Topics in Current Physics

31

Topics in Current Physics

Founded by Helmut K. V. Lotsch

1 **Beam-Foil Spectroscopy**
Editor: S. Bashkin

2 **Modern Three-Hadron Physics**
Editor: A. W. Thomas

3 **Dynamics of Solids and Liquids by Neutron Scattering**
Editors: S. W. Lovesey and T. Springer

4 **Electron Spectroscopy for Surface Analysis**
Editor: H. Ibach

5 **Structure and Collisions of Ions and Atoms**
Editor: I. A. Sellin

6 **Neutron Diffraction**
Editor: H. Dachs

7 **Monte Carlo Methods** in Statistical Physics
Editor: K. Binder

8 **Ocean Acoustics**
Editor: J. A. DeSanto

9 **Inverse Source Problems** in Optics
Editor: H. P. Baltes

10 **Synchrotron Radiation**
Techniques and Applications
Editor: C. Kunz

11 **Raman Spectroscopy** of Gases and Liquids
Editor: A. Weber

12 **Positrons in Solids**
Editor: P. Hautojärvi

13 **Computer Processing of Electron Microscope Images**
Editor: P. W. Hawkes

14 **Excitons** Editor: K. Cho

15 Physics of **Superionic Conductors**
Editor: M. B. Salamon

16 **Aerosol Microphysics I**
Particle Interactions
Editor: W. H. Marlow

17 **Solitons**
Editors: R. Bullough, P. Caudrey

18 **Magnetic Electron Lenses**
Editor: P. W. Hawkes

19 Theory of **Chemisorption**
Editor: J. R. Smith

20 **Inverse Scattering Problems** in Optics
Editor: H. P. Baltes

21 **Coherent Nonlinear Optics**
Recent Advances
Editors: M. S. Feld and V. S. Letokhov

22 **Electromagnetic Theory of Gratings**
Editor: R. Petit

23 **Structural Phase Transitions I**
Editors: K. A. Müller and H. Thomas

24 **Amorphous Solids**
Low-Temperature Properties
Editor: W. A. Phillips

25 **Mössbauer Spectroscopy II**
The Exotic Side of the Method
Editor: U. Gonser

26 **Crystal Cohesion and Conformational Energies** Editor: R. M. Metzger

27 **Dissipative Systems in Quantum Optics** Resonance Fluorescence
Optical Bistability, Superfluorescence
Editor: R. Bonifacio

28 **The Stratospheric Aerosol Layer**
Editor: R. C. Whitten

29 **Aerosol Microphysics II**
Chemical Physics of Microparticles
Editor: W. H. Marlow

30 **Real-Space Renormalization**
Editors: T. W. Burkhardt and J. M. J. van Leeuwen

31 **Hyperfine Interactions of Radioactive Nuclei**
Editor: J. Christiansen

32 **Superconductivity in Ternary Compounds I**
Structural, Electronic, and Lattice Properties
Editors: Ø. Fischer and M. B. Maple

33 **Molecular Collision Dynamics**
Editor: J. M. Bowman

34 **Superconductivity in Ternary Compounds II**
Superconductivity and Magnetism
Editors: M. B. Maple and Ø. Fischer

Hyperfine Interactions of Radioactive Nuclei

Edited by J. Christiansen

With Contributions by
H. Ackermann J. Christiansen W. Engel
O. Häusser P. Heitjans E. Recknagel
G. Schatz G. D. Sprouse H.-J. Stöckmann
I. S. Towner Th. Wichert W. Witthuhn

With 172 Figures

Springer-Verlag Berlin Heidelberg New York Tokyo 1983

Professor Dr. Jens Christiansen

Physikalisches Institut der Universität Erlangen-Nürnberg, Erwin-Rommel-Str. 1
D-8520 Erlangen, Fed. Rep. of Germany

ISBN-13:978-3-642-81971-1 e-ISBN-13:978-3-642-81969-8
DOI: 10.1007/978-3-642-81969-8

Library of Congress Cataloging in Publication Data. Main entry under title: Hyperfine interactions of radioactive nuclei. (Topics in current physics ; 31) Includes bibliographical references and index. 1. Hyperfine interactions. 2. Metals—Defects. I. Christiansen, J. (Jens), 1926–. II. Ackermann, M. III. Series. QC762.H94 1983 538'.3 82-19556

Softcover reprint of the hardcover 1st edition 1983

2153/3130-543210

Preface

This volume deals with the interaction between moments of excited or radioactive nuclei and electromagnetic fields. The experimental techniques developed for the observation of this hyperfine interaction are governed by the lifetime of the nuclear states in question. The dynamics of the interaction are reflected by the time dependence of the spatial distribution of the radioactive decay radiation. Basically, the experiments yield information on the energy shifts and/or splittings of the nuclear levels. These quantities are determined essentially by the product of the nuclear moment and the electromagnetic field acting at the site of the nucleus. Due to the strong decrease in the fields with distance, the measurements probe these fields within a highly localized region centered around the radioactive nuclei.

Detailed experimental methods with numerous ramifications were developed in the early sixties. In the period which followed, the main emphasis was on excitation of short-lived nuclear states by means of pulsed particle accelerators, implantation of radioactive nuclei, and production of polarized β-unstable nuclei by nuclear reactions with polarized neutrons or particles. The seventies were a period of fruitful applications directed to extensive studies of the moments of excited nuclear states on the one hand, and local internal fields on the other, resulting in far-reaching information on atomic and solid-state properties.

The organization of this *Topics* volume follows these main lines of research. Following the introduction, which summarizes some methodological and historical aspects, the second chapter discusses aspects of nuclear physics which are correlated with the measurements of magnetic dipole and electric quadrupole moments of excited nuclear states. The next chapter reviews the results of experiments on the electron shells of free atoms or ions. The three final chapters present current research on solid-state problems: lattice defects (Chap.4), electric field gradients in metals and alloys (Chap.5), and the applications of nuclear magnetic resonance experiments to β-unstable polarized nuclei (Chap.6).

The editor is grateful to the authors for their active cooperation, and to Dr. H. Lotsch for his encouragement and patience during the delayed maturing of this volume.

Erlangen, March 1983 *Jens Christiansen*

Contents

List of Contributors

Ackermann, Hans

Fachbereich Physik der Philipps-Universität, Renthof 5,
3550 Marburg, Fed. Rep. of Germany

Christiansen, Jens

Physikalisches Institut der Universität Erlangen-Nürnberg, Erwin-Rommel-Str. 1, 8520 Erlangen, Fed. Rep. of Germany

Engel, Wolfram

Physikalisches Institut der Universität Erlangen, Erwin-Rommel-Str. 1,
8520 Erlangen, Fed. Rep. of Germany

Häusser, Otto F.

Atomic Energy of Canada Limited, Chalk River Nuclear Laboratories, Chalk River, Ontario, Canada K0J 1J0

Heitjans, Paul

Fachbereich Physik der Philipps-Universität, Renthof 5,
3550 Marburg, Fed. Rep. of Germany

Recknagel, Ekkehard

Fakultät für Physik, Universität Konstanz, Bücklestr. 13,
7750 Konstanz, Fed. Rep. of Germany

Schatz, Günter

Fakultät für Physik, Universität Konstanz, Bücklestr. 13,
7750 Konstanz, Fed. Rep. of Germany

Sprouse, Gene D.

Department of Physics, State University of New York, Stony Brook,
New York 11794, USA

Stöckmann, Hans-Jürgen

Fachbereich Physik der Philipps-Universität, Renthof 5,
3550 Marburg, Fed. Rep. of Germany

Towner, Ian S.

Atomic Energy of Canada Limited, Chalk River Nuclear Laboratories,
Chalk River, Ontario, Canada KOJ 1J0

Wichert, Thomas

Fakultät für Physik, Universität Konstanz, Postfach 55 60 ,
7750 Konstanz, Fed. Rep. of Germany

Witthuhn, Wolfgang

Physikalisches Institut der Universität Erlangen, Erwin-Rommel-Str. 1,
8520 Erlangen, Fed. Rep. of Germany

1. Introduction

J. Christiansen and W. Witthuhn

With 11 Figures

The charge distribution of an atomic nucleus interacts with the electromagnetic fields produced by the extranuclear charges of the shell electrons or the constituents of the solid-state surroundings, or by totally external sources such as an applied magnetic field. This hyperfine interaction results in a shift and a splitting of the nuclear and atomic levels. The hyperfine interaction Hamiltonian H_{hf} can be expressed conveniently in terms of a multipole expansion:

$$H_{hf} = H(E0) + H(M1) + H(E2) + \ldots \quad , \tag{1.1}$$

where H(E0) denotes the electric monopole, H(M1) the magnetic dipole, H(E2) the electric quadrupole term, etc. They correspond to the moments of the nucleus not required by symmetry to vanish.

Each term of the expansion (1.1) can be written as the product of a nuclear moment and the corresponding term of the electromagnetic field. Thus, the investigation of this hyperfine interaction yields information on both the nuclear moments and the solid-state surroundings. However, extraction of one side out of the experimentally measured quantity – i.e., the level energy or/and splitting – requires detailed knowledge of the corresponding counterpart in the product term. Here one generally must rely on theoretical arguments. Only in the case of magnetic dipole interaction are external fields of the required strength available, allowing direct determination of nuclear dipole moments in many systems.

Success in the study of hyperfine interactions can be traced back to the beginning of modern atomic and nuclear physics. The determination of nuclear magnetic dipole moments was of fundamental importance in the development of the nuclear shell model and its modifications. Nuclear resonance methods yielded numerous results on extranuclear electromagnetic fields and significantly extended our knowledge of a variety of solid-state properties such as magnetism and the behaviour of spin systems in solids. Atomic and molecular beam methods and modern beam-foil spectroscopy were applied successfully in the investigation of interactions in free atoms and ions.

Hyperfine interactions can be investigated on stable or unstable nuclear ground states as well as on excited nuclear states. The classical methods – high-resolution optical spectroscopy, atomic beam resonance techniques and nuclear magnetic resonances – are restricted to stable or long-lived nuclear states with lifetimes of at least several minutes. In the fifties intense competition came from the nuclear methods – Mössbauer spectroscopy (ME) and perturbed angular correlation (PAC) – which systematically used nuclear radiation as the carrier of information on hyperfine interactions. These methods belong to the field of nuclear spectroscopy and therefore usually require expensive nuclear equipment, including particle accelerators and fairly complex radiation-detecting and radiation-analyzing systems. In the early phase the main emphasis was placed on the determination of nuclear properties such as magnetic moments, lifetimes, and spins of excited isomeric nuclear states. Today these techniques can be regarded as powerful tools for the investigation of unsolved problems in other fields of physics (primarily solid-state physics), chemistry, and even biology.

The characteristics of emitted nuclear radiation – energy, spatial distribution, polarization – are determined by the properties of the nuclear levels involved: level energies, spins, half-life. The radiation parameters are changed in a generally well-known manner under the influence of the hyperfine interaction between the moments of the nuclear level and the field acting at the position of the nucleus. The detection of the status of "single atoms" by their decay radiation accounts for some unique features of these methods.

The nuclear level in question can be populated by prior radioactive decay (source experiments, "off beam") or by a nuclear reaction ("in-beam" measurements). In the latter case, details of the nuclear reaction such as the nature of the incoming and outgoing particles, and their energies, directions and angular momenta, are additional parameters which determine the observed radiation.

This volume could have been divided in accordance with the two different ways of populating the isomeric level, with one part treating experiments on radioactive sources and another dealing with measurements following nuclear reactions. It could also have been divided on the basis of the methods applied, a course taken in the Mössbauer spectroscopy conferences and the conferences on perturbed angular correlations. We have, however, chosen to group the material according to the fields investigated, and have divided this volume into chapters on nuclear, atomic and solid-state physics, the latter being subdivided into three sections. This method has been shown by the most recent conferences on hyperfine interactions to be the most efficient one. The sections summarize the methods individually, and cover only the features relevant to the specific topic. In the following we present the basic ideas underlying the nuclear methods, together with some historical remarks.

The nuclear methods, which detect hyperfine interactions by means of their influence on nuclear radiation, can be grouped into three classes: the Mössbauer effect (ME), the perturbed angular correlation techniques (PAC: γ-γ and β-γ angular correlations), and the angular distribution methods (PAD: angular distribution of α-, β- or γ-radiation from aligned or polarized nuclei), the last of which is realized by low-temperature nuclear orientation or by nuclear reactions.

1.1 Perturbed Angular Distribution Techniques

In the following sections we summarize some of the basic ideas of the "in-beam" methods. The most simple experimental situation is the investigation of magnetic hyperfine interaction between nuclear magnetic moments and an applied external magnetic field.

1.1.1 Magnetic Hyperfine Interaction

Isomeric nuclear states with magnetic dipole moments and mean lifetimes τ are populated by a nuclear reaction. A typical experimental set-up is shown schematically in Fig.1.1. The pulsed particle beam of an accelerator is directed onto a target placed in a homogeneous magnetic field perpendicular to the beam axis. The nuclear reaction excites the isomeric levels and simultaneously orients the nuclear spins. The degree of orientation depends on the details of the reaction process itself, and can be deduced theoretically only in favourable cases. Assuming that the particle beam is unpolarized, an alignment is generally produced, i.e., substates with small $|m|$ quantum numbers are populated preferentially, with the beam axis chosen as the quantization axis. The angular distributions of the gamma radiation(s) emitted from these aligned excited states are anisotropic for a nuclear spin $I \geq 1$. The details of this angular distribution depend on the nuclear alignment, the spins of the nuclear levels involved, and the multipolarity of the γ radiation. Fairly reliable predictions of these distributions can be given for heavy-ion-induced reactions where the high transferred angular momentum allows sufficiently accurate estimates of the resulting alignment [1.1].

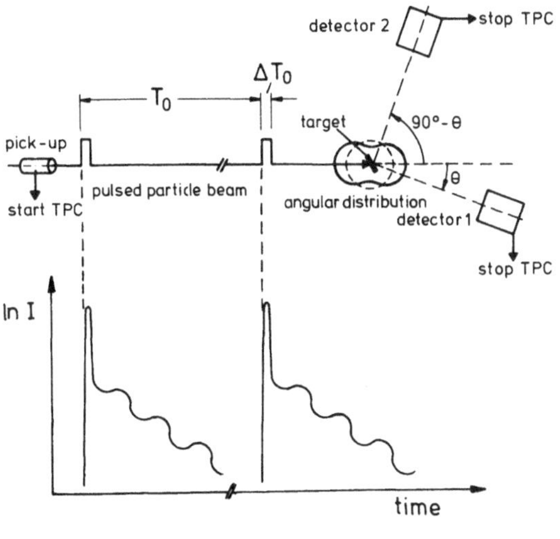

Fig.1.1. Schematic of the experimental set-up in a PAD experiment

The interaction of the nuclear magnetic moment with the applied external magnetic field[1] results in a change of the m-substate population. This repopulation is reflected in the angular distribution of the γ radiation. It can be taken into account by a perturbation factor. In a semi-classical picture the nuclear moments and hence the angular distribution precess around the magnetic field axis with the Larmor frequency

$$\omega_L = -\frac{\mu_N}{\hbar} g \cdot B \quad . \tag{1.2}$$

Here μ_N denotes the nuclear magneton and $g = (\mu/\mu_N)/(I/\hbar)$ the nuclear g factor of the isomeric nuclear state, with μ being the magnetic moment and I the nuclear spin of the level. The time-dependent γ intensity observed at an angle θ with respect to the beam axis is then given by

$$I(t,\theta,B) = I_0 \exp(-t/\tau) \cdot W(t,\theta,B) \quad , \tag{1.3}$$

where the mean life τ characterizes the exponential decay of the excited nuclei, and $W(t,\theta,B)$ reflects the oscillating intensity modulation due to the rotating angular distribution. The latter is given by (for details see [1.2]):

$$W(t,\theta,B) = \sum_{\substack{k \\ \text{even}}} A_k(1) A_k(2) P_k[\cos(\theta - \omega_L t)] \quad . \tag{1.4}$$

The functions P_k are the Legendre polynomials, the coefficients $A_k(1)$ depend on the degree of alignment produced by the nuclear reaction, and the coefficients $A_k(2)$ are determined by the parameters of the decay, i.e., the spins of the nuclear levels involved and the multipolarity of the γ radiation. In most cases – with the exception of Coulomb excitations – the coefficients $A_k(1)$ with $k \geq 4$ are negligible, and the angular distribution is characterized by the coefficients $A_2(1)$ and $A_2(2)$ only. The sign and magnitude of the nuclear g factor can be extracted conveniently from the intensity ratio of two measurements at angles θ and $\theta + 90^\circ$:

$$R(t,\theta,B) = \frac{I(t,\theta,B) - I(t,\theta + 90^\circ,B)}{I(t,\theta,B) + I(t,\theta + 90^\circ,B)}$$

$$= \frac{3A_{22}}{4 + A_{22}} \cdot \cos[2(\theta - \omega_L t)] \quad , \tag{1.5}$$

where the coefficient A_{22} stands for the product $A_2(1)\ A_2(2)$.

Typical intensity spectra of the decay of the $5/2^+$ level in ^{19}Ne, measured at two angles θ and $\theta + 90^\circ$, are shown in Fig.1.2a. The corresponding intensity ratio

1 We ignore here perturbations of the external field due to diamagnetics, paramagnetic shifts, ferromagnetism, etc. These effects are dealt with in detail in subsequent chapters.

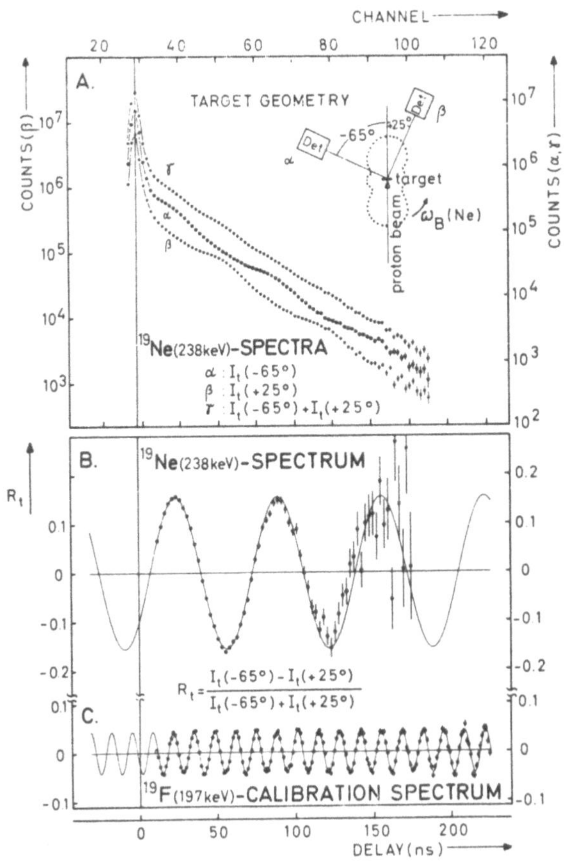

Fig.1.2. (a) Time spectra of the keV γ radiation emitted in the decay of the excited $5/2^+$ level of ^{19}Ne. The detectors are placed at angles θ and $\theta + 90°$ with respect to the beam axis as indicated in the inset. (b) Modulation spectrum of the γ ray intensities shown in (a), obtained from the data according to (1.5). The phase at t = 0 yields the sign of the nuclear magnetic moment [1.3]

$R(t,\theta,B)$, calculated from these data according to (1.5), is given in Fig.1.2b. The fit yields the magnitude and sign of the magnetic interaction and hence – assuming that the magnetic field acting at the nucleus is known – the corresponding parameters of the nuclear g factor. The PAD method was first applied by HRYNKIEWICZ [1.4], MATTHIAS and LUNDQUIST [1.5] and FREEMAN [1.6].

The application of this method requires that the mean lifetime of the excited nuclear state is short compared to the pulse repretition time T_0 of the particle accelerator. Otherwise the modulation amplitude is reduced as a result of incoherent superposition of successive decay intensities. Furthermore, the width of the beam pulses, ΔT_0, must be much smaller than the Larmor period: $\Delta T_0 \ll 1/\omega_L$. This condition ensures that the anisotropy is fully observed and is not smeared out (assuming that the time resolution of the equipment, Δt, is $<\Delta T_0$).

If the anisotropy is not attenuated by inhomogeneous or time-dependent fluctuating fields, the resolution of the method is limited by the uncertainty principle, i.e., by the mean lifetime τ of the isomeric level. Due to severe intensity problems, however, the conditions $\Delta T_0 \ll 1/\omega_L < \tau \ll T_0$ restrict the range of application to nuclear levels with mean lifetimes shorter than some 10^{-5} s, the lower limit

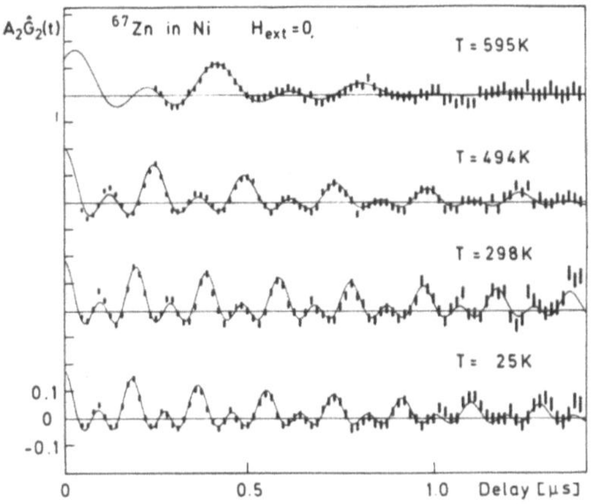

Fig.1.3. Spin precession spectra of ^{67}Zn nuclei in polarized ion and Ni-Fe alloys at different temperatures [1.8]

being of the order of 10^{-9} s. Within this range the PAD method has been applied extremely successfully in the determination of moments of excited nuclear states, as demonstrated by recent compilations of magnetic moments [1.7].

In addition to the measurement of magnetic moments, the method has yielded valuable information on internal fields. The high number of known isomeric states allows nearly every element to be investigated in a given host lattice, where the excited states can be populated by a nuclear reaction within the target material itself or be recoil-implanted into the material, e.g., a ferro- or antiferromagnetic lattice. An example is shown in Fig.1.3, where the modulation patterns of ^{67}Zn in Ni are shown at different target temperatures. In all these experiments one must be aware of the possibility of radiation damage perturbing the internal fields. A second problem which often arises relates to the question of whether the recoil comes to rest at a regular lattice site or an interstitial position.

1.1.2 Electric Quadrupole Hyperfine Interaction

The electric field gradient (EFG) acting on a nucleus at a lattice position in a non-cubic lattice results in a splitting of the nuclear levels due to the electric quadrupole hyperfine interaction between the nuclear quadrupole moment Q and this field gradient. This splitting can be observed with the same PAD apparatus as that used for magnetic interactions. However, the required strength of the EFG (of the order of 10^{15} V/cm^2) cannot be applied externally. Therefore, one must use internal field gradients of non-cubic crystalline solids and rely for the extraction of nuclear quadrupole moments on theoretical calculations of these gradients. The interpretation of numerous experimental results has thus been severely hampered.

The Hamiltonian of the quadrupole interaction has the form

$$H_Q = \frac{eV_{zz}Q}{4I(2I-1)}\left[3I_z^2 - I(I+1) + \frac{\eta}{2}\left(I_+^2 + I_-^2\right)\right] \quad , \qquad (1.6)$$

where I_z, $I_\pm$ are the conventional spin operators, and V_{zz} is the zz component of the electric field gradient. The asymmetry parameter η is defined by

$$\eta = (V_{xx} - V_{yy})/V_{zz} \quad . \tag{1.7}$$

In the case of an axially symmetric field gradient, i.e., $\eta = 0$, the level energies are given by

$$E_m = \frac{3cm^2 - I(I + 1)}{4I(2I - 1)} eQV_{zz} \quad . \tag{1.8}$$

In contrast to magnetic hyperfine splitting, the levels are still degenerated with respect to the sign, and the separation is not equidistant. For a vanishing asymmetry parameter η, e.g., in hexagonal or tetragonal lattice structures, the splittings are integer multiples of a basic energy difference

$$\Delta E_0 = \hbar\omega_0 = n \cdot eQV_{zz} \quad , \tag{1.9}$$

with n depending on the nuclear spin I.

The influence of this electric quadrupole hyperfine interaction on the angular distribution of the γ radiation is more complex than in the case of a magnetic interaction. One may still speak of a precession of the spin ensemble around the field axis, but the sense of the precession[2] is no longer defined, and the precession frequencies are different. In case of a unique field gradient with axial symmetry and random orientation within the target (polycrystalline sample) the angular distribution is given by

$$W(t,\theta) = \sum_{k \text{ even}} A_k(1)A_k(2)G_{kk}(t)P_k(\cos\theta) \quad , \tag{1.10}$$

where the coefficients $A_k(1)$ and $A_k(2)$ describe the γ ray anisotropy as in (1.4), and the P_k are the Legendre polynomials again. Information on the interaction is included in the perturbation factor $G_{kk}(t)$. Within the restrictions given above, this factor can be written in the form

$$G_{kk}(t) = \sum_n s_{kn} \cos(n\omega_0 t) \quad , \tag{1.11}$$

with the basic frequency ω_0 related directly to the interaction strength $\hbar\omega_Q$ [see (1.9)], and the amplitudes s_{kn} depending on the nuclear spin and the radiation parameter k; they are tabulated in [1.2]. The function (1.11) depends on the nuclear spin I (via n and s_{kn}). It therefore can be and has been used in a few cases to determine the spin of nuclear levels.

As an example of a quadrupole interaction measurement, Fig.1.4 shows the modulation pattern of ^{69}Ge in zinc (upper spectrum). The excited $9/2^+$ state of ^{69} Ge was

2 In the case of polarized nuclei this ambiguity can be resolved (Sect.1.2).

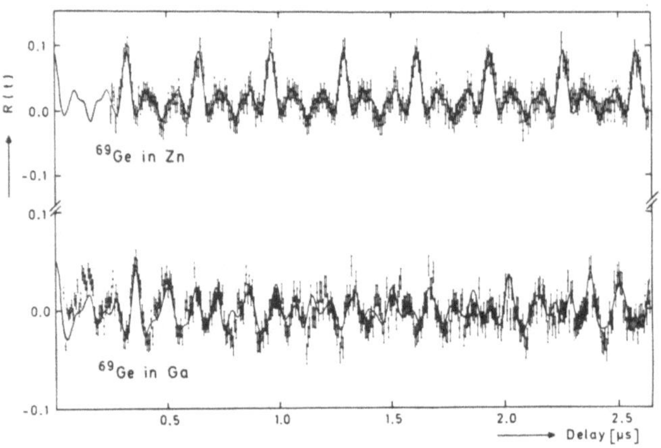

Fig.1.4. *Upper part.* Quadrupole interaction of ^{69}Ge in zinc. The isomeric $9/2^+$ level (τ = 4.2 μs) was populated by the nuclear reaction ^{66}Zn (α,n)^{69}Ge, and the modulation pattern was obtained in a polycrystalline sample at 575 K. *Lower part.* Quadrupole modulation pattern of ^{69}Ge nuclei in polycrystalline rhombohedral Ga at T = 253 K. The asymmetry parameter η = 0.697 results in a non-periodical modulation structure [1.9]

populated via the nuclear reaction ^{66}Zn(α,n)^{69}Ge in a target of enriched ^{66}Zn at a temperature close to the melting point. The interference due to radiation damage could be avoided (Chap.4), and the typical spin-$9/2^-$ interaction pattern can be followed clearly over many periods.

In non-axially symmetric crystals the perturbation function in (1.10) is in general a non-periodic function and cannot be written as an analytical expression. Here one has to diagonalize the Hamiltonian (1.6) numerically. However, leaving aside the more complex fitting procedure, even here the measurements yield unambiguous results on the V_{zz} component of the EFG and its asymmetry parameter η, as shown in the lower spectrum of Fig.1.4, where the ^{69}Ge nuclei are excited in a solid rhombohedral Ga host lattice at T = 253 K.

The first investigation of a quadrupole interaction using the PAD method was carried out by SUGIMOTO et al. [1.10]. They studied the field gradients of various insulating materials such as $(C_2F_4)_n$ and SF_6 at the second excited state of ^{19}F ($5/2^+$, 197 keV, τ = 128 ns). The wide field of this method was opened up and the first applications to the investigation of metals and alloys were made by the Berlin group [1.11,12].

1.1.3 The Stroboscopic Method

The time differential spin-precession method can be applied only to excited nuclear states with mean lifetimes considerably shorter than the pulse repetition time of the particle beam. This limitation can be overcome by stroboscopic observation of the perturbed angular distribution (SOPAD) [1.13,14]. The principle of this method is shown in Fig.1.5. The multiple excitations of the isomeric states during their lifetimes result in a superposition of the exponential decay curves. The γ ray intensity during a pulse interval is given by

$$I(t,\tau,T_0,\theta,B) = I_0 \sum_{n=0}^{\infty} \exp[-(t + nT_0)/\tau] \cdot W[\theta - \omega_L(t + nT_0)] \quad . \tag{1.12}$$

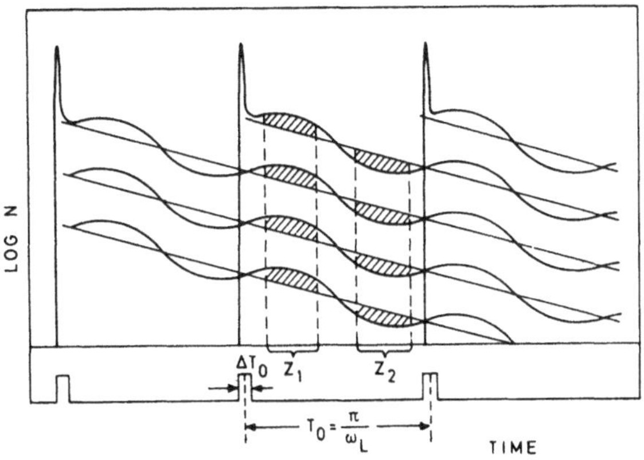

Fig.1.5. Principle of the SOPAD method. The counting rates within the marked time windows Z_1 and Z_2 depend on the anisotropy of the γ radiation and the ratio between the beam pulsing frequency and the Larmor frequency [1.14]

The counting rate measured within a fixed time interval (e.g., Z1 in Fig.1.5) shows resonance behaviour. The resonance maximum (or minimum) occurs if the Larmor frequency of the excited nuclei is equal to a multiple of half the pulse frequency. The resonance behaviour can be investigated by varying one of the two characteristic frequencies; it is most convenient to change the Larmor frequency by changing the applied magnetic field. The intensity measured at $t = t_0$ in the maxima (or minima) of the modulation, i.e., within small time windows Z1 (or Z2), is given by a Lorentzian:

$$I(t_0,\tau,T_0,\theta,B) \sim 1 \underset{(-)}{+} \frac{3A_{22}}{4 + A_{22}} \cdot \frac{(1/\tau)^2}{(1/\tau)^2 + 4(\omega_L - \omega_{L0})^2} \quad , \qquad (1.13)$$

where $k_{max} = 2$ is assumed for the angular distribution. The width of the resonance curve is given by

$$\Delta\omega = \frac{1}{\tau} \quad . \qquad (1.14)$$

This natural linewidth can be observed only if no additional relaxation process occurs during the lifetime of the nuclear state. Assuming an exponential decay of the anisotropy, with the relaxation time τ_{rel} being comparable to the mean lifetime τ, the resonance will be attenuated and broadened, and the width is then given by

$$\Delta\omega_{rel} = \frac{1}{\tau} + \frac{1}{\tau_{rel}} \quad . \qquad (1.15)$$

As an example we give the results of a stroboscopic experiment on the $9/2^+$ level of ^{71}Ge [1.15,17]. The 20-ms mean-lifetime of this level corresponds to a natural line width of about $3 \cdot 10^{-6}$ T. The relaxation time in the liquid Ga target is of the order of ms, thus increasing the linewidth but simultaneously reducing the height of the resonance. These difficulties were overcome by a special experimental setup, which artificially shortened the effective nuclear lifetime (Fig.1.6). By this means the linewidth was broadened without affecting the resonance amplitude (for details see [1.15,16]). Figure 1.7 shows three resonance curves obtained for three different effective lifetimes.

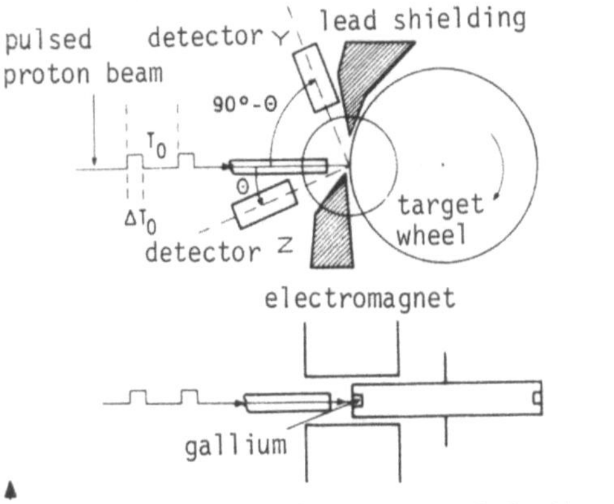

▲ Fig.1.6. Experimental arrangement in the ^{71}GeGa SOPAD experiment. The liquid Ga target is mounted on the periphery of the rotating wheel. By adjusting the wheel velocity the observable nuclear lifetime τ_{eff} could be reduced to 1 ms [1.16]

Fig.1.7. Stroboscopic resonances of the perturbed angular distribution of the 175 γ radiation of ^{71}Ge. The nuclear g factor obtained was g = -0.2304 ± 0.0005 [1.15] ▶

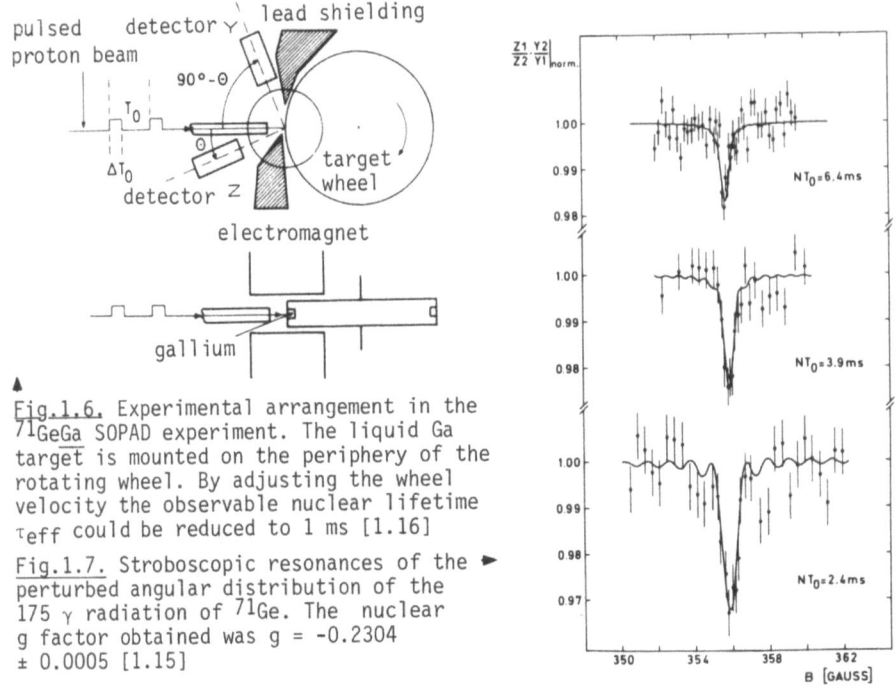

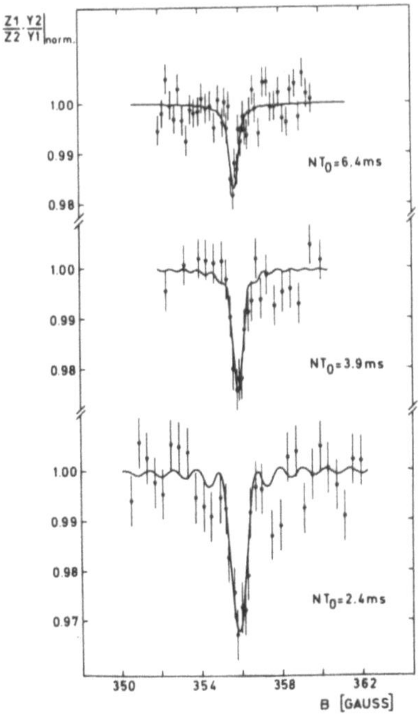

This method was used recently in high-precision measurements of the magnetic dipole moment of the muon [1.17].

1.2 Perturbed Angular Correlation Techniques

The "in-beam" methods discussed in the preceding sections required a nuclear reaction in order to produce the nuclear alignment needed for the observation of magnetic or electric hyperfine interactions. This alignment was characterized by the distribution coefficient $A_k(1)$ [see (1.4)]. A second possibility of generating an alignment is presented by the angular distribution of the radiation field itself, since this distribution depends in a well-known way on the nuclear spins, the m-sublevel population parameters, and the multipolarity of the radiation. Thus, the detection of gamma quanta in an arbitrary direction generally results in an alignment of the "observed nuclei". Any subsequent radiation emitted by the same nuclei shows a characteristic angular distribution with respect to the first radiation. In this method the nuclear level in question must be the intermediate level in a γ-γ cascade. The populating γ transition ("γ 1") selects a subensemble with a defined alignment. The angular distribution of γ 2 then reflects this m-sublevel population.

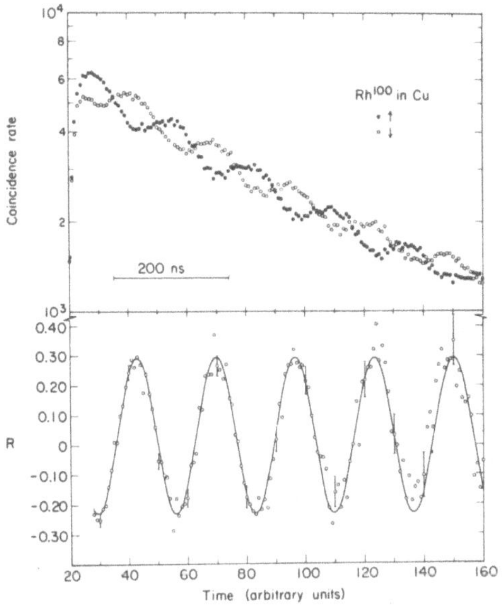

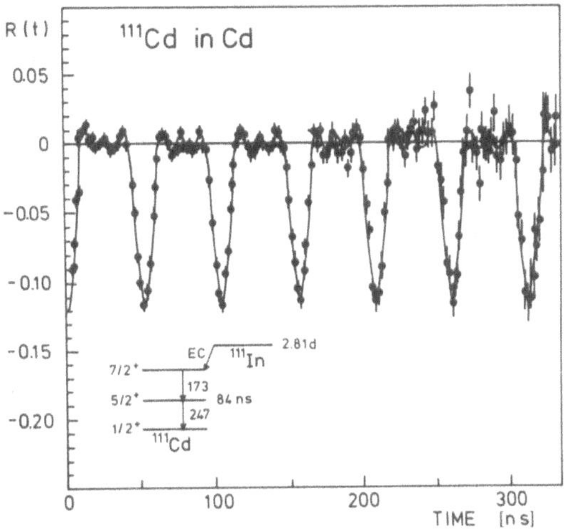

Fig.1.9 ▲

◄ Fig.1.8

Fig.1.8. PAC-intensity spectra (upper part) and the corresponding intensity ratio R(t,B) for the 2^+ level in ^{100}Rh [1.18]

Fig.1.9. Modulation spectra resulting from the quadrupole interaction of ^{111}Cd in hexagonal Cd metal at room temperature. The ^{111}Cd nuclei were populated via 2.8d-EC decay of ^{111}In. A 173-247 keV γ cascade was used in the PAC measurement

The angular correlation function can be described by an expression identical to that given for the PAD experiments (1.4). The coefficients $A_k(1)$ here depend on the initial and intermediate nuclear spins and on the radiation parameters of the preceding γ (γ1), and the coefficients $A_k(2)$ are given by the corresponding characteristics of the decay radiation (γ2). The hyperfine interaction between the nuclear moments of the intermediate state and the electromagnetic interaction fields again results in a characteristic perturbation of the angular correlation.

Figure 1.8 shows the result of a PAC experiment in which magnetic interaction was studied [1.18]. The excited 2^+ level of ^{100}Rh was populated by 84.0-keV γ-radiation. The "spin precession" of the nuclei in an external magnetic field of 0.222 T is reflected by the modulation of the 74.8-keV decay γ radiation.

A typical quadrupole interaction pattern is shown in Fig.1.9. Here the mother-activity ^{111}In was produced by a (d,n) nuclear reaction in the metallic Cd sample itself. The EC decay of the ^{111}In ($T_{1/2}$ = 2.8 d) populated the 173-247 keV γ cascade in ^{111}Cd which was used in the PAC measurement.

The periodic modulation is characteristic of a nuclear spin 5/2 and an axially symmetric electric field gradient with random orientation ("powder" sample). Because of the m^2 dependence of the level splitting (1.8), the sign of the interaction can generally not be deduced from the measured spectra. However, the inclusion of "polarization terms", i.e., a population $P(+m) \neq P(-m)$, resolves this ambiguity.

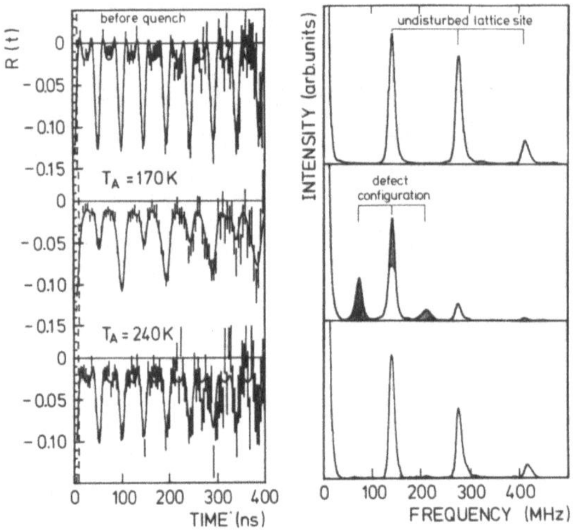

Fig.1.10. Trapping and detrapping of vacancies at In impurities in Zn. The spectra to the left give the modulation patterns observed at different annealing temperatures. The spectra on the right are the Fourier transforms, which clearly demonstrate the occurrence of a new, defect-specific EFG [1.19]

It can be realized experimentally by detecting the circular polarization of the γ radiation or by replacing one of the transitions by a β decay. Here the parity nonconservation then results in an occurrence of odd k terms in the angular correlation function (1.4). Details are given in Chap.5.

Quadrupole interaction has been investigated during the last decade in numerous substances, mainly by means of the isomeric level in ^{111}Cd. This state plays the same dominant role in PAC measurements as the first excited level in ^{57}Fe does in experiments involving the Mössbauer effect.

At present the PAC method is being applied intensively in the investigation of the EFG in metals and their alloys and in the study of local defect structures and their annealing behaviour in metals. These topics are dealt with in detail in Chaps. 4 and 5.

An example of the first field was the interaction in Cd metal mentioned above (Fig.1.9). The strong dependence of quadrupole hyperfine interaction on distance results in a unique sensitivity of the method to highly localized structures such as trapped point defects or small defect clusters. Here, in contrast to many "macroscopic methods", remote defects do not interfere and mask the measured pattern. Figure 1.10 illustrates this feature: the EFG of the undisturbed lattice (upper part of the figure) is changed significantly by a trapped vacancy, where the defect manifests itself by its characteristic defect EFG. At high temperatures the defect-probe-atom complex dissolves and the lattice EFG is restored again. Different defect configurations can be distinguished most simply by their different EFGs, and local symmetry can very often be deduced due to the tensor character of the EFG. In recent years this feature has proved extremely powerful and has yielded important structural data on defect configurations in cubic (fcc and bcc) and hexagonal metals.

1.3 Nuclear Magnetic Resonance on β-Emitting Nuclei

The PAD and PAC methods depend on the change in the population parameters of the excited level caused by the hyperfine interactions. The mean lifetimes of the isomeric states range from a few 10^{-9} s to several 10^{-3} s. In-beam β NMR detects the change in the β-ray asymmetry indicating a change of the nuclear polarization. Since the lifetimes of β-emitting nuclei are considerably longer (of the order of several 10^{-2} s to 10^{3} s), hyperfine transitions can be induced by classical nuclear magnetic resonance techniques. β NMR thus combines the high precision of a NMR experiment with high sensitivity based on the detection of "single atoms" via their β decay.

Nuclear polarization, which results in asymmetry of the β decay as a consequence of parity non-conservation, can be achieved by two different approaches. In the first, the β-unstable nuclei are produced by a nuclear reaction using unpolarized particle beams. Only the fraction of nuclei recoiling out of the target foil within a preselected angle in the reaction plane is implanted into the host lattice and investigated there (Fig.1.11). The first successful such experiment was carried out by SUGIMOTO et al. [1.20] using the experimental set-up shown in Fig.1.11.

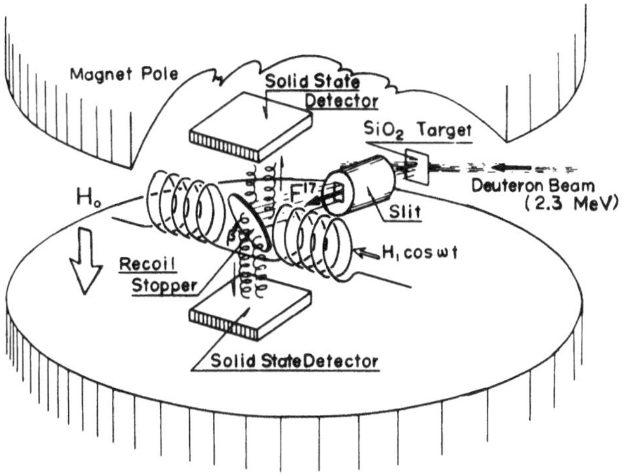

Fig.1.11. Experimental set-up of the β NMR experiment carried out by SUGIMOTO et al. [1.20]. The ^{17}F nuclei were produced via the nuclear reaction $^{16}O(d,n)^{17}F$, and the polarization of the ^{17}F nuclei was observed by detecting the asymmetry of the β^+ decay of ^{17}F [1.20]

The second possibility is the use of polarized particle beams to initiate the nuclear reaction. Here, a selected recoil angle is unnecessary, but the degree of polarization is generally smaller. The polarization of the β emitters produced can be calculated only in a very few favourable cases; like the alignment in PAD measurements, it usually has to be determined experimentally. The first experiment using polarized neutron capture was conducted by CONNOR [1.21], and polarized fast light-ion beams were first applied by MINAMINOSO et al. [1.22].

The polarization f_1^{nr} initially produced by the nuclear reaction results in an asymmetric angular distribution of the emitted electrons given by

$$W(\beta,\theta) = 1 + A_1 f_1^{nr} \cdot \frac{v}{c} \cos\theta \quad . \tag{1.16}$$

Here A_1 denotes the asymmetry coefficient, v the electron velocity, and θ the angle between $\underline{v}$ and the polarization $\underline{f}_1^{nr}$. The anisotropy can be destroyed in the event of resonance by rf-field-induced transitions. Experimental details and recent applications to solid-state problems are discussed in Chap.6.

References

1.1 T. Yamazaki: Nucl. Data *A3*, 1 (1967)
1.2 H. Frauenfelder, R.M. Steffen: "Angular Distribution of Nuclear Radiation", in *α-, β- and γ-Ray Spectroscopy*, ed. by K. Siegbahn (North-Holland, Amsterdam 1974) pp.997-1198
1.3 J. Bleck, D.W. Haag, W. Leitz, R. Michaelsen, W. Ribbe, F. Sichelschmidt: Nucl. Phys. *A123*, 65 (1969)
1.4 A.Z. Hrynkiewicz: Postepy Fiz. *11*, 521 (1960)
1.5 E. Matthias, T. Lundquist: Nucl. Instrum. Methods *13*, 356 (1961)
1.6 R. Freemann: Nucl. Phys. *25*, 446 (1961)
1.7 C.M. Lederer, V.S. Shirley: "Table of Nuclear Moments", in *Table of Isotopes*, ed. by C.M. Lederer, V.S. Shirley (Wiley, New York 1978) Appendix VII
1.8 H. Bertschat, H. Haas, F. Pleiter, E. Recknagel, E. Schlodder, B. Spellmeyer: Z. Phys. *267*, 299 (1974)
1.9 H. Haas, W. Leitz, H.-E. Mahnke, W. Semmler, R. Sielemann, T. Wichert: Phys. Rev. Lett. *30*, 656 (1973)
1.10 K. Sugimoto, A. Mizobuchi, K. Nakai: Phys. Rev. *B134*, 539 (1964)
1.11 J. Bleck, R. Butt, H. Haas, W. Ribbe, W. Zeitz: Phys. Rev. Lett. *29*, 1371 (1972)
1.12 H. Haas: Phys. Scr. *11*, 221 (1975)
1.13 J. Christiansen, H.-E. Mahnke, E. Recknagel, D. Riegel, G. Weyer, W. Witthuhn: Phys. Rev. Lett. *21*, 554 (1968)
1.14 J. Christiansen, H.-E. Mahnke, E. Recknagel, D. Riegel, G. Schatz, G. Weyer, W. Witthuhn: Phys. Rev. *C1*, 613 (1970)
1.15 H. Bertschat, J. Christiansen, H.-E. Mahnke, E. Recknagel, G. Schatz, R. Sielemann, W. Witthuhn: Nucl. Phys. *A150*, 282 (1970)
1.16 W. Witthuhn: "Das Magnetische Moment des 20 ms Zustandes im Kern ^{71}Ge", Ph.D. Thesis, University of Erlangen (1971)
1.17 M. Camani, F.N. Gygax, E. Klempt, P.D. Patterson, W. Rüegg, A. Schenck, R. Schulze, H. Wolf: Annual Report Swiss Institute for Nuclear Research (1976), E7
1.18 E. Matthias, D.A. Shirley, J.S. Evans, R.A. Naumann: Phys. Rev. *B140*, 264 (1965)
1.19 F. Simonato, W. Engel, S. Hoth, R. Keitel, R. Seeböck, W. Witthuhn: Phys. Lett. *A84*, 393 (1981)
1.20 K. Sugimoto, A. Mizobuchi, K. Nakai, K. Matuda: J. Phys. Soc. Jpn. *21*, 213 (1966)
1.21 D. Connor: Phys. Rev. Lett. *3*, 429 (1959)
1.22 T. Minaminoso, J. Hugg, D. Mavais, T. Saylor, S. Lazarus, H. Glavish, S. Hanna: Phys. Rev. Lett. *34*, 1465 (1975)

2. Hyperfine Interactions of Excited Nuclei in Atomic Systems

G.D. Sprouse

With 21 Figures

The idea of utilizing an excited nucleus as a probe to observe the hyperfine interaction in free atoms was evident from the very birth of the perturbed angular correlation method [2.1]. Since that time, there have been many developments in experimental methods and in our understanding of the different atomic systems in which measurements can be performed, and many interesting results have been obtained. There have been periodic reviews of the status of the field, the most recent of which was made by GOLDRING [2.2]. Together with his students and former students, Professor Goldring has made many of the important advances in this field of physics, and his review is an excellent source for more information on some of the topics covered in this chapter. Because of the existence of this excellent review by GOLDRING, which places strong emphasis on the time integral methods, I have tended in this chapter to emphasize more the techniques and methods for dealing with longer-lived nuclear levels where one can apply time differential methods.

Another excellent source for further information about hyperfine interaction in atomic systems is the Proceedings of the Haifa International Workshop [2.3], in which there are excellent articles on atomic systems by several different authors.

Previous reviews of the use of atomic systems for g-factor measurements of short-lived states have been made periodically at the conference series on hyperfine interaction [2.4-6]. Although many advances have been made since these earlier papers, I include them here for reference.

This chapter is a review of the methods and techniques that have been used for nuclear moment measurements in atomic systems, with emphasis placed on those techniques using an accelerator as a source. The advantages and limitations of the different methods are indicated, as well as future directions in some cases.

2.1 Free Atoms in Flight - The Recoil Distance Method

The quest for methods to measure nuclear moments of short-lived isomeric states has led to a very beautiful and elegant technique which utilizes the strong hyperfine interactions in highly ionized atoms [2.7]. This method, which we will call the recoil distance method, can best be understood by referring to Fig.2.1. A beam

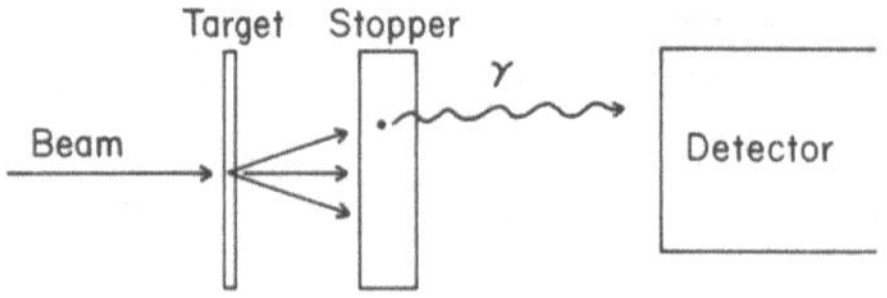

Fig.2.1. Schematic diagram of apparatus for recoil distance method

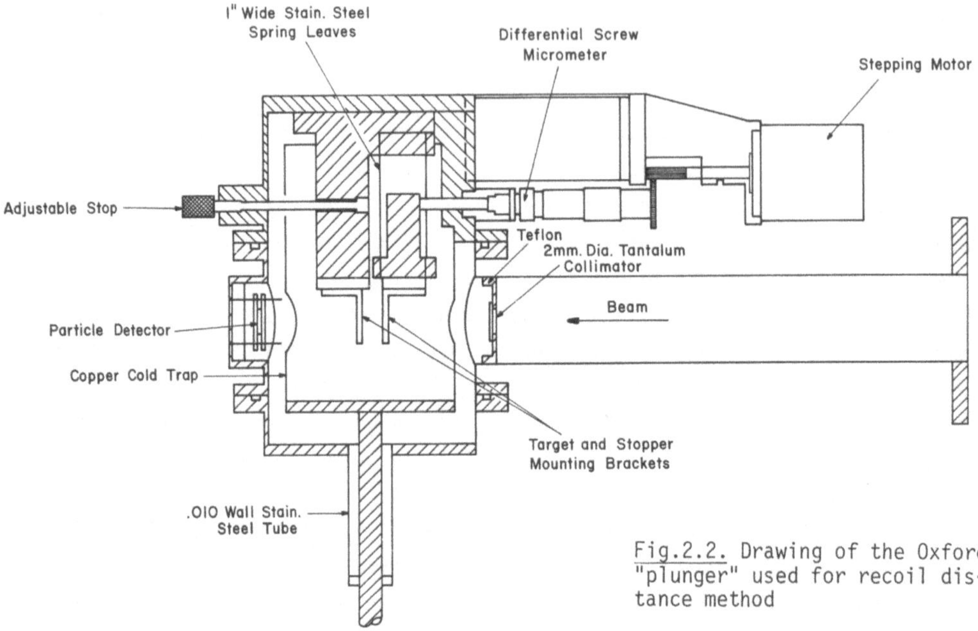

Fig.2.2. Drawing of the Oxford "plunger" used for recoil distance method

of projectile nuclei impinge on a thin target and produce nuclear reactions. Those product nuclei which are recoiling from the reaction in the forward direction are selected either by coincidence with another particle or by kinematic constraints. Upon emission from the target foil, the nucleus is in a free ion. The emission velocity is chosen so that some fraction of the ions will be in a charge state where the atomic configuration has a strong hyperfine interaction, and this interaction will modulate the anisotropy of decay γ rays at the hyperfine interaction frequency. The time of interaction is controlled by varying the distance to the recoil stopper by a precision-mechanical device. Upon hitting the stopper, the nucleus slows down, thereby regaining the outer electrons and turning off the strong hyperfine interaction. For lifetimes very long compared to the flight time, the decay in flight is usually neglected, and simple oscillations of the anisotropy are observed when the flight time is varied. For lifetimes comparable to the flight time, the stopped nuclear decays can be separated from the in-flight decays by the Doppler-shift technique with high-resolution Ge(Li) detectors, or the unseparated data can also be interpreted to obtain the hyperfine interaction frequencies.

In Fig.2.2 an example is shown of the apparatus used to perform the recoil distance measurement [2.8]. Extreme care must be taken in order to obtain precise parallelism of the target and stopper, which are usually thin foils stretched tightly across a hole which has been ground flat. The usual limit to the distance resolution with similar systems is of the order of a few microns. A very precise apparatus utilizing an interferometer for monitoring the system can achieve better resolution [2.9]. Beam currents on the foils must typically be limited to less than 100 nA before thermal heating and curling of the foils takes place.

2.1.1 Calculation of the Perturbation Function $G^{kk}(t)$

A necessary condition for the use of any perturbed angular distribution method is that an aligned ensemble of nuclear spins I be created. In the measurements described here, the alignment is produced by the nuclear reactions. In the flight region, the nuclei are coupled to the electronic spins J which are usually assumed to be randomly aligned. The possibility of the intermediate state propagating during the flight time in different hyperfine states F with different energies gives rise to interference at frequencies given by $\omega_{FF'} = (E_F - E_{F'})/\hbar$. There zero-field quantum beats were first given by ALDER [2.1], and the angular correlation function is:

$$G^{kk}(t) = \sum_{FF'} \frac{(2F+1)(2F'+1)}{(2J+1)} \begin{Bmatrix} F & F' & k \\ I & I & J \end{Bmatrix}^2 \exp(-i\omega_{FF'}t) \tag{2.1}$$

$$\hbar\omega_{FF'} = (2I+1)g_I\mu_N\langle B_J(0)\rangle \tag{2.2}$$

where $\langle B_J(0)\rangle$ is the expectation value of the magnetic field at the nucleus.

For the particularly simple case of J = 1/2, (2.1) reduces to:

$$G_k(t) = 1 - \frac{k(k+1)}{(2I+1)^2}(1 - \cos\omega_{FF'}t) \quad . \tag{2.3}$$

The observation of the oscillatory behavior allows a direct measurement of $\omega_{FF'}$. Note that the oscillations in time are always cosines, so that all information as to the sign of the magnetic moment is lost.

Several points should be made regarding these equations. Note that the perturbation frequency is proportional to the product $g_I(2I+1)$. In cases where the nuclear spin is unknown and the lifetime is long enough, a spin measurement could be made by comparison of a hyperfine frequency measurement with an external field measurement which only determines g_I. Another aspect is that the observed frequency is considerable raised for high spin states, which can either help or hinder measurements, depending on the values of other parameters. The g_I measured in an external field and in the internal hyperfine field will, once the spin is known, differ slightly due to a hyperfine anomaly [2.10], but the precision of measurements performed up until now has not been sufficient to consider this small effect.

In most angular correlation work, the higher k statistical tensors are smaller in magnitude than the lower k values, with most experiments utilizing only $k = 2$. This fact places a limitation on the use of the recoil distance method for high spin states, since the peak-to-peak amplitude of the oscillations is equal to $2k(k + 1)/(2I + 2)^2$. JAIN et al. [2.11] have utilized an angular correlation with a strong $k = 6$ term to circumvent this difficulty to measure $I = 3$ in ^{40}Ca. SPROUSE et al. [2.12] have developed a method utilizing a strong transverse magnetic field to measure a state with $I = 15/2$ in ^{41}Ca (Sect.2.3.2). Of course, large values of J could be utilized to generate in appreciable perturbation, but large J values are not usually ground-state atomic configurations.

2.1.2 Fine-Structure Beats

The observation of quantum beats from the "inside" with a nuclear probe is not limited to hyperfine structure. ANDRAÄ [2.13] points out that if a fine-structure multiplet and the hyperfine structure are fully unresolved in an experiment, and J and F are still good quantum numbers, then:

$$G^{kk}(t) = \sum_{JFJ'F'} \frac{(2F + 1)(2F' + 1)(2J + 1)(2J' + 1)}{(2S + 1)(2I + 1)} \times \begin{Bmatrix} J & J' & k \\ L & L & S \end{Bmatrix}^2 \begin{Bmatrix} F & F' & k \\ J' & J & I \end{Bmatrix}^2 \exp(-i\omega_{JFJ'F'}) \quad ; \tag{2.4}$$

where

$$\hbar\omega_{JFJ'F'} = E_{JF} - E_{J'F'} \quad . \tag{2.5}$$

In most experiments, however, these fine-structure beats are of too high frequency to be resolved, and the (2.4) reduces to (2.1) with an additional sum over the fine-structure levels.

2.1.3 Hyperfine Interactions in Free Atoms

In the absence of applied external fields, the electronic angular momentum, J, and the nuclear angular momentum, I, form a multiplet of levels with energies, E_F, given by:

$$E_F = A\,C/2 + B\,\frac{3/4C(C + 1) - I(I + 1)(J)(J + 1)}{2I(2I - 1)J(2J - 1)} \tag{2.6}$$

with

$$C = F(F + 1) - I(I + 1) - J(J + 1)$$

and

$$A = \frac{\mu_I \langle B_J(0)\rangle}{IJ} \tag{2.7}$$

$$B = eQ\langle\phi_{JJ}^{(0)}\rangle \quad . \tag{2.8}$$

The constants A and B are the magnetic dipole and electric quadrupole coupling constants familiar from resonance and optical spectroscopy [2.14]. For the recoil distance method, only energy differences between different members of the multiplet are important.

$$E_F - E_{F'} = A/2[F(F' + 1) - F'(F' + 1)] + \frac{3/4\ B[C(C + 1) - C'(C' + 1)]}{2I(2I - 1)J(2J - 1)} \quad . \tag{2.9}$$

Although the quadrupole term vanishes indentically only for $J = 1/2$ and $Q = 0$, it is in general much smaller than the magnetic term, and experimental observation of the quadrupole term has not been made with the recoil distance method.

For convenience, we will summarize in the next sections the expressions for hyperfine fields calculated for various simple atomic states.

2.1.4 Hydrogenlike Atoms

The simple one-electron atom has been utilized for most measurements to data. The hyperfine field for a hydrogenlike atom (including the Schwinger correction) is given by:

$$\langle B(0)\rangle = 0.167\ Z^3\ F_r(j,z) \times 10^6\ G \tag{2.10}$$

where $F_r(j,z)$ is a factor very nearly 1 which accounts for the small components of the Dirac wave function. Table 2.1 summarizes all measurements reported to data which have utilized the hydrogenlike atom.

In Fig.2.3, these measurements are plotted on a g-factor lifetime scale to indicate the range of Z and τ where the method has been applied. The smooth curve corresponds to a half cycle of the oscillation for the field in a hydrogenlike atom with a g factor of 1.

The strong dependence of the hyperfine field on Z restricts the range of usefulness of the hydrogenic atom for g-factor measurements to $Z \leq 20$, because the period of the oscillations becomes less than the achievable time resolution, which is indicated in the figure by the cross-hatching. The usefulness of other atomic species will be discussed in subsequent sections.

Equation (2.10) has been checked to better than 1% by MOORHOUSE et al. [2.24] by measuring the well-known g factor of $^{19}F(5/2^+)$ [2.29]. The data for the one-electron atom are shown in Fig.2.4, and illustrate the elegance and precision of this type of measurement.

One of the main advantages of the hydrogenic system is that because the atomic ground state is so much more tightly bound than the excited states, the velocity can be adjusted so that contamination with other excited states is not usually a

Table 2.1. Summary of g-factor measurements in hydrogenic atoms

Nucleus	E	I	τ		g	References
^{13}C	3850	$5/2^+$	12.4(8)	ps	0.59(5)	BEENE et al. [2.15]
^{14}C	6728	3^-	97(15)	ps	0.282(7)	ALEXANDER et al. [2.16]
^{14}N	5110	2^-	6.2(4)	ps	0.66(4)	MOORHOUSE et al. [2.17]
^{16}N	397	1^-	6.5(5)	ps	1.83(13)	ASHER et al. [2.18]
^{15}O	5240	$5/2^+$	3.25(30)	ps	0.260(28	BECK et al. [2.19]
^{16}O	6130	3^-	25(2)	ps	0.55(3)	RANDOLPH et al. [2.20]
^{18}O	1980	2^+	2.99(12)	ps	0.287(15)	ASHER et al. [2.21]
^{18}O	3550	4^+	27(5)	ps	0.62(12)	BERANT et al. [2.22]
^{20}O	1675	2^+	14.2(8)	ps	0.39(4)	BERANT et al. [2.23]
^{19}F	197	$5/2^+$	128.8(15)	ns	1.442(3)	MOORHOUSE et al. [2.24]
^{20}Ne	1634	2^+	0.8(2)	ps	0.54(4)	HORSTMAN et al. [2.9]
^{21}Ne	351	$5/2^+$	10.23(20)	ps	0.196(14)	ROWE et al. [2.25]
^{22}Ne	1275	2^+	5.4(4)	ps	0.36(3)	BOHM et al. [2.26]
^{21}Na	332	$5/2^+$	14(3)	ps	1.48(8)	BECK et al. [2.27]
^{22}Na	2210	1^-	22.0(8)	ps	0.36(7)	BECK et al. [2.28]
^{24}Mg	1369	2^+	2.09(13)	ps	0.51(2)	HORSTMAN et al. [2.9]

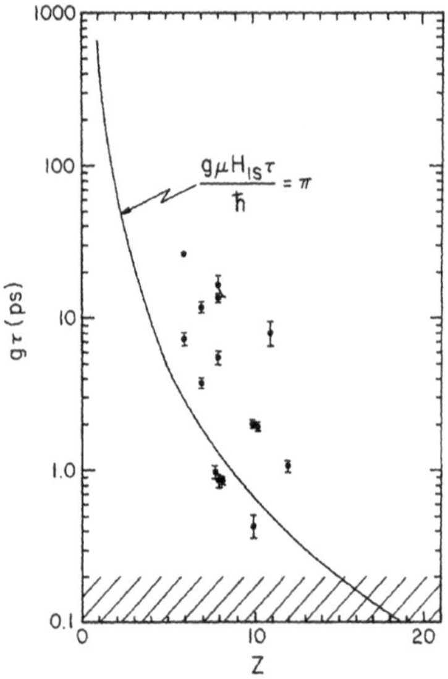

Fig.2.3. Summary of g factors measured with one-electron atoms. The smooth curve corresponds to a half cycle of the anisotropy oscillation in the nuclear lifetime. The shaded region indicates an approximate limit of the time resolution

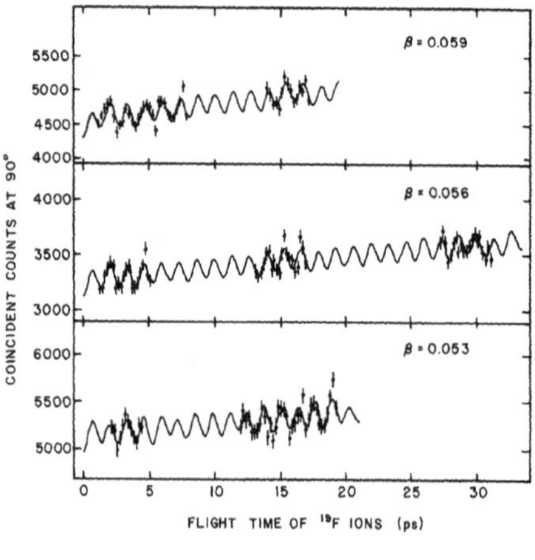

Fig.2.4. Data from [2.24] for $^{19}F(5/2^+)$ recoiling at different velocities

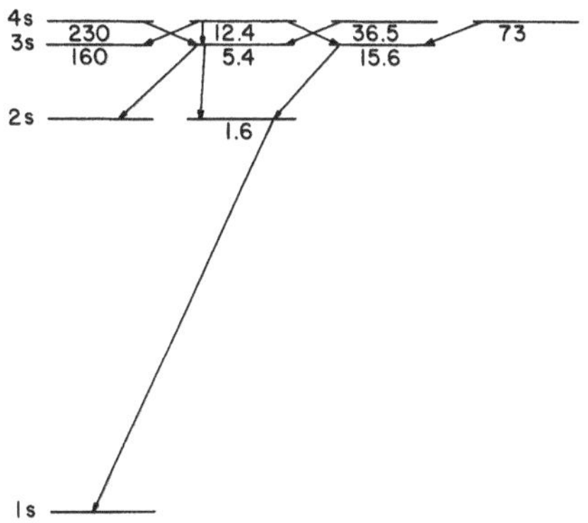

Fig.2.5. Lifetimes (in ns) for hydrogenic atoms. These lifetimes scale approximately as Z^{-4}

Table 2.2. Measured ground-state feeding times for hydrogenic systems

Nucleus	Feeding time	τ_{2p}	Reference
$^{14}C(3^-)$	1.8±0.4 ps[a]	1.23 ps	ALEXANDER et al. [2.16]
$^{20}Ne(2^+)$	1 ps	0.16 ps	FAESSLER et al. [2.31]
$^{24}Mg(2^+)$	0.15 ps	0.08 ps	HORSTMAN et al. [2.9]

[a]DOUBT [2.30] points out that the measured phase shift is inconsistent with the measured amplitude, and that a small shift in the time zero will account for the measured phase and amplitude.

problem. DOUBT [2.30] has summarized all of the reported data on excited-state population on one-electron systems and has shown that the measured fraction of ions which are in the ground state after emission from the foil is typically 90 ± 5%. The lifetimes of the excited atomic states are shown in Fig.2.5 for hydrogenic systems. These transitions are quite fast, e.g., for the 2p state in carbon, the lifetime is 1.23 ps. Therefore, if the 2p state is populated directly in the emission from the foil, it decays to the ground state in a time too short to disturb the measurement. If there is direct population of higher atomic levels, this will give rise to both a phase shift and attenuation of the time differential signal. These feeding times have been measured or a limit stated in some cases. These are summarized in Table 2.2, and it is seen that the measured times are of the order of the 2p lifetime.

The 2s state in the hydrogenic atom is metastable, as can be seen in Fig.2.5. The 2s hyperfine field is 8 times weaker than the 1s field, and is given by:

$$H(2s) = 2.09\ Z^3\ \text{tesla} \quad . \tag{2.11}$$

Populations of 2s metastable atoms are reported in [2.9] but with a very small fraction of the atoms in this state. In general, metastables are of little importance for these measurements because of their low population probability.

In summary, we see that the hydrogenic system is ideal for measuring hyperfine interaction frequencies by the recoil distance method because:

1. the ground state is either populated directly or with a very short feeding time
2. the hyperfine interaction constant is known precisely for the simple atom.

The major disadvantage of the hydrogenic system is that for $Z \geq 15$, the frequency of the oscillations is usually too high to measure with the plunger technique, and the low spin of the atom can give rise to small effects for high-spin nuclei.

2.1.5 He-Like Atoms

The two-electron atom can be used for g-factor measurements when the reaction kinematics do not allow sufficient velocity to populate the H-like species. In some cases, the larger spin of the electronic states gives larger amplitude oscillations of the anisotropy, making the measurements more sensitive. The important states to consider are the $^3S_1$, $^3P_2$, and $^3P_1$, since the $^1S_0$ ground state has no hyperfine field. The lifetimes of these states are given by MARRUS [2.32] and reproduced in Fig.2.6.

For heliumlike configurations with one electron in the n = 2 shell, the hyperfine interaction coefficients $A(^{2S+1}L_J)$ can be calculated approximately in terms of $A(^2S_{1/2})$ [2.28,33]. In the limit of pure *LS* coupling, the coupling constants for the important states are given as:

$$A(^3S_1) = 9/16\ A(^2S_{1/2}) \tag{2.12}$$

$$A(^3P_1) = 1/4\ A(^2S_{1/2}) \tag{2.13}$$

$$A(^3P_2) = 1/4\ A(^2S_{1/2}) \quad . \tag{2.14}$$

If more accuracy is required, relativistic Hartree-Fock calculations can be used, but in general the results differ by no more than 2%-3% from the formulae above.

The main difficulty confronting measurements with these He-like atoms is that the atomic population is split into many different configurations, including a sizeable fraction with no hyperfine interactions. Integral measurements are almost impossible to interpret unless strong assumptions are made concerning the relative populations, and differential measurements suffer because of the loss of effect from the fraction of the atoms with no field, and because the relative populations of the different active configurations require more parameters for the computer fit of the data. However, some information has been obtained about the relative population of excited states ($^3S_1$, $^3P_2$, $^3P_1$...) and the ground state ($^1S_0$). Again, this information has been summarized by DOUBT [2.30]. Figure 2.7 shows the measured ^3S and ^3P state population versus the ionic velocity parameter P_2 defined by DOUBT. This parameter is 1.0 at the peak of the heliumlike atomic charge state, 0.5 at the low-velocity half point and 1.5 at the high-velocity half point. Also shown as the dashed curve is the He-like total charge-state fraction. It can easily be seen that

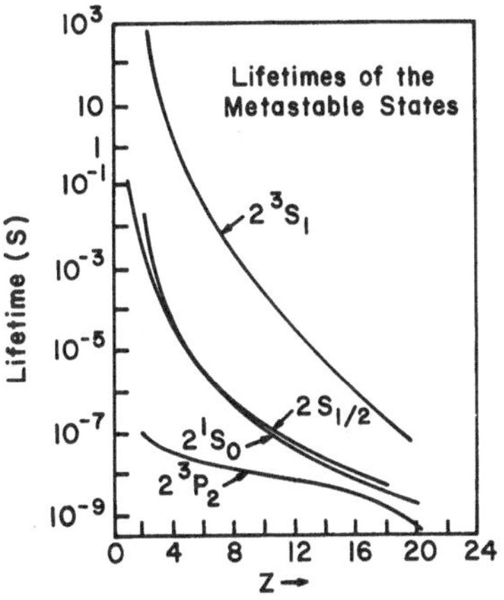

Fig.2.6. Lifetimes of metastable levels in H-like and He-like atoms vs Z [2.32]

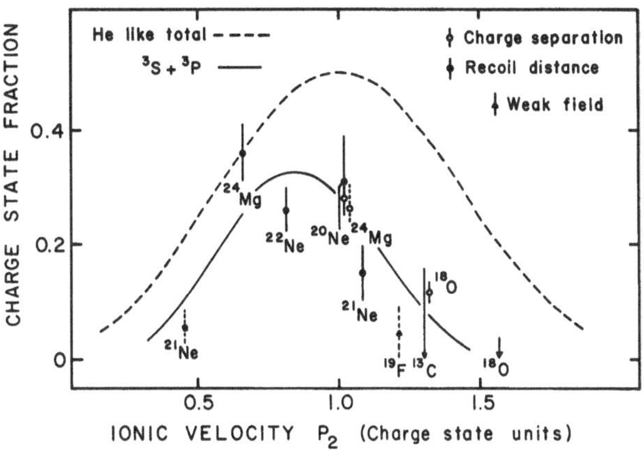

Fig.2.7. Measured excited-state population vs the the ionic velocity parameter P_2 [2.30]

the triplet-state population is shifted to the low-velocity side of the total distribution. This curve implies that the usual charge-state excitation curves should be further subdivided into many atomic-level excitation curves, and peak population of a particular level will not occur necessarily at the peak of the charge-state curve, but on the low-velocity side for weakly bound levels and on the high-velocity side for tightly bound levels of a particular charge state [2.2,30].

These ideas follow naturally from the earliest ideas of Bohr, which have been expanded by DMITRIEV [2.34]. Each electronic state is associated with the mean orbital velocity $v_I = (2E_I/m)^{\frac{1}{2}}$, where E_I is the binding energy of the state. The probability that an electron remains on the atom is then a universal function of the reduced velocity of the atom, v/v_I. This universal function is determined from the data for hydrogen. This model then predicts that the widths of each charge state, when plotted on a logarithmic velocity scale, will be very similar, since they result from the same universal function. The measured charge fractions bear this out. This width in velocity is called the "matching width" and has been extensively used by GOLDRING [2.2] and his collaborators to interpret their early integral measurements.

Figure 2.8 [2.30] shows the energy-level diagram of the different fluorine ions, with a velocity scale superimposed according to $v = \sqrt{2E/m}$. The braces indicate the measured full width half maximum (FWHM) of the charge distribution. The brackets correspond qualitatively with the positions of the actual energy levels of each charge state. From this kind of diagram, it is especially clear that the reason why the hydrogenic atomic is most ideal for achieving an essentially unique atomic population in the ground state is that it is so well separated from the excited states.

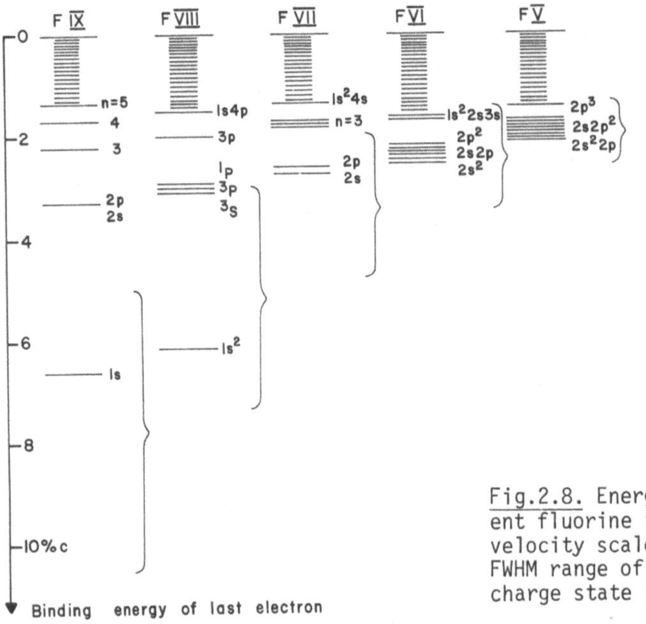

Fig.2.8. Energy level diagram of different fluorine ions, with a corresponding velocity scale. The brackets indicate the FWHM range of velocities over which the charge state is populated [2.30]

A further question remains as to whether the individual configurations are statistically populated. In particular, the questions remain whether states with the same n and ℓ, but with different total spin S are equally populated, and whether there is a dependence on the azimuthal quantum number ℓ if n is fixed. There is very little information on these questions for the more highly ionized systems, but beam-foil work [2.35] has shown that triplet-singlet ratios of between 3.2 and 2.3 are observed, indicating some spin dependence in the foil excitation. Populations of higher ℓ states of the same n are typically 1.5 times that of the s state, and not $2\ell + 1$. Of course, these results must depend strongly on the parameters of the system studied, but the message is that statistical population is most probably not a good assumption.

2.1.6 Li-Like Atoms

The Li-like system has been observed and utilized for hyperfine interaction measurements [2.9,36]. Table 2.3 gives an enumeration of the relevant atomic states which need to be considered in these measurements. The $^{19}F(5/2^+)$ level was used by RANDOLPH et al. [2.36] to test the theoretical predictions of the hyperfine interaction frequency in the $^2S_{1/2}$ ground state of the Li-like system. Although the quality of the data was not high, a value of 73.1(1.5) GHz was measured and agrees well with the calculation of Lindgren [Ref.2.36, Ref.10] of 71.6 GHz, which included relativistic and core-polarization corrections.

Table 2.3. Properties of Li-like atomic states

		ω/ω_{1s}	$\tau_m(^{20}Ne)$
$1s^2\ 2s\ 1s^22s$	$^2S_{1/2}$	~0.10	∞
$1s^2\ 2p\ 1s^22p$	$^2P_{1/2,3/2}$	~0.03	1.8 ns
$1s^1\ 2s^1\ 2p^1$	$^2P_{1/2,3/2}$	~0.03	
$1s^1\ 2s^1\ 2p^1$	$^4P_{1/2,3/2}$	~ 1	
$1s^1\ 2s^1\ 2p^1$	$^4P_{5/2}$	~ 1	10 ns

The excited quartet states have been observed extensively in beam-foil work, and at first sight should be useful for hyperfine interaction studies because of their long lifetime (the $^4P_{5/2} \rightarrow {}^2S_{1/2}$ transitions are *LS* forbidden and decay by M2 radiation) and because of their large hyperfine interaction strength (the fields from the unpaired 1s and 2s electrons add). However, the fraction of Li-like atoms in these states is usually small. HORSTMAN et al. [2.9] measured only about 1/4 of the Li-like population in all of the ^4P states. The beam-foil work can concentrate on the few atoms that are long lived in a sample, but the hyperfine interaction experiments require a large fraction of an atomic species to obtain a good signal.

2.1.7 Na-Like Atoms

Exploitation of the energy gap that exists between the ground state and excited states of other alkali-like systems becomes more and more difficult as the number of remaining electrons is increased. However, LIN et al. [2.37,38,12] have used a special technique (Sect.2.3.2) to increase the sensitivity of the recoil distance method in order to utilize the signal from a small fraction of atoms with a large hyperfine interaction strength. With this technique, they have performed measurements on the Na-like Ca system and measured hyperfine interaction constants of Na- and Mg-like atoms. Figure 2.9 gives the energy diagrams for several of the different charge states of Ca [2.39]. The measurements of LIN et al. for the populations are shown in Fig.2.10 and again indicate that the ground-state population is enhanced on the high-velocity side of the charge-state distribution, as would be expected from the matching velocity arguments. From energy-level diagrams such as that shown in Fig.2.9, it becomes apparent that as the possible level population becomes more dense, a statistical viewpoint becomes necessary in order to "understand" the measurements. However, this unfortunately results in a loss of the specificity and connection to known atomic properties which are part of the beauty of the recoil distance measurements.

Between the Na-like calcium atoms and the Li-like atoms is a wide range of charge states with a high density of atomic levels. The only experimenters to study

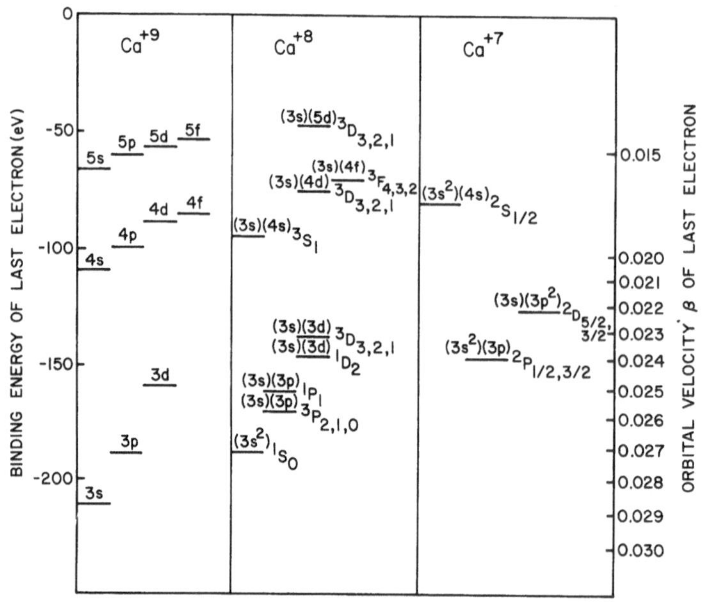

▲ Fig.2.9. Energy-level diagram of different calcium ions with a corresponding velocity compound scale

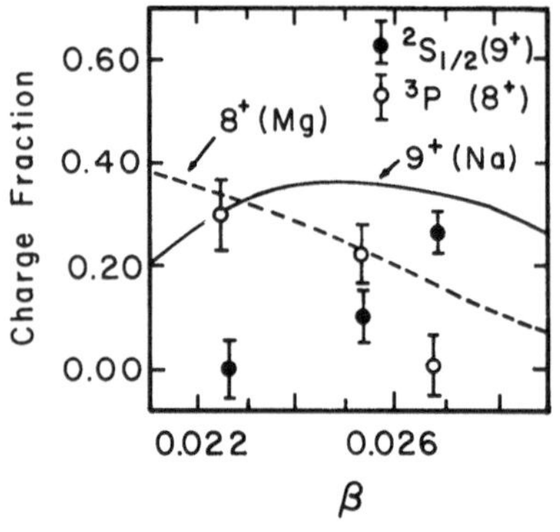

◀ Fig.2.10. Population of different atomic states vs velocity for Na- and Mg-like calcium ions [2.38]

this region so far have been LIEB et al. [2.40]. The high density of levels observed precluded connecting the observations with particular atomic states. A statistical method was used to interpret the data, and the average atomic spin of the states they observed was $J = 1.7 \pm 0.2$ for calcium and potassium ions with $v/c = 0.051$.

2.1.8 Other Atomic Systems in Vacuum

When the number of possible configurations that can be excited on emission from the foil becomes too large to consider in detail or there is too little information about these states, then a statistical approach similar to that used in the early experiments of GOLDRING and his collaborators [2.41] seems to be the only method of interpreting the data and extracting information. There has been considerable discussion

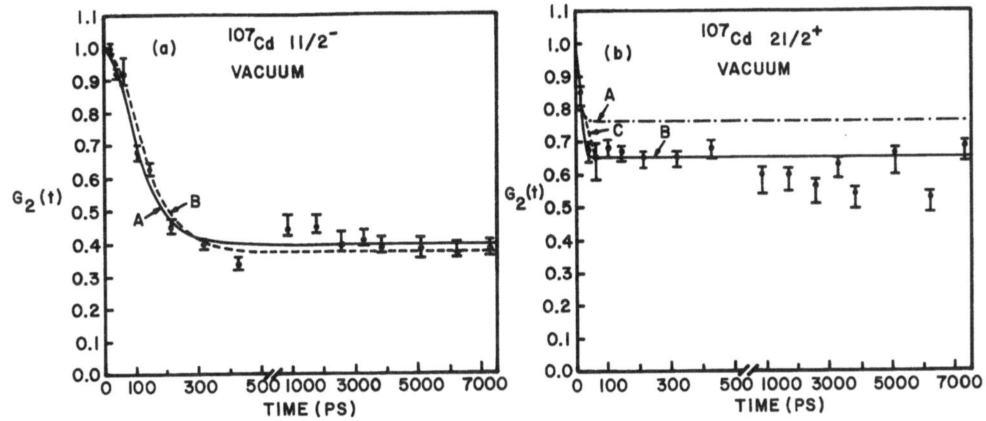

Fig.2.11a,b. $G_2(t)$ for two different ions in ^{107}Cd recoiling in vacuum at $v/c = 0.01$ [2.42]

over the years as to whether the interaction in these systems is static or fluctuating and whether the quadrupole interaction should be considered in addition to the magnetic interaction in analyzing measurements. Extremely good data have recently become available [2.42,43] that finally shed considerable light on these subjects.

The significant step taken by the Chalk River Group [2.42] was in finding a system with two long-lived isomers with different spins and known g factors which were populated under identical conditions of the atomic system. This has allowed for the first time a clear determination of the properties of the atomic system which caused the nuclear dealignment. Figure 2.11 shows the measured $G_2(t)$ for the two ^{107}Cd isomers recoiling in vacuum at 0.01c. Both curves exhibit a "hard core" at long times, which is indicative of static interactions. From (2.1), we can see that the time-independent term of the perturbation factor can be writen as:

$$G^{kk}(t) = \sum_F \frac{(2F+1)^2}{(2J+1)} \begin{Bmatrix} F & F & K \\ I & I & J \end{Bmatrix}^2 \quad . \tag{2.15}$$

This "hard-core" value is a function of the atomic spin J for a given nuclear spin I. The authors indicate that an average atomic spin of 3.5 can fit their data for both isomers. The preservation of alignment is greater for large I/J, so we see that high-spin nuclear levels can be transferred from a target to a catcher with the preservation of at least half of the alignment for $I > 11/2$ for flight times of the order of 10 ns.

The time dependence of the approach to the "hard core" is significantly different in the two systems. The $21/2^+$ state ($g = 0.866$) reaches the hard core in less than 50 ps, indicating that at this time the atom is no longer emitting photons and changing atomic states. On the other hand the $11/2^-$ state ($g = -0.195$) takes approximately 200 ps to reach the "hard core". The interpretation made by the authors is that a Gaussian distribution of static fields with a mean of 350T and standard deviation of 525T causes a dephasing of the static oscillations which appears as a Gaussian in time.

An exact mathematical treatment of the nonequilibrium atomic system is not possible at this time with currently available mathematical techniques. The Chalk River Group has analyzed their data with an Abragam-Pound treatment for the first 40 ps and deduced that about two atomic transitions take place in this time. Another alternative method is to model the system in two states, a fluctuating state and a static state, and to consider transitions between them [2.43], or to use a Monte Carlo method to model the system [2.44]. The discussion so far has been concerned mainly with medium mass atoms. The behavior of very heavy atoms recoiling into vacuum and gas has been studied to some extent with the recoil nuclei following α decay [2.45]. In addition, a recent experiment by METAG et al. [2.46] has observed very high charge states from Auger cascades for some atoms following emission from a foil after a nuclear reaction. The perturbations were studied using the conversion electron angular distribution. This work has led to the very elegant "charge-plunger" method to measure nuclear lifetimes [2.47,48].

2.2 Atoms in Gases

The use of a gaseous medium to study hyperfine interactions following nuclear reactions has a long history, with the first experiment being reported by TREACY [2.49] soon after the first work on γ-γ angular correlations. It was also soon realized that the randomizing collisions in a gas could be used to average motionally a hyperfine field so that the nuclear alignment could be preserved in flight [2.50]. Many other significant developments in both experimental technique and theoretical framework for understanding the results have taken place since these early experiments. In order to cover them, the discussion has been divided into three parts.

The first part is a discussion of the interaction of fast ions where the observation time is very short compared to the slowing-down time, so that a constant velocity can be assumed. In the second part, effects resulting from atoms that have reached thermal equilibrium will be discussed, and in the third, the intermediate velocity region, which is the most difficult to understand.

2.2.1 Fast Atoms in Gases

A fast atom moving through a gas will undergo collisions with the host gas and as a result can either gain or lose electrons, or be excited to a different state of the same ionic charge. The hyperfine interaction will change as the atom samples the

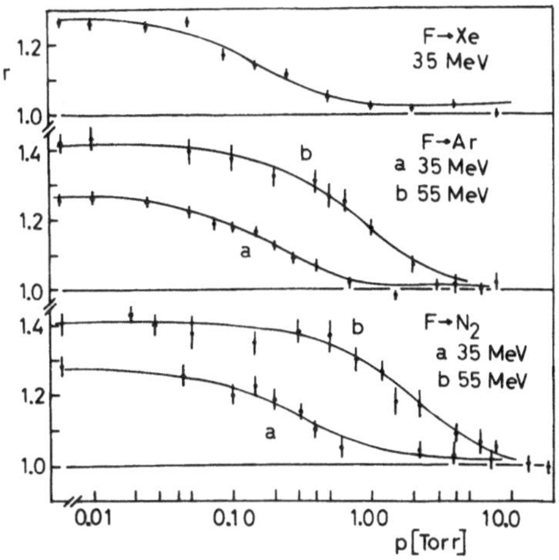

Fig.2.12. Ratio, r of γ-ray intensities at two orthogonal angles for $^{19}F(5/2^+)$ ions after traveling through a 12-cm path at different pressures [2.52]

many different possible atomic configurations of the system. In general, the cross sections for these processes are not known, and for heavier systems, the number of possible charge states and atomic levels involved is so large that a calculation of the expected perturbation from these cross sections is not feasible. However, for the lighter systems, much of the cross-section information is available from the extensive literature on x rays excited by particle beams. HAGEMEYER et al. [2.51] and BRENN et al. [2.52] have used these data to make a first-principles analysis and have tried to relate the observed dealignment at different pressures to the experimentally measured pickup and loss cross sections. Figure 2.12 shows the experimentally determined dealignment for $^{19}F(5/2^+)$ after flight through a gas cell of 12 cm length for different gases [2.52].

The gross dependence of the dealignment is that the heavier the host gas, the lower the pressure at which half of the alignment is lost. However, some deviations occur, as Xe is not as effective as Kr for dealignment. The gross structure and the deviations are well explained by a microscopic model which is based on the pickup and loss cross-section data, but only if one postulates significant inelastic excitation and collisional quenching of the He-like fluorine atom, which can produce a large dealignment due to the strong hyperfine field in the $^3S_1$ state. A similar conclusion was drawn from decoupling measurements by GOLDBERG et al. [2.53].

2.2.2 Statistical Approach

It is clear that the number of possible atomic states to be considered in calculating the perturbation factors is very large and that much of the necessary information is unknown, so that connecting the observed effects on the nucleus with specific properties of the atomic systems involved is not feasible in most cases. The initial parametrization of the problem [2.41] was to define an average effective hyperfine field $\langle H_{eff} \rangle$ and to utilize the Abragam-Pound expression:

$$G^{kk}(t) = \exp(-\lambda_k t) \qquad (2.16)$$

$$\lambda_k = \frac{k(k+1)}{3} \langle\omega^2\rangle\tau_c$$

$$\hbar\omega = g_I \mu H_{eff} \quad .$$

The conditions $\langle\omega^2\rangle^{\frac{1}{2}}\tau_c \ll 1$ and $t/\tau_c \gg 1$ have to be fulfilled for (2.16) to be valid. The correlation time τ_c (gas) for a random reorientation of the hyperfine field direction by gas collision was related to the collision time, and therefore was taken as inversely proportional to the pressure. The observation that some perturbation occurs at zero pressure required the introduction of a vacuum correlation time τ_c(vac), so that the total fluctuation rate was given by $1/\tau_c$(vac) + $1/\tau_c$(gas). The interpretation of the early results for Sm and Nd isotopes indicated τ_c(vac) $\simeq$ 3 ps. The value τ_c(vac) = 3 ps has received widespread usage for entirely different regions of the periodic table and for time ranges well outside the 100-ps times for which it was initially determined. When analyzed with the Abragram-Pound formalism, the time differential data of the Chalk River group [2.42] for the Cd isotopes give τ_c(vac) $\approx$ 20 ps, an order of magnitude larger than for the Sm system. It is clear that the stage of ionization and the atomic configuration involved will play an important role in the range of atomic lifetimes in a cascade to the ground state of the atomic system, and that large variations in this parameter will occur. In addition, more extensive recent work by the Weizmann Institute group [2.54] has led to an interpretation of their original data as a distribution of static interaction strengths which mimics an exponential decay. In this case, the dependence of the perturbation on the nuclear g factor in the short time region becomes linear rather than quadratic.

2.2.3 Nuclear Spin Dependence

The replacement of the hyperfine interaction with an effective field is only an approximation, and gives rise to erroneous results when levels of different spin are compared. In particular, nuclear spins I, large compared to the atomic spins J, are almost parallel to F, the free-atom angular momentum, which is fixed in space if no external fields are applied. Therefore, the constant term in the perturbation factor $G^{kk}(t)$ is large, and very little perturbation takes place [2.55]. In this case, replacement of J with a classical field is grossly incorrect and can lead to erroneous results. This topic has recently been discussed extensively by GOLDRING [2.56].

2.2.4 Relaxing the Condition $<\omega^2>^{\frac{1}{2}}\tau_c \ll 1$

The pressure dependence of the perturbation, especially at low pressure [2.43] shows some deviation that can be interpreted as a deviation from the Abragam-Pound formulae. A higher-order extension of the Abragam-Pound theory can be used in this situation, and the corresponding formulae are [2.57]:

$$G_{kk}(t) = (1 - \lambda_k\tau_c)\exp(-\lambda_k t)\tau/\tau_c \gg 1$$

$$\lambda_k = \frac{k(k+1)}{3}\left[\frac{a(J)}{h}\right]^2 \cdot J(J+1)\tau_c \times \left\{1 - 1/400\left[\frac{a(J)}{h}\right]^2 \right.$$

$$\left. \tau_c^2[(2I+1)^2 + 3k^2 + 3k - 1] \times (2J-1)(2J+3)\right\} \quad . \qquad (2.17)$$

The first-order theory gives a fixed ratio of $\lambda_2/\lambda_4 = 0.3$. Deviations from this value had been interpreted as evidence for some mixture of quadrupole interaction [2.58]. However, these second-order formulae allow for deviations of λ_2/λ_4 from 0.3 within the scope of purely magnetic interactions [2.43].

A nonperturbation approach to the study of hyperfine interactions in gas has been taken by SCHERER [2.59], with some extensions and corrections by BLUME [2.60]. Within their theory, the calculation of the attenuation coefficients involves the solution of a polynomial equation with the degree determined by the number of hyperfine frequencies. For simple atoms such as hydrogen or helium, the number of frequencies is small and the numerical work is tractable. However, when one considers an atomic system with a large number of possible levels, the Scherer-Blume theory is difficult to apply. ZEMEL and NIV [2.61] have analyzed the Scherer-Blume theory for a large number of contributing states, each with a small fraction of the total population, and have shown that the minimum in $G_k(t)$ versus τ_c occurs at $<\omega^2>^{\frac{1}{2}}\tau_c = 1$. This fact can be used to make relative g-factor measurements of different nuclear states in the same atomic system by measuring the pressure dependence of the perturbation.

2.2.5 The Approach to Thermal Equilibrium - Chemical Effects

The use of inert gases as a medium to preserve the nuclear alignment from a reaction for times greater than 10^{-3} holds great promise, since depolarization times longer than 10^{-2} have been observed in optical pumping experiments [2.62]. However, one of the important obstacles yet to be handled is how to preserve the alignment during the slowing-down time for a general atomic species. The work of BLECK et al. [2.63], which we will now discuss briefly, has shown that the chemical nature of the host gas is important in preserving the nuclear alignment. Figure 2.13 shows the observed attenuation coefficient $G_2(t)$ as a function of concentration of methane gas in a mixture of SF_6 and methane held at 1 atmosphere pressure. The slowing-down time to thermal energies was calculated as 0.9 ns. The measurement of the nuclear

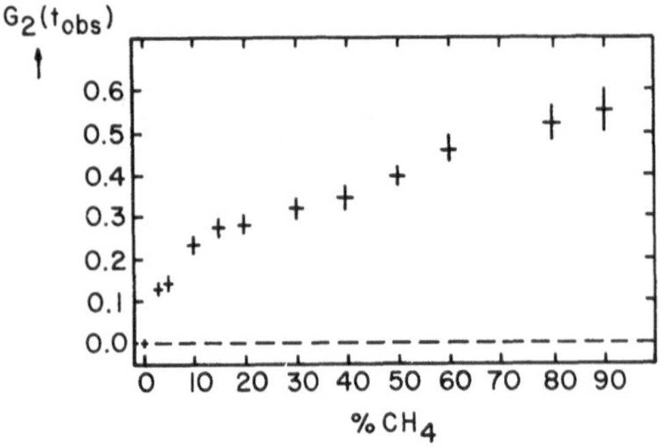

Fig.2.13. Attenuation of the γ-ray anisotropy 50 ns after excitation (relative to a thick CaF_2 target) as a function of the CH_4 concentration in the gaseous SF_6+CH_4 target of 1 atmosphere total pressure [2.63]

alignment starts only after about 40 ns, and a constant anisotropy was observed after this time, with the amount of anisotropy varying with CH_4 concentration as shown in the figure. The observed change in anisotropy versus pressure is attributed to processes which take place during the slowing-down time, because the amplitude remains constant at long times independent of the CH_4 concentration.

An exact treatment of the dealignment during the slowing-down time would require knowledge of the collision cross sections, the hyperfine interaction constants of the atomic states, and the collision times, all as a function of the energy of the slowing-down atom. An approximate analysis was first made by KLEPPER and SPEHL [2.64], who tried to explain the isotropic correlations they observed for $^{19}F(5/2^+)$ in all gases they measured. The correlation time, τ_c, should be approximately equal to the mean free time between collisions, which should be approximately inversely proportional to the velocity of the recoil. BLECK et al. [2.63] estimated for $^{19}F(5/2^+)$ that

$$\tau_c = \left(\frac{E_{thermal}}{E}\right)^{\frac{1}{2}} \times 0.1\ \text{ns} \quad . \tag{2.18}$$

The average hyperfine interaction constant for the free atomic $^{19}F(5/2^+)$ is $\omega \approx 9$ GHz. At high velocities, the atoms are more highly ionized and the hyperfine fields, corresponding to more tightly bound unpaired inner electrons, can be much larger. Therefore, as the atom slows down, the effective time probably becomes smaller and the correlation time longer, making the behavior of the relevant parameter of the Abragam-Pound theory, $\omega^2\tau_c$, difficult to predict. Near the end of the stopping time, it is clear for the $^{19}F(5/2^+)$ case of BLECK et al. that the parameter $\omega^2\tau_c$ becomes large because of the large hyperfine interaction frequency of the atomic ^{19}F, and the alignment is lost. The introduction of methane into the SF_6 target gas induces the chemical reactions:

$$F + CH_4 \rightarrow HF + CH_3$$

where the HF molecule provides a stable diamagnetic surrounding for the flourine nuclei during the final slowing down and thermalization, thus preserving the nuclear alignment. The variation of alignment observed with CH_4 pressure can then be explained as varying the fraction of atoms undergoing the above chemical reaction.

A similar investigation on the effect of impurity gas on the preservation of alignment was recently made by IKEZOE [2.65]. In it, an increase of the alignment of over a factor of two was observed for ^{86}Sr in Kr when the Kr was diluted to 50% with CO_2. The investigations were done with zero field and p = 1.6 atm. The chemical reactions taking place in this system probably inhibit the formation of Sr^+ atoms, which have a strong hyperfine field which can relax the nuclear alignment in a short time.

2.2.6 Thermalized Atoms in Rare-Gas Hosts

Several experiments have now been performed with the time differential method with nuclear probes in rare-gas hosts. Table 2.4 summarizes some of the experiments with longer-lived probes. In general, a fraction of the alignment is preserved during the slowing-down process and is apparently constant during the observation times, which are of the order of microseconds. The interpretation which is suggested for most systems is that the high charge states are neutralized as the ions slow down, until a stable charge state in which there is no hyperfine field is achieved for some fraction of the nuclei. For example, the stable state postulated for ^{85}Rb could be Rb^+, as the reaction:

$$Rb^+ + Kr \rightarrow Kr^+ + Rb - 9.83 \text{ eV}$$

is endothermic. Of course, the kinetic energy of the recoil could drive this reaction, but presumably some fraction of the atoms could be thermalized without the reaction taking place. BARTSCH et al. [2.66] indicated that 90% of the atoms are in the diamagnetic 1^+ state after slowing down.

Table 2.4. Summary of recoil into rare-gas hosts

Host gas	Nucleus	$t_{1/2}$	g	A_2[1 atm]	Reference
Ne	$^{22}Na(3^+)$	168 ns		0.01	BLECK et al. [2.63]
Ar	$^{37}Ar(7/2^-)$	3.2 ns	+0.365(20	0.10	BLECK et al. [2.63]
Kr	$^{85}Rb(9/2^+)$	1 μs	+1.350(14)	0.03	BARTSCH et al. [2.66]
Kr	$^{86}Sr(8^+)$	0.46 μs	-0.242(4)	0.06	IKEZOE et al. [2.67]
Xe	$^{138}Ba(6^+)$	0.77 μs	0.98(2)	0.02	IKEZOE et al. [2.67]

The same type of argument can be considered for Sr [2.67], where alignment is observed for the 8^+ state of ^{86}Sr in Kr gas. The reaction

$$Sr^{++} + Kr \rightarrow Sr^+ + Kr^+ - 2.97 \text{ eV}$$

is endothermic, and therefore some fraction of the Sr atoms should remain in the stable Kr-like Sr^{++} configuration. Indeed, it is this fraction which probably ac-

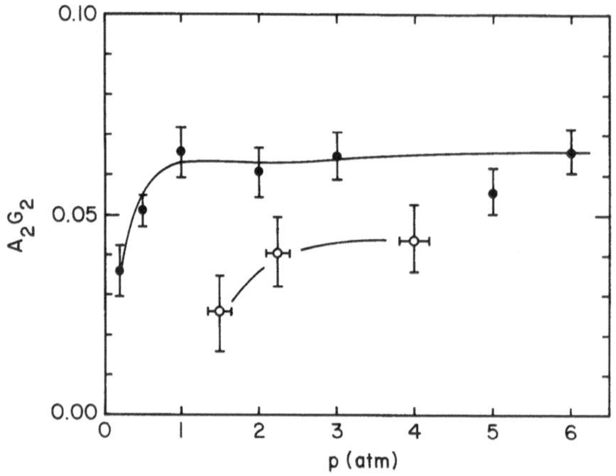

Fig.2.14. Pressure dependence of the nuclear alignment. The filled circles are for ^{86}Sr in Kr [2.68] and the open circles are for ^{85}Rb in Kr [2.67]

counts for the preserved alignment. Again, some fraction of the atoms should be converted to the Sr^+ by the above reaction, which is driven by the kinetic energy. The strong hyperfine interaction of the $^2S_{1/2}$ atomic state will couple to the large nuclear spin of the isomer and prevent the nuclei from precessing at the nuclear Larmor frequency. Instead, the atomic spin F will precess at the atomic Larmor frequency, which is much larger. The angular correlation will precess along with the coupled atom, and the spin rotation in lower fields can be used to determine this fraction and measure the atomic g factor [2.68,67].

2.2.7 Pressure Dependence of the Alignment

The pressure dependence of the alignment has been observed for ^{86}Sr in Kr and for ^{85}Rb in Kr. Very similar results are observed, as can be seen in Fig.2.14, where the results are summarized. The alignment is essentially constant above 1 atm for Sr and 2 atm for Kr, and below these values falls with decreasing pressure. The larger pressure required for saturation of the Rb isomer can qualitatively be understood in terms of the larger g factor of the nucleus, and further indicates that the dealignment is taking place during the slowing-down time, where the parameter $\omega^2\tau_c$ is important. For the Sr case, IKEZOE [2.65] estimated that if the active field during the slowing-down time were comparable to the 2.7 MG field of the ^2S state of the Sr^+ ion, then the time this field acts is approximately $T = (5(2) \times 10^{-10})/p$ s, where p is the pressure in atmospheres. As the excited-state fields are probably somewhat larger than 2.7 MG, this time should be viewed as an upper limit, i.e., at 1 atmosphere of pressure, the field is active for greater than 1/2 ns.

2.2.8 Magnetic Field Dependence

The transverse field dependence for the ^{86}Sr and ^{85}Kr isomers in Kr gas and the ^{138}Ba isomer in Xe gas are shown in Fig.2.15. Very little structure is observed in either case, except for a slow rise with decreasing field for ^{86}Sr. Since the

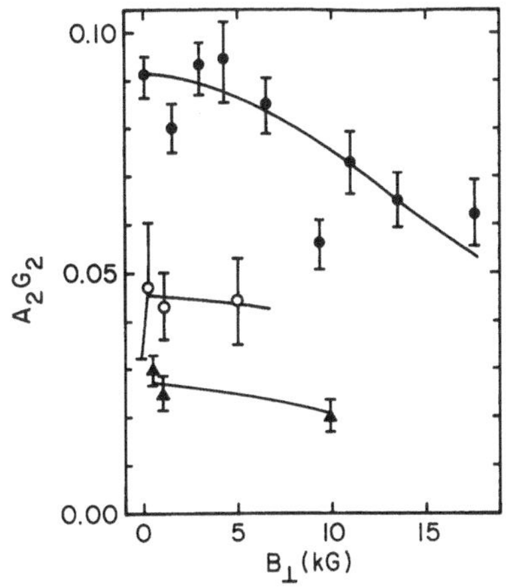

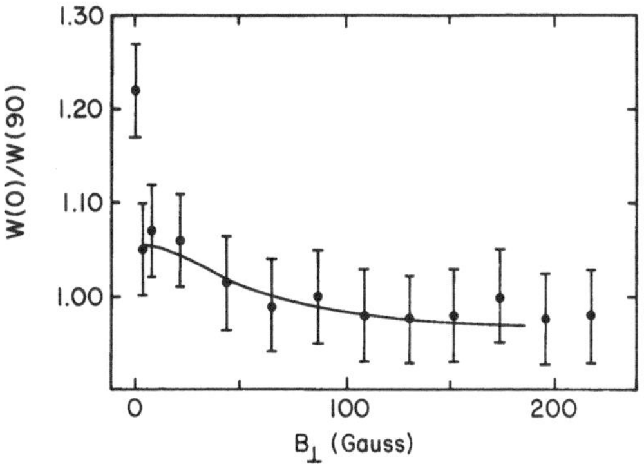

◄ Fig.2.15 ▲ Fig.2.16

Fig.2.15. Magnetic field dependence of the nuclear alignment. The filled circles are for ^{86}Sr in Kr [2.68], the open circles for ^{86}Rb in Kr [2.67], and the filled triangles for ^{138}Ba in Xe [2.68]

Fig.2.16. Magnetic field dependence at low fields for ^{85}Rb in Kr [2.67]

correlation times during the slowing down are too short to consider dealignment by a coupling of the external field to the nuclear g factor, one must consider another mechanism. An indirect coupling to the external field through the strong hyperfine field can take place, with the whole atom precessing with the much larger atomic g factor for a short time during the slowing-down process. Since these precessions are randomly oriented, an integral dealignment occurs and gives rise to the usual Lorentzian-shaped curve centered around zero field. For the data shown in Fig.2.15, only about 1/3 of the nuclei are evidently affected by this process, as the high-field data suggest a leveling off of the observed anisotropy.

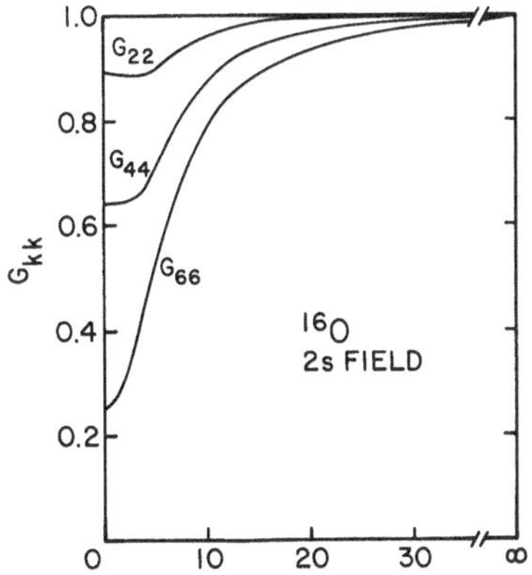

Fig.2.17. Decoupling curves for $^{16}O(3^-)$ in the 3-electron ground state [2.11]

At very low fields, BARTSCH [2.69] also observed a slight increase in the anisotropy with decreasing fields and a striking increase when the field was reduced below a few gauss. This dramatic increase is seen in Fig.2.16 and is a result of that fraction of the atoms in the neutral Rb atom. These atoms do not take part in the Larmor precession at the nuclear frequency, and their anisotropy is presumed "lost," but they are actually strongly coupled via the hyperfine interaction to the atomic spin and are stopped from precessing at the atomic Larmor frequency only when the field is reduced below several gauss.

2.3 Magnetic Decoupling Measurements in Vacuum

2.3.1 Fields Parallel to the Quantization (Beam) Axis

When an external magnetic field is applied along the quantization axis, the interaction Hamiltonian of the system is given by [2.14]:

$$a(I\cdot J) + \mu_B g_J H_a J_z + \mu_N g_I H_a I_z \quad .$$

In a sufficiently large external magnetic field, the atomic and nuclear angular momenta are decoupled (Paschen-Back effect), and as a consequence the angular correlation will no longer be perturbed. A calculation of the angular correlation at intermediate fields B [2.70] requires diagonalization of the Hamiltonian at each field, and typical curves are shown in Fig.2.17 [2.11]. The half point of the curve occurs where $g_I\mu_N H_{hf} \approx \mu g_J\ _B H_{ext}$, i.e., where the first two terms of the Hamiltonian are about equal in magnitude. Unfortunately, the rise with field is rather smooth, which lessens the sensitivity of the decoupling technique for precise measurements of g factors. The effect of the larger value of the "hard-core" anisotropy for smaller tensor orders k can also be seen in Fig.2.17. The calculation is for I = 3 and J = 1/2, and the maximum perturbation for k = 2 is only 10%. Even for a large value of $A_2 \approx 0.5$, the maximum observed change in intensity would be 5%. Therefore it is a necessary prerequisite that these kinds of measurements be performed to work with a large k = 4 or k = 6 correlation [2.11].

2.3.2 Strong Magnetic Fields Transverse to the Beam Direction

The effect of the hyperfine interaction becomes more and more difficult to observe by means of the perturbed angular correlation method as the spin of nucleus becomes larger. Since many of the nuclear states of current interest have high spin, methods to measure these interesting moments must be found. Another related difficulty in utilizing specific atomic states to perturb nuclei is that in some regions, the atomic systems have so many states populated that the perturbation factors in turn have so many Fourier components as to make it impossible to identify any one specific state. SPROUSE [2.12] and LIN et al. [2.37,38] have developed a method

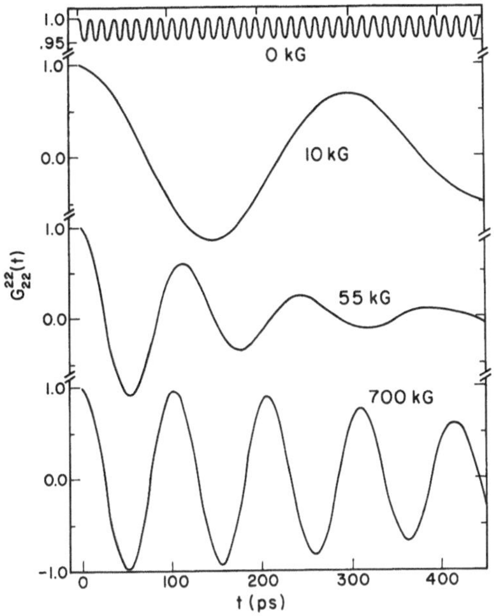

Fig.2.18. Perturbation factor $G_{22}^{22}(t)$ for Na-like $^{41}Ca(15/2^+)$ state in different magnetic fields transverse to the beam direction

of circumventing some of these difficulties by utilizing a strong transverse magnetic field from a superconducting magnet and by observing the anisotropy time differentially with the recoil distance method.

The experimental system chosen [2.71] to investigate the transverse decoupling effect was $^{41}Ca(15/2^+)$ in a Na-like atom. The $15/2^+$ isomer in ^{41}Ca with a mean life τ = 4.7(2) ns and a g factor of +0.29(2) [2.72] can be populated intensely by several heavy-ion fusion-evaporation reactions. Two different decay gamma rays of the isomer have large and opposite anisotropies, making the detection of the nuclear alignment particularly favorable.

The calculation of the perturbation factor in the strong magnetic field must be carried out numerically by diagonalizing the Hamiltonian at each value of the field. The results for $G_{22}^{22}(t)$ are shown for the $^2S_{1/2}$ ground state of the Na-like atom (H_{hf} = 21.3 MG) in Fig.2.18. The top of the figure shows the zero-field precession of I around J at the hyperfine interaction frequency. Because of the large nuclear spin, the amplitude of the effect is less than 3%, which would make it impossible to observe experimentally. At the intermediate field of 10 kG, the precession of the strongly coupled atom takes place at the frequency essentially given by $\hbar\omega_F = g_J\mu_B\{F(F+1) + J(J+1) - I(I+1)/[2F(F+1)]\}$. The only nuclear information available from this spectrum would be the value of the nuclear spin I.

At fields greater than about 25 kG the atom begins to decouple, and the oscillations shown for 55 kG correspond to the nuclei precessing in the Z component of the hyperfine field, which is strongly coupled to the external field. Notice that the period of the oscillations is essentially the same for the 55 kG case as it is

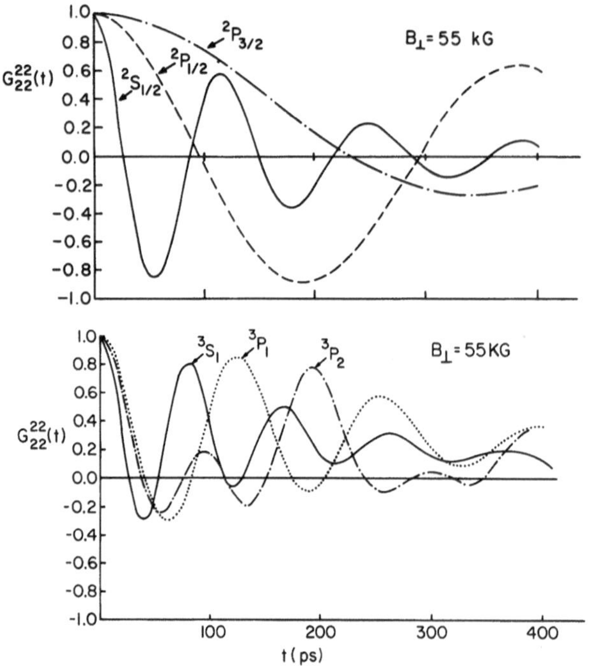

Fig.2.19a,b. Perturbation factors for different states of (a) the Na-like $^{41}Ca(15/2^+)$ atom and (b) Mg-like atoms at 55 kG transverse field

for the experimentally impossible field of 700 kG, but that a damping of the oscillations is observed at the lower field because the energies of the separate states are still not uniformly spaced.

The most important aspect of the calculation is that a large oscillation of the anisotropy can be achieved by the application of a strong transverse field. It should be noted that the period of the oscillation of the completely decoupled system is a factor of $|\vec{I}|/|\vec{J}|$ longer than the zero-field oscillation period. In both cases, the "torque" causing the precession is the hyperfine interaction. In the zero-field case, $\vec{J}$ precesses around the large $\vec{I}$, but in the decoupled case, $\vec{J}$ is "clamped" to the external field and the large $\vec{I}$ is forced to precess, but at a rate which is $|\vec{J}|/|\vec{I}|$ times slower. However, the decoupled situation produces a large effect on the γ-ray anisotropy. Figure 2.19 shows the perturbation factor for atomic configuration appropriate to Mg-like Ca atoms. The different angular momenta give rise to differently shaped curves. Note that all integer J values have a constant term added to $G_{22}^{22}(t)$, corresponding to the M = 0 state where the hyperfine field averages to zero.

Figure 2.20 shows the measured $G_{22}^{22}(t)$ in a transverse field of 55 kG for v/c = 2.25% (upper part) and v/c = 2.67% (lower part). The data have been fit with different fractions of the $^2S_{1/2}(9^+)$ and $^3P(8^+)$ time spectra as shown in Fig. 2.10. At the higher velocity, approximately 25% of the atoms are in the $^1S_0$ Na-like atomic state, and the $^2S_{1/2}$ patterns shown in Fig. 2.16 can easily be discerned in the data.

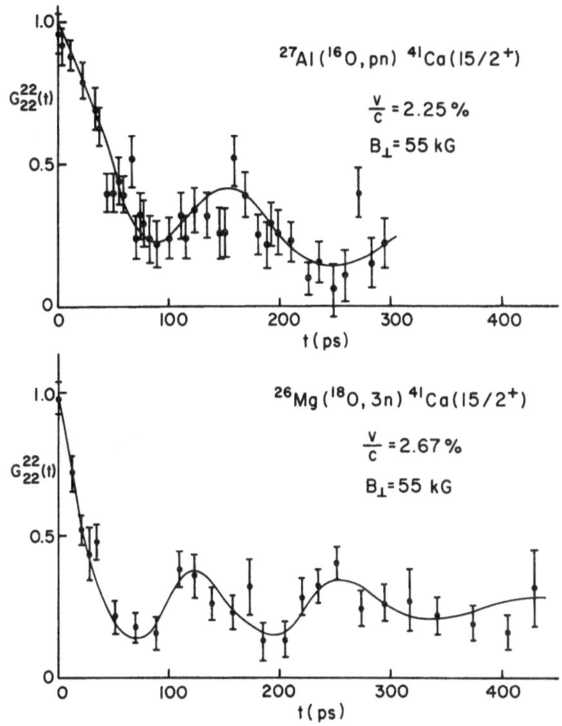

Fig.2.20. Time differential experimental data for $^{41}Ca(15/2^+)$ at different recoil velocities [2.71]

The method outlined here has been used to measure other g factors in the calcium region [2.71], but the requirements on the field intensity, recoil velocity, and atomic state population are quite restrictive, so that general application of this method is difficult. However, it will remain important for a few special cases where it can be applied.

2.4 New Methods and Future Directions

Almost all of the experiments described in this review have required the use of an accelerator to induce a nuclear reaction. As more time becomes available on various new heavy-ion accelerators, we can expect to see an extension of the recoil distance method to heavier nuclei than those shown in Fig.2.3. However, there are certain limitations on the time resolution obtainable with the "plunger." This will make it necessary to utilize the less desirable alkali-like systems, if the oscillations of the γ-ray anisotropy are to be resolved. Figure 2.21 shows the ion energy necessary to produce highly stripped ions in several representative charge states, along with the output energy curve of a modest heavy-ion accelerator (the Stony Brook FN Tandem and a superconducting post accelerator).

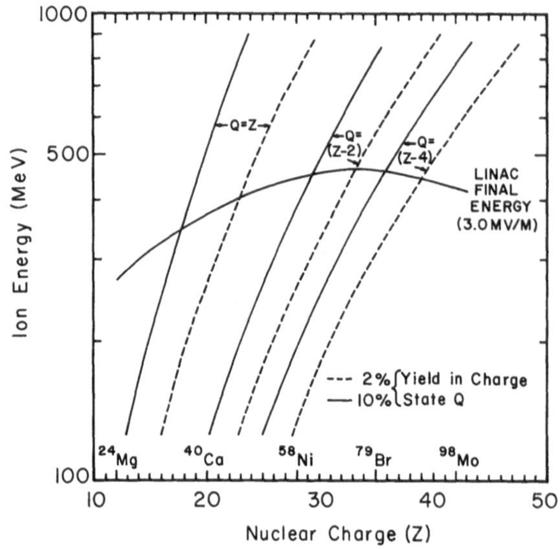

Fig.2.21. Ion energy necessary to produce highly stripped atoms for heavy ions. For reference, the output energy for the Stony Brook LINAC is included

Even this modest accelerator opens up a new mass region for precise g-factor measurements of short-lived low-spin levels, and the larger accelerators now available can extend this range even further.

The application of large electromagnetic fields, either from lasers or from resonant cavities, tuned to the resonance frequency of atomic or hyperfine transitions, is an area that is still in its infancy, and we can expect many interesting results in the future. Intensity is usually a problem in these experiments, as one is effectively using the beam of charged particles as a target for the laser beam, but these difficulties can be overcome in many cases.

The study of aligned nuclei stopped in rare-gas hosts may open up the large nuclear lifetime range from ms to minutes; this region is at present essentially inaccessible to nuclear moment measurements. The techniques for preserving the nuclear alignment during the slowing-down time and after stopping must first be developed, and then, when coupled with the usual NMR, a very powerful technique would be available for measuring many interesting and important moments.

One of the areas in which atomic systems were first used to gain understanding of nuclear systems was in the determination of the nuclear spin I by counting the number of different hyperfine components in an optical spectrum. A current problem in nuclear physics is the determination of the exact spin of large angular momentum isomeric levels, which are high in excitation energy in the nucleus. Imbedding these levels in an atomic system and measuring g_F by application of a weak field could allow an unambiguous determination of the nuclear spin.

Related to the determination of spins of the high-spin isomers is the determination of their moments. The recoil-into-gas method discussed above may be one of the best ways in which the moments of the very high-spin isomers can be determined. Much current interest and effort is being concentrated in this direction.

Finally, the behavior of paramagnetic ions in gases should be studied much more extensively, with the aim of developing methods for moment measurements in the rare-earth and actinide regions. These are at present essentially inaccessible except for a small "window" near the middle of the shell because of the strong hyperfine fields and consequent paramagnetic relaxation in solid or liquid hosts.

References

2.1 K. Alder: Helv. Phys. Acta *25*, 235 (1952)
2.2 G. Goldring: "Hyperfine Interactions in Isolated Ions", in *Heavy Ion Physics*, ed. by R. Bock (North-Holland, Amsterdam forthcoming) Chap.4
2.3 B. Rosner, R. Kalish (eds.): *Atomic Physics in Nuclear Experiments*, Annals of the Israel Physical Society, Vol.1 (Adam Hilger, Bristol 1977)
2.4 G. Goldring: "Internal Fields and Implantation Techniques", in *Hyperfine Structure and Nuclear Radiations*, ed. by E. Matthias, D. Shirley (North-Holland, Amsterdam 1968) pp.640-654
2.5 G. Sprouse: "Hyperfine Interactions of Fast Recoil Nuclei in Gas" in *Hyperfine Interactions of Excited Nuclei*, ed. by G. Goldring, R. Kalish, Proc. Int. Conf., Rehovot, 1970 (Gordon and Breach, New York 1971) pp.931-960
2.6 M.B. Goldberg: Phys. Scr. *11*, 84 (1975)
2.7 W.L. Randolph, N. Ayres de Campos, J.R. Beene, J. Burde, M.A. Grace, D.F.H. Start, R.E. Warner: Phys. Lett. B*44*, 36 (1973)
2.8 T.J. Moorhouse, J. Asher, H.A. Doubt, M.A. Grace, P.D. Johnston, W.L. Randolph, P.M. Rowe: Hyperfine Interact. *4*, 179 (1978)
2.9 R.E. Horstman, J.L. Eberhardt, H.A. Doubt, C.M.E. Otten, G. Van Middlekoop: Nucl. Phys. A*248*, 291 (1975)
2.10 G. Perlow: "Hyperfine Anomalies", in *Hyperfine Interactions of Excited Nuclei*, ed. by G. Goldring, R. Kalish, Proc. Int. Conf., Rehovot, 1970 (Gordon and Breach, New York 1971) p.651
2.11 H.C. Jain, W.A. Little, S.M. Lazarus, T.K. Saylor, B.B. Triplett, S.S. Hanna: Phys. Rev. C*14*, 2013 (1976);
W.A. Little: Ph.D. Thesis, Stanford University (1975)
2.12 G.D. Sprouse: Hyperfine Interact. *2*, 39-44 (1976)
2.13 H.J. Andraä: "Quantum Beats in Atomic and Nuclear Physics", in Ref.2.3, pp.373-440
2.14 H. Kopferman: *Nuclear Moments* (Academic, New York 1958)
2.15 J.R. Beene, J. Asher, N. Ayres de Campos, R.D. Gill, M.A. Grace, W.C. Randolph: Nucl. Phys. A*230*, 141 (1974)
2.16 T.K. Alexander, G.J. Costa, J.S. Forster, O. Häusser, A.B. McDonald, A. Olin, W. Wittuhn: Phys. Rev. C*9*, 1748 (1974)
2.17 T.J. Moorhouse, J. Asher, D.W. Bennett, H.A. Doubt, M.A. Grace, P.D. Johnston, W.L. Randolph, P.M. Rowe: J. Phys. G*4*, 1593-1605 (1978)
2.18 J. Asher, J.R. Beene, M.A. Grace, W.L. Randolph, D.F.H. Start: J. Phys. G*1*, 415 (1975)
2.19 F.A. Beck, T. Byrski, G.J. Costa, W.L. Randolph, J.P. Vivien: Hyperfine Interact. *4*, 181 (1978)
2.20 W.L. Randolph, N. Ayres de Campos, J.R. Beene, J. Burde, M.A. Grace, D.F.H. Start, R.E. Warner: Phys. Lett. B*44*, 36 (1973)
2.21 J. Asher, M.A. Grace, P.D. Johnston, J.W. Koen, P.M. Rowe, W.L. Randolph: J. Phys. G*2*, 477-486 (1976)
2.22 Z. Berant, C. Broude, S. Dima, G. Goldring, M. Hass, Z. Shkedi, D.F.H. Start, Y. Wolfson: Nucl. Phys. A*235*, 410 (1974)

2.23 Z. Berant, C. Broude, G. Engler, M. Hass, R. Levy, B. Richter: Nucl. Phys. A*243*, 519-527 (1975)
2.24 T.J. Moorhouse, J. Asher, H.A. Doubt, M.A. Grace, P.D. Johnston, P.M. Rowe: To be published
2.25 P.M. Rowe, J. Asher, H.A. Doubt, M.A. Grace, P.D. Johnston, T.J. Moorhouse: J. Phys. G*4*, 431 (1978)
2.26 R. Bohm, R. Keitel, W. Klinger, W. Wittuhn, W.L. Randolph, E. Matthias: Z. Phys A*278*, 133 (1976)
2.27 F.A. Beck, T. Byrski, G.J. Costa, J.P. Vivien: Hyperfine Interact. *4*, 188 (1978)
2.28 F.A. Beck, Y. Dar, M. Forterre, J.P. Vivien: Phys. Rev. C*13*, 895 (1976)
2.29 J. Bleck, D.W. Haag, W. Ribbe: Nucl. Instrum. Methods *67*, 169 (1969)
2.30 H.A. Doubt: "Excitation Regimes in Hydrogen- and Helium-Like Ions", in Ref.2.3, pp.537-575
2.31 M. Faessler, B. Povh, D. Schwalm: "Deorientation Measurement in ^{20}Ne", in Ref.2.3, pp.968-969
2.32 R. Marrus: Nucl. Instrum. Methods *110*, 333 (1973)
2.33 H.A. Bethe, E.E. Salpeter: *Quantum Mechanics of One- and Two-Electron Atoms* (Springer, Berlin, Göttingen, Heidelberg 1957)
2.34 I.S. Dmitriev: Zh. Eksp. Teor. Fiz. *32*, 570 (1957); Sov. Phys. JETP *5*, 473 (1957)
2.35 H.H. Bukow, H.v. Butlar, G. Heine, M. Reinke: In *Beam Foil Spectroscopy*, Vol.1, ed. by I.A. Sellin, D.J. Pegg (Plenum, New York 1967) p.263
2.36 W.L. Randolph, J. Asher, J.W. Koen, P. Rowe, E. Matthias: Hyperfine Interact. *1*, 145 (1975)
2.37 Y.C. Lin, W.A. Little, P.D. Bond, G.D. Sprouse: Hyperfine Interact. *2*, 388 (1976)
2.38 Y.C. Lin, W.A. Little, P.D. Bond, G.D. Sprouse: Hyperfine Interact. *4*, 105-109 (1978)
2.39 W.L. Wiese, M.W. Smith, B.M. Miles: *Atomic Transition Probabilities*, Vol.II (National Bureau of Standards, Washington, D.C. 1969)
2.40 K.P. Lieb, A.M. Nathan, J.W. Olness: Hyperfine Interact. *5*, 113 (1978)
2.41 I. Ben-Zvi, P. Gilad, M. Goldberg, G. Goldring, A. Schwarzschild, Z. Vager: Nucl. Phys. A*121* 592 (1968); A*122*, 73 (1968); A*151*, 401 (1970)
2.42 H.R. Andrews, R.L. Graham, J.S. Geiger, J.R. Beene, O. Häusser, D. Ward, D. Horn: Hyperfine Interact. *4*, 110-114 (1978)
2.43 R. Brenn, H. Spehl, A. Weckherlein, H.A. Doubt, G. van Middlekoop: At. Phys. A*281*, 219-227 (1977)
2.44 H.R. Andrews: Private communication (1978)
2.45 B. Orre, A. Linnfors, A. Falk, J.E. Thun, K. Johansson: Nucl. Phys. A*148*, 516-528 (1970)
2.46 V. Metag, D. Habs, H.J. Specht, G. Ulfert, C. Kozhuharov: Hyperfine Interact. *1*, 405 (1976)
2.47 G. Ulfert, D. Habs, V. Metag, H.J. Specht: Nucl. Instrum. Methods *148*, 369 (1978)
2.48 G. Ulfert, V. Metag, D. Habs, H.J. Specht: Phys. Rev. Lett. *42*, 1596 (1979)
2.49 P.B. Treacy: Nucl. Phys. *2*, 239 (1956)
2.50 L. Chase, G. Igo: Phys. Rev. *116*, 170 (1959)
2.51 K. Hagemeyer, M.B. Goldberg, G.J. Kumbartzki, K.H. Speidel, P.N. Taylor, M. Schramm, G. Kraft: Hyperfine Interact. *1*, 301-312 (1975)
2.52 R. Brenn, F. Hopkins, G.D. Sprouse: Phys. Rev. A*17*, 1837 (1979)
2.53 M.B. Goldberg, M. Hass, Z. Shkedi: Hyperfine Interact. *1*, 361-365 (1976)
2.54 G. Goldring, K. Hagemeyer, N. Benczer-Koller, R. Levy, Y. Lipshitz, B. Richter, Z. Shkedi, Y. Wolfson: Hyperfine Interact. *4*, 118-122 (1978)
2.55 R. Nordhagen, G. Goldring, R.M. Diamond, K. Nakai, F.S. Stephens: Nucl. Phys. A*142*, 577 (1970)
2.56 G. Goldring: Winter School on Heavy Ions, University of Witwatersrand, Johannesburg, South Africa, August 1978
2.57 F. Bosch, H. Spehl: Z. Phys. A*280*, 329 (1977)
2.58 H. Spehl, S. Steadman, A. Weckherlin, H.A. Doubt, K. Hagemeyer, G. Kumbartszki, K.H. Speidel: Nucl. Phys. A*215*, 446 (1973)

2.59 C. Scherer: Nucl. Phys. A*157*, 81 (1970)
2.60 M. Blume: Nucl. Phys. A*167*, 81 (1971)
2.61 A. Zemel, Y. Niv: Hyperfine Interact. *3*, 125-133 (1977)
2.62 E.W. Weber, H. Ackermann, N.S. Laulainen, G. zu Putlitz: Z. Phys. *259*, 371-390 (1973)
2.63 J. Bleck, D.W. Haag, W. Ribbe: Z. Phys. *233*, 65-73 (1970)
2.64 O. Klepper, H. Spehl: Nucl. Phys. *64*, 393-400 (1965)
2.65 H. Ikezoe: Private communication
2.66 W. Bartsch, W. Leitz, H.E. Mahnke, W. Semmler, R. Sielemann, Th. Wichert: Proc. Int. Meetings on Hyperfine Interactions, Leuven, Belgium, Sept. 1975, ed. by R. Coussement (University of Leuven, Belgium)
2.67 H. Ikezoe, G.D. Sprouse, H. Hamagaki, Y. Yamazaki, H. Nakayama, K. Nakai: Hyperfine Interact. *2*, 331 (1976)
2.68 G.D. Sprouse, R. Brenn, H.A. Calvin, H.J. Metcalf: Phys. Rev. Lett. *30*, 419 (1973)
2.69 W. Bartsch: Diploma Thesis, Hahn Meitner Institut, Berlin (1975)
2.70 C. Broude, M.-B. Goldberg, G. Goldring, M. Hass, M.J. Renan, B. Sharon, A. Shkedi, D.F.H. Start: Nucl. Phys. A*215*, 617-628 (1973)
2.71 Y.C. Lin: Ph.D. Thesis, State University of New York at Stony Brook (1978)
2.72 L.E. Young, G.D. Sprouse, D.E. Strottman: Phys. Rev. C*12*, 1358 (1975)

3. Hyperfine Interaction Studies in Nuclear Physics

O. Häusser and I.S. Towner

With 24 Figures

Hyperfine interaction studies have made, and continue to make, substantial contributions towards our understanding of nuclear structure. Following the overview, an elementary section is devoted to a discussion of the various quantities that describe the distributions of charge and magnetism in nuclei. As the sources of nuclear hyperfine fields we consider electronic and mesic atoms, and solid crystal lattices. Several topics which have had a significant impact on nuclear structure physics, and which appear promising for the near future, have then been selected for a more detailed description. A section on laser spectroscopy is linked to a discussion of the electromagnetic moments of unusual nuclear states, i.e. nuclei far off the stability line and superdeformed states in actinide nuclei. Recently, much has been learned about the structure and deformation of high-spin states. First experimental evidence for the unusually large oblate deformation of a high-spin state in ^{147}Gd has been obtained. The unexpectedly large transient magnetic fields for fast nuclear recoils in polarized ferromagnetics may be utilized to determine g factors of short-lived, deformed high-spin states. These measurements provide us with an improved understanding of the rotational alignment of particle spins in the direction of the collective angular momentum occurring under the stress of large rotational frequencies. In the last two sections we discuss how hyperfine interaction experiments may serve to probe hadronic currents and weak interactions. The evidence for mesonic exchange currents from magnetic moments of simple, nearly spherical nuclei is presented, and investigations of weak magnetism and of second-class currents in nuclear β decay are shown to rely crucially on the methods of hyperfine interaction physics.

3.1 Overview

Hyperfine interaction physics is concerned with the response of the atomic nucleus to electromagnetic fields produced by surrounding charged particles. Of the four fundamental interactions of the physical world, the electromagnetic interaction is by far the best understood, and thus is well suited to probe the structure of nuclei. Its strength is ideal in that it produces easily observable effects on nuclear charge and current distributions, yet constitutes only a small perturbation compared to the strong hadronic interactions. For this reason hyperfine interaction studies have made a very substantial contribution to our present understanding of nuclei.

Many of the early achievements were based on interference spectroscopy of optical hyperfine structure in atoms and molecules. Following the explanation of magnetic hyperfine structure by PAULI in 1924 [3.1], many nuclear spins and magnetic moments could be determined. From irregularities in the separation of hyperfine levels SCHÜLER and SCHMIDT [3.2] deduced the existence of nuclear quadrupole moments. The accuracy of the hyperfine structure observations in atoms could be greatly improved after Rabi and collaborators had developed the atomic beam radio frequency resonance method. Precise magnetic moments proved to be important for the development of the nuclear shell model. In rare-earth nuclei very large isotope shifts were observed [3.3], a finding which provided a starting point for the formulation of theories of collective motion in intrinsically deformed nuclei.

The perturbed angular correlation method as described by FRAUENFELDER and STEFFEN [3.4] became the single most powerful method for the measurement of nuclear moments of excited states. The first evidence for extranuclear perturbations of angular correlations was obtained in 1951 by the Zürich group [3.5] who studied the γ-γ correlation in ^{111}Cd in different chemical environments. In more recent years experimentalists pushed the limit of accessible nuclear lifetimes into the subpicosecond regime by harnessing the transient, fluctuating magnetic fields experienced by fast-moving, stripped ions recoiling in solids, gases, or in vacuo. Mössbauer's discovery of recoilless γ-ray absorption provided a new, complementary method of observing directly the hyperfine energy splittings of excited nuclear states. The impressive arsenal of experimental techniques has been further augmented through the use of mesonic probes (μ^-, π^-, K^-) which - as a consequence of their larger mass - cause much larger hyperfine effects than those common in electronic atoms.

From the preceding remarks it hardly seems surprising that these powerful experimental methods have provided a wealth of information on electromagnetic properties of nuclei which cannot be conveyed in the limited space of this chapter. Rather than commenting superficially on many aspects of hyperfine interactions that are important to nuclear physics, we have selected but a few examples. The topics chosen exemplify, in our opinion, some of the recent advances, or promise to come to the forefront in the near future. The latter category includes the use of lasers, which has already revitalized atomic spectroscopy. The development of the dye laser has supplied us with light beams of smooth frequency variability, of rapidly improving power density, and of unsurpassed energy resolution. The number of atomic species and the range of nuclear lifetimes that are accessible to measurement has already been expanded (Sect.3.3). During the last five years unexpectedly large transient magnetic fields have been observed for fast nuclear recoils in polarized ferromagnetics. This finding has opened up new possibilities for the measurement of g factors of very short-lived nuclear states, especially of high angular momentum states in intrinsically deformed nuclei. In less deformed, transitional nuclei, long-lived isomeric states of very high spin have been discovered, whose magnetic and quadrupole moments give valuable insight into their structure (Sect.3.4). During the last decade magnetic moments of simple shell model states near doubly closed shells have given theoretical physicists some incentive to reevaluate contributions to the magnetic moment from mesonic exchange currents and from polarization of the nuclear core. Our understanding of these effects has reached a considerable degree of sophistication, thanks to the quality of the experiments (Sect. 3.5). The well-understood electromagnetic hyperfine interaction may be used as a probe of fundamental aspects of hadronic and weak interactions. In Sect.3.6 it is shown how hyperfine interaction experiments have played a crucial role in investigations of weak magnetism and of second-class currents in nuclear β decay.

The subject matter in Sect.3.3-6 is preceded by an introductory discussion of the hyperfine Hamiltonian, of nuclear moments, and of hyperfine fields. This Sect. 3.2 is intended for the non-specialist, who may become familiar with the efforts of extracting nuclear moment information in the common situation where the hyperfine field is poorly known and has to be calibrated by a supplementary experiment. More important, this section should help to emphasize the basic unity that underlies the diverse branches of hyperfine interaction physics.

3.2 Hyperfine Hamiltonian and Nuclear Moments

The total Hamiltonian of a nucleus and surrounding charged particles (e.g. electrons in an atom) can be written as the sum of three terms (e.g. internucleon, interelectron, and nucleus-electron interactions)

$$\mathcal{H} = \mathcal{H}_{nn} + \mathcal{H}_{ee} + \mathcal{H}_{ne} \quad . \tag{3.1}$$

If we exclude for the moment mesic atoms from our consideration, then the energy differences of the three parts of the Hamiltonian differ by several orders of magnitude, being typically of the order of MeV for nuclear, eV for electronic, and μeV for hyperfine energy splittings. Perturbation theory is therefore adequate to solve subsequently $\mathcal{H}_{ee}$ and $\mathcal{H}_{ne}$.

The hyperfine interaction $\mathcal{H}_{ne}$ can traditionally be written as the products of hyperfine fields and nuclear moments (see the standard texts [3.6-8]). The moments may be derived in a more general context when considering the interaction between the nuclear current density $\underset{\sim}{j}$ and a plane electromagnetic wave $\underset{\sim}{A}$, i.e., $\mathcal{H}_{int} = -\int \underset{\sim}{j}\cdot\underset{\sim}{A}d\tau$. A multipole expansion of the vector potential $\underset{\sim}{A}$ leads to the definition of electric (Eλ) and magnetic (Mλ) tensor operators with parities of $(-)^L$ and $(-)^{L+1}$, respectively [3.9]. The nuclear moments can be viewed as diagonal matrix elements of the corresponding multipole operators sandwiched between the appropriate nuclear wave functions. Since nuclear states are also known to have good parity (i.e., they are invariant with respect to space inversion to an accuracy of $\sim 10^{-7}$), only even-rank electric and odd-rank magnetic moments are non-zero. In the following we confine our attention to electric monopole (E0), magnetic dipole (M1), and electric quadrupole (E2) moments, although M3 and E4 moments have been determined in a few cases [3.10].

The nuclear moments would be the only measurable nuclear quantities if the electromagnetic field did not vary appreciably over the nuclear volume, i.e., for point nuclei. The finite nuclear volume causes measurable effects in the electric monopole and the magnetic dipole components of $\mathcal{H}_{ne}$ that are manifest as the volume-dependent isomer shift and the hyperfine anomaly, respectively. These consequences of the variation of the electromagnetic field over the nuclear volume will be discussed in connection with the appropriate multipole moments.

The hyperfine fields in atoms and solids will now be discussed separately. For a description of the experimental methods of measuring hyperfine interactions for long-lived or stable nuclei (e.g., paramagnetic resonance, microwave spectroscopy, nuclear magnetic resonance, nuclear quadrupole resonance, atomic and molecular beam spectroscopy, optical spectroscopy), we refer to the standard texts [3.6-8]. The more recent perturbed angular distribution methods used to measure moments of very short-lived nuclear states will receive a very brief treatment.

3.2.1 Atomic Isotope Shifts and Nuclear Mean-Square Radii

The volume-dependent isomer shift (VS, frequently referred to as the field shift) is the measurable part of the very large electric monopole energy of the electron-nucleus interaction (i.e., the atomic binding energy). The origin of the volume shift may be illustrated by a simplified, non-relativistic estimate. Consider a nucleus whose charge (Ze) is distributed uniformly over a sphere with radius R. The electrostatic potential felt by an electron within the sphere, $V_i(r) = (Ze/2R)[3 - (r/R)^2]$, differs from that for a point nucleus, $V_0(r) = Ze/r$. The finite nuclear size causes an energy shift which is obtained by integrating $[V_i(r) - V_0(r)]$ over the nuclear volume, with the result

$$E_V = (2\pi/5)Ze^2R^2|\psi(0)|^2 = (2\pi/3)Ze^2\langle r^2\rangle|\psi(0)|^2 \quad . \tag{3.2}$$

Here $|\psi(0)|^2$ is the electron density at the nucleus, and $\langle r^2\rangle = (3/5)R^2$ is the mean-square charge radius of the sphere. The energy of an atomic transition i exhibits a shift, provided the electron density changes. The change $\Delta|\psi(0)|_i^2$ is large for transitions that either involve an s-electron jump or in which the screening of inner closed s-electron shells changes. The VS for transition i in two nuclear isotopes with mass numbers A and A' is then

$$VS^i_{AA'} = (2\pi/3)Ze^2\Delta|\psi(0)|_i^2\delta\langle r^2\rangle_{AA'} \quad . \tag{3.3}$$

Actually observed isotope shifts (IS) are difficult to interpret, because they contain, in addition to the VS, a mass shift in the frequency ν^i of the optical transition i

$$\delta\nu^i_{AA'} = \delta\nu^i_{AA'}(VS) + C_i(A' - A)/AA' \quad . \tag{3.4}$$

The mass shift accounts for the atomic recoil energy and has two parts corresponding to the non-relativistic kinematic energies of an atom, $T = (2\mu)^{-1}\sum \mathbf{p}_i^2 + (2M)^{-1}\sum \mathbf{p}_i\cdot\mathbf{p}_j)$, where $\mathbf{p}_i$ are the electron momenta and μ, M are reduced electron and nuclear mass, respectively. The normal mass shift, resulting from the one-electron part in T, is simply given by $C_i = \nu_i/1836$ cm, whereas the specific mass shift depends on the correlated motion of all of the electrons in the atom. The specific mass shift may exceed the normal mass shift by an order of magnitude, for example in rare-earth atoms, and atomic Hartree-Fock calculations are not yet capable of predicting it reliably [3.11]. The uncertainty in the specific mass shift may therefore seriously obscure the more interesting volume shift. The volume shift in optical transitions is frequently written as

$$\delta\nu^i_{AA'}(VS) = F_i\delta\langle r^2\rangle_{AA'} \tag{3.5}$$

where higher moments of $\delta\langle r^2\rangle_{AA'}$, whose contribution amounts to a few percent [3.12] have been neglected. It is customary to factorize F_i into $F_i = E_i f(Z)$, where E_i is an electronic factor which contains the change of the total non-relativistic charge density at the (point) nucleus, $\Delta|\psi(0)|^2_i$, in the transition i

$$E_i = \pi a_0^3\Delta|\psi(0)|^2_i/Z \quad . \tag{3.6}$$

The factor f(Z) includes the relativistic correction to E_i and the effects of the finite nuclear charge distribution. In a compilation of $\delta\langle r^2\rangle$ values by HEILIG and STEUDEL [3.13], f(Z) was calculated from analytical expressions given by BABUSHKIN [3.14].

The observed isotope shifts are dominated by the mass shift in light nuclei ($Z \lesssim 35$) and by the volume shift in heavy nuclei ($Z \gtrsim 60$). In any case, even if the isotope shift is known for a long series of isotopes, one may obtain $\delta\langle r^2\rangle$ only if the electron density change for transition i, $\Delta|\psi(0)|^2_i$, and the mass shift parameter C_i are known from other empirical data, usually on stable isotopes. For (core + ns) → (core' + np) transitions in alkali-like atoms one may write

$$\Delta|\psi(0)|^2_{ns-np} = \beta|\psi(0)|^2_{ns} \tag{3.7}$$

where the screening factor β,

$$\beta = (|\psi(0)|^2_{core} + |\psi(0)|^2_{ns} - |\psi(0)|^2_{core'})/|\psi(0)|^2_{ns} \quad ,$$

is reliably obtained from Hartree-Fock calculations. The ns electron density at the nucleus, $|\psi(0)|^2_{ns}$, is related to the magnetic hyperfine splitting resulting from the ns electron [3.7].

Independent methods for determining $\delta\langle r^2\rangle$, for example from mesic X rays and electron scattering, are becoming increasingly important in calibrating the optical isotope shift, although the results depend somewhat on the parametrization of the charge distribution. Most straightforward is the determination of both $\Delta|\psi(0)|^2_i$ and C_i from X ray isotope shifts [3.15], provided at least three isotopes have been studied. In Fig.3.1 we show a "King plot" of X ray versus optical isotope shift for several isotopic pairs of Sm, using the linear relation

$$\{2AA'/[A_1A_2(A'-A)]\}\delta\nu^i_{AA'} = 2AA'\Delta|\psi(0)|^2_i/[A_1A_2(A'-A)\Delta|\psi(0)|^2_{X\ ray}] + 2C_i/(A_1A_2)\ . \tag{3.8}$$

Here A_1 and A_2 are the mass numbers of the lightest pair $\Delta A = 2$ studied, and the factor $2/A_1A_2$ is introduced to normalize the A dependence to a value close to unity.

The $\delta\langle r^2\rangle$ values derived from experiment are best compared directly to theoretical prediction (e.g., nuclear Hartree-Fock calculations) to avoid model assumptions of the nuclear charge distributions. It is nevertheless useful to discuss briefly the effects that may influence $\delta\langle r^2\rangle$.

For any given charge distribution $\rho(r,\theta,\phi)$, let us define a volume-equivalent sphere with $\langle r^2\rangle = (3/5)R^2_{eq}$ and $R_{eq} = a_0A^{1/3}$ ($a_0 \sim 1.2\times 10^{-13}$ cm). One easily finds

$$\delta\langle r^2\rangle = 2\langle r^2\rangle\delta R_{eq}/R_{eq} \tag{3.9}$$

where $\delta R/R = \xi\eta\ \delta A/3A$ with

$$\xi = (3A/R)(\delta R/\delta A)\quad ,\quad \eta = (\delta R/\delta N)/(\delta R/\delta A)\quad .$$

For spherical, incompressible spheres ($\xi = \eta = 1$), one obtains the standard value of the volume-dependent isomer shift,

$$\delta\langle r^2\rangle_{std} = (2/5)a_0^2\ \delta A/A^{1/3}\quad . \tag{3.10}$$

In the region of spherical nuclei one finds volume shifts of about 0.6 times the standard value [3.16,17]. The parameter $\xi \sim 0.87$ arises from the diffuseness of the nuclear surface, which is well determined by elastic electron scattering data. The parameter η is introduced to account for the compression of the proton charge distribution as neutrons are added (δN). The compression is an effect of the increased proton binding energy, and has been discussed by BODMER [3.16]. The isomer shift data yield a value of $\eta \sim 0.7$ for the 'regular nuclear compressibility'.

The onset of nuclear deformation may cause strikingly large volume-dependent isomer shifts which may amount to several times the standard value [3.3]. Let us consider a nuclear charge distribution of the form

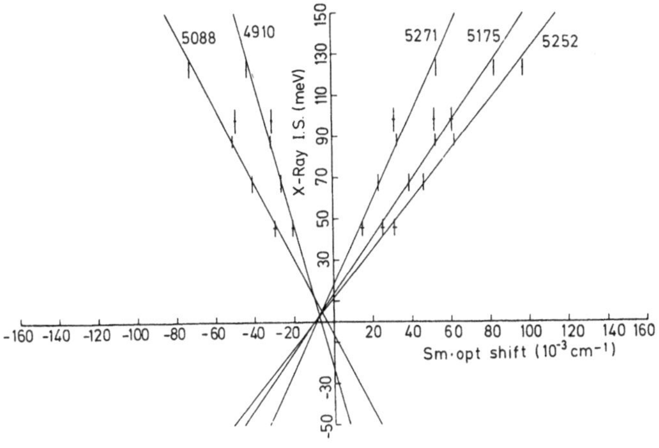

Fig.3.1. King plot of optical vs X ray isotope shifts for Sm [3.15]

$$\rho(r,\theta,\phi) = \rho_0 \, f(r/R) \tag{3.11}$$

where ρ_0 is the density at the centre of the nucleus, R is the radius to the nuclear surface, and f(r/R) specifies the radial density variation. For ellipsoidal deformations

$$R = \bar{R}\left[1 + \sum_{\mu} \alpha_{\mu} Y_{2\mu}(\theta,\phi)\right] \tag{3.12}$$

with $\alpha_0 = \beta\cos\gamma$, $\alpha_1 = \alpha_{-1} = 0$, and $\alpha_2 = \alpha_{-2} = \beta\sin\gamma/\sqrt{2}$. The average radius $\bar{R}$ may be expressed in terms of the radius of a volume-equivalent sphere, R_0, by

$$R_0^3 = \bar{R}^3[1 + (3/4\pi)\beta^2 + (5/4\pi)^{\frac{1}{2}}\beta^3 f(\gamma)/(14\pi)] \tag{3.13}$$

with $f(\gamma) = \cos\gamma(1 - 4\sin^2\gamma)$. The expression for $\langle r^2\rangle$ is of the form

$$\langle r^2\rangle = R_0^2 \, s[1 + (5/4\pi)\beta^2 + (25/42\pi)(5/4\pi)^{\frac{1}{2}}\beta^3 f(\gamma)] \tag{3.14}$$

where $s = \int f(x)x^4 \, dx / \int f(x)x^2 \, dx$ depends on the radial density distribution. For a uniform distribution, and dropping the β^3 term, one obtains the standard expression for deformed nuclei

$$\langle r^2\rangle = (3/5) \, R_{eq}^2[1 + (5/4\pi)\beta^2] \tag{3.15}$$

where the equivalent uniform charge distribution is considered to be incompressible, i.e. $R_{eq} = a_0 \, A^{1/3}$. In prolate, rotational nuclei the deformation parameter β may be calculated from data that determine intrinsic quadrupole moments (Sect.3.2.2.c.I) and inferred into (3.15) to calculate $\delta\langle r^2\rangle$. It is then found that (3.15) overestimates the observed changes in the mean-square radii. This discrepancy has led FRADKIN [3.17] to introduce a new parameter ξ, the 'compressibility under deformation'. He defines

$$R_0^2 s = (3/5)R_{eq}^2 = (3/5)R_{0\ eq}^2(1 + \xi\beta^2) \tag{3.16}$$

where the equivalence radius R_{eq} now depends on the deformation β, whereas the new radius, $R_{0\ eq} = (1.2 \times 10^{-13}\ \mathrm{cm})A^{1/3}$, does not. FRADKIN finds empirically a value $\xi \approx -(5/8\pi)$; consequently

$$\langle r^2 \rangle \approx (3/5)R_{0\ eq}^2[1 + (5/8\pi)\beta^2 + (25/42\pi)(5/4\pi)^{\frac{1}{2}}\beta^3 f(\gamma)] \quad . \tag{3.17}$$

The negative value of ξ can be interpreted as an increased proton concentration at the centre of the deformed nucleus as compared to spherical nuclei. A derivation of ξ has been given in terms of a harmonic oscillator model by CUSSON and CASTEL [3.18].

3.2.2 Quadrupole Interactions and Nuclear Quadrupole Moments

In an expansion of the interaction energy of the nuclear charge distribution ρ with an electrostatic potential V produced by external charges (electrons, ions), the most important orientation-dependent term is the quadrupole interaction [3.6-8]

$$\mathcal{H}_Q = \sum_q Q_2^{(q)} \mathcal{B}_2^{(-q)} \quad . \tag{3.18}$$

In this expression $Q_2^{(q)}$ are multipole tensors of the nuclear charge distribution, and $\mathcal{B}_2^{(-q)}$ are tensor components of the electromagnetic field. In a spherical representation the quadrupole tensor for a nuclear state with spin I is

$$\begin{aligned} Q_2^{(0)} &= eQ/[2I(2I-1)] \times [3I_z^2 - I(I+1)] \quad , \\ Q_2^{(\pm 1)} &= \pm(6)^{\frac{1}{2}} eQ/[4I(2I-1)] \times [I_z I_\pm + I_\pm I_z] \quad , \\ Q_2^{(\pm 2)} &= (6)^{\frac{1}{2}} eQ/[4I(2I-1)] \times I_\pm^2 \end{aligned} \tag{3.19}$$

where I_z and $I_\pm = I_x \pm iI_y$ are the usual spin operators. The nuclear quadrupole moment Q is conventionally defined as the matrix element

$$\begin{aligned} eQ &= \langle I\ m = I | \sum_i e_i(3z_i^2 - r^2) | I\ m = I \rangle \\ &= (16\pi/5)^{\frac{1}{2}} \langle II | \sum_i e_i r_i^2 Y_{20}(\theta,\phi) | II \rangle \\ &= (16\pi/5)^{\frac{1}{2}} \langle II | E2 | II \rangle \quad . \end{aligned} \tag{3.20}$$

This expression may be further reduced by applying the Wigner-Eckart theorem for tensor operators T_{LM} of rank L

$$\langle I_f m_f | T_{LM} | I_i m_i \rangle = (-)^{I_f m_f} \begin{pmatrix} I_f & L & I_i \\ -m_f & M & m_i \end{pmatrix} \langle I_f \| T_L \| I_i \rangle \tag{3.21}$$

where $\langle I_f \| T_L \| I_i \rangle$ is an m-independent reduced matrix element and the 3-j symbol is given by EDMONDS [3.19]. From (3.16,17) it follows that only states with $I \geq 1$ have a non-vanishing quadrupole moment, and furthermore

$$eQ = (16\pi/5)^{\frac{1}{2}}[I(2I - 1)/(I + 1)(2I + 1)(2I + 3)]^{\frac{1}{2}}\langle I \| E2 \| I \rangle \quad . \tag{3.22}$$

The tensor components of the electromagnetic field $\mathcal{B}_2^{(q)}$ can be obtained from the expression

$$\mathcal{B}_2^{(q)} = (4\pi/5)^{\frac{1}{2}} \int r_e^{-3} \rho_e Y_{2q}(\theta_e, \phi_e) d\tau_e \tag{3.23}$$

where ρ_e is the density of charges external to the nucleus. In the atomic case (3.23) is to be considered as a quantum-mechanical operator whose expectation value has to be evaluated over the atomic wave function. In solid-state physics (3.23) is usually replaced by a classical field gradient tensor. In terms of Cartesian field gradient operators (3.23) may be rewritten as

$$\mathcal{B}_2^{(0)} = (1/2)V_{z'z'} \quad ,$$

$$\mathcal{B}_2^{(\pm 1)} = \mp(6)^{-1/2}(V_{x'z'} \pm iV_{y'z'}) \quad ,$$

$$\mathcal{B}_2^{(\pm 2)} = (1/2)(6)^{-\frac{1}{2}}(V_{x'x'} - V_{y'y'} \pm iV_{x'y'}) \tag{3.24}$$

where $V_{x'y'} = \partial^2 V/\partial x'\partial y'$, etc. The expression (3.24) assumes the validity of the Laplace equation, $\langle V_{x'x'} + V_{y'y'} + V_{z'z'} \rangle = 0$, thus ignoring s electrons that have a non-vanishing overlap with the nucleus. It is nevertheless justified, because s electrons are spherically symmetric and hence produce a vanishing electric field gradient at the nucleus.

a) *Quadrupole Interactions in Atoms*

The atomic states in free atoms or molecules of non-zero angular momentum J exhibit rotational symmetry, i.e., the (2J + 1) manifold of electron wave functions is degenerate in the absence of hyperfine interactions. The tensor operators $\mathcal{B}_2^{(q)}$ can then be constructed in a fashion similar to that used to derive the nuclear quadrupole tensor (3.19). The quadrupole coupling in atoms is often written as [3.6-8]

$$\mathcal{H}_Q = B_J[2I(2I - 1)J(2J - 1)]^{-1}[3(\underset{\sim}{I}\cdot\underset{\sim}{J})^2 + (3/2)\underset{\sim}{I}\cdot\underset{\sim}{J} - I(I + 1)J(J + 1)] \tag{3.25}$$

where $B_J = e^2 q_J Q$. The electronic field gradient q_J is the expectation value for the m = J electronic substate

$$q_J = e^{-1}\langle JJ| \int \rho_e r_e^{-3}(3\cos^2\theta_e - 1)d\tau_e |JJ\rangle \quad . \tag{3.26}$$

In absence of external (magnetic) fields the quadrupolar hyperfine splitting is

$$E_Q(IJF) = B_J\ C(IJF)[8I(2I - 1)J(2J - 1)]^{-1} \tag{3.27}$$

where $\underset{\sim}{F} = \underset{\sim}{I} + \underset{\sim}{J}$ is the total angular momentum, $C(IJF) = [3K(K + 1) - I(I + 1)J(J + 1)]/[2I(2I - 1)J(2J - 1)]$, and $K = F(F + 1) - I(I + 1) - J(J + 1)$. From the observation of several transitions between hyperfine structure levels of known I, J and F it is usually possible to determine both magnitude and sign of B_J.

The electric field gradient q_J produced by a single, bound electron with quantum numbers n, ℓ and j is [3.7]:

$$q_j = R_r(\ell,j)\langle r^{-3}\rangle(2j - 1)/(2j + 2)/4\pi\varepsilon_0 \tag{3.28}$$

where $R_r(\ell,j)$ is a relativistic correction factor. In complex atoms the reliable evaluation of $\langle r^{-3}\rangle$ and q_J requires nothing less than an accurate solution of the relativistic many-body problem [3.20,21]. Only very few complete many-body calculations of atomic hyperfine structure have been carried out so far. The various theoretical approaches are discussed in the articles by LINDGREN and ROSÉN [3.21].

A conventional procedure for evaluating $\langle r^{-3}\rangle_Q$ is to use empirical values $\langle r^{-3}\rangle_0$, obtained, for example, from magnetic dipole interactions, to which a shielding correction is then applied

$$\langle r^{3}\rangle_Q = (1 - R_Q)\langle r^{-3}\rangle_0 \quad . \tag{3.29}$$

R_Q describes the additional field gradients from distortions of inner closed shells induced by non-spherical open-shell electrons, e.g., 4f electrons in rare-earth atoms. The procedure implied by (3.29) is only correct if R_Q denotes the *difference* between the polarization effects entering into $\langle r^{-3}\rangle_Q$ and $\langle r^{-3}\rangle_0$. The shielding correction R_Q arising from first-order core polarization has been calculated frequently by STERNHEIMER [3.22]. These calculations neglect higher-order polarization effects as well as all of the correlation effects associated with multiple electron excitations [3.21]. These problems are sometimes compounded by configuration mixing in the atomic state $|J\rangle$. For these reasons it is useful to measure the quadrupole splittings of several independent optical transitions to test whether the values for the derived nuclear quadrupole moment Q come out consistently. In a few cases, where Q can be measured reliably by independent methods (Sect.3.2.2.c), it is possible to assess the accuracy of calculated atomic field gradients.

b) *Quadrupole Interaction Non-Cubic Lattices*

The macroscopic electric field gradients that can be produced in the laboratory are too weak by orders of magnitude to cause measurable quadrupole interactions in nuclei. Sufficiently large field gradients may, however, be experienced by probe-nuclei substituting in solids with a regular lattice structure. The expectation value of the electric-field-gradient tensor (3.24) may then be evaluated over the non-degenerate wave functions for the external charges (electrons, ions), giving rise to a classical field gradient tensor. By a suitable rotation a principal axis system may always be found such that the mixed derivatives in (3.24) vanish, i.e.

$$B_0 = V_{xx}/2 \quad , \quad B_{\pm 1} = 0 \quad , \quad B_{\pm 2} = (24)^{-\frac{1}{2}} \eta V_{zz} \quad , \tag{3.30}$$

where $\eta = (V_{xx} - V_{yy})/V_{zz}$. The axes are usually labelled such that $|V_{zz}| \geq |V_{xx}| \geq |V_{yy}|$, which together with the Laplace equation limits η to values between 0 and 1. In crystals of cubic symmetry, $V_{xx} = V_{yy} = V_{zz} = 0$. The electric field gradient is axially symmetric ($\eta = 0$) if the point symmetry of the crystal is trigonal, hexagonal or tetragonal. In crystals of rhombic or lower symmetry $\eta \neq 0$.

If one defines the electric field gradient $V_{zz} = eq$ in analogy to the atomic case (3.25,26), the quadrupole Hamiltonian can be written

$$\mathcal{H}_Q = e^2qQ[3I_z^2 - I(I + 1) + (\eta/2)(I_+^2 + I_-^2)]/[4I(2I - 1)] \quad . \tag{3.31}$$

For non-axially symmetric electric field gradients ($\eta \neq 0$) the Hamiltonian (3.31) must be diagonalized numerically. In the axially symmetric case ($\eta = 0$) the Hamiltonian is diagonal in the $|Im\rangle$ representation with eigenvalues

$$E_m = e^2qQ[3m^2 - I(I + 1)]/[4I(2I - 1)] \quad . \tag{3.32}$$

It is seen that, in the absence of other perturbations, magnetic substates with $\pm m$ are degenerate, and that the energy splittings depend on the value of the spin I. Finite electric field gradients eq in solids are typically of the order of $\approx 10^{17}$ V/cm^2, with corresponding energy splittings of $\approx 10^{-7}$ eV and quadrupole interaction frequencies $\nu_Q = e^2q\,Q/h \approx 100$ MHz. The observation of nuclear quadrupole moments by static interactions is therefore limited to isomeric nuclear states with mean lifetimes $\tau \gtrsim 1$ ns. Very large, transient electric field gradients ($\approx 10^{30}$ V/cm^2) are produced in Coulomb scattering of a heavy-ion beam from a target nucleus, especially for backscatter trajectories. The time-dependent electric field gradients are exploited by the Coulomb reorientation method [3.23,24] to obtain quadrupole moments of short-lived low-lying nuclear states.

I) *Measurements of Quadrupolar Energy Splittings.* Several experimental methods are available to determine the product e^2qQ. One class of experiments determines the energy splitting directly, e.g. (3.32). One may detect the absorption of a radio-frequency signal when the applied frequency matches $(E_m - E_{m'})/h$. This nuclear-quadrupole-resonance (NQR) method [3.25] may only be used for abundant, stable nuclei.

Of wider applicability are the low-temperature nuclear orientation techniques if they are combined with the detection of radiation emitted from radioactive nuclei [3.26]. Because of the smallness of the quadrupolar energy splittings, temperatures in the milli-Kelvin range have to be applied to achieve non-uniform population of the nuclear sublevels through the Boltzmann factor, $\exp[(E_{m'} - E_m)/kT]$. The use of single crystals is necessary to provide a unique nuclear orientation in the laboratory frame. The anisotropy of the emitted radiation depends on whether the $|m| = I$ or $m = 0$ or 1/2 substates lie lowest in energy, and thus the sign of e^2qQ is also measured.

The sign of the quadrupole interaction may also be measured by the Mössbauer effect, provided the quadrupole splitting is comparable to or larger than the natural linewidth $\Delta E = \hbar/\tau$ [3.27]. The Mössbauer method is limited to stable isotopes having γ-emitting states at energies of $\lesssim 150$ keV.

II) *Perturbed Angular Distributions.* A second class of experiments exploits the fact that nuclear radiations (γ rays, α particles, etc.) emitted by an ensemble of oriented radioactive nuclei exhibit anisotropic angular distributions that are correlated with the nuclear-spin directions. The hyperfine interaction, i.e., the static quadrupole interaction considered here, causes a change of the spin orientation of the nuclear ensemble as a function of time elapsed between the formation of the oriented state and the time of emission of the radiation. The spin orientation of the state I is described by a statistical tensor $\rho_q^\lambda(I,t)$ whose initial value can usually be computed from the known orientation mechanism, e.g., preceding γ rays, Coulomb excitation and nuclear reactions. The time evolution of the statistical tensor is completely specified by a perturbation coefficient $G_{\lambda\bar{\lambda}}^{q\bar{q}}(t)$. (For an up-to-date formulation of perturbed angular distributions and correlations see [3.28].)

$$\rho_{\bar{q}}^{\bar{\lambda}}(I,t)_{S_k} = \sum_{qq'\lambda} \rho_{q'}^{\lambda}(I,t=0)_S \, D_{qq'}^{\lambda}(S \to S_k) G_{\lambda\bar{\lambda}}^{q\bar{q}}(t) \quad . \tag{3.33}$$

The D matrix [3.19] describes the rotation from the symmetry axis S of the orienting interaction, e.g., the beam direction in nuclear reactions, to the perturbation coordinate system S_K, e.g., the symmetry axis of a non-cubic crystal. The angular distribution of radiation X is then [3.28]

$$W(\underset{\sim}{k}\underset{\sim}{Q}t) = (d\Omega/4\pi)(2I+1) \sum_{\bar{\lambda}\bar{q}\bar{q}'} \rho_{\bar{q}}^{\bar{\lambda}*}(I,t)_{S_k} A_{\bar{\lambda}\bar{q}'}(X) D_{\bar{q}\bar{q}'}^{\bar{\lambda}}(S_k \to \underset{\sim}{k}\underset{\sim}{x}) \tag{3.34}$$

where $A_{\bar{\lambda}\bar{q}'}(X)$ is an angular distribution coefficient characteristic of the radiative transition $I \to I'$. The propagation direction $\underset{\sim}{k}$ and polarization $\underset{\sim}{Q}$ of radiation X are given with respect to the perturbation frame S_k through the rotation $S_k \to \underset{\sim}{k}\underset{\sim}{x}'$.

For axially symmetric perturbations, e.g. axially symmetric field gradients, the perturbation factor becomes [3.28]

$$G_{\lambda\bar{\lambda}}^{q\bar{q}}(t) = \delta_{q\bar{q}}[(2\lambda + 1)(2\bar{\lambda} + 1)]^{\frac{1}{2}} \sum_{mm'} \begin{pmatrix} I & I & \lambda \\ -m & m' & q \end{pmatrix} \begin{pmatrix} I & I & \bar{\lambda} \\ -m & m' & q \end{pmatrix} \exp[-i(E_m - E_{m'})t/\hbar] \quad . \tag{3.35}$$

The intensity variation of radiation X observed in a fixed detector contains frequency components $\omega = (E_m - E_{m'})/\hbar$ that are multiples of a fundamental frequency ω_0, see (3.32)

$$\omega_0 = 3\omega_Q \quad \text{for integer } I$$
$$\omega_0 = 6\omega_Q \quad \text{for half (odd) integer } I \tag{3.36}$$

where ω_Q is conventionally written as

$$\delta_Q = e^2qQ/[4I(2I - 1)/\hbar] = \pi\nu_Q/[2I(2I - 1)] \quad . \tag{3.37}$$

Inserting (3.35) into the expression for perturbed angular distributions (3.33), one finds that the real terms ($\lambda + \bar{\lambda}$ = even) correspond to a pure intensity modulation, whereas the imaginary temrs ($\lambda + \bar{\lambda}$ = odd) can be interpreted as an angular shift, and thus allow a measurement of the sign of e^2qQ. For unpolarized, *aligned* nuclei, having equal populations of +m and -m substates, only statistical tensors $\rho_q^\lambda(I,t)$ with $q = 0$ and λ = even survive; consequently, only the magnitude of $|e^2qQ|$ can be determined. This result can be understood if one considers the spin precession of a classical quadrupole about the symmetry axis of a non-uniform electric field. States with opposite signs of m precess in opposite directions [3.7], and thus no net precession is observable for aligned nuclei.

To determine the sign of e^2qQ the initial state must be polarized. This may be achieved by detecting the circular polarization of a preceeding γ ray [3.29], by populating state I by an appropriate β decay [3.30], or by direct nuclear reactions, e.g. the Coulomb excitation process [3.31].

III) *Electric Field Gradients in Non-Cubic Metals.* For most applications non-cubic metals provide the most suitable environment for the observation of a unique quadrupole interaction. The reservoir of conduction electrons in the metal brings the recoil atom, produced by the nuclear reaction, quickly into its equilibrium charge state. A second advantage of metals over insulators is the tendency of the former to form alloys with impurities such that the impurity atom occupies a substitutional site in the lattice. The solubility of the impurity is especially good if the ionic radii and electronegativity of impurity and host ions are similar [3.32].

It is usually difficult to calculate eq in metals with sufficient accuracy. A general expression for the electric field gradient in metals is

$$eq = (1 - R_Q')eq_{int} + (1 - \gamma_\infty')\, eq_{latt} + eq_{CE} \quad . \tag{3.38}$$

Here eq_{int} arises from partially closed shells in the probe atom, e.g., d or f electrons. The shielding factor R_Q' is modified from that for free ions, see (3.29), because of the presence of the conduction electrons [3.33]. The ionic contribution eq_{latt} can be reliably computed by point-charge lattice summations [3.34]. For a hexagonal close-packed (hcp) crystal structure with lattice constants c and a, DAS and POMERANTZ [3.35] found

$$eq_{latt} = (Ze/4\pi\varepsilon_0 a^3)[0.0065 - 4.3584\ (c/a - 1.633)] \quad . \tag{3.39}$$

Corrections resulting from higher multipole moments of the lattice ions have been evaluated by DAS and RAY [3.36] for hcp, and by TAYLOR [3.37] for tetragonal structures.

The term $(1 - \gamma_\infty')$ describes the polarization of the electronic shells of the probe ion by the perturbing external potential, an effect which may amount to a huge enhancement of eq_{latt} by factors between ~10 and ~400. The quantity γ_∞ is the Sternheimer antishielding ($\gamma_\infty < 0$) factor [3.22], which has been evaluated for free ions with a relativistic Hartree-Fock-Slater electron theory by FEIOCK and JOHNSON [3.38]. The conduction-electron density inside the ion is usually smaller than the average value, causing a larger enhancement factor in metals, $(1 - \gamma_\infty')$ [3.36]. Other conduction-electron contributions are collected in eq_{CE} and are discussed by WATSON et al. [3.39]. For a thorough discussion of recent developments in the understanding of electric field gradients in non-cubic metals we refer the reader to the chapter by CHRISTIANSEN and WITTHUHN and to a recent review article by KAUFMANN and VIANDEN [3.40].

c) *Calibration of Electric Field Gradients*

Although considerable progress has been made in theoretical calculations of electric field gradients in non-cubic metals, the accuracy obtainable is rarely comparable to that achieved in the calculation of atomic field gradients. In most cases, especially when the host is a transitional metal, and when probe and host ions are different, the electric field gradient has to be viewed as an unknown. Quadrupole moments of nuclear ground states determined from atomic hyperfine structure can sometimes be used to calibrate the electric field gradients. Successful NQR experiments for about eight different metals have been reported (see the compilation in [3.40]). In these experiments the metal is embedded as a fine powder in insulating materials to avoid the skin effect which would otherwise prohibit the penetration of the radio-frequency field into the metallic sample. In some instances, Mössbauer data may be used for calibration purposes, provided the nuclear ground state exhibits an observable quadrupole splitting. In the majority of cases, however, a suitable nuclear state in one of the neighbouring isotopes has to be found whose quadrupole moment can either be measured from the hyperfine structure of mesic X rays, or which can be calculated by well-applicable nuclear models. Such a state may then be used in a calibration experiment in which eq is deduced from the measured quadrupole frequency $\nu_Q = e^2qQ/h$.

I) *Quadrupole Moments of Deformed Nuclei.* In heavy, deformed nuclei of odd mass number and with spins $I \geq 3/2$, precise static quadrupole moments Q have been derived from measured X ray energies in mesic atoms. In such atoms (nucleus + μ^-, π^- or K^-) the magnetic dipole interactions are strongly suppressed compared to ordinary atoms, and the hyperfine interactions are often dominated by quadrupole effects. For pionic atoms $J = \ell$, and the quadrupole splitting can be written in close analogy to the atomic case (3.27) (a similar expression holds for muonic atoms)

$$E_Q(I\ell F) = Q_{eff}(A_2^P/Q)C(I\ell F) \quad . \tag{3.40}$$

In this expression (A_2^P/Q) is a well-known field gradient produced by the meson at a *point* nucleus [B_J/Q in (3.27)], and $C(I\ell F)$ is as defined in (3.27). From the observed hyperfine structure of mesic X rays, and after allowing in the analysis for vacuum polarization and magnetic dipole interactions, one derives an effective quadrupole moment which is related to Q by [3.41,42]

$$Q_{eff} = Q(1 + \delta_1 + \delta_2 - \varepsilon_2/A_2^P) \quad . \tag{3.41}$$

In muonic atoms the only correction necessary is δ_1, which accounts for the decrease of the electric quadrupole interaction as a result of the finite nuclear size. In deformed nuclei with $Z \sim 67$ (Ho), the overlap of a muon in its 3d state with the nucleus is only ~2%, and thus Q can be deduced from the 4f → 3d transitions in an essentially model-independent way. Figure 3.2 shows the spectrum of 4f → 3d transitions in muonic ^{165}Ho from the work of POWERS et al. [3.43]. From the fitted X ray line shapes a ~1% accurate value for the quadrupole moment of ^{165}Ho was obtained.

Strongly interacting probes (π^-, K^-) cause several effects resulting from the strong meson-nucleus interaction, e.g., a monopole energy shift ε_0, a line broadening Γ_0, and corrections δ_2 and ε_2 to the quadrupole hyperfine splitting [3.42]. Because of the strong interactions, the (π^-, K^-) wave functions are changed even at large distances from the nucleus, giving rise to an increased long-range electric quadrupole interaction (term δ_2). The strong quadrupole constant ε_2, which is related to the *matter deformation* of the nucleus, may be extracted, provided accurate muonic data exist that determine the quantity $Q(1 + \delta_1)$. The strong-interaction quadrupole effect for the 5g → 4f transition in ^{175}Lu has recently been observed by EBERSOLD et al. [3.41].

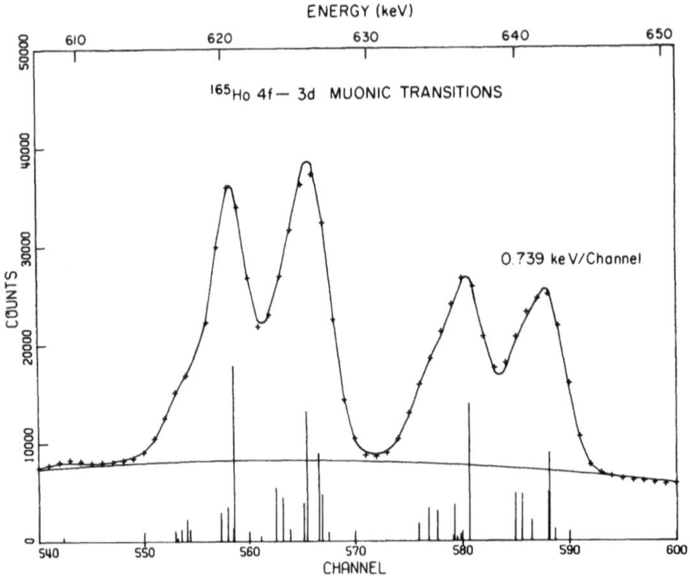

Fig.3.2. The spectrum of 4f → 3d muonic transitions of muonic ^{165}Ho. The solid curves indicate the fit to the data and the fitted background. The vertical bars indicate the best-fit line spectrum [3.43]

For lower muonic orbitals (e.g. 2p), the muonic fine structure is comparable to excitation energies of the ground-state rotational band in deformed nuclei. By the time the muon reaches the 1s level, the probability is large of finding the nucleus in an excited state. The energies and intensities of hyperfine components of muonic K or L X rays depend, then, on E2 transition moments between nuclear ground state and several excited states, in addition to the static quadrupole moments of these states. The static and transition quadrupole moments are related by the rotational model to a single quantity, the intrinsic quadrupole moment Q_0 of the deformed nucleus.

If $X_K(q)$ describes the internal motion of the nucleons that cause an axially symmetric deformation with quadrupole moment Q_0 and spin projection K onto the symmetry axis, the adiabatic rotation of $X_k(q)$ may be described by the wave function [3.44]

$$\psi^I_{MK} = [(2I + 1)/16\pi^2]^{\frac{1}{2}}[D^I_{MK}(\theta_i)X_k(q) + (-)^{I+K}\, D^I_{M-K}(\theta_i)X_{-k}(q)] \quad . \tag{3.42}$$

Inserting (3.42) into the equation for the quadrupole moment (3.20), one obtains

$$Q = Q_0[3K^2 - I(I + 1)]/[(I + 1)(2I + 3)] + \text{small terms} \quad . \tag{3.43}$$

The E2 transition probability B(E2) is defined by

$$B(E2,I_i \rightarrow I_f) = (2I_i + 1)^{-1}\langle I_f\|E2\|I_i\rangle^2 \quad . \tag{3.44}$$

With the wave function (3.42) one obtains for in-band transitions (K → K)

$$B(E2,I_iK \rightarrow I_fK) = (5/16\pi)e^2Q_0^2(2I_f + 1)\begin{pmatrix} I_i & 2 & I_f \\ K & 0 & -K \end{pmatrix}^2 \quad . \tag{3.45}$$

The data on different muonic transitions in ^{161}Dy [3.45] and ^{165}Ho [3.43] yield consistent values for Q_0 and provide excellent support for the applicability of the rotational model in these nuclei.

For nuclei in well-established regions of strong prolate deformation it seems, then, justified to deduce Q via (3.43,45) from measurements of Coulomb excitation yields and of partial E2 lifetimes. A compilation of Q_0 values from various methods has been given by LÖBNER et al. [3.46].

In Table 3.1 the quadrupole moments of strongly deformed nuclei obtained from mesic atoms and from Coulomb excitation yields of ELBEK [3.47] are compared to those from atomic beam resonance experiments. The latter values are deduced from measured B_J values using the $\langle r^{-3}\rangle_Q$ integrals of LINDGREN and ROSÉN [3.21]. The apparent discrepancies can be attributed to the Sternheimer shielding factor $(1 - R_Q)$, which amounts to ~1.25 for $4f^n$, and to ~0.78 for $5d^n$ configurations, respectively.

Table 3.1. Quadrupole moments of strongly deformed nuclei (1b = 10^{-28} m^2)

Nucleus	I^π	Atomic configuration	B_J[MHz]	Q_{atomic} [a] [b]	Ref.	meson	Q_{mesic} [b] [b]	Ref.	Q_{coulex} [c] [b]
^{161}Dy	$5/2^+$	$4f^{10}$ $^5I_8$	1091.6	2.01	3.48	μ^-	2.47(3)	3.45	2.64(8)
^{165}Ho	$7/2^-$	$4f^{11}$ $^4I_{15/2}$	-1668.1	2.65	3.49	μ^-	3.49(3)	3.43	3.58(6)
						π^-	3.53(8)	3.41	
^{175}Lu	$7/2^+$	5d $^2D_{3/2}$	1511.4	4.67	3.50	μ^-	3.49(2)	3.41	3.50(7)
						π^-	~3.49	3.41	
^{181}Ta	$7/2^+$	$5d^3$ $^4F_{3/2}$	-1012.3	4.15	3.51	μ^-	3.39(2)	3.52	3.25(10)
							3.18(3)	3.53	
						π^-	3.30(6)	3.54	
						K^-	3.45(10)	3.55	

[a]Derived from B_J using the optimized Hartree-Fock-Slater wave functions of LINDGREN and ROSÉN [3.21]. The values are uncorrected for Sternheimer shielding.
[b]Only the most accurate values are quoted except for muonic ^{181}Ta, where a large discrepancy exists.
[c]Derived from B(E2) values of ELBEK [3.47].

II) *Quadrupole Moments of Shell-Model Nuclei.* Nuclei in the vicinity of doubly closed shells possess very small quadrupole deformations. The resolution of Ge or Ge(Li) detectors is not sufficiently high for the quadrupole splittings of mesic X rays to be resolved with sufficient accuracy in these nuclei. Even in the Pb region the uncertainties ΔQ/Q become, then, comparable to those for calculating atomic field gradients q_J. As an alternative, one may deduce Q from an appropriate theoretical description of the nuclear states using the relationships of the nuclear shell model [3.56]. In the simplest case of a single valence nucleon with angular momentum j outside a doubly closed shell, the quadrupole moment Q_j is obtained from (3.20,22)

$$eQ_j = (16\pi/5)^{\frac{1}{2}}[j(2j - 1)/(j + 1)(2j + 1)(2j + 3)]^{\frac{1}{2}}\langle j\|E2\|j\rangle$$
$$= -\beta_e[(2j - 1)/)2j + 2)]\langle r_j^2\rangle \quad (3.46)$$

where $\langle r_j^2\rangle$ is an integral over the radial wave function of the valence nucleon, and the effective charge β_e differs from the bare nucleon charge to include polarization of the nuclear core resulting from core-nucleon interactions [3.57]. The reduced matrix element $\langle j\|E2\|j\rangle$ and Q_j may be derived from E2 transition probabilities in neighbouring nuclei. In the (A_{core} ± 2) nuclei, and for E2 transitions between two states of the $|j^2I\rangle$ multiplet, one obtains

$$[B(E2,j^2I \rightarrow j^2I'))]^{\frac{1}{2}} = 2[(2I' + 1)/(2I + 1)]^{\frac{1}{2}}W(jIjI';j2)\langle j\|E2\|j\rangle \quad (3.47)$$

where W is a Racah coefficient defined by EDMONDS [3.19]. The nucleon-nucleon interaction between like particles is such that the I = 2j-1 member of the multiplet lies highest in energy, with a small energy gap to the I' = 2j - 3 member (Fig.3.12). Because of the correspondingly long mean lifetime of the I = 2j - 1 state, perturbed angular distribution methods can frequently be applied (Sects. 3.2.2.b.II,3.4.1). The quadrupole moment of the $|j^2I\rangle$ state is related to $\langle j\|E2\|j\rangle$ via the expression

$$eQ(j^2I) = (64\pi/5)^{\frac{1}{2}}[I(2I - 1)(2I + 1)/(I + 1)(2I + 3)]^{\frac{1}{2}}W(jIjI,j2)\langle j\|E2\|j\rangle \quad . \quad (3.48)$$

It should be noted that (3.47,48) neglect the effects of configuration mixing and of two-body correlations [3.58].

As an illustration of the accuracy of the various approaches, the quadrupole moment of the ^{209}Bi ground state is shown in Table 3.2. In the simplest shell-model description, ^{209}Bi consists of a $1h_{9/2}$ proton outside a doubly closed ^{208}Pb core. The quadrupole moment may then be deduced from three E2 transition probabilities in the two-valence-proton nucleus ^{210}Po after applying small corrections due to configuration mixing [3.59]. Two-body contributions to B(E2) values in ^{210}Po were estimated by KHANNA (quoted in [3.59]), who found that they amount to at most a few percent. Unfortunately, the accuracy of the muonic X ray data [3.60,61] is unsufficient to decide whether the discrepancy between mesic X ray and $[B(E2)]^{\frac{1}{2}}$ results of ~15% is significant or not. The Sternheimer shielding factor has not yet been evaluated for the atomic configurations, and the calculations of $\langle r^{-3}\rangle_Q$ are somewhat uncertain because of atomic configuration mixing and relativistic effects [3.62].

Table 3.2. Quadrupole moment of the ^{209}Bi ground state (1b = 10^{-28} m^2)

Atomic configuration	Q_{atomic}[a] [b]	Q_{mesic} [b]	$\lvert Q_{E2}\rvert$[b] [b]	$I_i \to I_f$
$6p^3$	-0.46	-0.37±0.05[c]	0.43±0.01	$8^+ \to 6^+$
6p 7s	-0.40	-0.37±0.03[d]	0.43±0.02	$6^+ \to 4^+$
			0.42±0.02	$4^+ \to 2^+$

[a]Uncorrected for Sternheimer shielding. For further references see [3.62].
[b]Deduced from $B(E2,I_i \to I_f)$ in ^{210}Po. For further references see [3.59].
[c][3.60].
[d][3.61].

3.2.3 Magnetic Interactions and Nuclear Magnetic Moments

The magnetic moment $\underset{\sim}{\mu}$ of a nucleus is an axial vector that is parallel to the total angular momentum $\underset{\sim}{I}$, i.e.

$$\underset{\sim}{\mu} = g\mu_N \underset{\sim}{I} \tag{3.49}$$

where g is the gyromagnetic ratio, and the nuclear magneton $\mu_N = e\hbar/2m_pc$ is defined in terms of the proton mass m_p. The strong deviations of the nucleon g factors ($I = s = 1/2$, $g_s = 5.58555$ for proton and $g_s = -3.82608$ for neutron) from those of Dirac particles (2 and 0, respectively) indicate that hadrons are not elementary particles. In nuclei there are contributions from both spins and orbital motion of the nucleons. The most rudimentary shell-model operator for magnetic moments is (for a more general formulation see Sect.3.5)

$$\begin{aligned}\mu &= |II|\mu_z|II\rangle \\ &= \langle II|\sum_i (g_s^{(i)}s_z + g_\ell^{(i)}\ell_z)|II\rangle \\ &= (4\pi/3)^{\frac{1}{2}}[I/(2I+1)(I+1)]^{\frac{1}{2}}\langle I||M1||I\rangle \quad .\end{aligned} \tag{3.50}$$

The last equation, which may be derived from (3.21) by making use of the operator equality, $\langle\mu_z\rangle = (4\pi/3)^{\frac{1}{2}}\langle M1,0\rangle$, is often used in comparisons of μ with off-diagonal matrix elements from M1 transition probabilities, $B(M1,I \to I') = \langle I||M1||I'\rangle^2/(2I+1)$, to test nuclear models.

The interaction energy of the magnetic moment μ with an external magnetic field B is

$$\mathcal{H}_{mag} = -\underset{\sim}{\mu}\cdot\underset{\sim}{B} = -g\,\mu_N\underset{\sim}{I}\cdot\underset{\sim}{B} \quad . \tag{3.51}$$

The magnetic substates are good quantum numbers with corresponding energy eigenvalues

$$E_m = -g\mu_N Bm = -\hbar\omega_L m \tag{3.52}$$

where $\omega_L = g\mu_N B/\hbar$ is the Larmor precession frequency. Our knowledge of nuclear magnetic moments is considerably better than that of nuclear quadrupole moments, because magnetic fields produced in the laboratory ($\lesssim$15 tesla, 1T = 10^4 gauss) cause measurable effects. For an external magnetic field B = 10 T and a g factor g = 1, the energy spacing between two adjacent substates m, m' is $\Delta E_{mm'} = 3.1524 \times 10^{-7}$ eV, and the precession angle is 0.479 millirad/ps. Magnetic fields that are orders of magnitude larger can be exploited on a microscopic scale when recoil nuclei are implanted into ferromagnetics, allowing measurements of g factors for states with subpicosecond lifetimes. The large fields in ferromagnetics are operative when the recoils are in motion (transient fields) or at rest (static hyperfine fields). For sufficiently large external polarizing fields $\underset{\sim}{B}_0$ the induced hyperfine fields $\underset{\sim}{B}_{int}$ are parallel or antiparallel to $\underset{\sim}{B}_0$, and their sum may be considered as a classical field. This is not the case in atoms or molecules where the quantum-mechanical coupling between atomic and nuclear spins has to be evaluated. We shall only recall the features that are essential for further discussions of nuclear g-factor measurements in free atoms. Many interesting g factors of short-lived states, especially in light nuclei, have been determined from magnetic interactions in highly stripped, free atoms. These results are discussed in Chap.2.

a) *Magnetic Interactions in Atoms*

The non-relativistic Hamiltonian of the M1 interaction in an isolated atom was considered as a first-order perturbation by BETHE and SALPETER [3.63], who obtained

$$\begin{aligned}\mathcal{H}_{mag} &= -\underset{\sim}{\mu}\cdot\underset{\sim}{B}\\ &= 2\mu_B \sum_i \left\{(\underset{\sim}{\ell}_i\cdot\underset{\sim}{\mu})r_i^{-3} - [(\underset{\sim}{s}_i\cdot\underset{\sim}{\mu})r_i^{-3} - 3(\underset{\sim}{s}_i\cdot\underset{\sim}{r}_i)(\underset{\sim}{\mu}\cdot\underset{\sim}{r}_i)r_i^{-5}] + (8\pi/3)(\underset{\sim}{s}_i\cdot\underset{\sim}{\mu})\delta(r_i)\right\} .\end{aligned} \tag{3.53}$$

In this expression $\underset{\sim}{r}_i$ is the radius vector from the center of the nucleus to the ith electron and μ_B is the Bohr magneton. The first term represents the magnetic field created at the nucleus by the orbital motion of the electrons. The next term is the dipole-dipole interaction between nuclear and electronic magnetic moments. The last term is the Fermi contact interaction, which is large only for s electrons. It is customary to rewrite (3.53) as

$$\mathcal{H}_{mag} = A_J \underset{\sim}{I}\cdot\underset{\sim}{J} \tag{3.54}$$

where $A_J = g\mu_N B(0)/J$. The average field B(0) at the position of the nucleus is $B(0) = -\langle J||B||J\rangle J/\langle J||J||J\rangle$. The energy eigenvalues of the Hamiltonian (3.54) are

$$E_F = A_J/2[F(F+1) - I(I+1) - J(J+1)] \quad . \tag{3.55}$$

The magnetic field at a nucleus with atomic number Z produced by a single electron with quantum numbers $n\ell j$ (hydrogen-like atom) is in the non-relativistic approximation [3.7]

$$B(0) = 12.5\ Z^3/[n^3(\ell + \tfrac{1}{2})(j + 1)]\ \text{Tesla} \quad . \tag{3.56}$$

The strong Z^3/n^3 dependence of the magnetic field arises from the $\langle r^{-3}\rangle$ factor in the dipole interaction (3.53). Of great importance are the large magnetic fields from unpaired s electrons in atoms, which may be estimated by the approximate relationship [3.7]

$$B_{ns} = 16.7\ Z(Z - q)^2/n_{eff}^3\ \text{Tesla} \tag{3.57}$$

where q is the screening charge (0, 2, and 10 for n = 1, 2, and 3, respectively) and n_{eff} is the effective principle quantum number [1, 2 - 0.8/(Z - 1), and 2.67 for n = 1, 2, and 3, respectively].

In an applied external magnetic field $\underset{\sim}{B}_0$ the magnetic interaction of a free-atom-nucleus system becomes

$$\mathcal{H}_{mag} = A_J \underset{\sim}{I}\cdot\underset{\sim}{J} + g_J\mu_B\underset{\sim}{B}_0\cdot\underset{\sim}{J} - g\mu_N\underset{\sim}{B}_0\cdot\underset{\sim}{I} \quad . \tag{3.58}$$

In weak external fields the magnetic substates m_I are not good quantum numbers, and (3.58) has to be diagonalized to obtain the eigenvalues. A simple relationship exists for the energy spacing of two adjacent Zeeman sublevels ($\Delta F = 0$, $\Delta m_I = 1$) in weak external fields (~1 mT)

$$h\nu = g_F\mu_B B_0 \tag{3.59}$$

where $g_F = g_J[F(F + 1) + J(J + 1) - I(I + 1)]/[2F(F + 1)]$. The value of the nuclear spin I in g_F is frequently the only unknown parameter, and its determination by the atomic beam resonance method is straightforward [3.6,7] and Sect.3.3).

In the limit of very strong magnetic fields, m_J and m_I tend to become good quantum numbers, and the eigenvalues of (3.58) become

$$E_{mag} = A_J m_I m_J + g_J\mu_B B_0 m_J - g\mu_N B_0 m_I \quad . \tag{3.60}$$

In the common case of $|g_J\mu_B J| \gg |g\mu_N I|$, the eigenstates are separated into (2J + 1) manifolds of (2I + 1) states each. The use of strong magnetic fields in conjunction with atomic beam resonance methods usually enables one to determine highly accurate values of both A_J and g [3.6,7].

If g and A_J are known for two isotopes (1 and 2), one frequently finds the inequality $A_J(1)/A_J(2) \neq g(1)/g(2)$. This discrepancy is in fact expected, since in a direct magnetic moment measurement an external, *homogeneous* magnetic field $\underset{\sim}{B}_0$ is applied, whereas the hyperfine structure is caused by the interaction of the nuclear moment with the slightly *inhomogeneous* magnetic field produced by the electrons. The hyperfine constant A_J is therefore sensitive to the distribution of nuclear magnetization, especially for s (and in heavier nuclei also $p_{1/2}$) electrons which have a large probability distribution over the nuclear volume. The hyperfine anomaly ε is defined by

$$A_J = A_J^0(1 + \varepsilon)$$

where A_J^0 is the hyperfine constant for a point nucleus. The *hyperfine anomaly* is thus a finite nuclear size effect in close analogy to the volume-dependent isotope shift. The value of ε depends on the relative contributions of orbital motion and intrinsic spins (3.50). The nuclear intrinsic spin samples the electron-spin density at the nucleon's instantaneous position, whereas the nucleon's orbital motion can be regarded as a current loop that samples the electron-spin density at smaller radii. The differential hyperfine anomaly Δ_{12} for isotopes 1 and 2 is given by

$$\Delta_{12} = [A_J(1)/A_J(2)]/[g(1)/g(2)] - 1 \approx \varepsilon_1 - \varepsilon_2 \quad . \tag{3.61}$$

Typical values for Δ_{12} in atoms are 0%-1%, although a value as large as 10% has been observed. The hyperfine anomalies measured before 1969 have been discussed by FOLEY and STROKE [3.64]. The hyperfine anomaly for ^{161,163}Dy has recently been observed by CLARK et al. [3.65] in a high-resolution laser experiment. They find poor agreement with theoretical calculations based on Nilsson model wave functions. It is hoped that the hyperfine anomaly parameter will receive more theoretical attention as more experimental results become available.

b) *Magnetic Interactions in Solids*

I) *The Spin Precession Methods.* The g factors of short-lived nuclear states are usually determined by the perturbed angular distribution method (Sect.3.2.2.b.II) using strong external magnetic fields $\underset{\sim}{B}_0$. The excited nuclear recoils, populated and aligned by a nuclear reaction, are stopped in a solid, preferably a metal of cubic lattice structure to eliminate static quadrupole interactions. Because of the equidistant energy spacing of the magnetic substates, the perturbation factor (3.35) becomes simply

$$G_{\lambda\bar{\lambda}}^{q\bar{q}}(t) = \exp(-iq\omega_L t)\delta_{q\bar{q}}\delta_{\lambda\bar{\lambda}} = D_{qq}^{\lambda}(-\omega_L t,0,0) \tag{3.62}$$

where the Larmor frequency $\omega_L = -g\mu_N B_0/\hbar$. In the most commonly used geometry $\underset{\sim}{B}_0$ is applied perpendicular to both the direction of the orienting interaction (beam axis or - in correlation experiments - direction of the radiation populating level I) and the direction of observation, $\underset{\sim}{k}$, for the radiation emitted from level I. Introducing (3.61) into the general expression for perturbed angular distributions (3.34), one obtains

$$W(\theta,t,B) = (d\Omega/4\pi) \sum_{\lambda \text{even}} B_\lambda(I)A_\lambda(X)P_\lambda[\cos(\theta - \omega_L t)] \tag{3.63}$$

where θ is the angle between beam axis and $\underset{\sim}{k}$. The alignment parameter $B_\lambda(I)$ is related to the orientation tensor $\rho_0^\lambda(I)$ and population numbers p(m) by

$$B_\lambda(I) = (2I+1)^{\frac{1}{2}}\rho_0^\lambda(I) = \sum_m (-)^{I+m}[(2\lambda+1)(2I+1)]^{\frac{1}{2}}\begin{pmatrix} I & I & \lambda \\ -m & m & 0 \end{pmatrix} p(m) \quad . \tag{3.64}$$

Tabulations of $B_\lambda(I)$ and of $A_\lambda(\gamma \text{ ray})$ are, for example, given by ROSE and BRINK [3.9]. Equation (3.63) implies that the nuclear spins and the angular distribution of emitted radiation rotate around the applied magnetic field with frequency ω_L. The intensity variations in a detector at fixed angle θ usually contain a dominant frequency component of $2\omega_L$. This is more easily seen when we rewrite (3.63) as

$$W(\theta,t,B) = (d\Omega/4\pi)\sum_\lambda b_\lambda \cos[\lambda(\theta - \omega_L t)] \tag{3.65}$$

where, for $\lambda_{max} = 4$, and with the common definition $a_\lambda = B_\lambda(I)A_\lambda(X)$, the coefficients b_λ are given by

$$b_0 = 1 + (1/4)a_2 + (9/64)a_4 \quad , \quad b_2 = (3/4)a_2 + (5/16)a_4 \quad , \quad b_4 = (35/64)a_4 \quad .$$

The precession of the angular distribution is most easily observed by populating state I in a nuclear reaction by means of pulsed beams from an accelerator. The intensity of radiation X emitted from state I at time t with respect to the beam arrival time ($t = 0$) is simply

$$Y(\theta,t,B) = (-Y_0/\tau)e^{-t/\tau}W(\theta,t,B) \quad . \tag{3.66}$$

Usually two detectors are used at angles θ_1 and $\theta_2 = \theta_1 + 90^\circ$, and the intensity ratio $R(t)$ is calculated

$$R(\theta,t,B) = [Y(\theta,t,B) - Y(\theta+90^\circ,t,B)]/[Y(\theta,t,B) + Y(\theta+90^\circ,t,B)] \quad . \tag{3.67}$$

Frequently, e.g., when state I is populated by heavy-ion-induced reactions, $|a_4| \ll |a_2|$, and (3.67) becomes simply

$$R(\theta,t,B) \sim [3a_2/(4+a_2)]\cos[2(\theta - \omega_L t)] \quad . \tag{3.68}$$

With a suitably chosen angle θ (typically 45° or 135°, but avoiding 0° and 90°), both sign and magnitude of the g factor can be easily determined.

In cases where the mean lifetime τ exceeds greatly the pulsed beam repetition time ΔT, one may determine ω_L by the elegant "stroboscopic" method [3.66]. In this technique the applied magnetic field is swept, and counting rates in suitably chosen time windows exhibit a resonance when the Larmor frequency $\omega_L/2\pi$ is an even multiple of the beam pulsing frequency. A detailed discussion of this and other methods of determining g factors is given by RECKNAGEL [3.67].

If the mean lifetime τ of the isomeric state is short compared to the Larmor frequency, one may obtain an average precession angle from the time-integrated perturbed angular distribution (3.66)

$$Y(\theta,B)\sum_\lambda \{b_\lambda/[1 + (\lambda\omega_L\tau)^2]^{\frac{1}{2}}\}\cos[\lambda(\theta - \Delta\theta)] \tag{3.69}$$

with $\tan(\lambda\Delta\theta) = \lambda\omega_L\tau$. In addition to the average rotation angle $\Delta\theta$, an attenuation factor $[1 + (\lambda\omega_L\tau)^2]^{-\frac{1}{2}}$ has to be considered. To extract the g factor from a measurement of $Y(\theta,B)$ the mean lifetime τ has to be known. Small precession angles $\Delta\theta$ are usually measured from the change in yield when the external field $\underset{\sim}{B}_0$ is reversed. It is convenient to define the yield ratio ε

$$\varepsilon = [Y(\theta,B) - Y(\theta,-B)]/[Y(\theta,B) + Y(\theta,-B)] \quad . \tag{3.70}$$

Uncertainties in normalizing the yields can be avoided when pairs of detectors are placed at $\pm\theta$, symmetric to the beam direction. The double ratio

$$R = [Y(\theta_1,B)/Y(\theta_1,-B)]/[Y(\theta_2,B)/Y(\theta_2,-B)] \tag{3.71}$$

is then related to ε by

$$\varepsilon = (\sqrt{R} - 1)/(\sqrt{R} + 1) \quad . \tag{3.72}$$

For small rotations $\Delta\theta$, and in the absence of delayed feeding of state I, one may write

$$\theta = S\Delta\theta \tag{3.73}$$

where $S = (dW/d\theta)/W(\theta)$ is the slope of the angular distribution. Large slopes for $I \to I - 2$ E2 transitions are observed near $\theta \sim 70^\circ$, and for $2^+ \to 0^+$ transitions near $\theta \sim 18^\circ$.

<u>II) *Diamagnetic, Paramagnetic, and Knight-Shift Corrections.*</u> The magnetic field at the nuclear site is generally the sum of the externally applied field $\underset{\sim}{B}_0$ and an induced internal field $\underset{\sim}{B}_{int}$ which is usually collinear with $\underset{\sim}{B}_0$. The g factors deduced from the measured Larmor frequency using the expression $\omega_L = -g_0\mu_N B_0/\hbar$ have therefore to be multiplied by a correction factor, i.e. $g_{corrected} = g_0(1 + B_{int}/B_0)^{-1}$. The most common internal field is that resulting from the diamagnetic current density induced by the external field in the atomic electron cloud. The most reliable calculations of this effect use relativistic Hartree-Fock electron wave functions. A table of diamagnetic correction factors $(1 - \sigma)^{-1}$ for neutral atoms, adapted from the unpublished work of LIN et al. (cf. [3.38]), can be found in FULLER's compilation [3.10]. Compared to the diamagnetic correction, the chemical shift associated with the induced field in the molecular environment [3.8] can frequently be neglected.

Much larger shifts may arise in paramagnetic atoms having unfilled inner electronic shells with large orbital momentum. Paramagnetic atoms are grouped into transition elements (3d, 4d and 5d shells unfilled), rare earths (lanthanides, 4f shell unfilled) and actinides (5f unfilled). When an external field is applied to such ions the degeneracy of the $(2J + 1)$ ionic states is removed. At thermal equilibrium the Boltzmann distribution leaves the magnetic substates unequally populated, with a resulting induced field

$$B_{int}(T) = -B_{int}(0)\langle J_Z\rangle_T/J \quad . \tag{3.74}$$

Here $\langle J_Z\rangle/J$ describes the average polarization of the ion parallel to the external field at temperature T. For pure Zeeman splitting and $g_J\mu_B J_Z B_0 \ll kT$, one obtains

$$B = B_0\beta(T) = B_0[1 + B_{int}(0)g_J\mu_B(J + 1)/(3kT)] \tag{3.75}$$

where $\beta(T)$ is the paramagnetic enhancement factor, and k is Boltzmann's constant. Equation (3.75) is derived with the assumption that L, S, J and J_Z are good quantum numbers, and that atomic states other than the ground state J manifold can be ignored. The fully aligned magnetic hyperfine field $B_{int}(T = 0)$ may be as large as 700 T (7 megagauss) in the rare earths, and the corresponding enhancement factor $\beta(T)$ may then assume values as large as ~7 at T = 300 K. Measurements of g factors are difficult in such extreme situations, because thermal fluctuations will cause a fluctuating field at the nucleus which tends to randomize rapidly the original spin orientation at t = 0. Calculations of paramagnetic enhancement factors have been performed by GÜNTHER and LINDGREN [3.68b] for the rare earths, and by KALISH et al. [3.69] for the actinides.

In NMR experiments one finds a large shift in the resonance frequency when comparing insulators and metals. The shift in metals is generally known as the Knight shift [3.70], and arises from the polarization of conduction-electron spins at the nucleus. The dominant contribution to the shift usually comes from the Fermi contact interaction between the nucleus and the electronic spins (3.53) and is given by

$$K_s = B_{int}/B_0 = (8\pi/3)M(0)/B_0 \quad . \tag{3.76}$$

Here M(0) is the electron-spin magnetic-moment density at the nucleus. If one assumes that the external field B_0 does not alter the spatial character of the electronic band states but simply redistributes the occupation of these states at the Fermi level, one may express the Knight shift in its most commonly used form

$$K_s = (8\pi/3)\chi_s\langle|\chi(0)|^2\rangle_F \tag{3.77}$$

where χ_s is the paramagnetic spin susceptibility per atom, and $\langle|\chi(0)|^2\rangle_F$ is the probability amplitude for electrons at the nucleus, averaged over all electronic states at the Fermi level. The Knight shift is always positive, and may amount to several percent in heavy nuclei. It cannot be calculated reliably and is best determined experimentally.

Two distinct experimental techniques have been employed to measure the Knight shift. One involves the determination of spin precession frequencies in both the metal host and in an environment free of conduction electrons. Successful spin rotation experiments after recoil implantation into cubic insulators have been reported by BEENE et al. [3.71] for Fr in a TℓCℓ host, and by DEWEY et al. [3.72] for Cs in a CsI host. Recoil thermalization in buffer gases has also been successfully employed in spin rotation experiments, e.g. for Fr recoils in 800 Torr (103 kPa) argon gas [3.71]. It is essential in such experiments that a large fraction of the recoils attain a hyperfine interaction-free noble-gas configuration (e.g. Fr^+) before thermalization. The observed long spin-relaxation times (e.g. ~1 μs for Fr in Ar) imply that the electron-capture cross section ($Fr^+ \rightarrow Fr^0$) is strongly inhibited as a result of the mismatch in electron binding energies for recoil (~4 eV for Fr^0) and buffer gas (~16 eV in Ar).

A second technique involves measurements of the spin-lattice relaxation times T_{rel} under conditions in which they can be related to the Knight shift via the KORRINGA relation [3.73]. The relaxation of the nuclear alignment coefficient, $a_2(t) = a_2(0)\exp(-t/\tau_2)$, is described by the relaxation rate τ_2^{-1}, which contains dipolar and quadrupolar terms, $\tau_2^{-1} = \tau_M^{-1} + \tau_Q^{-1}$. For non transition metals the KORRINGA relation implies

$$\tau_M^{-1} = 366.2\ g^2 K(\alpha) K^2 T\ \text{ms}^{-1} \tag{3.78}$$

where $K(\alpha)$ is an electron-electron correlation factor [3.73] and T is the host temperature. The Knight shift K may be derived from the observed relaxation rate τ_2^{-1}, provided the effects of both τ_M^{-1} and τ_Q^{-1} can be separated. This can be done either by exploiting their different temperature dependence [3.74] or by choosing isomeric states whose quadrupole moments are very small ($\tau_Q^{-1} \ll \tau_M^{-1}$). The latter option was chosen for Fr in Tℓ metal by BEENE et al. [3.71], who found excellent agreement between the Knight shifts from this and from two direct measurements.

III) *Static and Transient Magnetic Fields in Ferromagnets.* Several options exist for the measurement of g factors of very short-lived nuclear states ($\tau \ll 0.1$ ns). In a first class of experiments one exploits the dynamic magnetic field in isolated, fast-moving, and highly stripped atoms. The time evolution of the fields at the nucleus is governed either by atomic transition rates (recoils in vacuum) or by controllable collision rates (recoils in a buffer gas). A discussion of this topic can be found in Chap.2. A second class of experiments makes use of the microscopic hyperfine fields experienced by recoils in ferromagnetics (Fe, Co, Ni, Gd, etc.).

The *static* magnetic fields for dilute impurities substituting in ferromagnetic lattices have been studied extensively [3.75]. The effective field at the nucleus is

$$\boldsymbol{B} = \boldsymbol{B}_0 + \boldsymbol{B}_{stat} - D\boldsymbol{M} \quad . \tag{3.79}$$

Here, $\boldsymbol{M}$ is the macroscopic magnetic moment resulting from the strong spin-spin exchange forces in the ferromagnet, and the demagnetization factor D depends on the dimensions of the sample and its orientation with respect to $\boldsymbol{B}_0$ [3.76]. The sample is magnetized if $\underset{\sim}{B}_0$ exceeds $D\underset{\sim}{M}$ appreciably. The static hyperfine field $\underset{\sim}{B}_{stat}$ is caused by several competing mechanisms which are not sufficiently well understood to warrant a discussion here. However, $\underset{\sim}{B}_{stat}$ can often be calibrated from known magnetic moments using NMR, the Mössbauer effect, and perturbed angular correlations. In Fig.3.3 the static hyperfine fields are shown for various solutes in iron (from the compilation of RAO [3.75]). The most striking feature is the reversal of the sign of $\underset{\sim}{B}_{stat}$ in the lanthanides. The lanthanides are unique because of the validity of Russel-Saunders coupling (L, S and J are good quantum numbers). Since $\underset{\sim}{S}$ determines the orientation of the atomic moment and $\underset{\sim}{L}$ dominates the contribution to $\underset{\sim}{B}_{stat}$, the hyperfine fields are positive in the first half of the series ($\underset{\sim}{L}$ and $\underset{\sim}{S}$ antiparallel) and negative in the second half ($\underset{\sim}{L}$ and $\underset{\sim}{S}$ parallel).

The large *transient fields* experienced by nuclei slowing down in polarized ferromagnetics were first observed in the pioneering study by BORCHERS et al. [3.77]. The transient magnetic fields (TMF) operate only while the recoils are in motion. For a state with mean lifetime τ, the TMF precession effect may be evaluated by (3.69-73) if we neglect the attenuation factor and rewrite the precession angle $\Delta\theta$ as

$$\phi_{TMF}(t) = -0.00479\ g \int_0^t B(t)\ \exp(-t/\tau)\ dt\ \text{mrad Tesla}^{-1}\ \text{ps}^{-1} \quad . \tag{3.80}$$

This expression is usually transformed into an integral over velocity, viz.

$$\phi_{TMF}(t_0) = 1.487(gA/\rho c) \int_{v_i}^{v(t_0)} B(v)\ e^{-\lambda t(v)}(dE/\rho dx)^{-1}\ dv \tag{3.81}$$

where v_i is the initial velocity, t_0 is the exit time from (or stopping time in) the ferromagnetic layer of density ρ [g cm^{-3}], and ($dE/\rho dx$) is the stopping power [MeV mg^{-1} cm^2]. Two approximations are often made to integrate (3.81). First, the stopping power is assumed to be proportional to velocity; consequently $v = v_i \exp(-t/\alpha)$, with α being the slowing-down time. Secondly, the transient field is assumed to be linear with velocity, i.e.,

$$B(v) = C(Z)v/v_0 \tag{3.82}$$

where $v_0 = c/137$ is the Bohr velocity, and the constant C(Z) describes the atomic number dependence of the TMF. With these two assumptions one easily obtains

$$\phi_{TMF}(t) = -0.00479gC(Z)(v_i/v_0)(1/\alpha + 1/\tau)^{-1}\{1 - \exp[-t(1/\alpha + 1/\tau)]\} \quad . \tag{3.83}$$

If the mean life τ is much longer than the slowing-down time α, (3.83) simplifies to

$$\phi_{TMF}(t) \to -0.00479gC(Z)D/v_0 \quad , \tag{3.84}$$

i.e., the precession angle is proportional to the thickness D of the traversed ferromagnetic layer, independent of the initial velocity v_i and of the slowing-down time α.

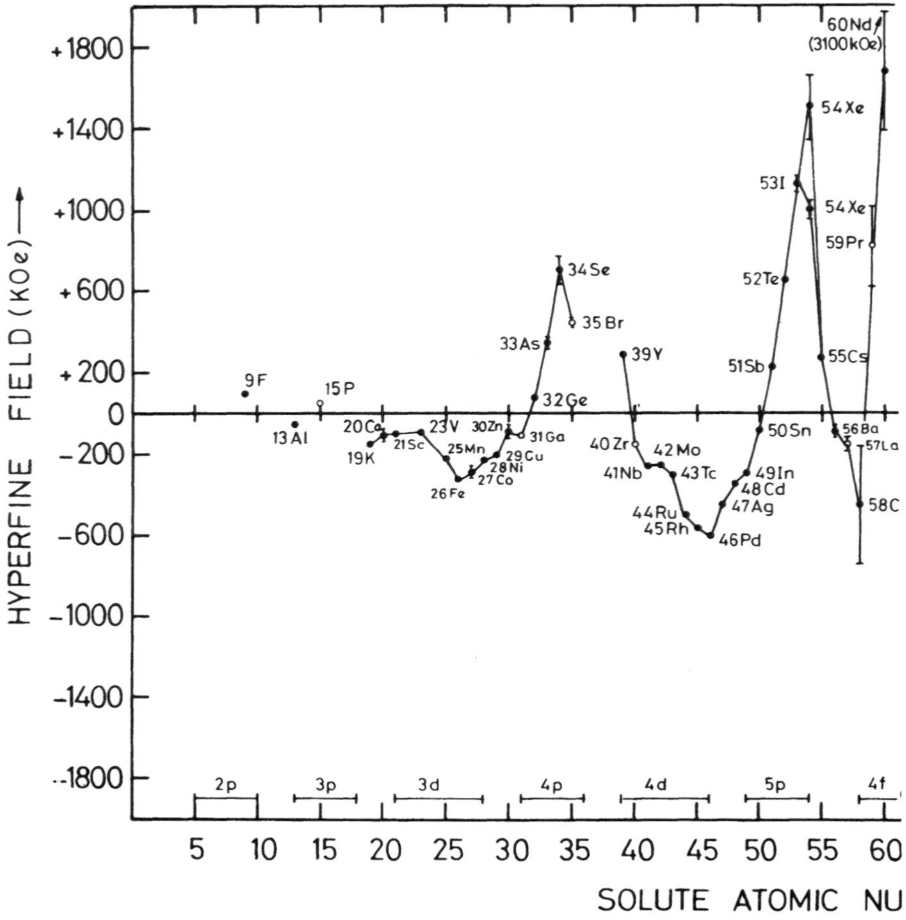

Fig.3.3. Static magnetic hyperfine fields of various solutes in iron (1 kO

Until about 1975 the experiments were analyzed by the theory of LINDHARD and WINTHER [3.78] (LW) which evaluates the consequences of Coulomb scattering of *unbound*, polarized electrons by the moving recoils. The LW field is proportional to 1/v [in contrast to (3.82)] and becomes largest at the end of the recoil path. Precession angles much larger than predicted by the LW theory were observed in 1975 in experiments with high-velocity light ions [3.79,80]. Since then, considerable experimental effort has been directed toward mapping out and exploiting the large TMF at high recoil velocities. Detailed discussions of these developments can be found in articles by van MIDDELKOOP [3.81] and by BENCZER-KOLLER et al. [3.82].

The *mechanism* that produces the large TMF has been most clearly elucidated by experiments in light nuclei. EBERHARDT et al. [3.83] have suggested that deep holes in ns shells are created by promotion of electrons from the moving recoil into empty orbitals of the combined molecular state of recoil and host atom. The average TMF is then

$$B(v) = \sum_n \xi_1^{ns}(v) F_1^{ns}(v) B_{ns} \qquad (3.85)$$

where $F_1^{ns}(v)$ is the fraction of ions carrying an unpaired ns electron, $\xi_1^{ns}(v)$ is its degree of polarization, and B_{ns} is the magnetic contact field (3.57). The polarized electrons are replaced so frequently during the slowing-down process that the transient field can be regarded as parallel to the external polarizing field. The basic correctness of this model is supported by measurements of the K-vacancy fractions for recoils leaving the ferromagnetic layer [3.84,85]. The measured K-vacancy fractions are strongly correlated with the observed velocity dependence of B(v) and the observed discontinuity of B(v) at $Z = 9$, provided one assumes a constant polarization $\xi \sim 0.13$. A possible mechanism for the polarization is spin exchange between the unpaired ns electron and polarized electrons in higher shells which in turn are polarized by capture-loss and spin-flip collisions with polarized (M and N) electrons of the ferromagnet (Fe).

The gross dependence of the TMF on velocity and atomic number of the recoil is now roughly known. Unfortunately, different parametrizations of B(v,Z) have been used by different authors, making detailed comparisons cumbersome. The TMF in different ferromagnetics is found to be approximately proportional to the magnetization per unit volume, i.e. B(v) is roughly in the ratio of 1:0.8:0.3:1.2 for fully magnetized Fe, Co, Ni and Gd, respectively [3.82]. Over a limited velocity range, and provided the electrons are not fully stripped ($v < Zv_0$), B is generally found to be proportional to v (3.82). More recent results taken over a wider range of velocities [3.82,112] indicate a weaker than linear increase of the TMF with recoil velocity. There is yet no conclusive experimental proof that the full LW contribution has to be included, partly because of large uncertainties in the energy loss ($dE/\rho dx$) at low recoil velocities. Experiments using thin ferromagnetic layers

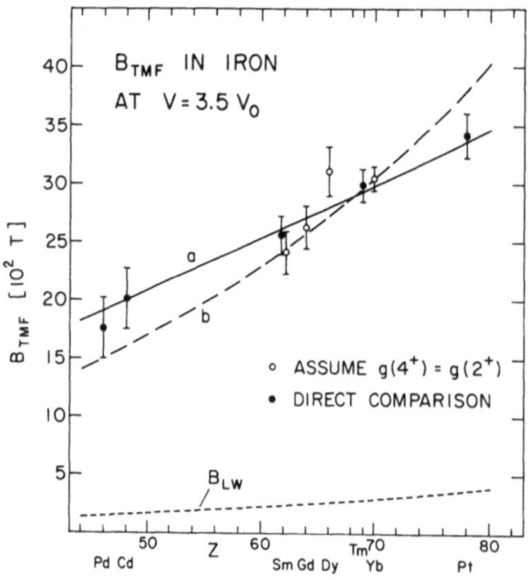

Fig.3.4. Transient hyperfine fields in iron at a recoil velocity v = 3.5 v_0. All data points were obtained with thin iron in the velocity range. $1.4 \leq v/v_0 \leq 5.6$. The parametrizations a and b of the transient field are explained in the text. The weak Lindhard-Winther field (B_{LW}) contributes at most 10% to the total field. Adapted from [3.86]

and high initial recoil velocities are advantageous because one may thus avoid the main regime of the LW field. Furthermore, even in light nuclei the pronounced shell effects observed at low recoil velocities ($v/v_0 \lesssim 2$) tend to be washed out at higher recoil velocities [3.85], and, finally, the recoils can be stopped in a suitable cubic environment, thus avoiding an additional spin precession due to the static field B_{stat}.

When considering the atomic-number dependence of B(v,Z) it is important to compare data from a similar velocity region. In Fig.3.4 we show the TMF in Fe for various heavy nuclei between Pd and Pt [3.86]. The data were all obtained in the same velocity region $1.4 \leq v/v_0 \leq 5.6$, assuming (3.82). The results are best fit by a linear Z dependence (solid line), C(Z) = 11.0 Z tesla. The existence of a relativistic factor familiar from the LW theory [3.78] cannot be ruled out, although the corresponding best fit (dashed line), $C(Z) = 6.89\, Z[1 + (Z/84)^{5/2}]$ Tesla, is worse. It is remarkable that the TMF behaves in a rather smooth fashion. For acurate g factor measurements it is, however, not advisable to rely on empirical parametrizations of the TMF. It is at present much safer to calibrate the TMF in the relevant velocity region using states of known g factor in neighbouring nuclei.

Of the recently determined properties of the TMF, three may be singled out to have important consequences for the measurements of nuclear g factors. Firstly, the TMF depends rather smoothly on v and Z. From a comparison of the Z dependence of static and transient fields (Figs.3.3,4), it is obvious that the latter is more readily applicable to a larger variety of nuclei. Secondly, at suitably high recoil velocities the TMF is larger than both static and LW fields. This makes measurements on states with subpicosecond lifetimes possible. Finally, and most importantly, the TMF is largest at the beginning of the recoil path. This property is essential when considering short-lived high-spin states which can only be formed in heavy-ion-induced reactions with consequently large initial recoil velocities and long slowing-down times. It is in this latter area of nuclear structure physics where qualitatively new results can be expected.

3.3 Laser Spectroscopy and Hyperfine Structure of Exotic Nuclear States

The development of single-mode tunable lasers has made it possible to scan the hyperfine structure of atomic transitions by very sharp, tunable optical excitation. Because of the excellent energy resolution (10^{-7} - 10^{-10}) and the high power ($\lesssim$1 W cw power corresponding to $\lesssim 3\times10^{17}$ photons/s) of laser light, and because of the very large cross sections for atomic excitations ($\sim10^{-16}$ cm^2), extremely low atomic densities and even single atoms may be detected. Another important advantage of laser beams is their narrow divergence and large energy density. A variety of applications of laser techniques to nuclear physics has recently been described by MURNICK and FELD [3.87]. We confine ourselves here to a description of two different experiments using the optical pumping technique to determine spins, isotope shifts and moments of exotic (this term is used here as a synonym for uncommon) nuclear states.

In optical pumping experiments the atoms are placed in a magnetic field which is sufficiently weak ($\sim$1 mT) that the Zeeman splitting is smaller than the natural linewidth of the optical transition. The atoms are then irradiated along the direction of the magnetic field (quantization axis) by circularly polarized (laser) light. If we consider the circular polarization to be left-handed (σ^- light), then the selection rule $\Delta m_F = -1$ applies to the absorption, whereas the spontaneous radiative decay from the excited state occurs according to statistical probabilities, with the selection rule $\Delta m_F = 0, \pm1$. Consequently, after the first pumping cycle the centroid of magnetic substates in the atomic ground state, $\bar{m}_F$, will be negative. If the process of absorption and reemission can be repeated many times without loss of polarization resulting from spin relaxation, the atoms will eventually be pumped into their lowest substate, $m_F = -F = -I-J$, with complete polarization of both atomic and nuclear spins. The equilibrium polarization $\bar{P}$ is approximately given by

$$\bar{P} \approx T_\ell/(T_\ell + p_c T_p) \tag{3.86}$$

where T_ℓ is the (longitudinal) relaxation time, T_p is the average time between subsequent absorptions of a light quantum per atom, and p_c is the polarization after one pumping cycle. In high-resolution studies the power of the laser light has to be chosen carefully. It should be large enough that $T_p \ll T_\ell$, and at the same time low enough to avoid power broadening of the optical transition. The onset of optical

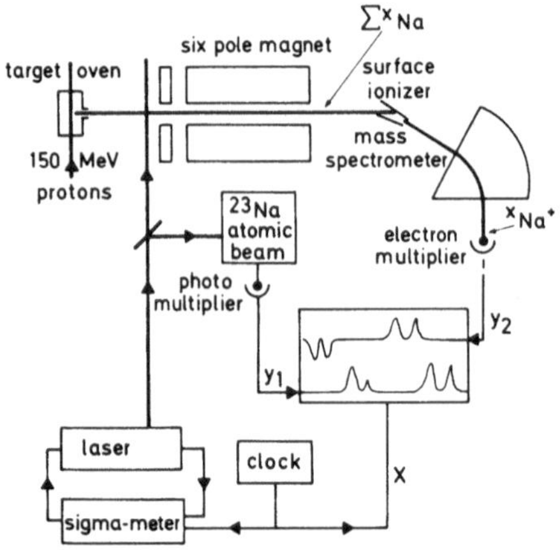

◄Fig.3.5

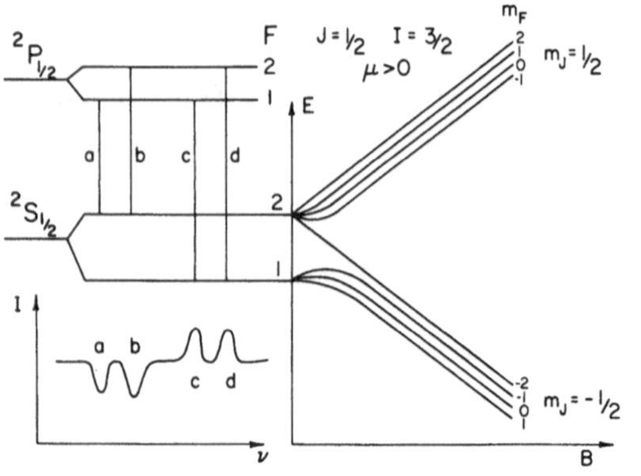

▲Fig.3.6

Fig.3.5. Schematic view of the experimental setup used for atomic beam laser spectroscopic studies [3.88]

Fig.3.6. Breit-Rabi diagram of a $^2S_{1/2}$ state with I = 3/2, showing the hyperfine structure as a function of the magnetic field (right-hand side). The hyperfine components a-d of the sodium D_1 line observed with the experimental setup of Fig.3.5 are shown schematically on the left

pumping can best be detected either by the anisotropy of emitted radiation (if the nuclear state is very short lived, $\tau \lesssim 0.01$ s), or by deflection of the atoms in an atomic beam (if the nuclei are long-lived, $\tau \gtrsim 0.01$ s).

In the first example we describe the measurement of spins, isotope shifts and magnetic moments in a series of sodium isotopes [3.88]. The principle of the experiment is shown schematically in Figs.3.5,6. The sodium isotopes are produced by the interaction of a high-energy proton beam (150 MeV, 2 GeV) with several $^{238}UO_2$ targets mounted in a stack on graphite backings. The target assembly is heated to 2000°C to allow quick diffusion of the sodium through the graphite. In the region of a weak magnetic field, the collimated beam of sodium atoms is intersected at right angles by the σ^- light beam from a tunable dye laser. The laser frequency is scanned across the sodium D_1 line (Fig.3.6), and the stability of the system is monitored by detecting the fluorescence signal from a reference atomic beam of stable ^{23}Na that is illuminated by the same laser beam. The atomic beam passes then adiabatically from the weak-field region to the strong-field region inside a six-pole magnet. In the six-pole magnet the magnetic field is zero on the axis and increases with the distance from the axis. As a consequence the magnet has focussing properties for atoms with $m_j > 0$ ($|+\rangle$ states) and defocussing properties for atoms with $m_j < 0$ ($|-\rangle$ states) [see the Breit-Rabi diagram in Fig.3.6 for the ^{23}Na ground state with I = 3/2, $^2S_{1/2}$, and also (3.60)]. The use of a six-pole magnet has the advantage of increased transmission over the standard Rabi-type atomic beam experiment [3.6,7]. A signal proportional to the number of focussed atoms of the sodium isotope studied is obtained by counting them with an electron multiplier

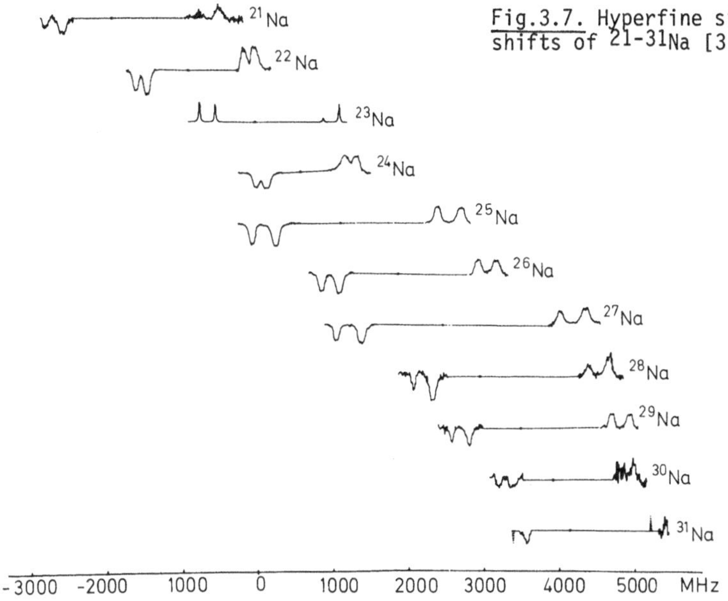

Fig.3.7. Hyperfine structure and isotope shifts of $^{21-31}$Na [3.88]

after they have been ionized and passed through a mass spectrometer. As the frequency of the σ^- laser light matches one of the hyperfine transitions (a-d in Fig. 3.6) the number of detected atoms decreases (increases), depending on whether atoms are transferred from $|+\rangle$ to $|-\rangle$ states (components a,b) or vice versa (components c,d). Details of the signals a-d depend on the transition probabilities and the product of power density times the interaction time of the σ^- laser light. Signals observed by HUBER et al. [3.88] for the isotopes $^{21-31}$Na are shown in Fig.3.7.

The spins I of the sodium isotopes have been determined in the same experiment by using the atomic beam resonance method with weak magnetic fields (Sect.3.2.3.a). For this purpose the σ^- laser light is frequency locked to the $^2S_{1/2}$, $F = I + 1/2 \rightarrow {}^2P_{3/2}$, $F = I + 3/2$ hyperfine transition of the sodium D_2 line. The atoms involved in this transition are optically pumped into the Zeeman sublevel $m_F = -(I + 1/2)$ of the $F = I + 1/2$ hyperfine ground state and become subsequently defocussed by the six-pole magnet. A radio-frequency field is applied perpendicular to the weak magnetic field $\underset{\sim}{B}_0$. When the radio frequency matches the frequency of the transition between two adjacent Zeeman levels ($\Delta F = 0$, $\Delta m_F = 1$), the RF field tends to equalize the populations of the Zeeman sublevels, and the number of detected atoms increases. The resonance condition is in the present case [$J = 1/2$, $g_J = 2$; see (3.59)]

$$\nu_{RF} = 2\mu_B B_0/[(2I + 1)h] \qquad (3.87)$$

with the nuclear spin I being the only unknown.

The magnetic moments are derived from the centroids of the four hyperfine components of the D_1 line (Fig.3.7). The A_J factors and average energies for both the

Table 3.3. Spins, magnetic moments and isotope shifts for $^{21-31}Na^{a}$

A	$T_{1/2}$	I^{b}	$\mu_I[\mu_N]^{b}$	$IS^{23,A}$ [MHz]
21	22.5 s	3/2	2.3861(1)	-1596.7(2.3)
22	2.60 yr	3	1.746(3)	- 758.5(7)
23	stable	3/2	2.217520(2)	0
24	15.02 h	4	1.6902(5)	706.4(6.2)
25	60.0 s	5/2	3.683(4)	1347.2(1.3)
26	1.07 s	3	2.851(2)	1397.5(9)
27	290 ms	5/2	3.895(5)	2481.3(2.0)
28	30.5 ms	1	2.426(3)	2985.8(2.7)
29	43 ms	3/2	2.449(8)	3446.2(3.8)
30	53 ms	2	2.083(10)	3883.5(6.0)
31	17 ms	3/2	2.283(38)	2486(16)

[a]From [3.88]. The quoted errors are one standard deviations.
[b]The spins and magnetic moments for $^{21-25}Na$ were previously known. For further references see [3.88].

$^2S_{1/2}$ and $^2P_{1/2}$ states are obtained from the centroids using (3.55). The magnetic moments are then calculated relative to the well-known magnetic moment of the reference isotope ^{23}Na via the relationship $\mu = \mu_{23}(I/I_{23})[A_J/A_J(23)]$. These magnetic moments, which do not include corrections for hyperfine anomaly effects, are shown in Table 3.3 together with spins, half-lives and isotope shifts.

Of particular interest are the isotope shifts which deviate strongly from the expected dependence for a pure mass shift, i.e., $\delta\nu_{23,A} = C(A-23)/(23A)$, see (3.4). Sodium is presently the lightest element for which a volume shift has been unambiguously observed. The mass-shift constant C can be estimated from the isotope shifts of $^{25-27}Na$ after calculating the volume shifts of these isotopes according to (3.5-7). The change in the mean-square radii, $\delta\langle r^2\rangle_{AA'}$, for these nearly spherical nuclei is assumed to be 0.5 times the 'standard' value (3.10), as is suggested by the deformed Hartree-Fock calculations of CAMPI et al. [3.89]. With the derived mass-shift constant C = 385 GHz, one may then deduce $\delta\langle r^2\rangle_{25,A}$ for the complete series of isotopes (see Fig.3.8). The reduced data agree well with the Hartree-Fock calculations [3.89] for $^{23-28}Na$ and ^{31}Na, and tend to support the predicted strong prolate deformation for the heavy isotopes $^{31-35}Na$. This analysis of nuclear deformation is somewhat dependent on the assumptions for the mass shift. It could be confirmed more directly by measurements of the hyperfine structure of the D_2 line which would allow a determination of the static quadrupole moments (Sect. 3.2.2.a).

Techniques similar to the one just described have been applied in extensive on-line measurements of isotope shifts in isotopes of mercury [3.90] and of cesium [3.91].

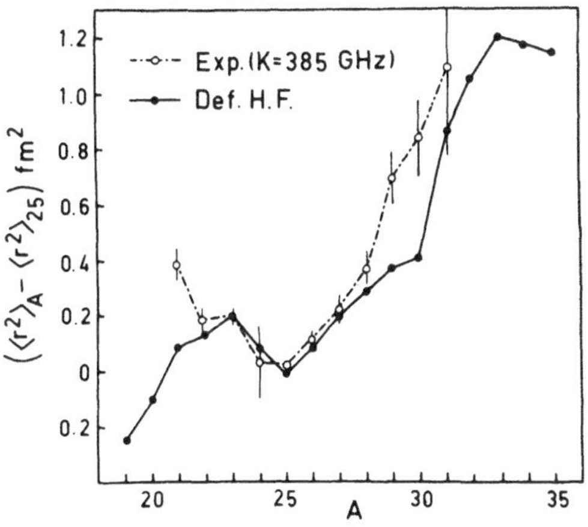

Fig.3.8. Comparison of the change in the mean square radius of sodium isotopes from experiment (assuming a mass shift constant C = 385 GHz [3.88] and from a deformed Hartree-Fock calculation [3.89]

The combination of laser spectroscopy with the atomic beam method is clearly a very powerful one. The strong suppression of the Doppler width, the lack of collisional line broadening, and the very high collection efficiency are distinct advantages that can no longer be retained when studying shorter-lived nuclear states ($\tau \lesssim 0.01$ s). In the second example we describe the recent experiment of BEMIS et al. [3.92], who observed optical pumping of a fissioning isomer in ^{240}Am. This isomer has an excitation energy of approximately 2.6 MeV and a half-life of ~0.94 ms [3.93]. The atomic ground state for americium, $^8S_{7/2}(5f^77s^2)$, is suitable for optical pumping because of its sphericity, which suppresses collisional relaxation. The lifetime of the $^{10}P_{7/2}(5f^77s7p)$ excited state (~2 μs) is much shorter than the isomer lifetime, and the hyperfine components of the $^8S_{7/2} \rightarrow {}^{10}P_{7/2}$ transition are all contained within a wavelength band of ~0.5 Å (0.5×10^{-8} cm). Laser optical pumping with σ^+ laser light of ~0.5 Å bandwidth will then result in the population of the $F_{max} = I + 7/2$ Zeeman level, with complete polarization of both atomic and nuclear spins.

The experimental arrangement is shown schematically in Fig.3.9. The ^{240}Am isomer was produced in the $^{238}U(^7Li,5n)$ reaction at an effective bombarding energy $E(^7Li) \sim 47.5$ MeV and with a cross section of ~1.8 μb (1.8×10^{-30} cm^2). The isomeric recoils were ejected from the UO_2 target and thermalized in 180 torr (24 kPa) of helium gas. The peak of the thermalized recoil distribution was located about 15 cm downstream of the target, and fission fragments from the decay of the isomer were observed in two position-sensitive gas proportional detectors. The ^{7}Li beam was mechanically chopped to provide equal 2-ms-long beam-on and beam-off periods, and data were recorded during the beam-off period. A time distribution of fission fragments is shown in Fig.3.10b. The fitted half-life of 0.942 ± 0.038 ms ensures the correct assignment of the fission events to ^{240m}Am (see half-lives of other fissioning isomers in [3.93]).

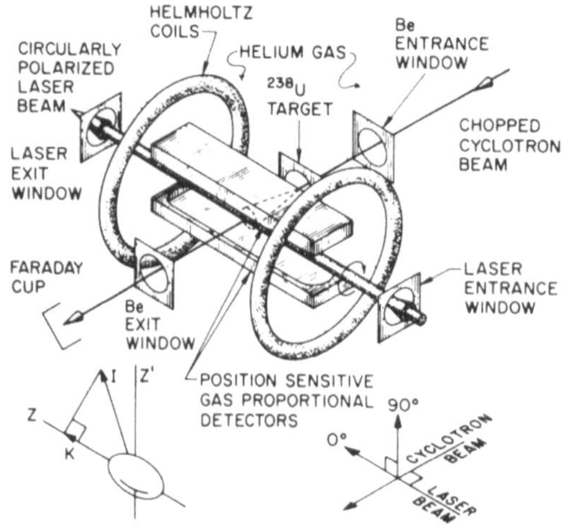

Fig.3.9. Schematic view of the experimental setup used by BEMIS et al. [3.92] to observe the isomer shift of the ^{240}Am fissioning isomer

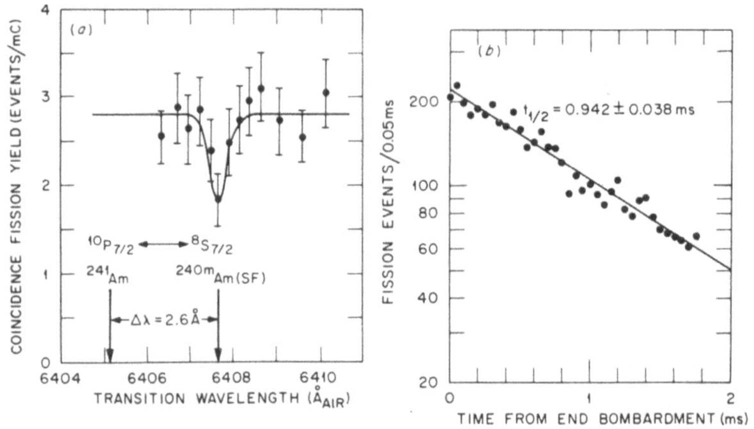

Fig.3.10. (a) Coincidence fission yield from the ^{240}Am fissioning isomer vs the wavelength of the scanning laser light. The dip in the yield indicates the onset of optical pumping; (b) time distribution of fission fragments from ^{240}Am [3.92]

A σ^+ continuous-wave laser beam, directed perpendicular to the Li^7 particle beam and along a weak magnetic field (~0.3 mT), was focussed at the location of the thermalized recoils. The laser output wavelength λ was scanned in the interval 6406 Å - 6410 Å, and absolute values of λ were established relative to the 6402.246 Å line in NeI. When optical pumping occurs, the angular distribution of the fission fragments becomes anisotropic. For the oriented nuclei with $K = I = |m_I|$, the fission fragments are emitted preferentially along the laser beam direction at 0°, the anisotropy being ideally $W(0^\circ)/W(90^\circ) = 2^{2K-1}$. The yield of coincident fission fragments shows a dip at $\lambda = 6407.7 \pm 0.2$ Å (see Fig. 3.10a). This result implies an extremely large isomer shift of 2.6 ± 0.2 Å relative to the known wavelength of the $^8S_{7/2} \rightarrow {}^{10}P_{7/2}$ transition in ^{241}Am, $\lambda = 6405.105$ Å.

The observed large isomer shift is a clear indication of a substantial change in nuclear shapes between ^{240,241}Am and ^{240m}Am. One may deduce the deformation parameter β of the fissioning isomer with a number of reasonable assumptions. The scaling factor F between frequency shift and $\delta\langle r^2\rangle$ [see (3.5)] can be determined from the known isotope shift between ^{243,241}Am, $\Delta\lambda = 0.09685$ Å. Assuming that the difference in wavelength is attributable to the change in radius of the volume-equivalent sphere [i.e. ignoring the small mass shift in (3.4) and assuming that the deformation parameter, $\beta = 0.24$, is the same for both ^{241}Am and ^{243}Am], a value $\delta\langle r^2\rangle = (5.1 \pm 0.4) \times 10^{-26}$ cm^2 is derived for the change in the mean-square radius of ^{240}Am and ^{240m}Am. With the 'standard' expression for $\delta\langle r^2\rangle$ in deformed nuclei (3.15), this result implies a deformation parameter $\beta = 0.66 \pm 0.04$, and an intrinsic quadrupole moment $Q_0 = (32.7 \pm 2.4) \times 10^{-24}$ cm^2 for the fissioning isomer in ^{240}Am. If one includes the effect of 'compressibility under deformation' and retains terms of the order of β^3 [see Sect.3.2.1 and (3.17)], one obtains the values $\beta = 0.73 \pm 0.04$ and $Q_0 = (36.9 \pm 2.4) \times 10^{-24}$ cm^2. In any case, the extremely large isomer shift for ^{240m}Am provides at present the most direct confirmation of the theoretically expected unusually large deformation of fissioning isomers.

Variations of the optical pumping method of BEMIS et al. [3.92] can in principle probe the hyperfine structure of nuclear states with half-lives even shorter than 1 ms. From (3.86) it is seen that significant polarizations can be obtained, provided a suitable excited atomic state can be found such that the average time between subsequent absorptions of light quanta per atom, T_p, is shorter than the mean lifetime of the isomer, τ, and the spin-relaxation time T_ℓ. It is hoped that future studies with laser beams will make a substantial contribution to hyperfine structure investigations of short-lived nuclear states.

3.4 Nuclear Moments of High-Spin States

The development of heavy-ion accelerators has made it possible to transfer large amounts of angular momentum to the nucleus ($\lesssim 30\hbar$ in Coulomb excitation with heavy ions, $\lesssim 80\hbar$ in fusion-evaporation reactions). During the last decade rapid progress has been made in establishing the main features that govern the behaviour of nuclei at high spin. Of particular interest is the structure of nuclear states at the yrast line, i.e. the ensemble of states whose energy is lowest for any chosen value $I\hbar$ of the angular momentum.

Two basically different types of yrast structures have so far been identified. The nuclei whose neutron and proton numbers are far removed from the closed-shell (magic) values (many stable rare-earth nuclei, actinides, etc.) prefer strongly deformed, intrinsic deformations with collective adiabatic rotations occurring around an axis perpendicular to the prolate symmetry axis. The detailed properties of high-spin yrast states in these nuclei depend on the underlying microscopic shell structure and its response to the strong Coriolis forces [3.94]. The new phenomena at spins $I \gtrsim 20$ include a partial alignment of the spins of individual particles in the direction of the axis of rotation, and a gradual disappearance of superfluidity, with the moment of inertia approaching that for rotation of a rigid body. The yrast states decay by collective E2 transitions whose extremely short lifetimes ($\lesssim 1$ ps) preclude measurements of static quadrupole moments by presently known techniques (Sect.3.2.2.b). The use of transient magnetic fields, however, has opened up the possibility of testing the microscopic shell effects that are manifest in the g factors of collective yrast states, as will be discussed in Sect.3.4.2.

A second class of discrete high-spin states ($I \gtrsim 20$), whose properties differ radically from those of collectively rotating nuclei, has recently been identified in several nuclei at the beginning [3.95-97] or towards the end [3.98] of major (jj) shells. These nuclei are characterized by irregular energy spacings between adjacent yrast levels, by γ decays of various multipolarities (M1, E2, E3), and by the existence of long-lived isomers ($\tau \gtrsim 1$ ns). Standard perturbed angular distribution methods (see Sects.3.2.2.b.II, 3.2.3.b.I, and 3.5) are then applicable to determine the g factors and in favorable cases the static quadrupole moments of the isomeric states, to deduce the structure and deformation of yrast states at isolated spin values. We restrict our discussion to states at the beginning of (jj) shells having an oblate deformation.

3.4.1 The Deformation of High-Spin Yrast Isomers

The nucleus ^{212}Rn was the first in which discrete yrast states have been observed to spin values as high as 30. In Fig.3.11 we show the yrast energies from the work of HORN et al. [3.95] plotted versus $I(I+1)$. The yrast states contain a total of nine isomeric states with half-lives $T_{1/2}$ longer than 1 ns. The existence of isomers can usually be attributed to microscopic shell effects. This is well established for the states in ^{212}Rn with $I \leq 18$, for which the shell model interaction of Blomqvist (quoted in [3.95]) between valence protons ($1h_{9/2}$, $2f_{7/2}$, $1i_{13/2}$) produces exceptionally good agreement with experiment, the deviations being typically 10 keV. This interpretation is further supported by the measured g factors of the $I^{\pi} = 8^{+}$ and 17^{-} isomers, which agree with calculations using the valence proton configurations of Blomqvist (see lower part of Fig.3.11). Above $I = 18$ the yrast states exhibit a number of features that have meanwhile been observed in several other nuclei [3.96,97]: a) The yrast energies, when plotted versus $I(I+1)$, follow closely a straight line characterized by an effective moment of inertia close to that for the rotation of a rigid sphere ($2\mathcal{J}/\hbar^2 \sim 207$ MeV^{-1} in ^{212}Rn); b) the E2 transition probabilities are severely inhibited when compared to collective transitions; c) the g factors approach a value $g \sim 0.5$ (see Fig.3.11), indi-

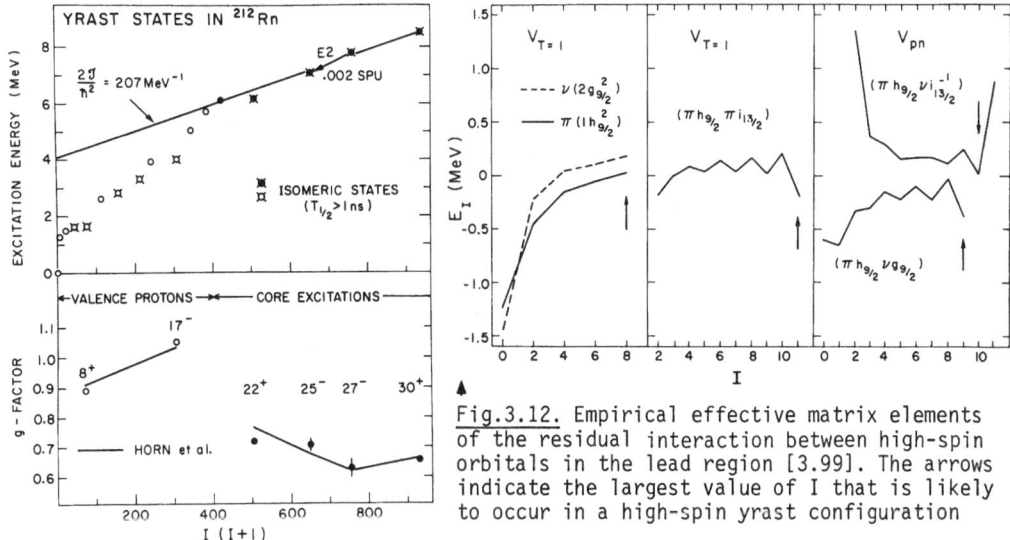

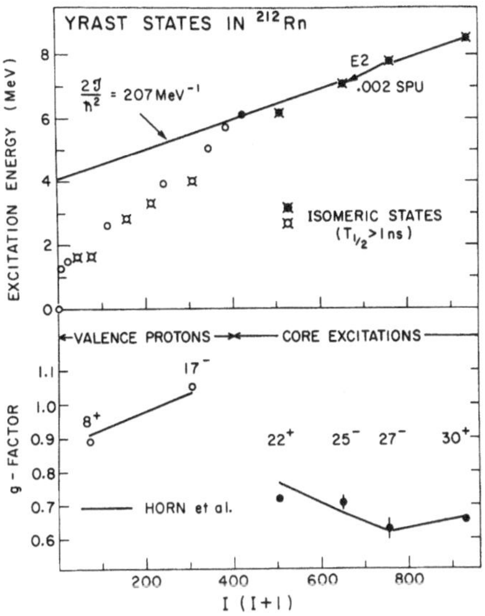

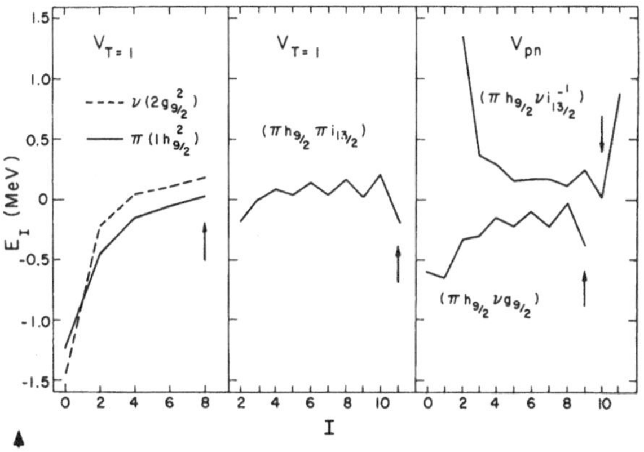

Fig.3.12. Empirical effective matrix elements of the residual interaction between high-spin orbitals in the lead region [3.99]. The arrows indicate the largest value of I that is likely to occur in a high-spin yrast configuration

Fig.3.11. Rotational plot of the energies of yrast levels in ^{212}Rn vs I(I + 1) (top half). The g factors shown below illustrate the transition from a pure valence proton structure to a mixed proton-neutron structure [3.95]. The $27^- \rightarrow 25^-$ E2 transition probability is known to be extremely weak (0.002 single particle units)

cating that the closed (N = 126) neutron shell is broken, with both protons and neutrons contributing to the angular momentum; d) some states, the so-called yrast traps, may be lowered sufficiently in energy that they can only decay by γ rays of multipolarity $\lambda > 2$ ($I^\pi = 25^-$ and 30^+ in ^{212}Rn).

The reason for the occurrence of yrast isomers can be traced to the general properties of the nucleon-nucleon interaction. In Fig.3.12 we show several examples of the effective interaction between pairs of high-j particles (holes) in the Pb region (see [3.99]). It is usually energetically favorable to form high-spin states by maximally aligning the valence particles, causing an oblate deformation of the nucleus [$Q < 0$; see (3.46)]. This trend towards oblate deformation is not substantially altered when also considering the (neutron) hole orbitals which - for a chosen substate m_j - produce a deformation of opposite sign to that of the equivalent particle orbital. It is seen from Fig.3.12 that maximally aligned particle-hole pairs experience a strong repulsion and are thus unlikely to occur in yrast configurations. The hole orbitals tend to contribute to the total angular momentum with small m projections which may even add to the oblate deformation produced by the valence particles.

These qualitative considerations are confirmed in two types of calculations, one using the cranked Hartree-Bogolyubov approximation [3.100], and one using cranked Nilsson potentials with Strutinsky normalization to the liquid drop [3.101,102]. In both approaches the β_2 deformation for each value I of the yrast spin is found by an energy minimization procedure. The calculations predict that the valence particles induce a sizeable oblate deformation of the core which adds to the shell model deformation.

The g factors of high-spin yrast isomers can be measured relatively easily by the spin rotation method; however, their values turn out to be remarkably insensitive to the core deformation [3.103]. In a deformed model one may assume $I = K$, where $K = \sum \Omega_i$ is the projection of all the particle angular momenta on the symmetry axis. The g factor is then

$$g = \left\{g_R + \sum_i \left[g_\ell^{(i)}\Omega_i + (g_s^{(i)} - g_\ell^{(i)})\langle S_3^{(i)}\rangle\right]\right\}/(I + 1) \tag{3.88}$$

where $g_\ell^{(i)}$ and $g_s^{(i)}$ are the orbital and intrinsic g factors of the valence particles and holes, g_R is the gyromagnetic ratio associated with the collective flow of the rotating core, and $\langle S_3^{(i)}\rangle$ is the expectation value of the intrinsic spins along the symmetry axis. The latter spin term is the only one which depends on the core deformation. For high-spin configurations it may frequently be neglected, and one obtains

$$g \approx (g_R + g_\ell^P\Omega_p)/(I + 1) \approx \Omega_p/(\Omega_p + \Omega_n) \quad , \tag{3.89}$$

i.e. the g factors of high-spin yrast states measure the fraction of the angular momentum supplied by the protons, largely independent of the core deformation.

The static quadrupole moment and deformation of a high-spin yrast state has recently been determined [3.104] in ^{147}Gd, one of several transitional rare-earth nuclei in which high-spin isomers have been found. Of the rare-earth metals, gadolinium (and europium) occupy a special role, since in their common $3^+(2^+)$ charge states they have an almost pure, spherical $^8S_{7/2}$ atomic ground state. For this reason the enormous orbital contribution to the magnetic hyperfine field B_{int} vanishes, and only the small, negative spin contribution remains. Consequently, long relaxation times ($\gtrsim 0.1$ μs) are observed, which make sensitive perturbed angular distribution experiments feasible [3.105,106].

The high-spin isomer in ^{147}Gd was populated and aligned by the ^{124}Sn (^{28}Si,5n) reaction using pulsed beams of ^{28}Si at a bombarding energy of 144 MeV. The beam bursts were $\lesssim 2$ ns wide and had a repetition time of 3.2 μs, much longer than the isomer half-life $T_{1/2}$ = 510 ns. The excited ^{147}Gd recoils traversed the 2 mg/cm^2-thick target of enriched ^{124}Sn and were implanted into 1-mm-thick single crystals of hexagonal Gd. The target assembly was heated to elevated temperatures between 324 K and 469 K, well above the Curie temperature for ferromagnetic Gd (293 K).

The static quadrupole interaction of the quadrupole moment of the isomeric state and the axially symmetric ($\eta = 0$) electric field gradient of the Gd host was observed using the quadrupole modulation technique for aligned nuclear states (Sect.3.2.2.b.II). Figure 3.13 shows examples of calculated modulation patterns for

3641-A

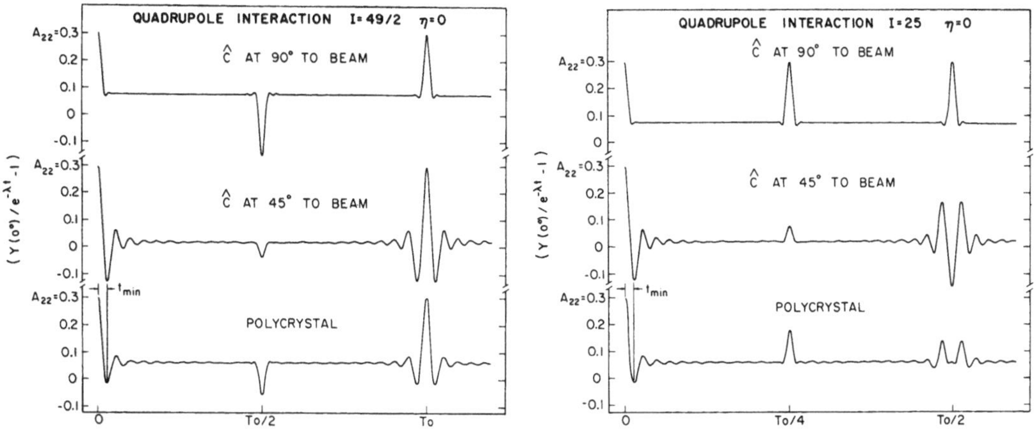

Fig.3.13. Quadrupole modulation patterns for an axially symmetric field gradient ($\eta = 0$) and integer (right panel) and half-integer (left panel) spin values

γ rays with positive a_2 angular distribution coefficients ($a_4 = 0$ is assumed) in an even-A nucleus (I = 25, right panel) and an odd-A nucleus (I = 49/2, left panel). The γ detector was assumed to be at 0° to the beam, and both polycrystalline hosts and single crystals at two different orientations are considered. The time T_0 at which the pattern repeats itself [see (3.36,37)],

$$T_0 = \nu_0^{-1} = \begin{cases} 4I(2I-1)h/3|e^2qQ| & \text{for } I \text{ integer} \\ 2I(2I-1)h/3|e^2qQ| & \text{for } I \text{ half integer} \end{cases} \quad , \qquad (3.90)$$

is proportional to I^2, a most unfavorable dependence when considering high-spin states. One is, however, less restricted if one is content with observation of only a small part of the complete modulation pattern near time t = 0. There, the geometry with the symmetry ($\hat{c}$) axis at 45° to the beam yields the most pronounced interference feature, with the minimum occurring at a time

$$t_{min} \sim 4(2I-1)h/9|e^2qQ| \quad . \qquad (3.91)$$

From this relationship, which is only accurate provided $a_4 = 0$, one may then determine the ratio $I/|e^2qQ|$.

In the favored geometry ($\hat{c}$ at 45° to the beam), time spectra for deexcitation γ rays from the ^{147}Gd isomer were obtained in two Ge(Li) detectors placed at 0° and 90° to the beam. Ratios of normalized γ-ray yields, $R(t) = [Y(0^\circ,t) - Y(90^\circ,t)]/[Y(0^\circ,t) + Y(90^\circ,t)]$, are shown in Fig.3.14 for a 254 keV γ ray ($a_2 > 0$, middle panel) and a doublet of 339 keV γ rays (combined $a_2 < 0$, top panel). The fitted curves correspond to a quadrupole coupling constant, $\nu_Q = |e^2qQ|/h = 250 \pm 7$ MHz at 413 K, assuming I = 49/2 for the spin of the isomer.

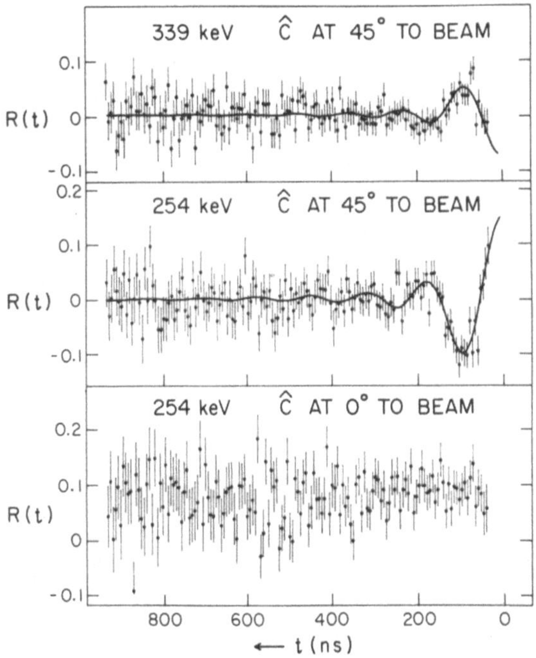

Fig.3.14. Modulation patterns resulting from quadrupole interaction of the 510 ns isomer in ^{147}Gd in a single crystal of Gd at 413 K. The reduced R(t = 0) in the lower panel arises from the fact that one of the γ detectors was placed at 68° rather than at 90° to the beam to avoid absorption of low-energy γ rays [3.104]

The use of single crystals has the further advantage that the effect of the static quadrupole interaction can be effectively switched off. If the symmetry axis of the perturbation is turned to coincide either with the direction of the orienting interaction (beam axis) or with the direction of observation for the γ rays, the unperturbed angular distribution yield is obtained [this follows from (3.33-35)]. The former option was chosen to obtain the data in Fig.3.14 (lower panel), which show that much of the initial alignment is preserved up to $\sim 1\ \mu s$. This implies that most ^{147}Gd recoils substitute in the Gd lattice, and that the observed modulation patterns result from static quadrupole interaction with a unique electric field gradient.

The value of the electric field gradient was calibrated by using ^{4}He Coulomb excitation of rotational 2^+ states in ^{156,158,160}Gd whose quadrupole moments can be reliably calibrated from the 2^+ lifetimes via (3.43,45). The result, $|eq| = (3.43 \pm 0.14)10^{17}\ V/cm^2$, allows one to deduce the quadrupole moment of the ^{147}Gd high-spin isomer and of two other isomers of lower spin in the same nucleus. In Table 3.4 the experimental quadrupole moments [3.104] and g factors [3.106] in ^{147}Gd are compared to theoretical calculations with the deformed single-particle model [3.107]. The (oblate) deformation ($\beta < 0$) is seen to increase with spin, acquiring a substantial value ($|\beta| \sim 0.2$) for the 510 ns isomer with a probable spin of 49/2. The data thus provide direct evidence for the core deformation effect in oblate high-spin isomers.

Table 3.4. Isomeric states in ^{147}Gd

I^{π}	Main configuration	$T_{1/2}$ [ns]	E_x[MeV][a]	$g^{a,b}$
$13/2^+$	$\nu i_{13/2}$	22.5±1.5	0.997 [0.8]	-0.037
$27/2^-$	$\pi\{(d^{-2}_{5/2})_0(h^2_{11/2})_{10}\}\nu f_{7/2}$	26.8±0.7	3.582 [4.2]	0.840
$(49/2^+)$	$\pi\{(d^{-2}_{5/2})_0(h^2_{11/2})_{10}\}$ $\nu\{(d^{-2}_{3/2})_0 f_{7/2} i_{13/2} h_{9/2}\}$	510 ± 20	~7.6 [8.1]	0.446

[a]Values in brackets are from [3.107].
[b]Experimental g factors [3.106].
[c]Experimental quadrupole moments [3.104]; $1b = 10^{-28}\ m^2$.

The spin of the 510 ns isomer can in principle be obtained by observing the complete quadrupole modulation pattern. With increasing spin I the interference features become sharper because of the larger number of quadrupole frequencies. From (3.90, 91) one obtains $T_0/t_{min} \sim 3I/2$. The expected large value for T_0 (~3 μs), which exceeds considerably the half-life (0.5 μs), and the spin-relaxation time (~1.5 μs) make such an experiment very difficult.

3.4.2 The g Factors of Collective High-Spin States

The most striking feature in the yrast energies of deformed nuclei is the failure of the energy spacings to conform to a constant moment of inertia. In some of the less deformed rare-earth nuclei the moments of inertia exhibit a dramatic increase (backbending), indicating that a new structure can carry angular momentum more favorably than a rotation of the ground state. It is now generally thought that the alignment of the spins of high-j orbitals with the axis of rotation is the dominant effect [3.94], whereas a gradual weakening of the pairing gap for either neutrons (Δ_n) or protons (Δ_p) may contribute as a background effect. In Fig.3.15 we show a calculation by SANO et al. [3.108] of the moments of inertia, of pairing gaps, B(E2) values and collective g factors (g_R) in ^{158}Dy. This calculation may not be correct in detail, but serves as an illustration of the phenomena that one may observe in prolate rotors at high spins. It is apparent that the g factors are very sensitive to the sharing of angular momentum between collective and single-particle degrees of freedom.

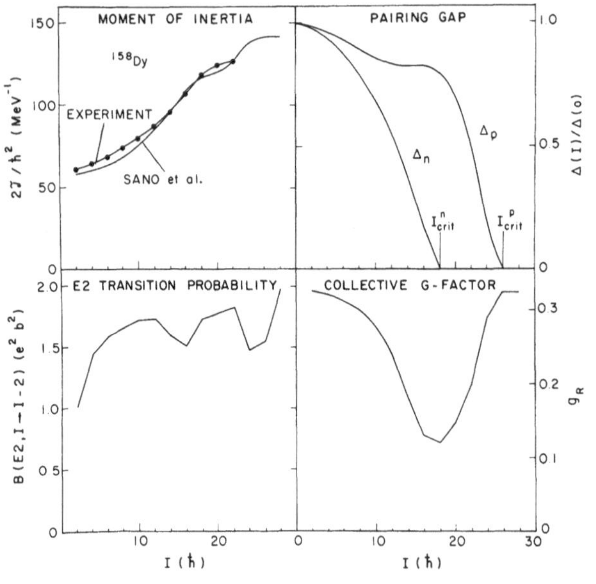

Fig.3.15. Moments of inertia, pairing gaps for protons (Δ_p) and neutrons (Δ_n), B(E2) transition probabilities and collective g factors in ^{158}Dy [3.108]

The first experiments using transient magnetic fields to probe g factors of high-spin states were attempted by KALISH et al. [3.109] and by SKAALI et al. [3.110] using (heavy-ion, xn) reactions. Such experiments are difficult to interpret, because the feeding times encountered in the population of the yrast states of interest are comparable to the slowing-down time of the recoils. Therefore, assumptions about the g factors of the continuum of levels involved in the feeding cascade have to be made to analyze the observed precession angles.

It is clearly advantageous to populate the levels under investigation instantaneously by a direct reaction process such as multiple Coulomb excitation. This reaction has the added bonus of populating the high-spin states at high recoil velocities and correspondingly large transient magnetic fields (Sect.3.2.3.b.III). The instantaneous level populations and the subsequent time evolutions of populations and precession angles can be calculated precisely, with the nuclear g factors as the only unknowns. It was shown by HÄUSSER et al. [3.86] that this can be done conveniently using the exponential expression for the precession angle (3.83). The experiments make use of thin polarized iron foils to sample the g factors of states near those populated instantaneously. The intense γ-ray transitions between lower-spin states carry information on the g factors of higher-lying states where the initial population is centered, because the accumulated precessions are passed on from level to level via the γ cascade.

Experiments along these lines have been carried out in the rotational nuclei ^{170,172,174}Yb using beams of ^{40}Ca [3.111], and in ^{170,174}Yb and ^{160}Dy using beams of ^{86}Kr [3.112]. The velocity dependence of the transient field was calibrated against the well-known g factors in ^{169}Tm [3.113]. The ^{40}Ca experiment which samples levels with $I \leq 10$ is consistent with a constant rotational g factor, $g(I) = g(2) = \text{const}$. In the ^{86}Kr experiment which samples levels with $I \leq 14$, the precession data were analyzed with the expression $g(I) = g(2)(1 + \alpha I^2)/(1 + 4\alpha)$, where g(2) is determined by Mössbauer experiments [3.10]. This relationship reproduces well the g factors below the backbend as predicted in the calculation by SANO et al. [3.108]. In Fig.3.16 we show the average precession angles $\bar{\phi} = \varepsilon/\bar{S}$ [see (3.73)] for various E2 transitions in ^{170,174}Yb and ^{160}Dy (middle panel), together with the percentage contributions of the various spins to the measured precession angles (lower panel) [3.112]. The precession results are best fitted by $\alpha = (-1.5 \pm 1.3) \times 10^{-3}$, $(-2.4 \pm 1.3) \times 10^{-3}$, and $(-1.3 \pm 0.8) \times 10^{-3}$, for ^{160}Dy, ^{170}Yb and ^{174}Yb, respectively (solid lines), and are nearly compatible with a constant collective g factor (dashed lines). The results for ^{160}Dy and ^{170}Yb, which exhibit a rapid increase of the moment of inertia at $I \sim 16$, and for ^{174}Yb, which exhibits a smooth variation of the moment of inertia, are qualitatively similar. There are clearly no dramatic changes in the g factors below and close to the critical spin value $I = 16$ in both ^{160}Dy and ^{170}Yb. The consistently negative sign for α may indicate that the pairing gap for neutrons decreases more rapidly that that for protons as the spin increases close to the critical value.

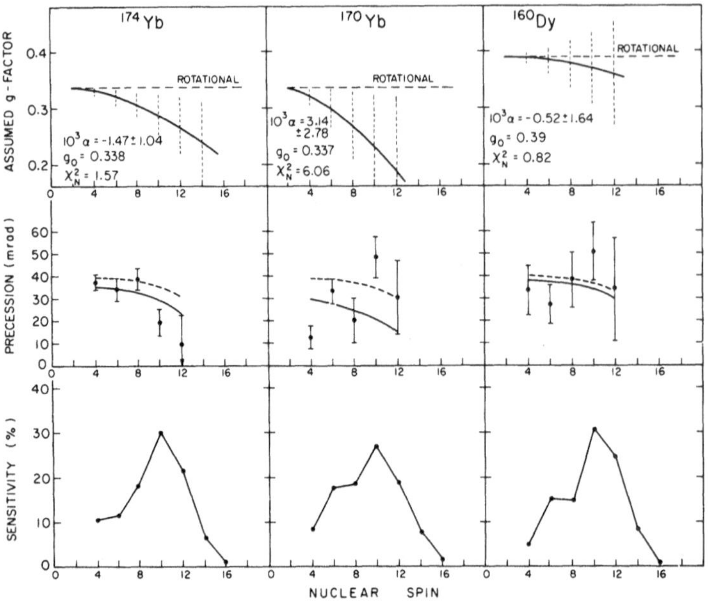

Fig.3.16. Average transient field precession angles ϕ for various E2 transitions in ^{170,174}Yb and ^{160}Dy. The theoretical lines represent various assumptions of the dependence of g factors vs spin [3.112]

In a more recent experiment at GSI Darmstadt [3.114] ^{158}Dy was Coulomb excited to even higher spins using a Pb beam from the UNILAC accelerator. The precession results are presently being analyzed. One may hope that similar experiments will eventually provide a more complete understanding of the microscopic shell effects that cause the anomalies in the moment of inertia in collectively rotating nuclei.

3.5 Magnetic Moments of Simple Shell-Model Configurations

Our discussion of magnetic moments in this section will be largely theoretical; only where an experimental measurement illustrates the theoretical concept will we pause to discuss the experimental technique.

We start by defining again the magnetic moment operator, this time deliberately introducing the concept of an electromagnetic current. A charged particle interacts with an external magnetic field both via its internal magnetic moment and via the current it produces. Its electromagnetic interaction may be written in the form [3.115]

$$\mathcal{H} = -\int j_\mu(\underset{\sim}{r},t)A_\mu(\underset{\sim}{r},t)d^3\underset{\sim}{r}$$
$$\equiv \int \phi(\underset{\sim}{r},t)\rho(\underset{\sim}{r},t)d^3\underset{\sim}{r} - \frac{1}{c}\int \underset{\sim}{j}(\underset{\sim}{r},t)\cdot\underset{\sim}{A}(\underset{\sim}{r},t)d^3\underset{\sim}{r} \quad , \qquad (3.92)$$

representing a local coupling between the four-vector potential $A_\mu = (\phi, \mathbf{A})$ and the four-vector charge density $\mathbf{j}_\mu = (\rho, \mathbf{j}/c)$, where ϕ and ρ are the time-like components of the corresponding four vectors. The conservation of electric charge is expressed by the continuity equation

$$c \frac{\partial}{\partial x_\mu} j_\mu(\mathbf{r},t) \equiv \nabla \cdot \mathbf{j}(\mathbf{r},t) + \frac{\partial}{\partial t} \rho(\mathbf{r},t) = 0 \quad . \tag{3.93}$$

As a first approximation, we write the charge-current density in a finite nucleus as that due to a collection of point particles each having a charge and a magnetic moment

$$\rho(\mathbf{r}) = \frac{e}{2} \sum_k [1 - \tau_z(k)] \delta(\mathbf{r} - \mathbf{r}_k) \quad ,$$

$$\begin{aligned} \mathbf{j}(\mathbf{r}) &= \mathbf{j}^c(\mathbf{r}) + \mathbf{j}^m(\mathbf{r}) \\ &= \frac{e}{2M} \sum_k [1 - \tau_z(k)] \frac{1}{2} [\mathbf{p}_k \delta(\mathbf{r} - \mathbf{r}_k) + \delta(\mathbf{r} - \mathbf{r}_k) \mathbf{p}_k] \\ &\quad + \frac{e\hbar}{2M} \sum_k g_s(k) \nabla \wedge \mathbf{S}_k \delta(\mathbf{r} - \mathbf{r}_k) \end{aligned} \tag{3.94}$$

where the first term is called the convection current and the second the magnetization current, and where $\tau_z = +1$ for neutrons and -1 for protons. The spin g factor is

$$g_s = \frac{1}{2} [g_s(\nu) + g_s(\pi)] + \frac{1}{2} \tau_z [g_s(\nu) - g_s(\pi)] \tag{3.95}$$

in terms of the neutron and proton g factors. The sum k is over all nucleons in the nucleus. The only parameters in the charge-current density, apart from the electric charge, are the spin g factors, which if they are determined from the observed neutron and proton magnetic moments, $g_s(\pi) = 5.586$ and $g_s(\nu) = -3.826$, differ considerably from the values $g_s(\pi) = 2.0$ and $g_s(\nu) = 0.0$, expected for a point particle satisfying the Dirac equation. This anomaly, first discovered by FRISCH and STERN [3.116] in 1933, merely says that the nucleon is not a point-like object, but has some intrinsic structure.

The magnetic moment operator may then be written as a sum of one-body operators

$$\boldsymbol{\mu} \equiv \frac{1}{2c} \int \mathbf{r} \wedge \mathbf{j}(\mathbf{r}) d^3\mathbf{r} = \sum_k \boldsymbol{\mu}_k$$

where

$$\begin{aligned} \boldsymbol{\mu}_k &= \frac{e}{2Mc} g_\ell(k) \int \mathbf{r} \wedge \mathbf{p}_k \delta(\mathbf{r} - \mathbf{r}_k) d^3\mathbf{r} + \frac{e\hbar}{2Mc} g_s(k) \int \mathbf{r} \wedge (\nabla \wedge \mathbf{S}_k) \delta(\mathbf{r} - \mathbf{r}_k) d^3\mathbf{r} \\ &= \frac{e\hbar}{2Mc} [g_\ell(\mathbf{k}) \mathbf{L}_k + g_s(\mathbf{k}) \mathbf{S}_k] \end{aligned} \tag{3.96}$$

Table 3.5. Single-particle expectation values in the m = j substate for three rank-one tensors characterizing an effective magnetic moment operator

	$\langle L \rangle$	$\langle S \rangle$	$(8\pi)^{1/2}\langle Y_2 \times S \rangle$
$j = \ell + \frac{1}{2}$	ℓ	$\frac{1}{2}$	$\frac{\ell}{2\ell + 3}$
$j = \ell - \frac{1}{2}$	$\frac{(\ell + 1)(2\ell - 1)}{2\ell + 1}$	$-\frac{(2\ell - 1)}{2(2\ell + 1)}$	$-\frac{(\ell + 1)}{2\ell + 1}$

with $\hbar L_k = r_k \wedge p_k$ and $g_\ell(k) = \frac{1}{2}[1 - \tau_z(k)]$, viz. $g_\ell(\pi) = 1$ and $g_\ell(\nu) = 0$.

Specifying the g factors from the experimental anomalous proton and neutron magnetic moments and using them unmodified in the one-body operator in finite nuclei is commonly called *impulse approximation*. The magnetic moment is defined as the expectation value of the operator in the M = J magnetic substate. Closed shells naturally give no contribution. Thus the moment is determined (as is, of course, the nuclear spin) largely from the valence (unpaired) nucleons near the Fermi surface. In the extreme case, in which all but the last particle are coupled in zero angular momentum pairs, the calculation is trivial and leads to the Schmidt value of the magnetic moment for odd-mass nuclei. In Table 3.5 we list the single-particle expectation values of operators L, S and for future reference $[Y_2, S]^{(1)}$, a spherical harmonic of rank two in orbital space coupled to the spin operator to form a tensor of rank one.

The experimental magnetic moments for nuclei whose configuration consists of a closed shell plus (or minus) one nucleon show departures from the Schmidt values. The deviation is small in light nuclei (typically 10% in A = 15, and 2% in A = 17) and large in heavy nuclei (36% in ^{209}Bi). This is qualitatively understood as arising from two principal causes:

1. *Meson Exchange Corrections.* The first approximation, in which we wrote the charge-current density as that due to point nucleons, could be inadequate. The charge and current densities can be modified to take the finite extension of the nucleons into account by replacing the δ functions in (3.96) with electric and magnetic form factors. The effect is not large, and has little influence on static properties such as magnetic moments. Of more concern is the existence of two-body currents such as the electromagnetic interaction of a charge pion exchanged between two nucleons. These meson exchange currents are absent in a single proton or neutron and therefore are not contained in the impulse approximation.

2. *Core Polarization.* The configuration assignment of closed shells plus a single nucleon is too simplistic. Additional, but small, terms arise from the breakup of the closed shells, e.g. two particle-one hole (2p-1h) and three particle-two hole

(3p-2h) configurations, whose influence on the magnetic moment can be evaluated in perturbation theory.

Considerable progress has been made in recent years in the understanding of meson exchange currents in nuclei, but positive experimental identification is hard to achieve, because core-polarization corrections mask the effect sought.

3.5.1 Meson Exchange Currents

We begin by asking why we are so certain that meson exchange currents exist. Consider for a moment the equation of continuity (3.93)

$$\nabla\cdot\mathbf{j} + \frac{i}{\hbar}[H,\rho] = 0 \qquad (3.97)$$

re-expressed as the commutator of the Hamiltonian, $H = T + V$, where $T = \sum_k p_k^2/2M$ is the kinetic energy and $V = \sum_{k<\ell} V_{k\ell}$ the two-body potential-energy operator. Then it is quite straightforward to show that for the point-nucleus current densities,

$$\nabla\cdot\mathbf{j}^{c} = -\frac{i}{\hbar}[T,\rho] \quad ,$$
$$\nabla\cdot\mathbf{j}^{m} = 0 \quad . \qquad (3.98)$$

Thus if the equation of continuity is to be satisfied, a third term must be added to the current such that

$$\nabla\cdot\mathbf{j}^{ex} = -\frac{i}{\hbar}[V,\rho] \quad . \qquad (3.99)$$

This extra current $\mathbf{j}^{ex}$, which we identify as the meson exchange current, will be two body in character, since V is a two-body operator. Furthermore, if we identify V with the one-pion exchange potential, whose isospin dependence is $\boldsymbol{\tau}(k)\cdot\boldsymbol{\tau}(\ell)$, then the commutator

$$\nabla\cdot\mathbf{j}^{ex} = -\frac{i}{\hbar}\left[\sum_{k<\ell} V_{k\ell}\boldsymbol{\tau}(k)\cdot\boldsymbol{\tau}(\ell) \quad , \quad \sum_m \frac{e}{2}[1-\tau_z(m)]\delta(\mathbf{r}-\mathbf{r}_m)\right]$$

$$= -\frac{e}{\hbar}\sum_{k<\ell} V_{k\ell}[\boldsymbol{\tau}(k)\wedge\boldsymbol{\tau}(\ell)]_z[\delta(\mathbf{r}-\mathbf{r}_k)-\delta(\mathbf{r}-\mathbf{r}_\ell)] \quad , \qquad (3.100)$$

which is non-zero. Thus, *current continuity and the isospin dependence of the nuclear force guarantee the existence of exchange currents.*

The magnetic moment due to the exchange current

$$\boldsymbol{\mu}^{ex} = \sum_{k<\ell}\boldsymbol{\mu}^{ex}(k,\ell) = \sum_{k<\ell}\frac{1}{2c}\int \mathbf{r}\wedge\mathbf{j}^{ex}(k,\ell)d^3\mathbf{r} \qquad (3.101)$$

can be broken into two pieces [3.117-119] in the following way:

$$\int \underset{\sim}{r} \wedge \underset{\sim}{j}^{ex}(k,\ell) d^3\underset{\sim}{r} = \underset{\sim}{R}_{k\ell} \wedge \int \underset{\sim}{j}^{ex}(k,\ell) d^3\underset{\sim}{r} + \int \left[\underset{\sim}{r} - \frac{1}{2}(\underset{\sim}{r}_k + \underset{\sim}{r}_\ell)\right] \wedge \underset{\sim}{j}^{ex}(k,\ell) d^3\underset{\sim}{r} \quad (3.102)$$

where $\underset{\sim}{R}_{k\ell} = (\underset{\sim}{r}_k + \underset{\sim}{r}_\ell)/2$. Since $\underset{\sim}{j}^{ex}$ depends only on the relative coordinates $\underset{\sim}{r} - \underset{\sim}{r}_k$, $\underset{\sim}{r} - \underset{\sim}{r}_\ell$, the second term in (3.102) is translationally invariant. By constrast, the first term depends on the choice of origin and is *not* translationally invariant. Using Green's theorem and the physical fact that currents $\underset{\sim}{j}^{ex}$ vanish exponentially at large distances, the integral in the first term of (3.102) becomes

$$\int \underset{\sim}{j}^{ex} d^3\underset{\sim}{r} = - \int \underset{\sim}{r}(\underset{\sim}{\nabla} \cdot \underset{\sim}{j}^{ex}) d^3\underset{\sim}{r} \quad (3.103)$$

and hence, according to the equation of continuity, is known unambiguously. Therefore the magnetic-moment terms which are translationally non-invariant are uniquely determined

$$\underset{\sim}{\mu}^{ex} = \frac{M}{\hbar^2} \sum_{k<\ell} (\underset{\sim}{R}_{k\ell} \wedge \underset{\sim}{r}_{k\ell})[\underset{\sim}{\tau}(k) \wedge \underset{\sim}{\tau}(\ell)]_z V_{k\ell} \quad (3.104)$$

where $\underset{\sim}{r}_{k\ell} = \underset{\sim}{r}_k - \underset{\sim}{r}_\ell$. This is known as the *Sachs moment* and is completely model independent, inasmuch as it is determined by the nucleon-nucleon potential $V_{k\ell}$. The Sachs moment is the dominant contribution to the meson exchange correction to the magnetic moment. Note, in particular, that this correction is isovector in character.

This phenomenological approach, however, tells us nothing about the translationally invariant and hence model-dependent second term in (3.102). For this reason exchange-current calculations have tended to follow the approach used in ab initio calculations of the nuclear force: to concentrate on single and multiple meson exchanges of the lightest mass, which generate the long-range parts of the potential. The short-range parts of the potential are not understood, and are generally approached phenomenologically [3.120]. Most exchange-current calculations have concentrated on one-pion exchange, [3.121,122] although single ρ- and ω-meson exchanges are common [3.123]. Two-pion exchange processes were considered by HYUGA et al. [3.119]. Some of the physical processes which contribute to the current in a two-body system are shown in Fig.3.17. The impulse approximation (one-body nucleon operators) is illustrated in (a), with the blobs depicting initial and final wave functions. This is in general the most important contribution. The "pair" diagram is shown in (b), the gauge term needed for current conservation in (c), and the true pion exchange graph in (d). This latter graph contains the Sachs moment. These diagrams give the more important corrections to impulse approximation, and are all isovector in character. Somewhat less important are the isobar diagrams (e) and the meson decay graphs, ($\rho\pi\gamma$) and ($\omega\pi\delta$) currents, in diagram (f).

The decomposition of the exchange magnetic moment into a translationally non-invariant and a translationally invariant part corresponds to the decomposition of the one-body magnetic moment into the orbital magnetic moment (which is trans-

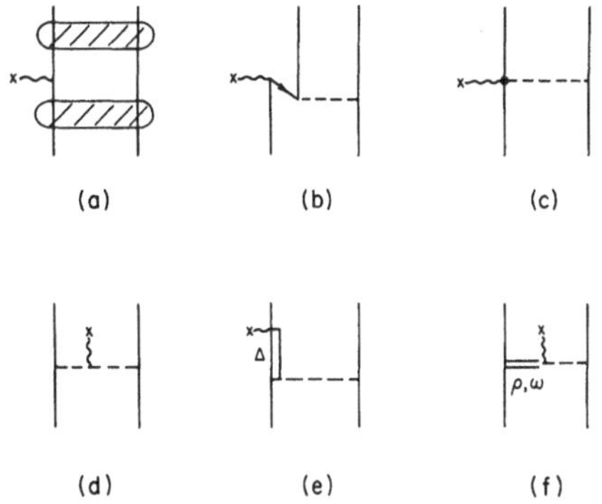

Fig.3.17a-f. Feynman diagrams which contribute to meson exchange current corrections to magnetic moments

lationally non-invariant) and the spin magnetic moment (translationally invariant). This analogy may be carried further [3.118].

Consider a configuration consisting of a closed shell plus (or minus) one nucleon, with the closed shells closed in LS coupling. An effective one-body magnetic dipole operator can be derived from the exchange magnetic moment by summing over all the two-body interactions between the valence nucleon and the nucleons in the LS core, viz.

$$<j|\underset{\sim}{\mu}^{eff}|j> = \sum_{h} <jh|\underset{\sim}{\mu}^{ex}|jh> \tag{3.105}$$

where j is the valence nucleon, h a core nucleon, $\underset{\sim}{\mu}^{ex}$ the two-body meson exchange operator, and $\underset{\sim}{\mu}^{eff}$ the deduced effective one-body operator. Indeed this operator can be re-expressed in terms of the three available first-rank operators [3.124]

$$\underset{\sim}{\mu}^{eff} = \delta g_{\ell}\underset{\sim}{L} + \delta g_{s}\underset{\sim}{S} + \delta g_{p}[Y^{(2)},\underset{\sim}{S}]^{(1)} \tag{3.106}$$

where the coefficients δg_{ℓ}, δg_{s} and δg_{p} are all uniquely determined. Because the closed shells have zero angular momentum (L = 0, S = 0), the tensorial character in orbital and spin space of the two-body operator is directly matched by the tensorial character of the one-body operator. For example, the orbital g factor δg_{ℓ} is determined exclusively from the Sachs moment, as this is the only meson exchange term which is axial vector in orbital space. Calculations with a one-pion exchange potential typically lead to a $\delta g_{\ell} \sim \pm 0.1$, plus for proton, minus for neutron. Incorporating heavy-meson exchange or two-pion exchange through such phenomenological potentials as the Hamada-Johnston potential enhances δg_{ℓ} by an additional 50%.

3.5.2 First-Order Core Polarization

The most important effect [3.125] is the excitation of the core to 1^+ excited states (M1 giant-resonance states), as formulated by ARIMA and HORIE [3.126]. In general, a closed-shell core, in which a single-particle orbital $j_1 = \ell_1 + 1/2$, is fully occupied and its spin-orbit partner $j_2 = \ell_1 - 1/2$ is empty can be excited to a $(j_1^{-1} j_2)1^+$ state by the presence of an extra particle $j = \ell \pm 1/2$. Physically, the core polarization is associated with the presence of unsaturated spins in closed shells, which can be partially aligned by interaction with the spin of the valence nucleon. This spin-dependent interaction between the valence nucleon and closed shells gives rise to the excitations of spin-orbit partners. For example, the ^{208}Pb core has two 1^+ excitation modes: $\pi(h_{11/2}^{-1} h_{9/2})1^+$ and $\nu(i^{-1}_{13/2} i_{11/2})1^+$. Then the single-particle state j is modified

$$|j, I_c = 0; I = j\rangle + A|j , (j_1^{-1} j_2)1^+; I = j\rangle \qquad (3.107)$$

where I_c is the spin of the closed-shell core. The admixture amplitude A may be evaluated in first-order perturbation theory, leading to a correction to the magnetic moment

$$\delta\mu = 2 \sum_\alpha \langle j|\underset{\sim}{\mu}|\alpha\rangle\langle\alpha|V|j\rangle/(\varepsilon_j - \varepsilon_\alpha) \qquad (3.108)$$

where j is the single-particle and α the 2p-1h configuration, and $(\varepsilon_j - \varepsilon_\alpha)$ their unperturbed energy separation. Such first-order corrections have been evaluated by many authors [3.126-130] with a wide range of results, depending on the two-body residual interaction selected. The trend, however, is clear. If the correction is expressed, as was the case for meson exchange corrections (3.106), in terms of an effective one-body operator, the trend is to reduce the magnitude of g_s. (There is no correction to g_ℓ if the residual interaction is purely central.) The magnitude of the correction is state dependent.

If one restricts the configurations of the admixtures to just the two 2p-1h states of (3.107), then there is no reason why the calculation should be limited to first-order perturbation theory. Diagonalization gives an all-orders result (Tamm-Dancoff approximation, TDA), while 2p-2h correlations in the closed shell can be incorporated in the random-phase approximation (RPA). Calculations by TOWNER et al. [3.130] show that these all-orders results quench the first-order results by anywhere from 30% to 50%, depending on the residual interaction.

In taking the calculation beyond first order, however, other configurations can be important. For example, in ^{208}Pb there are low-lying collective (vibrational) excited states with spins $J^\pi = 2^+, 4^+, 3^-, 5^- \ldots$. A valence nucleon may couple with these collective states, giving a second-order correction to the magnetic moment. The 3^- vibration plays a particularly important role in high-spin single-particle states ($\pi i_{13/2}$ and $\nu j_{15/2}$). Calculations of these effects have been given by HAMAMOTO [3.131].

In Table 3.6 we have assembled some sample calculations for magnetic moments in the Pb region. Note that the meson exchange correction is positive for proton states and negative for neutron states, with the magnitude increasing with increasing orbital angular momentum. For $j = \ell - 1/2$ states, the mesonic correction

Table 3.6. Corrections to the Schmidt values for the magnetic moments of single-particle (hole) states in the Pb region

Nucleus	Orbit	Schmidt	RPA [3.130]	Vibration [3.131]	Mesonic[a]	Sum	Expt[b]	Bauer et al.[c]
^{209}Bi	$0h_{9/2}$	2.62	0.53	0.00	0.40	0.93	1.50±0.01	1.44
	$0i_{13/2}$	8.79	-0.66	-0.68	0.45	-0.89	-0.78±0.10	-0.67
^{207}Tl	$2s^{-1}_{1/2}$	2.79	-0.68	-0.03	0.04	-0.67	-0.96±0.18	-0.85
^{209}Pb	$1g_{9/2}$	-1.91	0.51	-0.02	-0.07	0.42	0.58±0.06	0.59
^{207}Pb	$2p^{-1}_{1/2}$	0.64	-0.07	0.00	-0.05	-0.12	-0.05±0.00	-0.09
	$1f^{-1}_{5/2}$	1.37	-0.34	0.02	-0.21	-0.53	-0.67±0.03	-0.52
	$0i^{-1}_{13/2}$	-1.91	0.60	0.16	-0.21	0.55	0.90±0.03	0.77

[a] One-pion meson exchange correction [3.118].
[b] Experimental bibliography is listed in [3.130].
[c] [3.132]; a similar RPA calculation with vibrational and mesonic corrections incorporated.

enhances the core-polarization correction, whereas for $j = \ell + 1/2$ states, the reverse is true.

Table 3.6 considers second-order corrections from low excitation. ARIMA and HYUGA [3.118] argued, however, that the correction from high excitations must be included as well. As will be discussed later on, the tensor force component of the residual interaction has a very strong coupling with states of high excitation energy. Another correction still to be added is given by the first-order expression, (3.108), but now $\underset{\sim}{\mu}$ is the two-body meson exchange operator. Both of these effects have been estimated by ARIMA and HYUGA [3.118], and we quote their results in Table 3.7.

The core-polarization correction has been deliberately separated into that derived from the easily calculated first-order correction, which in this case is taken from ARIMA and HUANG-LIN [3.129], and that estimated from all second-order corrections. The meson exchange correction has likewise been separated into that due to one-pion exchange processes and that due to heavy-meson and/or multi-pion exchanges. The point of these separations, as stressed by ARIMA and HYUGA [3.118], is the observation that the sum of the contributions labelled C2, $n\pi$ and M1 in Table 3.7 is almost zero. All come from second-order processes involving at least two pions. This suggests, then, that a reasonable understanding can be achieved from a calculation of one-pion exchange processes and first-order core polarization. At least it warns of the dangers of calculating only some of the second-order effects and not all of them.

Table 3.7. Corrections to the Schmidt value for the magnetic moments of single-particle (hole) states in the Pb region [3.118]

Nucleus	Orbit	C1[a]	Vibration [3.131]	1π[b]	C2[c]	$n\pi$[d]	M1[e]	Sum	Expt[f]
^{209}Bi	$0h_{9/2}$	0.79	0.00	0.40	-0.30	0.18	0.45	1.52	1.50±0.01
	$0i_{13/2}$	-0.60	-0.68	0.45	-1.17	0.41	0.74	-0.85	-0.78±0.10
^{207}Tl	$2s_{1/2}^{-1}$	-0.77	-0.03	0.04	-0.60	0.15	0.22	-0.99	-0.96±0.18
^{209}Pb	$1g_{9/2}$	0.59	-0.02	-0.07	0.53	-0.20	-0.39	0.44	0.58±0.06
^{207}Pb	$2p_{1/2}^{-1}$	-0.11	0.00	-0.05	-0.01	0.00	-0.01	-0.18	-0.05±0.00
	$1f_{5/2}^{-1}$	-0.46	0.02	-0.21	0.13	-0.05	-0.14	-0.71	-0.58±0.03
	$0i_{13/2}^{-1}$	0.85	0.16	-0.21	0.66	-0.25	-0.45	0.76	0.90±0.03

[a]First-order core polarization [3.129].
[b]One-pion meson exchange correction.
[c]Second-order core polarization [3.119].
[d]Multi-pion and heavy-meson exchange correction.
[e]First-order core polarization with one-pion meson exchange operator.
[f]Experimental bibliography listed in TOWNER et al. [3.130].

3.5.3 High-Spin Isomers of Two-Particle Configuration

Is there any experimental evidence which irrefutably confirms the presence of a meson exchange current contribution? We recall that the largest meson exchange component arises from the Sachs moment, which from (3.106) is seen to modify the orbital g factor. By contrast, core polarization largely modifies the spin g factor. Thus the beautiful idea proposed by YAMAZAKI [3.125] was to examine a two-particle configuration $(j_1j_2)I$ for which $j_1 = \ell_1 + 1/2$ and $j_2 = \ell_2 - 1/2$. The magnetic moment of such configurations is dominated by the orbital moments, because the spin moments from the two particles are opposite in sign and cancelling.

For strong singlet forces, the lowest member of the multiplet $(j_1j_2)I$ is the stretched configuration $I = j_1 + j_2 = \ell_1 + \ell_2$. Furthermore, this state is isomeric because of a fortunate spin gap between this level and any of the low-lying levels. Thus its magnetic moment is amenable to experiment by measuring the time differential perturbed angular distribution of the delayed γ rays from the isomeric state. As an example consider the $(\pi h_{9/2}\, i_{13/2})11^-$ isomeric state in ^{210}Po which feeds the $(\pi h_{9/2}^2)8^+$ state. This 24-ns isomer was found with the 8^+ isomer in ^{208}Pb(α, 2n)^{210}Po reaction by YAMAZAKI and EWAN [3.133], and its magnetic moment was measured by YAMAZAKI et al. [3.134] and confirmed by HÄUSSER et al. [3.59].

In Yamazaki's experiment a thick metallic ^{208}Pb target was bombarded with a pulsed 30-MeV α-particle beam to populate the 11^- isomeric state. The recoiling Po nuclei were stopped in the thick target. No static quadrupole interaction is expected, since the product nuclei are stopped ultimately at a site of cubic symmetry. If neither the beam of incident particles nor the target Pb nuclei are polarized, then there will be an alignment induced in the product Po nuclei with respect to the beam direction. Between beam bursts the de-excitation γ rays are detected in a Ge(Li) counter placed at 135° to the beam.

Ideally, the time-integrated angular distribution of γ rays is given by [3.9]:

$$W(\theta) = \sum_{\lambda,\text{even}} a_\lambda P_\lambda(\cos\theta) \equiv \sum_{\lambda,\text{even}} b_\lambda \cos(\lambda\theta)$$

$$= \sum_{\lambda,\text{even}} B_\lambda(I) R_\lambda(I,I') P_\lambda(\cos\theta) \tag{3.109}$$

where θ is the angle between the beam direction and the γ counter. The coefficients $B_\lambda(I)$ describe the nuclear alignment and depend on the population parameters $p(m)$ of the magnetic substates of the initial state I (in this case $I = 11^-$) according to [see also (3.64)]:

$$B_\lambda(I) = \sum_m p(m)(2I + 1)\langle I\ m\ I\ -m|K\ 0\rangle\langle I\ m\ I\ -m|0\ 0\rangle \quad . \tag{3.110}$$

The populations are normalized such that $\sum_m p(m) = 1$. If there is no polarization, all magnetic states being equally populated, then $B_\lambda(I) = \delta_{\lambda,0}$, and $a_2 = a_4 = \cdots = 0$. The second factor, $R_\lambda(I,I')$, depends only on the nuclear matrix elements characterizing the transition $I \to I'$ (in this case $11^- \to 8^+$).

The time differential angular distribution, by contrast, is influenced by two factors. First, the population of the isomeric state is decreasing: $dN/dt = -N/\tau$, where τ is the mean lifetime. Second, in the presence of a magnetic field $\underset{\sim}{B}$, the spin vector (for the aligned nucleus) precesses about the magnetic field direction with characteristic frequency ω_L, the Larmor frequency, where $\omega_L = g\mu_N B/\hbar$ and g is the g factor of the isomeric state. The angular distribution is then (3.66):

$$Y(t) \equiv \frac{dW(\theta,t)}{dt} = -\frac{1}{\tau} e^{-t/\tau} \sum_{\lambda,\text{even}} b_\lambda \cos[\lambda(\theta - \omega_L t)] \quad . \tag{3.111}$$

Two other considerations give rise to corrections to this formula. First, the alignment is preserved in the presence of a strong magnetic field and assumed to be time independent. In fact, spin-relaxation processes in the metallic Pb gradually destroy the alignment. This effect can be included by multiplying the alignment parameters a_λ (or equivalently b_λ) with exponential factors containing the characteristic relaxation time. Second, the zero for time is taken at the beam burst, and counting of the de-excitation γ rays continues until the next beam burst. However, the contribution of the preceding bursts should be taken into account. The modified function [3.135] for periodic events with interval T is

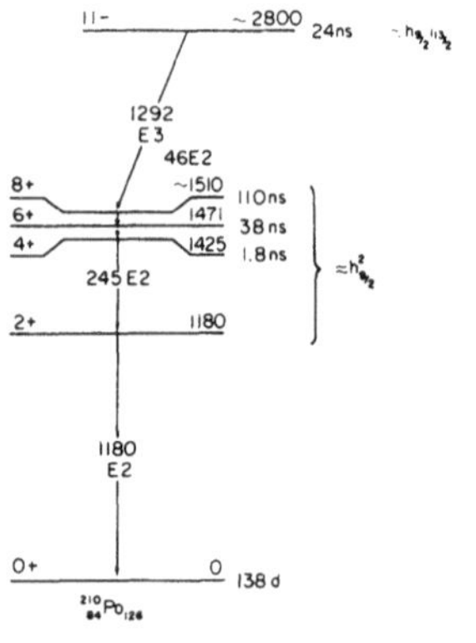

Fig.3.18a ▲

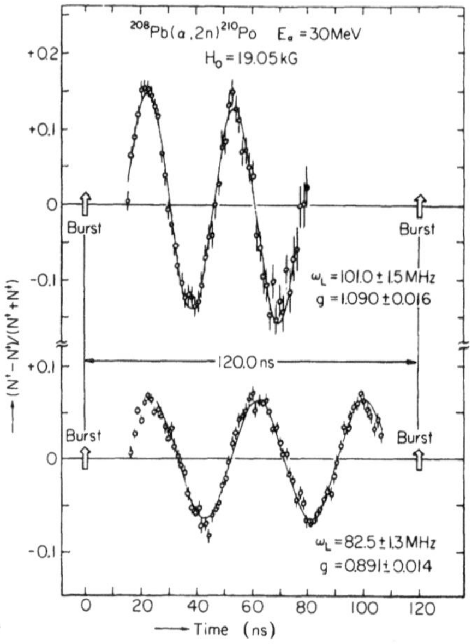

Fig.3.18b ►

Fig.3.18. (a) Level scheme of ^{210}Po associated with the high-spin isomeric states, populated in the reaction ^{208}Pb(α,2n)^{210}Po. (b) Time differential spin-rotation patterns for the 24-ns 11$^-$ state of ^{210}Po (upper) and for the 110-ns 8$^+$ state of ^{210}Po (lower). Counts N↑ and N↓ correspond to the external magnetic field up and down, respectively. Because of the long-lived tail in the time spectrum the quantity N↑ + N↓ is replaced by a decay function of 24 ns half-life in the upper case. The solid curves are best-fitted ones [3.134]

$$\bar{Y}(t) = \sum_{n=0}^{\infty} Y(t + nT)$$

$$= -\frac{1}{\tau}\frac{e^{-t/\tau}}{1 - e^{-T/\tau}}\{b_0 + \alpha b_2 \cos[2(\theta - \omega_L t) - \phi] + \dots\} \tag{3.112}$$

where α represents an attenuation factor due to the addition of preceding events, and ϕ a phase angle. The formula can be further generalized [3.59] to deal with a cascade of γ rays proceeding through a series of isomers, each with their own decay constant and characteristic Larmor frequency.

In the experiment of YAMAZAKI et al. [3.134] an external magnetic field of 19.05 kG was applied up and down perpendicular to the beam-detector plane. The difference in counts, with field up and down, respectively, is given by

$$R(t) = 2\frac{\bar{Y}(\uparrow,t) - \bar{Y}(\downarrow,t)}{\bar{Y}(\uparrow,t) + \bar{Y}(\downarrow,t)} = 2\alpha b_2 \sin(2\omega_L t + \phi) \quad , \tag{3.113}$$

and a plot of R(t) is shown in Fig.3.18. A fit to the frequency of oscillation yields ω_L and hence the g factor (uncorrected for diamagnetism and Knight shift) of 1.107 ± 0.016 from YAMAZAKI et al. [3.134] and 1.102 ± 0.008 from HÄUSSER et al. [3.59].

Theoretically, the g factor for the two-particle state $(j_1j_2)I$ is

$$g = \frac{1}{2}(g_1 + g_2) + \frac{j_1(j_1+1) - j_2(j_2+1)}{I(I+1)}\frac{1}{2}(g_1 - g_2) \tag{3.114}$$

where $g_1(g_2)$ is the g factor for the single-particle state $j_1(j_2)$,

$$g_1 = g_\ell^{eff} \pm (g_s^{eff} - g_\ell^{eff})/(2\ell_1 + 1) \quad \text{as} \quad j_1 = \ell_1 \pm \frac{1}{2} \quad , \tag{3.115}$$

and where $g_\ell^{eff} = g_\ell + \delta g_\ell^{ex}$ and $g_s^{eff} = g_s + \delta g_s^{ex}$ (neglect δg_p^{ex}) are effective g factors which include a meson exchange correction [see (3.106)]. Assuming these effective g factors are state independent, then the g factor for the isomer with $j_1 = \ell_1 + 1/2$, $j_2 = \ell_2 - 1/2$, and $I = \ell_1 + \ell_2$ is

$$g = g_\ell^{eff} + \frac{1}{I(2\ell_2 + 1)}(g_s^{eff} - g_\ell^{eff}) + \frac{1}{I}[\delta\mu(j_1) + \delta\mu(j_2)] \tag{3.116}$$

where $\delta\mu$ are core-polarization corrections yet to be applied. The key here is that the second and third terms are small corrections, so a measurement of g leads directly to g_ℓ^{eff}.

The experimentally observed 11^- state in ^{210}Po is not likely to be a pure two-particle state, but can conceivably have components $(h_{9/2}^2)8^+$ and $(h_{9/2}f_{7/2})8^+$ each coupled to a 3^- vibration in ^{208}Pb. These configurations are suggested by the strong $B(E3; 11^- \rightarrow 8^+)$ transitions to the two known 8^+ states in ^{210}Po, each of which is strongly enhanced over the single-particle estimate, and each of which is characteristic of the strog $B(E3; 3^- \rightarrow 0^+)$ known in ^{208}Pb. YAMAZAKI [3.125] estimated the isomeric state wave function to be

$$|11^-\rangle = 76\%|h_{9/2}\, i_{13/2}\rangle + 20\%|h_{9/2}f_{7/2} \otimes 3^-\rangle + 4\%|h_{9/2}^2 \otimes 3^-\rangle \quad . \tag{3.117}$$

Knowing or estimating the g factors of the 8^+ two-particle configuration and the 3^- vibration, the experimental g factor for the 11^- state can be corrected for these admixtures to yield a net g factor for the two-particle configuration:

$$g[(h_{9/2}i_{13/2})11^-] = 1.18 \pm 0.01 \quad . \tag{3.118}$$

Referring to the theoretical expression for g, (3.116), the second term here is 0.04 ± 0.01 if we take $g_s^{eff} - g_\ell^{eff} = 4.6$ and allow an ambiguity of ± 1.0. For the core-polarization correction, we can look to the RPA calculations of TOWNER et al. [3.130] (Table 3.6), which give -0.012, or to the first-order calculations of ARIMA and HUANG-LIN [3.129] (Table 3.7), which give 0.017. We will assume that these values spread the likely range and take for the third term in (3.116) a value 0.0 ± 0.2. Therefore, we obtain

$$g_\ell^{eff} = 1.14 \pm 0.02 \tag{3.119}$$

or $\delta g_\ell^{ex} = 0.14 \pm 0.02$, which is a clear indication of the meson exchange effect on g_ℓ.

3.5.4 Single-Particle States in Other Mass Regions

It is not our intention to give an extensive literature survey; however it is worth recording a few of the recent experiments using the same time differential perturbed angular distribution method to measure magnetic moments of isomeric states in nuclei with either neutron or proton closed shells. In ^{91}Nb with closed neutron shell (N = 50), the magnetic moments of the $13/2^-$ isomer of configuration $|(\pi g_{9/2})^2 6^+, (\pi p_{1/2}); 13/2^-\rangle$ and of the $17/2^-$ isomer, $|(\pi g_{9/2})^2 8^+, (\pi p_{1/2}); 17/2^-\rangle$, have been measured by BABA et al. [3.136] and by HÄUSSER et al. [3.137], respectively. Both moments are consistent with the shell-model assignment using empirical single-particle g factors taken from known moments in ^{89}Y and ^{90}Zr of $g[\pi g_{9/2}] = 1.38$ and $g[\pi p_{1/2}] = -0.275$. The Schmidt values are 1.51 and -0.53, respectively. HÄUSSER et al. [3.137] have calculated the first-order core-polarization and one-pion meson exchange corrections to the Schmidt value, obtaining $g(\pi g_{9/2}) = 1.38$ and $g(\pi p_{1/2}) = -0.446$, which agrees with the empirical value for the $g_{9/2}$-orbit, but failing to obtain sufficient inhibition for the $p_{1/2}$-orbit. In ^{91}Nb there are also some two-body, first-order core-polarization corrections, which we will discuss in the next section.

For the N = 82 shell closure, the $\pi h_{11/2}$ single-proton state magnetic moments in ^{141}Pr and in ^{145}Eu have been measured by EJIRI et al. [3.138] and by KLINGER et al. [3.139], who obtained g factors of 1.30 ± 0.08 and 1.356 ± 0.008, respectively. The Schmidt single-particle g factor for an $h_{11/2}$ proton is 1.42. In this case, the core-polarization and meson exchange current corrections are opposite in sign. We are not aware of any detailed meson exchange calculations for these nuclei, but an estimate based on the dominance of the Sachs moment, namely $\delta g_\ell^{mes} = 0.1$, gives $\delta g_{mes} = (10/11)\delta g_\ell^{mes} = 0.09$ for the meson correction. KLINGER et al. [3.139] calculated a core-polarization correction of $\delta g_{c.p.} = -0.15$ for ^{145}Eu, so that the summed result of $g_{Schmidt} + \delta g_{mes} + \delta g_{c.p.} = 1.36$ is in good agreement with experiment.

Finally, for the proton-shell closure at Z = 50, the $\nu h_{11/2}$ single-neutron state magnetic moments have been measured in odd-mass tin isotopes ^{111}Sn, ^{113}Sn, ^{115}Sn and ^{119}Sn [3.140-144], and in two $(\nu h_{11/2})^2\ 10^+$ states in ^{116}Sn and ^{118}Sn [3.145]. The general trend is for all of these g factors to be approximately constant; $g(\nu h_{11/2}) \sim -0.24$, compared with a Schmidt value of -0.35. BRENN et al. [3.140] found that this near constancy of g factor as a function of mass number is reproduced by a first-order core-polarization calculation using pairing model wave functions and experimentally determined energy denominators and occupancy parameters of the other partially filled neutron orbits $d_{5/2}$, $s_{1/2}$, $d_{3/2}$ and $g_{7/2}$.

3.5.5 Core-Polarization Blocking

Another less direct way to glean evidence for meson exchange processes comes from core-polarization blocking. As we will shortly explain, blocking gives a unique signal for the core-polarization correction (or more correctly, a portion of it), and as such differentiates between different model calculations. Thus with this additional constraint on the calculated core-polarization correction, experiment and theory can be compared and a meson exchange effect indirectly inferred.

The best example of core-polarization blocking is obtained from the N = 126 isotones with proton configuration $[(h_{9/2})^n, I]$, n = 1-6. Magnetic moments for high-spin isomers have been measured [3.146] in ^{210}Po (6^+ and 8^+), ^{211}At($21/2^-$), ^{212}Rn(8^+), ^{213}Fr($21/2^-$) and ^{214}Ra(8^+). The shell model prediction, based on the simple $h^n_{9/2}$ configuration, is that the g factors, $g = \mu/I$, for all of these states should be equal to each other and equal to the ^{209}Bi ground-state value. Experimentally there is a small but significant breakdown in this additivity rule, and this breakdown can be understood qualitatively as a blocking effect in the core-polarization correction.

As has already been mentioned, the first-order correction arises from the spin polarization of the ^{208}Pb core from the 1p-1h excitations: $\pi(h^{-1}_{11/2}h_{9/2})1^+$ and $\nu(i^{-1}_{13/2}i_{11/2})1^+$. However, as the occupancy of the proton $h_{9/2}$ orbit increases, so the probability of creating the $\pi(h^{-1}_{11/2}h_{9/2})$ excitation decreases (a manifestation of the Pauli exclusion principle), with a resulting reduction in the core-polarization correction. The g factor is given approximately by the formula [3.147]:

$$g(h^n_{9/2}) = g_{\text{Schmidt}} + g_{\text{meson}} + g^{\nu}_{\text{c.p.}} + \frac{1}{8}(9-n)g^{\pi}_{\text{c.p.}} \tag{3.120}$$

where $g^{\nu}_{\text{c.p.}}$ is the core-polarization correction from the neutron p-h excitation and $g^{\pi}_{\text{c.p.}}$ that from the proton excitation. In Table 3.8 we list some sample calculations of the core-polarization correction in ^{209}Bi in first order and all orders (TDA and RPA) in the 1p-1h model space, for a variety of residual interactions. The first three are all phenomenological central particle-hole interactions (fitted to energy levels of negative parity states in closed-shell nuclei), and the fourth is the more realistic Hamada-Johnston interaction. Note that for the realistic, the Gillet, and to some extent the Perez interaction, the contribution from the neutron p-h states is very small. This was pointed out by ARITA [3.147], who observed that a strong repulsive singlet-odd force is required to enhance the neutron contribution $g^{\nu}_{\text{c.p.}}$. ARITA increased the range of the Rosenfeld interaction from the more usual values in the neighbourhood of 1.4 to 1.6 fm to a long-range value of 2.07 fm to achieve this effect.

For the blocking effect, we define

$$\begin{aligned}\Delta_{\text{expt}} &= g(^{213}\text{Fr};21/2^-) - g(^{209}\text{Bi};9/2^-)\\ &= -0.027 \pm 0.004\end{aligned} \tag{3.121}$$

Table 3.8. Core-polarization corrections for the ^{209}Bi g factor and the core-polarization blocking calculated for four choices of residual interaction [3.130]

Source		$g^{\pi}_{c.p.}$	$g^{\nu}_{c.p.}$	Sum	Δ^{a}
Perez [3.148]	first	0.128	0.075	0.203	-0.064
	TDA	0.096	0.052	0.148	-0.048
	RPA	0.078	0.040	0.118	-0.039
Gillet [3.149]	first	0.100	-0.003	0.097	-0.050
	TDA	0.082	-0.004	0.078	-0.041
	RPA	0.069	-0.001	0.068	-0.035
Rosenfeld	first	0.136	0.147	0.283	-0.068
[3.147]	TDA	0.087	0.102	0.189	-0.044
	RPA	0.061	0.079	0.139	-0.031
Hamada-	first	0.115	-0.001	0.114	-0.058
Johnston	TDA	0.098	-0.003	0.094	-0.049
[3.150]	RPA	0.086	0.000	0.087	-0.043
expt				$\sim 0.24^{b}$	-0.027±0.004

[a] $\Delta \equiv g(^{213}Fr) - g(^{209}Bi) \simeq -\frac{1}{2} g^{\pi}_{c.p.}$

[b] Estimate: $g = g_{exp} - g_{Schmidt} - g_{meson}$ with $g_{meson} = 0.09$.

as the difference in g factors of ^{213}Fr and ^{209}Bi, the two most accurate data. Assuming that the meson exchange correction is nuclear mass independent, then the theoretical prediction from (3.120) is

$$\Delta \simeq -\frac{1}{2} g^{\pi}_{c.p.} \quad . \tag{3.122}$$

This value is also listed in Table 3.8. The first-order perturbation results for Δ are in all cases a factor of two larger than experiment. Higher order TDA and RPA approximations reduce the discrepancy.

From the results in Table 3.8 a basic dilemma is evident. In order to explain the absolute g-factor measurement in ^{209}Bi, the calculated core-polarization correction needs to be enhanced; yet in order to explain the blocking data, the proton contribution must be reduced. TOWNER et al. [3.130] gave a partial resolution of this problem by further considering 2p-2h excitations among the proton h orbitals and the neutron i orbitals and by estimating the configuration mixing with other proton orbitals. Their results are shown graphically in Fig.3.19, where each of these additional corrections is seen to reduce the degree of blocking; as a consequence, they permit larger values for $g_{c.p.}$ in ^{209}Bi, which are necessary to reconcile theory and experiment. The important facet to note from this is that the

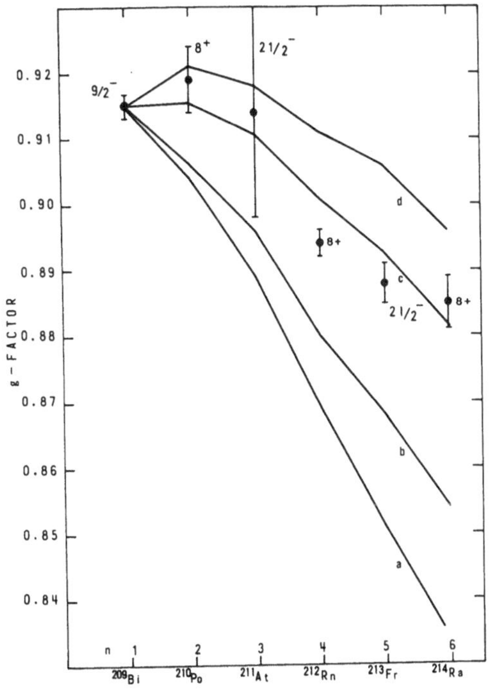

Fig.3.19. The g factors for $[(h_{9/2})^n;I]$ proton states. The solid lines illustrate the blocking in the core-polarization calculations taken (a) to first order in perturbation theory; (b) to all orders for 1p-1h intermediate states; (c) to all orders for 1p-1h and 2p-2h intermediate states; and (d) to include, in addition, configuration mixing as discussed in the text. All curves were shifted separately to fit the ^{209}Bi datum ([3.130], experimental data from [3.146])

blocking data give a powerful constraint to the core-polarization calculations.

While the N = 126 isotopes represent the best example of core-polarization blocking, the phenomenon is evident in other mass regions. In the N = 50 isotones, for example, g factors have been measured for 8^+ states in ^{90}Zr, ^{92}Mo and ^{94}Ru, and for $17/2^-$ states in ^{91}Nb and ^{93}Tc [3.137]. The configuration for these states is assumed to be $\pi(p_{1/2}^{n_1} g_{9/2}^{n_2})$, where $n_1 + n_2$ are the number of protons outside a ^{88}Sr closed-shell core. The experimental g factors show systematic departures from Schmidt values, depending on the mass number. First-order core-polarization corrections arise from the 1^+ particle-hole excitations in the core: $\nu(g_{9/2}^{-1} g_{7/2})$, $\pi(p_{3/2}^{-1} p_{1/2})$, and $\pi(g_{9/2}^{-1} g_{7/2})$. The latter two excitations should more correctly be written $\pi(p_{3/2}^{-1} p_{1/2}^{n_1+1})$ and $\pi(g_{9/2}^{n_2-1} g_{7/2})$, and as such depend on the orbital occupancy n_1 and n_2. This again is the core-polarization blocking.

Other examples are the 2^+ and $5/2^-$ states in the Ni isotopes [3.151], and isotopes and isotones of $f_{7/2}^n$ configuration [3.152]. When there is more than one valence nucleon, $n > 1$, first-order core-polarization generates a two-body effective operator. Indeed a linear dependence of g factors on the valence nucleon number can be expected if a simple configuration such as j^n is involved and seniority is a good quantum number [3.153].

3.5.6 Second-Order Core Polarization

We turn now to light nuclei and consider explicit configurations consisting of a closed shell plus (or minus) one nucleon in A = 15, 17, 39 and 41 nuclei. With the measurement of the magnetic moment for the β-emitting ^{39}Ca (to be discussed later), all of the moments for the mirror pairs ^{15}N-^{15}O, ^{17}F-^{17}O, ^{39}Ca-^{39}K and ^{41}Sc-^{41}Ca are available for systematic analysis. The deviations from the Schmidt values are relatively small, which is to be expected, since for closed LS shells first-order core-polarization corrections vanish.

In second order the correction to the magnetic moment can be written

$$\delta\mu = \sum_{\alpha\beta} \frac{\langle j|V|\alpha\rangle\langle\alpha|\underset{\sim}{\mu}|\beta\rangle\langle\beta|V|j\rangle}{(\varepsilon_j - \varepsilon_\alpha)(\varepsilon_j - \varepsilon_\beta)} - \sum_{\alpha} \frac{\langle j|V|\alpha\rangle\langle\alpha|V|j\rangle\langle j|\underset{\sim}{\mu}|j\rangle}{(\varepsilon_j - \varepsilon_\alpha)^2}$$

$$= \langle j|V \frac{Q}{e} [\underset{\sim}{\mu}, \frac{Q}{e} V]|j\rangle \qquad (3.123)$$

where j is the single-particle and α the 2p-1h or 3p-2h intermediate-state configurations, and $(\varepsilon_j - \varepsilon_\alpha)$ is their unperturbed energy separation. First-order corrections are zero because $\langle j|\underset{\sim}{\mu}|\alpha\rangle$ has no non-zero matrix elements for closed LS shells. The second expression in (3.123) was derived by ARIMA and HYUGA [3.118] and involves the commutator $[\underset{\sim}{\mu}, \frac{Q}{e} V]$, where Q is a projection operator projecting onto intermediate states α of 2p-1h or 3p-2h configuration, and e is the operator form of the energy denominator. Thus if $\underset{\sim}{\mu}$ commutes with e and the residual interaction V, the second-order correction to the magnetic moment will vanish.

It is useful in discussing light nuclei to consider an isoscalar one-body operator

$$\underset{\sim}{\mu}^S = \frac{1}{2} [g_\ell(\nu) + g_\ell(\pi)] \sum_k \underset{\sim}{L}_k + \frac{1}{2} [g_s(\nu) + g_s(\pi)] \sum_k \underset{\sim}{S}_k \qquad (3.124)$$

and an isovector one-body operator

$$\underset{\sim}{\mu}^V = \frac{1}{2} [g_\ell(\nu) - g_\ell(\pi)] \sum_k \underset{\sim}{L}_k \underset{\sim}{\tau}_k + \frac{1}{2} [g_s(\nu) - g_s(\pi)] \sum_k \underset{\sim}{S}_k \underset{\sim}{\tau}_k \qquad (3.125)$$

where the isospin convention is for $\tau_z = +1$ for neutrons and $\tau_z = -1$ for protons. If there is no spin-orbit interaction in the unperturbed Hamiltonian, and if the residual interaction is central, then the commutator vanishes for the isoscalar magnetic moment operator. Only non-central and in particular tensor interactions can generate a second-order correction to the Schmidt isoscalar moment. A similar argument holds for $\sum_k \underset{\sim}{S}_k \underset{\sim}{\tau}_k$, which commutes with any central interaction with Wigner and Majorana exchange character. Any realistic interaction contains only small Bartlett and Heisenberg spin exchange components. Thus one can expect that the second-order configuration mixing due to the central interaction produces a very small change in the matrix element of $\sum_k \underset{\sim}{S}_k \underset{\sim}{\tau}_k$. The same argument, however, cannot

Table 3.9. Correction to the Schmidt value in nuclear magnetons for the isoscalar and isovector magnetic moments of single-particle (hole) states near LS closed shells, A = 16 and A = 40

Orbit	C2[a]	Mesonic[b]	M1[c]	Sum	Expt.[d]
		Isoscalar magnetic moments			
$0p_{1/2}^{-1}$	0.040	0.006		0.034	0.031
$0d_{5/2}$	-0.035	0.005		-0.030	-0.026
$0d_{3/2}^{-1}$	0.063	-0.003		0.060	0.071
$0f_{7/2}$	-0.036	0.005		-0.029	-0.022
		Isovector magnetic moments			
$0p_{1/2}^{-1}$	0.03	0.01	-0.07	-0.03	0.05
$0d_{5/2}$	0.48	-0.26	-0.40	-0.17	0.05
$0d_{3/2}^{-1}$	0.23	-0.16	-0.12	-0.05	-0.20
$0f_{7/2}$	0.76	-0.36	-0.35	0.05	0.34[e]

[a]Second-order core polarization [3.157].
[b]Meson exchange correction [3.118].
[c]First-order core polarization with one-pion meson exchange operator [3.118].
[d][3.10].
[e]An estimate of -0.03 can be deduced from the experimental moments for spin $19/2^-$ states in mass 43; see text.

be applied to the matrix element of $\sum_k \underset{\sim}{L}_k \underset{\sim}{\tau}_k$, which does not commute with any central interaction.

Second-order core-polarization calculations have been reported by ICHIMURA and YAZAKI [3.154], MAVROMATIS and ZAMICK [3.155], and BERTSCH [3.156]. However, we choose to quote results from the more recent work of SHIMIZU, ICHIMURA and ARIMA [3.157], because these authors demonstrated explicitly the importance of highly excited configurations which are strongly coupled to low-lying configurations through the tensor force. Using a G matrix [3.158,159] derived from the Hamada-Johnston potential, these authors calculated the correction to the magnetic moment from low-lying states, defined as those configurations with $2\hbar\omega$ harmonic-oscillator energy above the single-particle state. For highly excited configurations with up to 10 or $12\hbar\omega$ energy, just the tensor force component of the bare Hamada-Johnston potential was used. The results of these computations are given in Table 3.9.

We must also consider the two-body meson exchange interaction between the valence nucleon and the core, and the first-order core-polarization correction involving the meson exchange operator. Both corrections have been evaluated by ARIMA and HYUGA [3.118], and their results are reproduced in Table 3.9. For isoscalar magnetic moments, the meson exchange correction is small, all one-pion exchange processes being purely isovector. The only isoscalar pieces come from the $\rho\pi\gamma$-diagram and heavy-meson exchanges. The agreement with experiment is quite good for isoscalar moments, and the feature that the correction is larger for hole states than for particle states is successfully reproduced. For isovector moments, by contrast, the agreement is very poor due to the very strong cancellation among the three main contributions.

One may worry about low-lying core-deformed states. They appear to be very important for the isovector moments in mass 41 and mass 39, and less so in masses 17 and 15. In the calculation of ERIKSON [3.160] in which 3p-2h deformed states were admixed into the ground states of ^{41}Ca and ^{41}Sc, and 3h-2p into the ground states of ^{39}K and ^{39}Ca, the correction is purely isovector. By contrast, high-spin states in neighbouring nuclei are largely free of core-deformed contaminants. For example, HÄUSSER et al. [3.161] argued that the $19/2^-$ states in ^{43}Sc and ^{43}Ti might furnish a more reliable value for the single-particle $f_{7/2}$ magnetic moment than that provided by the ground states of ^{41}Ca and ^{41}Sc. Indeed, following this example a little further, the wave function for the $19/2^-$ state might be

$$|19/2^-\rangle = 0.978|f^3_{7/2}\rangle - 0.208|f^2_{7/2}f_{5/2}\rangle \qquad (3.126)$$

where the 4% admixture of $|f^2_{7/2}f_{5/2}\rangle$ was estimated in a standard shell-model calculation. Then an experimental $f_{7/2}$ magnetic moment can be deduced from the experimental $19/2^-$ moments after a correction, calculated in the shell model, is applied for the 4% admixture. The result is that the isovector $f_{7/2}$ single-particle moment (after subtracting the Schmidt value) is -0.03 nuclear magneton compared to the value 0.34 nuclear magneton recorded in Table 3.9. This much smaller value is closer to theoretical expectation, indicating that second-order core-polarization and meson exchange corrections are opposite in sign and close to being mutually cancelling.

The $19/2^-$ states in mass 43 are isomeric [3.162], and their magnetic moments are measured by the time differential perturbed angular distribution method [3.161] described earlier. The isomers are populated in (α,n) and (α,p) reactions on calcium targets with a pulsed ^{4}He beam. The pulsed beam repetition time of 6.4 μs was chosen to exceed the isomer half-lives by at least a factor of ten. The delayed γ-ray spectra were observed with Ge(Li) spectrometers at $\theta_\gamma = 135^\circ$. Low-energy γ rays ($E_\gamma < 500$ keV) were strongly suppressed by the use of Pb absorbers, and the counting rate for the cascade higher-energy $15/2^- \rightarrow 11/2^-$ and $11/2^- \rightarrow 7/2^-$ transitions could thus be improved considerably by increasing the beam intensity.

Time differential perturbed angular distributions for these γ-ray transitions were measured by using an external magnetic field whose magnitude is determined in a proton NMR measurement. In Fig.3.20 the yield ratios $R(t) = [\bar{Y}(135^\circ) - \bar{Y}(-135^\circ)]/[\bar{Y}(135^\circ) + \bar{Y}(-135^\circ)]$ are shown. The isomer's magnetic moments are determined from the fitted Larmor frequencies after correcting for diamagnetism and Knight shift.

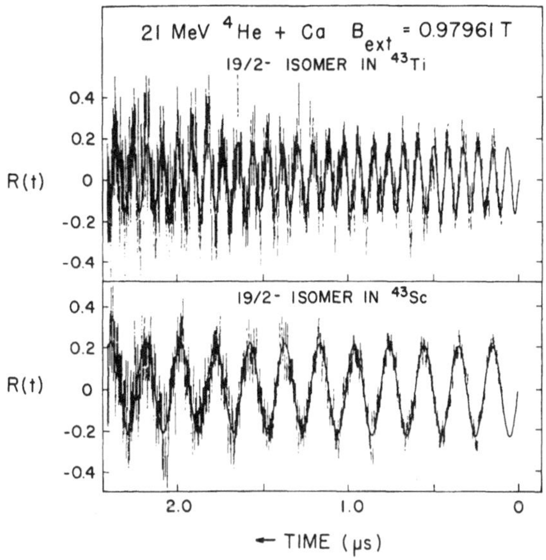

Fig.3.20. Time differentially-observed spin-rotation patterns for the sum of the $15/2^- \to 11/2^-$ and $11/2^- \to 7/2^-$ time distributions in ^{43}Ti (upper half) and ^{43}Sc (lower half) [3.161]

3.5.7 Magnetic Moments of β Emitters

The last of the magnetic moments to be measured in configurations consisting of a closed shell plus (or minus) one nucleon in the mirror pairs ^{15}N-^{15}O, ^{17}F-^{17}O, ^{39}Ca-^{39}K and ^{41}Sc-^{41}Ca was that of the β emitter ^{39}Ca ($T_{1/2}$ = 0.87 s). In a novel experiment by MINAMISONO et al. [3.163] a net polarization was detected in ^{39}Ca and then destroyed in a NMR measurement yielding the magnetic moment.

The β-emitting ^{39}Ca nuclei were polarized in the reaction $^{39}K(\vec{p},n)^{39}Ca$ initiated with polarized protons on thick polycrystalline targets such as KBr, which also served as host for the recoiling ^{39}Ca nuclei. A holding magnetic field $\underset{\sim}{B}_0$ of 6.5 kG was applied parallel to the beam polarization axis. The polarization of the ^{39}Ca nuclei was detected by means of two particle telescopes placed at 0° and 180° to $\underset{\sim}{B}_0$. The incident polarized beam was pulsed and β particles only counted during the beam-off period. The angular distribution of β's is $W(\theta) \sim 1 + P\cos\theta$, where P is the polarization and θ the angle between the polarization direction (parallel to $\underset{\sim}{B}_0$) and the β detector. The sign of P could also be reversed by reversing the beam polarization direction. The observed β asymmetry is defined as $4(r-1)/(r+1) = 4P$, where $r = [W_u(0)W_d(\pi)/W_u(\pi)W_d(0)]^{\frac{1}{2}}$ and the subscripts u and d refer to the beam polarization direction up or down.

The NMR spectrum of ^{39}Ca in the KBr target sample was observed by applying an rf magnetic field $\underset{\sim}{B}_1 \sim 3$G perpendicular to the field $\underset{\sim}{B}_0$ and measuring the β asymmetry as a function of frequency. At the resonant frequency the β asymmetry should be (partially) destroyed. A sharp NMR spectrum (Fig.3.21) was observed, and from the frequency, the Zeeman level splitting and hence the magnetic moment are determined. A correction for diamagnetism is applied.

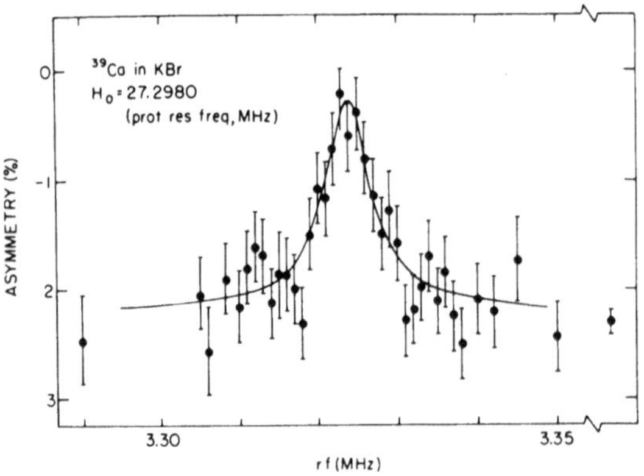

Fig.3.21. NMR spectrum of ^{39}Ca produced in a KBr target. The magnetic field H_0 was fixed, and an unmodulated rf field was varied to observe the resonance. The solid curve is a Lorentzian fit to the data [3.163]

In the next section, we shall discuss in more detail NMR experiments involving β emitters in connection with the measurement of β-decay properties.

3.6 Hyperfine Interactions in Nuclear β Decay

In assessing the impact of hyperfine interactions on our understanding of nuclear β decay - our task in this section - we decided to concentrate on a single topic and afford it extensive treatment. We do this deliberately (and apologetically with regard to those whose work is therefore not covered) because the experiments are novel and the techniques ingenious. The topic is a series of measurements of the correlation between nuclear alignment and the direction of the emitted electron following nuclear β decay. These measurements

a) provide a test of the conserved vector current (CVC) hypothesis and the presence of weak magnetism;

b) provide the best evidence (limits) for the absence of a second-class axial-vector current; and

c) suggest an indirect method of identifying meson exchange currents in the time component of the axial-vector current.

3.6.1 Theoretical Position

Let us briefly recall the theoretical position. In the customary theory of nuclear β decay, the weak-interaction Hamiltonian is taken as that of a collection of mutually isolated physical nucleons, while the initial and final nuclear states are described by wave functions dependent on the position, spin, and isospin of these nucleons. As a consequence, an impulse approximation is employed to relate the transition matrix elements in nuclear and nucleon β decay. Moreover the calculated matrix elements are in general rather sensitive to the details of the wave functions used, and this feature is often exploited in tests of nuclear spectroscopy. Beyond

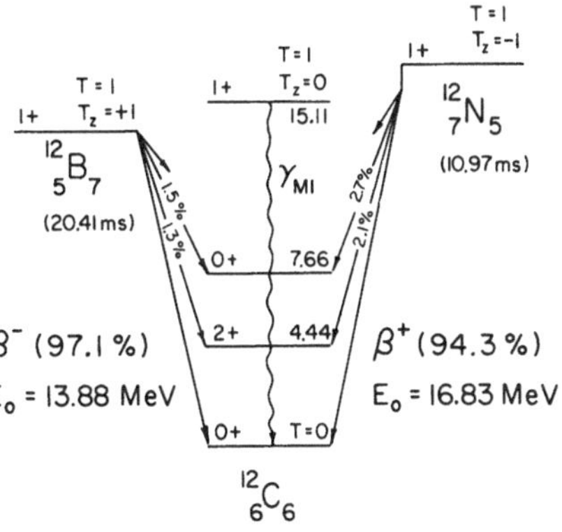

Fig.3.22. Decay scheme for the isospin triplet (T = 1) to the ground state in ^{12}C [3.171]

certain accuracy, however, one should expect deviation from the impulse approximation due to the fact that nucleons inside the nucleus are interacting with each other. For example, one or more π mesons might be exchanged between two nucleons in a nucleus undergoing a β transition. This process, which is absent in the neutron decay and consequently not included in impulse approximation, gives rise to extra two-body currents, the so-called meson exchange currents.

An alternative formulation is the elementary-particle treatment [3.164-170]. Here the nuclear matrix elements characterizing the transitions are expressed in terms of known kinematic quantities and of nuclear form factors which describe the relevant structural aspects of the nuclei involved. No explicit consideration is given to the dependence of the wave functions on the coordinates of the constituent nucleons, nor is the weak current explicitly decomposed into contributions from individual nucleons and contributions from meson exchanges. The elementary-particle treatment is essentially independent of nuclear models and as such is useful in analyses which explore the consequences of symmetry properties and conservation hypotheses.

We will concentrate exclusively on the allowed Gamow-Teller mirror transitions in the A = 12 nuclei: $^{12}B \to {}^{12}C + e^- + \bar{\nu}_e$ and $^{12}N \to {}^{12}C + e^+ + \nu_e$ (Fig.3.22). The matrix elements of the vector and axial currents for these decays (upper sign for ^{12}N decay and lower sign for ^{12}B decay) are expressed in terms of nuclear form factors [3.170]:

$$<0^+;p_1|V_\alpha^\pm(0)|1^+;p_2> = \varepsilon_{\alpha\beta\gamma\delta}\,\xi_\beta\,\frac{q_\gamma}{2m_p}\,\frac{Q_\delta}{2M}\,F_M^\pm(q^2)$$

$$<0^+;p_1|A_\alpha^\pm(0)|1^+;p_2> = \xi_\alpha F_A^\pm(q^2) + q_\alpha(q\cdot\xi)F_P^\pm(q^2) - \frac{Q_\alpha}{2M}\,\frac{q\cdot\xi}{2m_p}\,F_E^\pm(q^2) \quad , \qquad (3.127)$$

and the matrix element of the electromagnetic current for the γ transition from the analogue state in ^{12}C is likewise expressed in terms of a nuclear magnetic transition form factor:

$$<0^+;p_1|J_\alpha^{em}(0)|1^+;p_2> \equiv \varepsilon_{\alpha\beta\gamma\delta}\, \xi_\beta \frac{q_\gamma}{2m_p} \frac{Q_\delta}{2M} \mu(q^2) \qquad (3.128)$$

where $q_\alpha = (p_2 - p_1)_\alpha$, $Q_\alpha = (p_2 + p_1)_\alpha$, $M = (1/2)(M_1 + M_2) = (1/2)[M(^{12}C) + M(^{12}N, ^{12}B)]$, $\Delta^\pm = M(^{12}N, ^{12}B) - M(^{12}C)$, m_p is the nucleon mass, and ξ the polarization four-vector of the spin-one nuclei. Further, $F^\pm_{M,A,P,E}(q^2)$ are, respectively, the nuclear weak magnetism, axial, pseudoscalar and pseudotensor (weak electricity) form factors, which we decompose into first-class and second-class components

$$F^\pm_{M,A,P,E}(q^2) = F^{\pm(I)}_{M,A,P,E}(q^2) \pm F^{\pm(II)}_{M,A,P,E}(q^2) \quad . \qquad (3.129)$$

According to WEINBERG [3.172], who first introduced the classification, currents are first class if they transform as $G\, V_\alpha^{(I)} G^{-1} = V_\alpha^{(I)}$ and $G\, A_\alpha^{(I)} G^{-1} = -A_\alpha^{(I)}$, and second class (with superscript II) if they transform with the opposite sign. Here G is the product of charge conjugation and a rotation of π about the isospin y axis. Strong interactions are believed to be invariant under G-parity transformation. If it is hypothesized that weak interactions are likewise invariant, then second-class currents would not exist, and $F^{(II)}_{M,A,P,E}(q^2) = 0$. The search for second-class currents, therefore, concentrates on the isolation of form factors $F^{(II)}(q^2)$, which because they change sign in going from electron decay to positron decay can be detected as interference phenomena in comparing transition probabilities and angular distributions in mirror β decays. (However, the search is spoiled by internucleon electromagnetic interactions and the neutron-proton mass difference, which violate isospin symmetry and induce small amounts of G non-invariance).

The conserved vector current (CVC) hypothesis, besides requiring that the vector current be divergenceless, i.e., $\partial_\alpha V_\alpha = 0$, which is automatically satisfied, by construction, in the above matrix elements, also identifies the weak vector current with that obtained from the electromagnetic current by means of isotopic spin rotation: $V_\alpha^\pm = \mp[I^\pm, J_\alpha^{em}]$. Neglecting any electromagnetic interaction, which would break the isospin symmetry, the CVC hypothesis implies

$$F_M^{(I)}(q^2) = \sqrt{2}\mu(q^2) \quad ,$$

$$F_M^{(II)}(q^2) = 0 \quad , \qquad (3.130)$$

i.e., no second-class component in the vector current.

It is useful to make contact with the more familiar impulse approximation results. In this case it is necessary to write down the matrix elements of the vector and axial currents for *nucleon* decay:

$$\langle \tfrac{1}{2}^+;p_p|V^{\pm}_{\alpha}(0)|\tfrac{1}{2}^+;p_n\rangle = i\bar{u}_p(p_p)\left[\gamma_\alpha f^{\pm}_V(q^2) + \frac{\sigma_{\alpha\beta}q_\beta}{2m_p} f^{\pm}_M(q^2) + i\,\frac{(m_p+m_n)}{m_\pi^2} q_\alpha f^{\pm}_S(q^2)\right]u_n(p_n)$$

$$\langle \tfrac{1}{2}^+;p_p|A^{\pm}_{\alpha}(0)|\tfrac{1}{2}^+;p_n\rangle = i\bar{u}_p(p_p)\left[\gamma_\alpha\gamma_5 f^{\pm}_A(q^2) - i\,\frac{(m_p+m_n)}{m_\pi^2} q_\alpha\gamma_5 f^{\pm}_P(q^2) - \frac{\sigma_{\alpha\beta}q_\beta\gamma_5}{2m_p} f^{\pm}_E(q^2)\right] u_n(p_n) \quad (3.131)$$

where $q_\alpha = (p_n - p_p)_\alpha$, $\sigma_{\alpha\beta} = [\gamma_\alpha,\gamma_\beta]/2i$, and u_p and u_n the usual Dirac spinors. Here, $f^{\pm}_{V,M,S,A,P,E}(q^2)$ are, respectively, the vector, induced tensor, induced scalar, axial, pseudoscalar and pseudotensor form factors, which as before can be decomposed into first-class and second-class components

$$f^{\pm}_{V,M,S,A,P,E}(q^2) = f^{\pm(I)}_{V,M,S,A,P,E}(q^2) \pm f^{\pm(II)}_{V,M,S,A,P,E}(q^2) \quad . \qquad (3.132)$$

However, in the limit within which both neutron and proton are members of the same isospin doublet, we have $f_V^{(II)} = f_M^{(II)} = f_A^{(II)} = f_P^{(II)} = 0$ and $f_S^{(I)} = f_E^{(I)} = 0$. Furthermore, under the CVC hypothesis ($\partial_\alpha V_\alpha = 0$), the induced scalar form factor is required to vanish, $f_S = 0$, while the weak magnetism term is related to the vector term:

$$f_M = (\mu_p - \mu_n)f_V = 3.7\ f_V \qquad (3.133)$$

where μ_p and μ_n are the proton and neutron magnetic moments. Note, therefore, that the only second-class current form factor to survive is contained in the induced pseudotensor term f_E.

Impulse approximation uses these single-nucleon currents (reduced to their non-relativistic forms) as the one-body operators to be sandwiched between the many-body nuclear wave functions in forming the nuclear matrix elements. The form factors $f(q^2) \approx f(0)$ are treated as constants unchanged from nucleus to nucleus. The connection with the form factors of the elementary-particle approach is given by HWANG and PRIMAKOFF [3.170]

$$F^{\pm}_M(q^2) \to -\left[f_V^{(I)} + f_M^{(I)}\right]\langle\underset{\sim}{\sigma}\rangle - f_V^{(I)}\langle\underset{\sim}{L}\rangle \quad ,$$

$$F^{\pm}_A(q^2) \to -\left[f_A^{(I)} + \frac{\Delta^{\pm}}{2m_p} f_E^{(II)}\right]\langle\underset{\sim}{\sigma}\rangle \quad ,$$

$$F^{\pm}_P(q^2) \to -\left[f_P^{(I)} - \left(\frac{m_\pi}{2m_p}\right)^2 f_E^{(II)}\right]\langle\underset{\sim}{\sigma}\rangle \quad ,$$

$$F^{\pm}_E(q^2) \to -\left[f_A^{(I)} - f_E^{(II)}\right]\langle\underset{\sim}{\sigma}\rangle - 2f_A^{(I)}\langle i\underset{\sim}{r}(\underset{\sim}{\sigma}\cdot\underset{\sim}{p})\rangle \qquad (3.134)$$

where, for example, $\langle\underset{\sim}{\sigma}\rangle$ is the nuclear matrix element

$$\langle\sigma\rangle = \langle\psi_{12_C}|\sum_{a=1}^{12} \tau_+^{(a)}\sigma_z^{(a)}|\psi_{12_B}\rangle \quad .$$

Note, however, that the second-class form factors may be particularly sensitive to deviations from impulse approximation (effects from nucleons off the mass shell, and from meson exchange), so that $F^{(II)}_{M,A,P,E}$ could be quite different from the values indicated above. Nonetheless, HWANG and PRIMAKOFF [3.170] believe that if $f_E^{(II)} \simeq f_A^{(I)}$, the conclusion that only $F_E^{(II)}$ is important is likely to be correct.

The β decay rate, angular distributions and corresponding asymmetry coefficients $\mathscr{A}^\pm$ are given in terms of these nuclear form factors [3.170]:

$$d^3\Gamma(e^\pm) = \frac{1}{\hbar}\frac{G^2}{8\pi^4}|F_A^\pm(0)|^2 F_\pm(Z,E_e)p_e E_e(\Delta^\pm - E_e)^2(1 + \eta_\pm + a_\pm E_e)dE_e d\Omega_e$$

$$\times\ [1 \pm (h_1 - h_{-1})(1 + \alpha_\pm E_e)P_1(\cos\theta_e) + (1 - 3h_0)\alpha_\pm E_e P_2(\cos\theta_e)]$$

$$\mathscr{A}^\pm \equiv \frac{d^3\Gamma(e^\pm)_{\theta_e=0} - d^3\Gamma(e^\pm)_{\theta_e=\pi}}{d^3\Gamma(e^\pm)_{\theta_e=0} + d^3\Gamma(e^\pm)_{\theta_e=\pi}}$$

$$\simeq \pm (h_1 - h_{-1})\{1 + \alpha_\pm[1 - (1 - 3h_0)]E_e\} \quad . \tag{3.135}$$

Here G is the weak-interaction vector coupling constant[1], $G = (1.1476 \pm 0.0004) \times 10^{-11}\ \mathrm{MeV}^{-2}$; $F_\pm(Z,E_e)$ is the Fermi function for the $e^\pm$ decays (the ratio of the electron density at the nucleus to that at infinity); p_e and E_e are the electron momentum and energy; $\Delta^\pm$ is the maximum electron energy; h_1, h_{-1}, h_0 are the populations of the m = 1, -1, 0 states of ^{12}N or ^{12}B, normalized such that $h_1 + h_{-1} + h_0 = 1$; $\cos\theta_e = \hat{p}_e\cdot\hat{z}$; and $\Gamma = \ln 2/t_{1/2}$, where $t_{1/2}$ is the decay half-life. The coefficients $\eta_\pm$, $a_\pm$ and $\alpha_\pm$ are similarly given in terms of the nuclear form factors:

$$\eta_\pm = \frac{\Delta^\pm}{3m_p}\left(\pm 2\frac{F_M^\pm(0)}{F_A^\pm(0)} + \frac{F_E^\pm(0)}{F_A^\pm(0)}\right) \quad ,$$

$$a_\pm = \mp\frac{4}{3m_p}\frac{F_M^\pm(0)}{F_A^\pm(0)} \quad ,$$

$$\alpha_\pm = \frac{1}{3m_p}\left(\mp\frac{F_M^\pm(0)}{F_A^\pm(0)} - \frac{F_E^\pm(0)}{F_A^\pm(0)}\right) \quad . \tag{3.136}$$

---

1 The coupling constant of HWANG and PRIMAKOFF [3.170], $G = G_V'/(\hbar c)^3$, can be determined from $0^+ \to 0^+$ superallowed β-decay data, which in the analysis of TOWNER et al. [3.173] leads to a value $G_V' = (1.4128 \pm 0.0005) \times 10^{-49}\ \mathrm{erg\ cm}^3$.

Small residual Coulomb corrections to these coefficients have been omitted.

Let us consider a number of applications of the basic formula.

Case 1: The total decay rate. For non-aligned ^{12}B and ^{12}N nuclei, integrating over electron directions and energies gives for the ratio of decay rates in the mirror transitions the result:

$$\frac{(ft_{\frac{1}{2}})^+}{(ft_{\frac{1}{2}})^-} = \frac{|F_A^-(0)|^2(1 + \eta_- + a_-\Delta^-/2)}{|F_A^+(0)|^2(1 + \eta_+ + a_+\Delta^+/2)} \tag{3.137}$$

where f is the integrated Fermi function

$$f = \int F(Z,E_e)p_eE_e(\Delta - E_e)^2\, dE_e \quad , \tag{3.138}$$

which in general will include radiative and Coulombic corrections. The energy-dependent term aE_e has been replaced by its average value. Expanding binomially and neglecting terms proportional to $(\Delta^+ - \Delta^-)$, the ratio becomes

$$\frac{(ft_{\frac{1}{2}})^+}{(ft_{\frac{1}{2}})^-} \approx \left|\frac{F_A^{(I)-}}{F_A^{(I)+}}\right|^2 \left\{1 - 2\left[\frac{F_A^{(II)-}}{F_A^{(I)-}} + \frac{F_A^{(II)+}}{F_A^{(I)+}}\right] - \frac{2}{3}\,\frac{\Delta^+ + \Delta^-}{2m_p}\,\frac{F_E^{(II)}}{F_A^{(I)}}\right\}$$

$$\to 1 - \frac{4}{3}\,\frac{\Delta^+ + \Delta^-}{2m_p} \cdot \frac{f_E^{(II)}}{f_A^{(I)}} + \delta_{nucl.} \quad , \tag{3.139}$$

the last line representing the impulse approximation limit. It would seem that a comparison of ft values in mirror transitions would be a profitable way to search for second-class currents. This, indeed, was proposed by WILKINSON [3.174], and first results appeared encouraging. However, a definitive experiment was performed by WILKINSON and ALBURGER [3.175] in which the asymmetry of ft values was measured as a function of $(\Delta^+ + \Delta^-)$ in the β decays of ^8Li and ^8B. This is a unique case where the daughter state has a broad width, decaying by α emission and effectively permitting the end-point energy to be varied within a single decay. Their results showed the ratio of ft values to be almost independent of $(\Delta^+ + \Delta^-)$, giving little or no evidence for second-class currents.

Another difficulty with this type of analysis concerns the existence of the additional term $\delta_{nucl.}$ in (3.139). It arises from the fact that $\langle\underset{\sim}{\sigma}\rangle^+ \neq \langle\underset{\sim}{\sigma}\rangle^-$ $(|\langle\sigma\rangle^-/\langle\sigma\rangle^+|^2 = 1 + \delta_{nucl.})$, because $|^{12}B\rangle$, $|^{12}N\rangle$ and $|^{12}C\rangle$ are not exact isospin eigenstates; this is because the internucleon electromagnetic interaction and the neutron-proton mass difference violate isospin symmetry. Nuclear structure calculations [3.176,177] indeed find computed values for $\delta_{nucl.}$ to be comparable with experiment, and thus confuse the evidence, if any, for the presence of second-class currents.

Case 2: Electron shape. Again for non-aligned ^{12}B and ^{12}N nuclei, integrating over electron directions and measuring the electron energy spectrum makes it possible to determine the coefficient of the energy-dependent term $a_\pm E_e$. The coefficient $a_\pm$

$$a_\pm = \mp \frac{4}{3m_p} \frac{F_M^\pm}{F_A^\pm} \to \mp \frac{4}{3m_p} \left(4.7 + \frac{\langle \underset{\sim}{L} \rangle}{\langle \underset{\sim}{\sigma} \rangle}\right) \frac{f_V}{f_A} \tag{3.140}$$

can be predicted in advance. The form factor F_A is determined from the total rate, and F_M from the analogous electromagnetic M1 transition in ^{12}C using the CVC hypothesis, (3.130). Indeed the measurement of $a_\pm$ and its confirmation of expectation are viewed as a vindication of the CVC hypothesis. This is the classic experiment of LEE et al. [3.178].

Recently CALAPRICE and HOLSTEIN [3.179] have questioned the original analysis of the experiment and as a result thrown some doubt on the validity of CVC. The critique was discussed in a new analysis by the original group of experimenters (WU et al. [3.180]) of the data from the experiment, with the conclusion that these data still support CVC. A second experiment by KAINA et al. [3.181] is only in moderately good agreement with the first one, but these authors adopted a different philosophy in the analysis. WU et al. retained only terms linear in the electron energy, while KAINA et al. considered a quadratic term, E_e^2, important. Since the quadratic terms cancel in the difference $a_- - a_+$, and both experiments agree on this quantity, this difference is preferred in the extraction of F_M/F_A. The results are in agreement with theory [3.182], although small discrepancies remain when individual $a_\pm$ values are discussed.

Case 3: Polarization and alignment. We now turn to experiments with oriented nuclei and measurements of the decay electron's angular distribution. By either measuring the energy dependence of the polarization term, or the alignment correlation directly, the interesting coefficients $\alpha_\pm$ can be determined. In particular, the difference in coefficients is

$$\alpha_- - \alpha_+ \simeq \frac{4}{3}\frac{1}{2m_p}\left(\frac{F_M}{F_A^{(I)}} + \frac{F_E^{(II)}}{F_A^{(I)}}\right)$$

$$\to \frac{4}{3}\frac{1}{2m_p}\left[\left(4.7 + \frac{\langle L\rangle}{\langle \sigma\rangle}\right)\frac{f_V}{f_A} - \frac{f_E^{(II)}}{f_A}\right] \quad , \tag{3.141}$$

from which two approaches to the analysis are possible. Either one can assume that the CVC hypothesis is valid, in which case $F_M/F_A^{(I)}$ is known and experiment gives a measure of the second-class form factor $F_E^{(II)}$; or one can assume that second-class terms are absent and use the experiment as a validation of the CVC hypothesis. As we shall see, the data are consistent with no second-class currents and with the CVC hypothesis intact.

Also of interest is the sum of coefficients

$$\alpha_- + \alpha_+ \simeq -\frac{4}{3}\frac{1}{2m_p}\frac{F_E^{(I)}}{F_A^{(I)}}$$

$$\to -\frac{4}{3}\frac{1}{2m_p}\left[1 + 2\,\frac{\langle i\underset{\sim}{r}(\underset{\sim}{\sigma}\cdot\underset{\sim}{p})\rangle}{\langle\underset{\sim}{\sigma}\rangle}\right] \quad , \tag{3.142}$$

which is seen to depend exclusively on the small first-class component of the 'weak electricity' form factor. The impulse approximation limit shown in the second line is derived from the time component ($\alpha = 4$) of the axial-vector current A_α.

To see the relevance of this remark, we must digress for a moment. As was mentioned earlier, impulse approximation derives from one-body currents evident in nucleon decays; any two-body currents arising from meson exchange processes, for example, must be added subsequently as a correction. Of the possible meson exchange processes, the one-pion exchange process is believed to be dominant. Furthermore, to the extent that soft-pion theorems [3.122] are valid, these pion exchange processes may be calculated in a model-independent way. The point that we wish to make is that for the two-body meson exchange part of the vector electromagnetic current, J_α^{em}, the time ($\alpha = 4$) and space ($\alpha = 1,2,3$) components have quite different magnitudes: the time component is $\sim O(\underset{\sim}{p}/m_p)$ and the space component is $\sim O(1)$. In contrast, the one-body impulse approximation operator behaves non-relativistically like $\sim O(1)$ for time and $\sim O(\underset{\sim}{p}/m_p)$ for space components. Thus for the space components the meson exchange operator is intrinsically enhanced relative to the single-particle operator. This does not tell us how large two-body operators are relative to single-particle operators; it merely suggests that it would be more profitable to look at the space components of the vector electromagnetic current in searching for evidence for meson exchange processes. Indeed, in the best evidence to date, the thermal neutron capture by proton [3.183], meson exchange currents were identified in an M1 transition.

The argument is reversed for axial currents. KUBODERA et al. [3.184], who first advanced this idea, pointed out that the meson exchange operator is $\sim O(1)$ for $\alpha = 4$ and $\sim O(\underset{\sim}{p}/m_p)$ for $\alpha = 1,2,3$. By contrast, the single-particle axial-current operator behaves like $O(\underset{\sim}{p}/m_p)$ for $\alpha = 4$ and $O(1)$ for $\alpha = 1,2,3$. Thus, isolating time components of the axial current is important in the search for meson exchange currents, and for this reason the quantity $\alpha_- + \alpha_+$ has renewed importance.

KUBODERA et al. [3.184] estimated that exchange currents could account for as much as 40% of the experimentally observed $\alpha_- + \alpha_+$. By contrast MORITA et al. [3.185] have calculated the one-body impulse approximation matrix elements and found them to be in agreement with the experimental value of $\alpha_- + \alpha_+$, leaving little room for a substantial meson exchange contribution.

3.6.2 Experimental NMR Techniques for β Emitters

The experimental techniques for measuring the asymmetry parameters $\alpha_\pm$ all follow a common theme. Polarized recoil nuclei are implanted into monocrystalline metal foils for which the degeneracy of the Zeeman levels is broken by the quadrupole couplings. By a variety of novel means, an alignment can be induced in the recoil nucleus which is detected in the subsequent β decays.

The standard experimental arrangement is that discussed by MINAMISONO [3.186], HASKELL and MADANSKY [3.187] and HASKELL et al. [3.188]. Polarized ^{12}N and ^{12}B recoils are produced through the reactions $^{10}B(^3He,n)^{12}N$ and $^{11}B(d,p)^{12}B$, with the precise magnetic substate populations of the product nuclei depending on the incident beam energy and the recoil angle. The β^- unstable nuclei decay according to the schemes $^{12}N \to {}^{12}C + e^+ + \nu$ and $^{12}B \to {}^{12}C + e^- + \bar{\nu}$, the half-lives being 11.0 and 20.4 ms, respectively. The angular momenta and parities of the initial and final states are 1^+ and 0^+ in both decays; hence both are pure Gamow-Teller transitions. Thus any polarization of the ^{12}N and ^{12}B nuclei induced in the nuclear reaction will result in an asymmetry in the emission direction of the decay betas. From (3.135) the angular distribution is

$$W(\theta) \sim 1 \pm P(1 + \alpha_\pm E)P_1(\cos\theta) + A\alpha_\pm EP_2(\cos\theta) \tag{3.143}$$

where the polarization $P = (h_1 - h_{-1})$; the alignment $A = (h_1 + h_{-1} - 2h_0)$; h_1, h_{-1}, h_0 are the populations of the m = 1, -1, 0 states of ^{12}N or ^{12}B, normalized such that $h_1 + h_{-1} + h_0 = 1$; $\cos\theta = \hat{\underset{\sim}{p}}\cdot\hat{\underset{\sim}{z}}$; $\underset{\sim}{p}$ is the electron momentum and E the electron energy. The upper sign refers to ^{12}N decay, the lower to ^{12}B decay.

The polarized recoiling nuclei are collimated and stopped in a single crystal of Mg. The nuclear orientation produced by the reaction is preserved during the flight of recoil ions in vacuum by applying a strong holding magnetic field $\underset{\sim}{B}_0$ normal to the reaction plane. This defines the z direction. The subsequent decay betas were detected by two coincidence telescopes, one positioned directly above and one directly below the metal stopping crystal. Since the recoils are polarized perpendicular to the reaction plane, this polarization is detected by observing the asymmetry in the counting rates of the telescopes. Specifically

$$\mathscr{A}^\pm \equiv \frac{W(0) - W(\pi)}{W(0) + W(\pi)} = \frac{\pm P(1 + \alpha_\pm E)}{1 + A\alpha_\pm E}$$

$$\approx \pm P[1 + \alpha_{\pm}E(1 - A)]$$

$$\approx \pm P \quad , \tag{3.144}$$

since $\alpha_{\pm}E \ll 1$. A weak RF magnetic field $\tilde{B}_1$ is applied at the recoil stopper perpendicular to the static magnetic field to induce the nuclear magnetic resonance. The NMR effect is detected by the change in the counting rates produced by the change in the asymmetry, $\mathcal{A}^{\pm}$, of the β decay. We refer to this as the 'standard' arrangement in our subsequent discussion.

Ideally, the interaction of a strong magnetic field $\tilde{B}_0$ along the z direction with the recoil nucleus is described by the simple Hamiltonian $\mathcal{H} = -g\hbar B_0 I_z$, where g is the nuclear gyromagnetic ratio (magnetic moment divided by spin I). An alternating magnetic field of amplitude $\tilde{B}_1$, applied perpendicular to the static field, say in the x direction, introduces a perturbing term in the Hamiltonian of $\mathcal{H}_{pert} = -g\hbar B_1 I_x \cos\omega t$. The operator I_x has matrix elements between states m and m' which vanish unless $m' = m \pm 1$. Resonance occurs when the frequency of the alternating magnetic field matches the energy spacing (3.52)

$$\hbar\omega = \Delta E = g\hbar B_0 \tag{3.145}$$

inducing transitions. At saturation the magnetic-state populations have all become equalized, thus destroying any polarization in the recoil nucleus produced in the nuclear reaction. As all energy spacings are degenerate, there is a single resonance, the Zeeman resonance, in the response of the β asymmetry, $\mathcal{A}^{\pm}$, when measured as a function of the RF frequency. Off resonance, there is no redistribution of magnetic-state populations.

Implanting the recoil nucleus in a single crystal of Mg perturbs the situation. The magnesium crystal lattice is hexagonal close packed and possesses electric field gradients at both lattice and interstitial positions. The interaction of the electric quadrupole moments of the implanted recoil nuclei with these electric field gradients can be treated as a perturbation of the Zeeman splitting. The energy of the nuclear magnetic sublevel m is then ([3.189] and Sect.3.2.2.b):

$$E_m = -g\hbar B_0 m + \frac{e^2 qQ}{4I(2I - 1)} \frac{1}{2} (3 \cos^2\theta - 1)[3m^2 - I(I + 1)] \tag{3.146}$$

where q is the electric field gradient at the nucleus, Q is the nuclear electric quadrupole moment, I the nuclear spin (I = 1 for ^{12}N and ^{12}B) and θ the angle between the direction of the electric field gradient, i.e., the crystalline $\underline{c}$ axis, and the direction of the external magnetic field $\tilde{B}_0$. The frequencies of the resonance lines satisfy the relation

$$\hbar\omega = g\hbar B_0 \pm \frac{3}{4} e^2 qQ \frac{1}{2} (3 \cos^2\theta - 1) \quad . \tag{3.147}$$

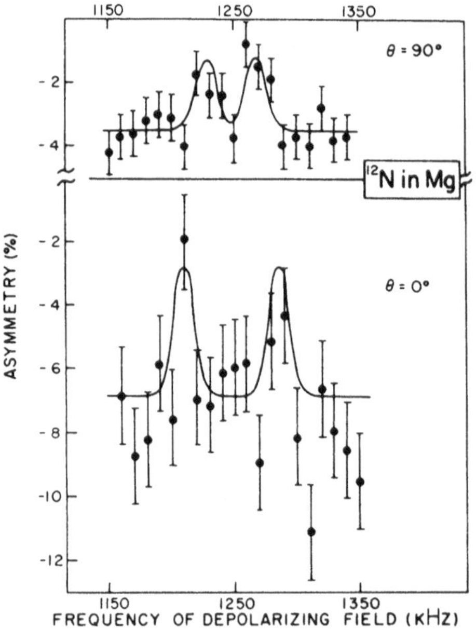

Fig.3.23. Resonance scans of ^{12}N implanted in a single crystal of Mg. The quadrupole splittings exhibit the proper dependence on the angle between the crystalline c axis and the applied magnetic field, the splitting at $\theta = 0^{\circ}$ being twice the splitting at $\theta = 90^{\circ}$ [3.188]

Instead of a single Zeeman resonance line, one obtains two quadrupole resonance lines equally split from the Zeeman frequency. The quadrupole splitting is a maximum when $\theta = 0$, i.e. when the crystalline $\underline{c}$ axis is parallel to the direction of the external magnetic field (Fig.3.23), and this condition is adopted in most cases.

If the RF frequency is matched to the energy spacing $E_0 - E_1$, i.e., $\hbar\omega_1 = g\hbar B_0 - \frac{3}{4} e^2qQ$, then on saturation the populations of the m = 1 and m = 0 magnetic substates become equalized. We shall call this condition 1. If we denote the initial populations with superscripts zero, h_1^0, h_{-1}^0, h_0^0, and the initial polarization and alignment P^0 and A^0, then after equalization

$$h_1 = h_0 = \frac{1}{2}(h_1^0 + h_0^0)$$

$$= \frac{1}{12}(4 + 3P^0 - A^0) \quad ,$$

$$h_{-1} = \frac{1}{12}(4 - 6P^0 + 2A^0) \quad , \tag{3.148}$$

while h_{-1} remains at its initial value h_{-1}^0. The polarization and alignment, under condition 1, are

$$P(1) = \frac{3}{4}P^0 - \frac{1}{4}A^0 \quad ,$$

$$A(1) = -P(1) = -\frac{3}{4}P^0 + \frac{1}{4}A^0 \quad . \tag{3.149}$$

Equally well, the RF frequency can be matched to the energy spacing $E_{-1} - E_0$ (condition $\bar{1}$), $\hbar\omega_{-1} = g\hbar B_0 + \frac{3}{4} e^2qQ$; then after equalization the magnetic populations become

$$h_{-1} = h_0 = \frac{1}{12}(4 - 3P^0 - A^0) \quad ,$$

$$h_1 = \frac{1}{12}(4 + 6P^0 + 2A^0) \tag{3.150}$$

and the polarization and alignment

$$P(\bar{1}) = \frac{3}{4}P^0 + \frac{1}{4}A^0 \quad ,$$

$$A(\bar{1}) = P(\bar{1}) = \frac{3}{4}P^0 + \frac{1}{4}A^0 \quad . \tag{3.151}$$

Experimentally, then, the initial polarization and alignment P^0 and A^0 are determined by the standard technique of saturating each of the two Zeeman resonances in turn and observing the difference in the β counting rates, $\mathscr{A}^{\pm}$, on and off each resonance. The sum of the appropriately defined signals, $P(\bar{1}) + P(1)$, gives $\frac{3}{2}P^0$, and the difference, $P(\bar{1}) - P(1)$, gives $\frac{1}{2}A^0$. (Actually the sign of A^0 is not determined unless the sign of the quadrupole coupling e^2qQ is known.)

Another technique which leads essentially to the same conditions but which has some inherent experimental advantage is known as "spontaneous level mixing" [3.190]. The energy of the nuclear magnetic sublevel,

$$E_m = -g\hbar B_0 m + \frac{1}{4}e^2qQ(3m^2 - 2) \equiv h\nu_m \quad , \tag{3.152}$$

when plotted as in Fig.3.24 as a function of the strength of the vertical static magnetic field $\underset{\sim}{B}_0$, shows a crossing of the sublevels E_1 and E_0 when $|g\hbar B_0| = |\frac{3}{4}e^2qQ|$. At this crossing point, the magnetic populations of the m = 1 and m = 0 sublevels "spontaneously" mix and become equal, thus producing condition 1 discussed above. Reversing the direction of the magnetic field $\underset{\sim}{B}_0$ exchanges h_1 and h_{-1} for fixed spatial orientation of the nuclear spin, and leads to condition $\bar{1}$ at the crossing point. Thus by complete analogy with the above, the initial polarization and alignment are determined by observing the difference in β counting rates, $\mathscr{A}^{\pm}$, on and off resonance. The resonance in this case is not obtained by varying the RF frequency of the alternating magnetic field, but by adjusting the strength of the static vertical field until the crossing point is reached.

A third technique, known as "adiabatic fast passage" [3.192], permits the magnetic populations of two states to be reversed. As used by SUGIMOTO et al. [3.171], adiabatic inversion is achieved by sweeping the RF frequency of the alternating magnetic field $\underset{\sim}{B}_1$ from a value below the resonance value to a value above, all in the presence of a strong static magnetic field $\underset{\sim}{B}_0$. The condition for a complete reversal of the magnetization is [3.192] $d\omega/dt \ll |gB_1^2|$. SUGIMOTO et al. [3.171] first set the RF frequency exactly on the resonance ω_1 (condition 1 discussed above), thereby equalizing the magnetic populations h_1 and h_0. This frequency, naturally, is off resonance for the second Zeeman line ($h\omega_{-1} = E_{-1} - E_0$). The RF frequency is

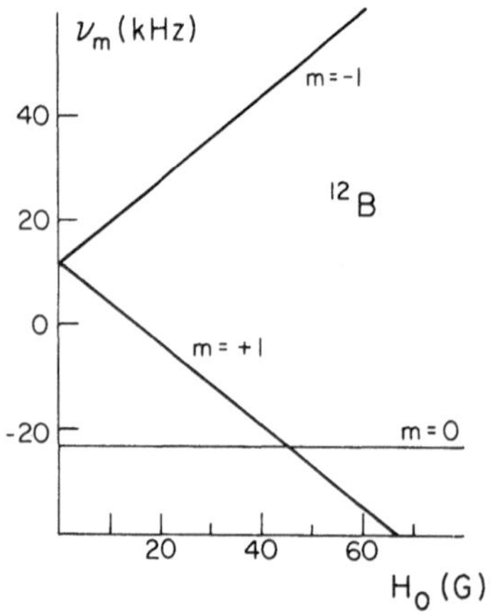

Fig.3.24. Zeeman level diagram of ^{12}B recoils implanted into a Mg crystal with the crystalline c axis parallel to the applied magnetic field H_0. The quadrupole interaction of the implanted recoil with the crystalline field produces an offset of the m = 0 level from the |m| = 1 levels, so that the m = 0 and m = 1 levels cross at H_0 = 45 G [3.191]

then swept from its value above ω_{-1} through this resonance value to a value below. The magnetic substate populations h_{-1} and h_0 are thereby reversed, inducing a net alignment and reducing the polarization (ideally) to zero. We call this condition 2.

The magnetic populations at condition 2 are

$$h_1 = h_{-1} = \frac{1}{12}(4 + 3P^0 - A^0) \quad ,$$

$$h_0 = \frac{1}{12}(4 - 6P^0 + 2A^0) \quad , \tag{3.153}$$

leading to a polarization and alignment

$$P(2) = 0 \quad ,$$

$$A(2) = \frac{3}{2}P^0 - \frac{1}{2}A^0 \quad . \tag{3.154}$$

Equally well, the RF frequency could first be set at ω_{-1} (condition $\bar{1}$) and then swept through the resonance ω_1 (condition $\bar{2}$), leading to a polarization and alignment

$$P(\bar{2}) = 0 \quad ,$$

$$A(\bar{2}) = -\frac{3}{2}P^0 - \frac{1}{2}A^0 \quad . \tag{3.155}$$

Again, the initial polarization and alignment can be determined by observing the difference in β counting rates, $\mathscr{A}^{\pm}$, both before and after sweeping the RF frequency.

3.6.3 Experimental Results

Following the general discussion of the last section, we now indicate how the various techniques of "saturating Zeeman resonances", "spontaneous level mixing", and "adiabatic inversion" have been used to measure the β decay asymmetry parameters, $\alpha_{\pm}$.

Example 1 [3.193]. Pulsed polarized ^{12}B and ^{12}N nuclei were produced in (d,p) and (^3He,n) reactions at the Osaka University Van de Graaff and implanted into an Aℓ foil. In a fcc metal, such as aluminium, only a single Zeeman resonance is observed, indicative of a high-symmetry crystallographic site with little or no quadrupole splitting of the magnetic substates. Off resonance the decay betas were detected by two coincident telescopes, one above and one below the stopping foil. The standard signal (3.144),

$$\mathscr{A}^{\pm} \equiv \frac{W(0) - W(\pi)}{W(0) + W(\pi)} \approx \pm P^0[1 + \alpha_{\pm}E(1 - A^0)] \quad , \tag{3.156}$$

was recorded as a function of β energy and the data fitted by a linear function of E: $\mathscr{A}^{\pm} = X(1 + YE)$, where X and Y are fitted parameters. Some systematic uncertainties can be eliminated by reversing the sign of the polarization once each beam burst using the "adiabatic fast passage" (AFP) technique of sweeping the radio frequency through the Zeeman resonance value. From a determination of the slope Y a value of $\alpha_{\pm}(1 - A^0)$ is obtained.

The alignment A^0 was then measured in a separate experiment in which recoil nuclei were implanted into a thin Mg single crystal. In this case, the Zeeman splitting was perturbed by the quadrupole interaction. Matching the radio frequency to ω_1 and ω_{-1} in turn (conditions 1 and $\bar{1}$), the standard signal yielded P(1) and $P(\bar{1})$, and from the difference $P(\bar{1}) - P(1) = A^0/2$, the alignment was determined. The alignments were small, and the coefficients $\alpha_{\pm}$ as measured are listed in Table 3.10.

A determination of $\alpha_{\pm}$ such as this from the energy dependence of the polarization term in the decay electron's angular distribution

$$W(\theta) = 1 \pm P(1 + \alpha_{\pm}E)P_1(\cos\theta) + A\alpha_{\pm}EP_2(\cos\theta) \tag{3.157}$$

is fraught with experimental difficulties (primarily from backscattering), since a small effect is being isolated against a large leading term ($\alpha E \ll 1$). The result for α_- is in serious disagreement with subsequent work and is probably in error. Later experiments all concentrate on isolating the alignment correlation in (3.157).

Example 2 [3.194]. In the Louvain's group work, ^{12}B recoils produced in the (d,p) reaction were implanted into a Mg single crystal oriented with its $\underline{c}$ axis parallel to the static magnetic field $\underset{\sim}{B}_0$. The deuteron energy and recoil angle were chosen to maximize the initial polarization P^0 and minimize the alignment A^0. Their values were determined by saturating the two Zeeman resonances ω_1 and ω_{-1} (conditions 1 and $\bar{1}$) and observing the changes in the β-counting rates, $\mathscr{A}^{\pm}$. To obtain the coefficients $\alpha_{\pm}$, one can form a signal from the β-counting rates under conditions 1 and $\bar{1}$:

$$S_{\pm}(\theta,E) = 2[W(\theta,\bar{1}) - W(\theta,1)]/[W(\theta,\bar{1}) + W(\theta,1)]$$

$$\simeq \frac{\frac{3}{2}P^0\alpha_\pm EP_2(\cos\theta) \pm \frac{1}{2}A^0P_1(\cos\theta)}{1 \pm \frac{3}{4}P^0P_1(\cos\theta)} \tag{3.158}$$

and hence obtain

$$\alpha_\pm(E) = [(1 \pm \tfrac{3}{4}P^0)S_\pm(0,E) + (1 \mp \tfrac{3}{4}P^0)S_\pm(\pi,E)]/3P^0E \quad . \tag{3.159}$$

The resulting $\alpha_-(E)$ is found to be almost independent of electron energy E, and the Louvain group quoted their result at the maximum electron energy.

Example 3 [3.191]. The SIN group in Zurich followed almost the same procedure as LEBRUN et al. [3.194], except that conditions 1 and $\bar{1}$ were produced by the "spontaneous level mixing" technique. This offers a number of advantages in that: (a) there are no problems of RF saturation and field modulation, and (b) the characteristic mixing time for the substate populations at the critical magnetic field strength where the levels cross is 30 μs, i.e. negligible on the time scale of relaxation, ms. Steady-state operation thus becomes possible without excessive RF requirements, a fact which is statistically quite advantageous. The derived result for α_-, listed in Table 3.10, is obtained by forming, as before, the signal $S(\theta,E)$ (3.158). The SIN group's result is fully consistent with the Louvain value. A subsequent experiment along the same lines [3.195] measured $|\alpha_+|$.

Example 4 [3.171]. Pulsed beams of polarized ^{12}B and ^{12}N produced in (d,p) and (^{3}He,n) reactions were implanted in a thin Mg single crystal, with the $\underline{c}$ axis set as before parallel to the external magnetic field. Immediately after an on-beam period, an equalization RF was applied at resonance frequency ω_1 (condition 1). In the next step, a sweeping RF was applied through the resonance frequency ω_{-1} for the adiabatic fast passage to reverse the populations of the m = 0 and m = -1 substates (condition 2). This induced a positive alignment A(2) and a small (ideally zero) residual polarization P(2). The alignment was re-examined by again applying a sweeping RF through the resonance ω_{-1}, reproducing a polarization P(3), ideally the same as that obtained at condition 1, P(1). In the next beam burst the roles were reversed. An equalization RF first achieved condition $\bar{1}$, a sweeping RF through ω_1 produced negative alignment $A(\bar{2})$, and a further sweeping re-examined $P(\bar{3})$ ideally equal to $P(\bar{1})$.

The required coefficients $\alpha_\pm$ are obtained from the β-counting rates under conditions 2 and $\bar{2}$ by forming a signal

$$R(\theta,E) \equiv \frac{W(\theta,2)}{W(\theta,\bar{2})}$$

$$\simeq 1 + [A(2) - A(\bar{2})]\alpha_\pm EP_2(\cos\theta) \pm [P(2) - P(\bar{2})]\,(1 + \alpha_\pm E)P_1(\cos\theta) \quad . \tag{3.160}$$

Then on averaging

$$R(E) = \frac{1}{2}[R(0,E) + R(\pi,E)]$$

$$= 1 + [A(2) - A(\bar{2})]\alpha_{\pm}E \tag{3.161}$$

and plotting as a function of energy E, a straight-line fit determines the slope $[A(2) - A(\bar{2})]\alpha_{\pm}$. The alignments $A(2) - A(\bar{2})$ are determined by observing the changes in β-counting rates of the standard signal, $\mathscr{A}^{\pm}$, under conditions 1 and $\bar{1}$, or equivalently 3 and $\bar{3}$, viz.

$$A(2) - A(\bar{2}) = 3P^0 = 2[P(1) + P(\bar{1})] = 2[P(3) + P(\bar{3})] \quad . \tag{3.162}$$

Corrections are applied for attenuation due to the spin-lattice relaxation and for the efficiency of population inversion in the adiabatic fast passage. The resulting values of $\alpha_{\pm}$ are listed in Table 3.10. Note, for the case of ^{12}N, that the sign of α_+ is not determined, because the sign of the quadrupole coupling constant e^2qQ is not known. A subsequent experiment [3.196] determined the sign by measuring a β-γ correlation in aligned ^{12}N.

Table 3.10. β decay asymmetry coefficients, $\alpha_{\pm}$ [% MeV^{-1}]

$\alpha_-(^{12}B)$	$\alpha_+(^{12}N)$	Reference
+(0.31 ± 0.06)	-(0.21 ± 0.07)	SUGIMOTO et al. [3.193]
-(0.007 ± 0.020)		LEBRUN et al. [3.194]
+(0.024 ± 0.044)		BRÄNDLE et al. [3.191]
+(0.025 ± 0.034)	-(0.277 ± 0.052)	SUGIMOTO et al. [3.171]
+(0.010 ± 0.030)	-(0.273 ± 0.039)	BRÄNDLE et al. [3.195]
-(0.001 ± 0.021)	-(0.267 ± 0.056)	MASUDA et al. [3.196]
+(0.003 ± 0.012)	-(0.273 ± 0.027)	Average values[a]
+(0.276 ± 0.030)		$\alpha_- - \alpha_+$
-(0.270 ± 0.030)		$\alpha_- + \alpha_+$

[a]Excluding [3.193]; see text.

3.6.4 Discussion

We summarize in Table 3.10 the experimental values of the β decay asymmetry coefficients, $\alpha_{\pm}$, obtained from the Osaka group and the Louvain-SIN collaboration. In obtaining the averages, we omitted the measurement from the energy dependence of the polarization term [3.193].

The theoretical prediction for the difference in coefficients is, in the elementary-particle treatment (3.141),

$$\alpha_- - \alpha_+ = \frac{4}{3}\frac{1}{2m_p}\left(\frac{F_M}{F_A^{(I)}} + \frac{F_E^{(II)}}{F_A^{(I)}}\right) \tag{3.163}$$

where $F_A^{(I)}$ is the average axial-vector form factor in ^{12}N and ^{12}B decay, viz. $\frac{1}{2}(F_A^{(I)+} + F_A^{(I)-})$. If we assume the validity of CVC, a numerical value for F_M can immediately be determined. The transition rate for the γ decay $^{12}C^* \to {}^{12}C + \gamma$ is specified by

$$\Gamma(^{12}C^* \to {}^{12}C) = \frac{\alpha}{3}\frac{E_\gamma^3}{m_p^2}|\sqrt{2}\mu(0)|^2 \tag{3.164}$$

where α is the fine-structure constant and E_γ the photon energy, $E_\gamma = 15.10$ MeV. Using the experimental value [3.197] of $\Gamma = 37.0 \pm 1.1$ eV, we obtain [3.170]:

$$F_M^\pm(0) = \sqrt{2}\mu(0) = 1.97 \pm 0.02 \quad . \tag{3.165}$$

Similarly $F_A^{(I)}$ is derived from the experimental half-lives [3.198] $(t_{1/2})_-$ $= 21.02 \pm 0.06$ ms, $(t_{1/2})_+ = 11.63 \pm 0.04$ ms to be [3.170]:

$$F_A^{(I)+} = 0.488 \pm 0.001 \quad , \quad F_A^{(I)-} = 0.516 \pm 0.001 \quad ,$$

$$F_A^{(I)} = 0.502 \quad . \tag{3.166}$$

Thus if the second-class current term $F_E^{(II)}/F_A^{(I)}$ is ignored for the moment, we have a prediction for $\alpha_- - \alpha_+$ of

$$\alpha_- - \alpha_+ = +(0.279 \pm 0.003)\ \%\ \mathrm{MeV}^{-1} \tag{3.167}$$

in excellent agreement with experiment. Clearly there is no evidence for second-class currents if we accept the validity of CVC.

The same conclusion is drawn from impulse approximation for which

$$\alpha_- - \alpha_+ \simeq \frac{4}{3}\frac{1}{2m_p}\left(4.706\ x\ \frac{f_V}{f_A} - \frac{f_E^{(II)}}{f_A}\right) \tag{3.168}$$

where x is a ratio of nuclear matrix elements,

$$x = 1 + \frac{1}{4.706}\frac{\langle \underset{\sim}{L} \rangle}{\langle \underset{\sim}{\sigma} \rangle} \quad . \tag{3.169}$$

MORITA et al. [3.168] have calculated these matrix elements using shell-model wave functions, assuming the ground states of $^{12}B(^{12}N)$ and ^{12}C can be described by four-hole p-shell configurations in a closed-shell nucleus ^{16}O. The wave function for ^{12}N is assumed to be identical with that for ^{12}B, neglecting any nuclear charge differences. Harmonic-oscillator radial wave functions are assumed. The mixing between the possible four-hole configurations is specified in the effective interaction of COHEN and KURATH [3.199]. These authors have three variants which lead to values of $x = 0.947, 0.981, 0.996$. We adopt $x = 0.98 \pm 0.03$. Again neglecting second-class currents and taking from neutron decay a value $f_A/f_V = 1.2605 \pm 0.0075$ [3.200], we predict

$$\alpha_- - \alpha_+ = +(0.260 \pm 0.008)\ \%\ \text{MeV}^{-1} \tag{3.170}$$

in agreement with experiment.

Finally we turn to the prediction for the sum of coefficients, which in impulse approximation is

$$\alpha_- + \alpha_+ = -\frac{4}{3}\frac{1}{2m_p}y \quad ,$$

$$y = 1 + \frac{2\langle i\underset{\sim}{r}(\underset{\sim}{\sigma}\cdot\underset{\sim}{p})\rangle}{\langle\underset{\sim}{\sigma}\rangle} \quad . \tag{3.171}$$

MORITA et al. [3.168] have calculated the ratio of matrix elements using the three variants of the COHEN and KURATH wave functions: $y = 3.72, 3.55, 3.56$. We adopt 3.6 ± 0.1 and hence predict

$$\alpha_- + \alpha_+ = -(0.256 \pm 0.007)\ \%\ \text{MeV}^{-1} \tag{3.172}$$

in agreement with experiment. There is little evidence for a significant correction to impulse approximation arising from two-body meson exchange currents.

This result poses a dilemma. Meson exchange theories [3.184] predict a substantial 40% enhancement to the matrix element $\langle i\underset{\sim}{r}(\underset{\sim}{\sigma}\cdot\underset{\sim}{p})\rangle$ arising from the timelike part of the axial-vector current. The prediction is largely model independent, relying only on the assumption that one-pion exchange processes be dominant over other shorter-ranged processes such as multipion and heavier-meson exchanges, and that soft-pion theorems be applicable.

The dilemma was resolved recently [3.201,202], when it was realized that extending the shell-model space beyond that spanned by p-shell configurations gives a significant correction to the timelike matrix element contained in the quantity y. KOSHIGIRI et al. [3.201] evaluated this core-polarization effect using first-order perturbation theory and found that it depended principally on the tensor part of the residual interaction. Using the Hamada-Johnston interaction [3.150], they computed a correction to y of -1.2. It should be noted, in parenthesis, that first-order perturbation theory gives no correction to the matrix elements contained in x, so the discussion on $\alpha_- - \alpha_+$ is unaffected. Meson exchange currents, on the other hand, give a strong enhancement to y, adding $+1.3$ in the calculation of KOSHIGIRI et al. [3.201]. Thus the resultant value of y is $3.6 - 1.2 + 1.3 = 3.7$, giving

$$\alpha_- + \alpha_+ = -(0.263 \pm 0.007)\ \%\ \text{MeV}^{-1}$$

in good agreement with experiment; the core-polarization and meson exchange current effects largely cancel each other.

References

3.1 W. Pauli: Naturwissenschaften *12*, 741 (1924)
3.2 H. Schüler, T. Schmidt: Z. Phys. *94*, 457 (1935)
3.3 P. Brix, H. Kopfermann: Rev. Mod. Phys. *30*, 1189 (1958)
3.4 H. Frauenfelder, R.M. Steffen: *α-, β-, and γ-Ray Spectroscopy*, Vol.2, ed. by K. Siegbahn (North-Holland, Amsterdam 1965) pp.997-1198
3.5 H. Aeppli, A.S. Bishop, H. Frauenfelder, M. Walter, W. Zünti: Phys. Rev. *82*, 550 (1951)
3.6 N.F. Ramsey: *Molecular Beams* (Oxford Univ. Press, London 1955)
3.7 H. Kopfermann: *Nuclear Moments* (Academic, New York 1958)
3.8 A. Abragam: *Nuclear Magnetism* (Oxford Univ. Press, London 1961)
3.9 H.J. Rose, D.M. Brink: Rev. Mod. Phys. *39*, 306 (1967)
3.10 G.H. Fuller: J. Phys. Chem. Ref. Data *5*, 835 (1976)
3.11 J. Bauche: J. Phys. *35*, 19 (1974)
3.12 E.C. Seltzer: Phys. Rev. *188*, 1916 (1969)
3.13 K. Heilig, A. Steudel: At. Data Nucl. Data Tables *14*, 613 (1974)
3.14 F.A. Babushkin: Zh. Eksp. Teor. Fiz. *44*, 1661 (1963) [English transl.: Sov. Phys.-JETP *17*, 1118 (1963)]
3.15 P.L. Lee, F. Boehm: Phys. Rev. C *8*, 819 (1973)
3.16 A.R. Bodmer: Proc. Phys. Soc. A *66*, 1041 (1953); Nucl. Phys. *9*, 371 (1958/1959)
3.17 E.E. Fradkin: Zh. Eksp. Teor. Fiz. *42*, 787 (1962) [English transl.: Sov. Phys.-JETP *15*, 550 (1962)]
3.18 R.Y. Cusson, B. Castel: Can. J. Phys. *47*, 1189 (1969)
3.19 A.R. Edmonds: *Angular Momentum in Quantum Mechanics* (Univ. Press, Princeton 1957)
3.20 P.G.H. Sandars, J. Beck: Proc. R. Soc. (London) A *289*, 97 (1965)
3.21 I. Lindgren, A. Rosén: Case Stud. in At. Phys. *4*, 93,197 (1974)
3.22 R.M. Sternheimer: Phys. Rev. *80*, 102 (1950); *84*, 244 (1951); *95*, 736 (1954); *105*, 158 (1957); *164*, 10 (1967)
3.23 J. de Boer, J. Eichler: "The Reorientation Effect", in *Advances in Nuclear Physics*, ed. by M. Baranger, E. Vogt (Plenum, New York 1968) pp.1-65
3.24 O. Häusser: "Coulomb Reorientation", in *Nuclear Spectroscopy and Reactions*, ed. by J. Cerny (Academic, New York 1974) pp.55-91
3.25 W.D. Knight, R.R. Hewitt, M. Pomerantz: Phys. Rev. *104*, 271 (1956)
3.26 W.D. Hamilton: "Gamma-Ray Angular Distribution and Correlation Measurements", in *The Electromagnetic Interaction in Nuclear Spectroscopy*, ed. by W.D. Hamilton (North-Holland, Amsterdam 1975) Chap.14, pp.645-700
3.27 G.K. Wertheim: *Mössbauer Effect; Principles and Applications* (Academic, New York 1964)
3.28 R.M. Steffen, K. Alder: "Extranuclear Perturbations of Angular Distributions and Correlations", in *The Electromagnetic Interaction in Nuclear Spectroscopy*, ed. by W.D. Hamilton (North-Holland, Amsterdam 1975) Chap.13, pp.583-644
3.29 H.J. Behrend, D. Budnick: Z. Physik *168*, 155 (1962)
3.30 R.S. Raghavan, P. Raghavan, E.N. Kaufmann: Phys. Rev. Lett. *31*, 111 (1973)
3.31 L. Grodzins, O. Klepper: Phys. Rev. C *3*, 1019 (1971)
3.32 L. Darken, R.W. Gurry: *Physical Chemistry of Metals* (McGraw-Hill, New York 1953)
3.33 J. Pelzl: Z. Phys. *251*, 13 (1972)
3.34 F.W. DeWette: Phys. Rev. *123*, 103 (1961)
3.35 K.C. Das, M. Pomerantz: Phys. Rev. *123*, 2070 (1961)
3.36 K.C. Das, R.K. Ray: Phys. Rev. *187*, 777 (1969)
3.37 T.T. Taylor: Phys. Rev. *127*, 120 (1962)
3.38 F.D. Feiock, W.R. Johnson: Phys. Rev. *187*, 39 (1969)
3.39 R.E. Watson, A.C. Gossard, Y. Yafet: Phys. Rev. A *140*, 375 (1965)
3.40 E.N. Kaufmann, R.J. Vianden: Rev. Mod. Phys. *51*, 161 (1979)
3.41 P. Ebersold, B. Aas, W. Dey, R. Eichler, H.J. Leisi, W.W. Sapp, F. Scheck: Nucl. Phys. A *296*, 493 (1978)
3.42 F. Scheck: Nucl. Phys. B *42*, 573 (1972)

3.43 R.J. Powers, F. Boehm, P. Vogel, A. Zehnder, T. King, A.R. Kunselmann, P. Roberson, P. Martin, G.H. Miller, R.E. Welsh, D.A. Jenkins: Nucl. Phys. A *262*, 493 (1976)
3.44 J.D. Rodgers: Annu. Rev. Nucl. Sci. *15*, 241 (1965)
3.45 R.J. Powers, F. Boehm, A.A. Hahn, J.P. Miller, J.-L. Vuilleumier, K.C. Wang, A. Zehnder: Nucl. Phys. A *292*, 487 (1977)
3.46 K.E.G. Löbner, M. Vetter, V. Hönig: Nucl. Data Tables A *7*, 495 (1970)
3.47 B. Elbek: Thesis, University of Copenhagen (1963)
3.48 J. Ferch, W. Dankwort, H. Gebauer: Phys. Lett. A *49*, 287 (1974)
3.49 W. Dankwort, J. Ferch, H. Gebauer: Z. Phys. *267*, 229 (1974)
3.50 H. Figger, G. Wolber, S. Penselin: Phys. Lett. A *34*, 21 (1971)
3.51 S. Büttgenbach, G. Meisel: Z. Phys. *244*, 149 (1971)
3.52 D. McLoughlin, S. Raboy, E. Deci, D. Adler, R. Sutton, A. Thompson: Phys. Rev. C *13*, 1644 (1976)
3.53 R.J. Powers, F. Boehm, A. Zehnder, A.R. Kunselman, P. Roberson: Nucl. Phys. A *278*, 477 (1977)
3.54 R. Beetz, F.W.N. DeBoer, K. Fransson, J. Konijn, J.K. Panman, L. Tauscher, G. Tibell: Nucl. Phys. A *300*, 369 (1978)
3.55 C.J. Batty, S.F. Biagi, R.A.J. Riddle, A. Roberts, B.L. Roberts, D.H. Worledge, N. Berovis, G.J. Pyle, G.T.A. Squier, A.S. Clough, P. Coddington, R.E. Hawkins: 6th Int. Conf. High-Energy Physics and Nuclear Structure, Santa Fe and Los Alamos, June 1975
3.56 A. de Shalit, I. Talmi: *Nuclear Shell Theory* (Academic, New York 1963)
3.57 J. Blomqvist: J. Phys. Soc. Jpn. *34*, 223 (1973)
3.58 M. Harvey, F.C. Khanna: Nucl. Phys. A *152*, 588 (1970); A *155*, 337 (1970)
3.59 O. Häusser, T.K. Alexander, J.R. Beene, E.D. Earle, A.B. McDonald, F.C. Khanna, I.S. Towner: Nucl. Phys. A *273*, 253 (1976)
3.60 R.J. Powers, V.L. Telegdi: Z. Phys. *202*, 214 (1967);
R.J. Powers: Phys. Rev. *169*, 1 (1968)
3.61 W.Y. Lee, M.Y. Chen, S.C. Cheng, E.R. Macagno, A.M. Rushton, C.S. Wu: Nucl. Phys. A *181*, 14 (1972)
3.62 L. Holmgren, A. Rosen: Phys. Scr. *6*, 37 (1972)
3.63 H.A. Bethe, E.E. Salpeter: *Quantum Mechanics of One- and Two-Electron Atoms* (Springer, Berlin, Heidelberg, New York 1957)
3.64 H.M. Foley: In *Atomic Physics*, ed. by B. Bederson, V.W. Cohen, F.M.J. Pichanic (Plenum, New York 1969) pp.509-522;
H.H. Stroke: In *Atomic Physics*, ed. by B. Bederson, V.W. Cohen, F.M.J. Pichanic (Plenum, New York 1969) pp.523-550
3.65 D.L. Clark, M.E. Cage, G.W. Greenless: Hyperfine Interact. *4*, 83 (1978)
3.66 J. Christiansen, H.E. Mahnke, E. Recknagel, D. Riegel, G. Weyer, W. Witthuhn: Phys. Rev. Lett. *21*, 554 (1968)
3.67 E. Recknagel: "Magnetic Moments of Excited States", in *Nuclear Spectroscopy and Reactions, Part C*, ed. by J. Cerny (Academic, New York 1974) pp.93-141
3.68 C. Günther, I. Lindgren: "Paramagnetic Effects", in *Perturbed Angular Correlations*, ed. by E. Karlsson, E. Matthias, K. Siegbahn (North-Holland, Amsterdam 1964) pp.357-375
3.69 R. Kalish, U. Shreter, J. Grunzweig-Genossar: Hyperfine Interact. *1*, 65 (1975)
3.70 W.D. Knight: Phys. Rev. *76*, 1259 (1947); Solid State Phys. *2*, 97 (1956)
3.71 J.R. Beene, O. Häusser, A.B. McDonald, T.K. Alexander, A.J. Ferguson: Hyperfine Interact. *3*, 397 (1977)
3.72 M.S. Dewey, H.E. Mahnke, P. Choudhury, U. Garg, T.P. Sjoreen, D.B. Fossan: Phys. Rev. C *18*, 2061 (1978)
3.73 J. Korringa: Physica *16*, 601 (1950);
F.A. Rossini, W.D. Knight: Phys. Rev. *178*, 641 (1968)
3.74 D. Riegel: Phys. Scr. *11*, 228 (1975)
3.75 G.N. Rao: At. Data Nucl. Data Tables *15*, 553 (1975)
3.76 C. Kittel: *Introduction to Solid-State Physics* (Wiley, New York 1971)
3.77 R.R. Borchers, B. Herskind, J.D. Bronson, L. Grodzins, R. Kalish, D.E. Murnick: Phys. Rev. Lett. *20*, 424 (1968)
3.78 J. Lindhard, A. Winther: Nucl. Phys. A *166*, 413 (1971)
3.79 J.L. Eberhardt, G. van Middelkoop, R.E. Horstmann, H.A. Doubt: Phys. Lett. B *56*, 329 (1975)

3.80 M.B. Goldberg, G.J. Kumbartzki, K.H. Speidel, M. Forterre, J. Gerber: Hyperfine Interact. *1*, 429 (1976)
3.81 G. van Middelkoop: Hyperfine Interact. *4*, 238 (1977)
3.82 N. Benczer-Koller, M. Hass, J. Sak: Annu. Rev. Nucl. Part. Sci. *30*, 53 (1980)
3.83 J.L. Eberhardt, R.E. Horstman, P.C. Zalm, H.A. Doubt, G. van Middelkoop: Hyperfine Interact. *3*, 195 (1977)
3.84 K. Dybdal, J.L. Eberhardt, N. Rud: Phys. Rev. Lett. *42*, 592 (1979)
3.85 K. Dybdal, J.S. Forster, N. Rud: Phys. Rev. Lett. *43*, 1711 (1979)
3.86 O. Häusser, D. Ward, H.R. Andrews, N. Rud, P. Skensved, C. Broude: Nucl. Instrum. Methods *169*, 539 (1980)
3.87 D.E. Murnick, M.S. Feld: Annu. Rev. Nucl. Part. Sci. *29*, 411 (1979)
3.88 G. Huber, F. Touchard, S. Büttgenbach, C. Thibault, R. Klapisch, H.T. Duong, S. Liberman, J. Pinard, J.L. Vialle, P. Juncar, P. Jacquinot: Phys. Rev. C *18*, 2342 (1978)
3.89 X. Campi, H. Flocard, A.K. Kerman, S. Koonin: Nucl. Phys. A *151*, 193 (1975)
3.90 T. Kühl, P. Dabkiewicz, C. Duke, H. Fischer, H.-J. Kluge, H. Kremmling, E.W. Otten: Phys. Rev. Lett. *39*, 180 (1977)
3.91 G. Huber, F. Touchard, S. Büttgenbach, C. Thibault, R. Klapisch, S. Liberman, J. Pinard, H.T. Duong, P. Juncar, J.L. Vialle, P. Jacquinot, A. Pesnelle: Phys. Rev. Lett. *41*, 459 (1978)
3.92 C.E. Bemis, J.R. Beene, J.P. Young, S.D. Kramer: Phys. Rev. Lett. *43*, 1854 (1979)
3.93 H.C. Britt: At. Data Nucl. Data Tables *12*, 407 (1973)
3.94 F.S. Stephens: Rev. Mod. Phys. *47*, 43 (1975)
3.95 D. Horn, O. Häusser, T. Faestermann, A.B. McDonald, T.K. Alexander, J.R. Beene, C.J. Herrlander: Phys. Rev. Lett. *39*, 389 (1977)
3.96 T.L. Khoo, R.K. Smither, B. Haas, O. Häusser, H.R. Andrews, D. Horn, D. Ward: Phys. Rev. Lett. *41*, 1027 (1978)
3.97 C. Baktash, E. ter Mateosian, O.C. Kistner, A.W. Sunyar: Phys. Rev. Lett. *42*, 637 (1979)
3.98 T.L. Khoo, F.M. Bernthal, R.G.H. Robertson, R.A. Warner: Phys. Rev. Lett. *37*, 823 (1976)
3.99 J.P. Schiffer, W.W. True: Rev. Mod. Phys. *48*, 191 (1976)
3.100 A. Faessler, M. Ploszajczak, K.R.S. Devi: Phys. Rev. Lett. *36*, 1028 (1976)
3.101 C.G. Andersson, G. Hellström, G. Leander, I. Ragnarsson, S. Aberg, J. Krumlinde, S.G. Nilsson, Z. Szymanski: Nucl. Phys. A *309*, 141 (1978)
3.102 K. Matsuyanagi, T. Dossing, K. Neergaard: Nucl. Phys. A *307*, 253 (1978)
3.103 O. Häusser: Proc. Symposium on High Spin Phenomena in Nuclei, Argonne (1979)
3.104 O. Häusser, H.E. Mahnke, J.F. Sharpey-Schafer, M.L. Swanson, P. Taras, D. Ward, H.R. Andrews, T.K. Alexander: Phys. Rev. Lett. *44*, 132 (1980)
3.105 W. Klinger, R. Böhm, W. Engel, W. Sandner, R. Seeböck, W. Witthuhn: Z. Phys. A *290*, 227 (1979)
3.106 O. Häusser, P. Taras, W. Trautmann, D. Ward, T.K. Alexander, H.R. Andrews, B. Haas, D. Horn: Phys. Rev. Lett. *42*, 1451 (1979)
3.107 T. Døssing, K. Neergaard, H. Sagawa: Int. Conf. Nuclear Behaviour at High Angular Momentum, Strasbourg (1980); Phys. Lett. B *99*, 191 (1981)
3.108 M. Sano, T. Takemasa, W. Wakai: J. Phys. Soc. Jpn. Suppl. *34*, 365 (1975)
3.109 R. Kalish, B. Herskind, G.B. Hagemann: Phys. Rev. C *8*, 757 (1973)
3.110 B. Skaali, R. Kalish, J. Eriksen, B. Herskind: Nucl. Phys. A *238*, 159 (1975)
3.111 D. Ward, O. Häusser, H.R. Andrews, P. Taras, P. Skensved, N. Rud, C. Broude: Nucl. Phys. A *330*, 225 (1979)
3.112 H.R. Andrews, O. Häusser, D. Ward, P. Taras, N. Rud, B. Haas, R.M. Diamond, D. Fossan, H. Kluge, M. Nieman, C. Roulet, F.S. Stephens: Phys. Rev. Lett. *45*, 1835 (1980)
3.113 P. Taras, D. Ward, H.R. Andrews, J.S. Geiger, R.L. Graham, W. McLatchie: Nucl. Phys. A *289*, 165 (1977)
3.114 D. Schwalm: Private communication
3.115 A. Bohr, B.R. Mottelson: *Nuclear Structure*, Vol.1 (Benjamin, New York 1969)
3.116 R. Frisch, O. Stern: Z. Phys. *85*, 4 (1933)
3.117 R.H. Dalitz: Phys. Rev. *95*, 799 (1954)
3.118 A. Arima, H. Hyuga: In *Mesons in Nuclei*, ed. by M. Rho, D.H. Wilkinson (North-Holland, Amsterdam 1969) pp.683-720

3.119 H. Hyuga, A. Arima, K. Shimizu: Nucl. Phys. A *336*, 363 (1980)
3.120 J.L. Fiar: Int. Conf. on Nuclear Physics with Electromagnetic Interactions, Mainz 1979
3.121 M. Chemtob: Nucl. Phys. A *123*, 449 (1969)
3.122 M. Chemtob, M. Rho: Nucl. Phys. A *163*, 1 (1971)
3.123 G. Konopka, M. Gari, J.G. Zabolitzky: Nucl. Phys. A *290*, 360 (1977)
3.124 A. Arima: J. Phys. Soc. Jpn. Suppl. *34*, 205 (1973)
3.125 T. Yamazaki: In *Mesons in Nuclei*, ed. by M. Rho, D.H. Wilkinson (North-Holland, Amsterdam 1969) pp.651-682
3.126 A. Arima, H. Horie: Prog. Theor. Phys. *12*, 623 (1954)
3.127 J. Blomqvist, N. Freed, H.O. Zetterstrom: Phys. Lett. *18*, 47 (1965)
3.128 H.A. Mavromatis, L. Zamick, G.E. Brown: Nucl. Phys. *80*, 545 (1966)
3.129 A. Arima, L.J. Huang-Lin: Phys. Lett. B *41*, 435 (1972)
3.130 I.S. Towner, F.C. Khanna, O. Häusser: Nucl. Phys. A *277*, 285 (1977)
3.131 I. Hamamoto: Nucl. Phys. A *277*, 285 (1977)
3.132 R. Bauer, J. Speth, V. Klemt, P. Ring, E. Werner, T. Yamazaki: Nucl. Phys. A *209*, 535 (1973)
3.133 T. Yamazaki, G.T. Ewan: Phys. Lett. B *24*, 278 (1967)
3.134 T. Yamazaki, T. Nomura, S. Nagamiya, T. Katou: Phys. Rev. Lett. *25*, 547 (1970)
3.135 T. Yamazaki, E. Matthias: Phys. Rev. *175*, 1476 (1968)
3.136 C.V.K. Baba, D.B. Fossan, T. Faestermann, F. Feilitzsch, M.R. Maier, P. Raghavan, R.S. Raghavan, C. Signorini: Nucl. Phys. A *257*, 135 (1976)
3.137 O. Häusser, I.S. Towner, T. Faestermann, H.R. Andrews, J.R. Beene, D. Horn, D. Ward, C. Broude: Nucl. Phys. A *293*, 248 (1977)
3.138 H. Ejiri, T. Shibata, M. Takeda: Nucl. Phys. A *221*, 211 (1974)
3.139 W. Klinger, R. Böhm, W. Engel, W. Sandner, R. Seeböck, W. Witthuhn: Z. Phys. A *290*, 227 (1979)
3.140 R. Brenn, S.K. Bhattacherjee, G.D. Sprouse, L.E. Young: Phys. Rev. C *10*, 1414 (1974)
3.141 D. Riegel, N. Bräuer, F. Dimmling, B. Focke, K. Nishiyama: Phys. Lett. B *46*, 170 (1973)
3.142 F. Dimmling, N. Bräuer, B. Focke, T. Kornrumpf, K. Nishiyama, D. Riegel: Z. Phys. *271*, 103 (1974)
3.143 N. Bräuer, B. Focke, B. Lehmann, E. Matthias, D. Riegel: Phys. Lett. B *34*, 54 (1971)
3.144 D.F. Gumprecht, T.E. Katila, L.C. Moberg: Phys. Lett. A *40*, 297 (1972)
3.145 M. Ishibara, R. Broda, B. Herskind: Proc. Int. Conf. on Nuclear Physics, Munich 1973, ed. by J. de Boer, H.J. Mang (North-Holland, Amsterdam 1973) Vol.1, p.256
3.146 O. Häusser, J.R. Beene, T. Faestermann, T.K. Alexander, D. Horn, A.B. McDonald, A.J. Ferguson: Hyperfine Interact. *4*, 219 (1978)
3.147 K. Arita: Proc. Int. Conf. on Nuclear Physics, Munich 1973, ed. by J. De Boer, H.J. Mang (North-Holland, Amsterdam 1973) Vol.1, p.264
3.148 S.M. Perez: Phys. Lett. B *33*, 317 (1970)
3.149 V. Gillet, A.M. Green, E.A. Sanderson: Phys. Lett. *11*, 44 (1964)
3.150 T. Hamada, I.D. Johnston: Nucl. Phys. *34*, 383 (1962)
3.151 M. Hass, N. Benczer-Koller, J.M. Brennan, H.T. King, P. Goode: Phys. Rev. C *17*, 997 (1978)
3.152 P. Goode, L. Zamick: Part. Nucl. *3*, 125 (1972)
3.153 A. Arima: Proc. of the Fourth Int. Conf. on Hyperfine Interactions, Madison, 1977, ed. by R.S. Raghavan, D.E. Murnick (North-Holland, Amsterdam 1978)
3.154 M. Ichimura, K. Yazaki: Nucl. Phys. *63*, 401 (1965)
3.155 H.A. Mavromatis, L. Zamick: Nucl. Phys. A *104*, 17 (1967)
3.156 G.F. Bertsch: Phys. Lett. B *28*, 302 (1968)
3.157 K. Shimizu, M. Ichimura, A. Arima: Nucl. Phys. A *226*, 282 (1974)
3.158 T.T.S. Kuo: Nucl. Phys. A *103*, 71 (1967)
3.159 T.T.S. Kuo, G.E. Brown: Nucl. Phys. A *114*, 241 (1968)
3.160 T. Erikson: Nucl. Phys. A *205*, 593 (1973)
3.161 O. Häusser, T.K. Alexander, T. Faestermann, D. Horn, D. Ward, H.R. Andrews, I.S. Towner: Phys. Lett. B *73*, 127 (1978)

3.162 L. Meyer-Schützmeister, A.J. Elwyn, K.E. Rehm, G. Hardie: Phys. Rev. C *17*, 1299 (1978)
3.163 T. Minamisono, J.W. Hugg, D.G. Mavis, T.K. Saylor, H.F. Glavish, S.S. Hanna: Phys. Lett. B *61*, 155 (1976)
3.164 C.W. Kim, H. Primakoff: Phys. Rev. B *139*, 1447 (1965)
3.165 W. Dreschler, B. Stech: Z. Phys. *178*, 1 (1964)
3.166 A. Fujii, Y. Yamaguchi: Prog. Theor. Phys. *31*, 107 (1964)
3.167 B.R. Holstein: Rev. Mod. Phys. *46*, 789 (1974)
3.168 M. Morita, M. Nishimura, A. Shimizu, H. Ohtsubo, K. Kubodera: Prog. Theor. Phys. Suppl. *60*, 1 (1976)
3.169 C.W. Kim, H. Primakoff: In *Mesons in Nuclei*, ed. by M. Rho, D.H. Wilkinson (North-Holland, Amsterdam 1979)
3.170 W.Y.P. Hwang, H. Primakoff: Phys. Rev. C *16*, 397 (1977)
3.171 K. Sugimoto, T. Minamisono, Y. Nojiri, Y. Masuda: J. Phys. Soc. Jpn. Suppl. *44*, 801 (1978)
3.172 S. Weinberg: Phys. Rev. *112*, 1375 (1958)
3.173 I.S. Towner, J.C. Hardy, M. Harvey: Nucl. Phys. A *284*, 269 (1977)
3.174 D.H. Wilkinson: Phys. Lett. B *31*, 447 (1970)
3.175 D.H. Wilkinson, D.E. Alburger: Phys. Rev. Lett. *26*, 1127 (1971)
3.176 D.H. Wilkinson: Phys. Rev. Lett. *27*, 1018 (1971)
3.177 I.S. Towner: Nucl. Phys. A *216*, 589 (1973)
3.178 Y.K. Lee, L. Mo, C.S. Wu: Phys. Rev. Lett. *10*, 253 (1963)
3.179 F.P. Calaprice, B.R. Holstein: Nucl. Phys. A *273*, 301 (1976)
3.180 C.W. Wu, Y.K. Lee, L. Mo: Phys. Rev. Lett. *39*, 72 (1977)
3.181 W. Kaina, V. Soergel, H. Thies, W. Trost: Phys. Lett. B *70*, 411 (1977)
3.182 K. Koshigiri, M. Nishimura, H. Ohtsubo, M. Morita: Nucl. Phys. A *319*, 301 (1979)
3.183 D.O. Riska, G.E. Brown: Phys. Lett. B *38*, 193 (1972)
3.184 K. Kubodera, J. Delorme, M. Rho: Phys. Rev. Lett. *40*, 755 (1978)
3.185 M. Morita, M. Nishimura, H. Ohtsubo: Phys. Lett. B *73*, 17 (1978)
3.186 T. Minamisono: J. Phys. Soc. Jpn. Suppl. *34*, 324 (1973)
3.187 R.C. Haskell, L. Madansky: J. Phys. Soc. Jpn. Suppl. *34*, 167 (1973)
3.188 R.C. Haskell, F.D. Correll, L. Madansky: Phys. Rev. B *11*, 3268 (1975)
3.189 C.P. Slichter: *Principles of Magnetic Resonance*, 2nd ed., Springer Ser. Solid-State Sci.,Vol.1 (Springer, Berlin, Heidelberg, New York 1980)
3.190 I. Tanihata, S. Kogo, K. Sugimoto: Phys. Lett. B *67*, 392 (1977)
3.191 H. Brändle, L. Grenacs, J. Lang, L.Ph. Roesch, V.L. Telegdi, P. Truttmann, A. Weiss, A. Zehnder: Phys. Rev. Lett. *40*, 306 (1978)
3.192 A. Abragam, B. Bleaney: *Electron Paramagnetic Resonance of Transition Ions* (Clarendon Press, Oxford 1970)
3.193 K. Sugimoto, I. Tanihata, J. Goring: Phys. Rev. Lett. *34*, 1533 (1975)
3.194 P. Lebrun, Ph. Deschepper, L. Grenacs, J. Lehmann, C. Leroy, L. Palffy, A. Possoz, A. Maio: Phys. Rev. Lett. *40*, 302 (1978)
3.195 H. Brändle, G. Miklos, L.Ph. Roesch, V.L. Telegdi, P. Truttmann, A. Zehnder, L. Grenacs, P. Lebrun, J. Lehmann: Phys. Rev. Lett. *41*, 299 (1978)
3.196 Y. Masuda, T. Minamisono, Y. Nojiri, K. Sugimoto: Phys. Rev. Lett. *43*, 1083 (1979)
3.197 B.T. Chertok, C. Sheffield, J.W. Lightbody, S. Penner, D. Blum: Phys. Rev. C *8*, 23 (1973)
3.198 F. Ajzenberg-Selove: Nucl. Phys. A *248*, 1 (1975)
3.199 S. Cohen, D. Kurath: Nucl. Phys. *73*, 1 (1965)
3.200 D.H. Wilkinson: Nucl. Phys. A *377*, 474 (1982)
3.201 K. Koshigiri, H. Ohtsubo, M. Morita: Prog. Theor. Phys. *66*, 358 (1981)
3.202 P.A.M. Guichon, C. Samour: Nucl. Phys. A *382*, 461 (1982)

4. Hyperfine Interactions of Defects in Metals

E. Recknagel, G. Schatz, and Th. Wichert

With 47 Figures

Hyperfine methods are well established for investigations of solid state properties. There are fields of application, which most recently have developed themselves rather impetously, among them the subject of defects in metals. Information on defects have so far been extracted from macroscopic physical properties like electric resistivity or elastic modulus; in contrast to this the hyperfine methods permit a microscopic glance onto the defect behaviour. Structural and dynamical defect aspects originating from the close vicinity to the radioactive probe atoms as well as the identification of defects on an atomic scale are the main thrust of hyperfine investigations. Originally a handicap for the study of hyperfine interactions following ion implantation or nuclear reactions, the influence of lattice defects became more and more the subject of investigations using hyperfine methods. During the last few years, the method has proved to be a new sensitive tool for defect studies as indicated by the recent proceedings of the international conferences on hyperfine interactions (Uppsala, Sweden, 1974 [4.1]; Leuven, Belgium 1975 [4.2]; Madison, USA, 1977 [4.3]; Berlin, Germany, 1980 [4.4]). Though the defect characteristic hyperfine interaction can be measured by several techniques, the main body of results is obtained by the nuclear techniques like the Mössbauer effect, the perturbed angular correlation and perturbed angular distribution technique.

This article, therefore, shall try to present a review of the great amount of experiments done by these techniques and it will also summarize the state of interpretation. For readers coming from the hyperfine interaction field, a short introduction to the current knowledge of defect physics is presented (Sect.4.1.1). The necessary elements of hyperfine physics which are needed for the understanding of the experimental facts are discussed in Sect.4.1.2; more details can be found in other contributions to this book. The main part – the discussion of defect investigations with radioactive probe atoms – is separated into experiments with radioactive sources (Sect.4.2.1) and nuclear reaction experiments (Sect.4.2.2). There the great body of existing information collected by conventional methods can only be touched on in direct connection with the microscopic result; excellent reviews and recent results of macroscopic data are available for more detailed instructions [4.5-7]. An attempt was made to include all relevant references published before June 1981. During the publication process these were then updated.

4.1 Relevant Solid State and Nuclear Physics Aspects

4.1.1 Defects in Metals

a) *Defects After Irradiation*

Radiation penetrating in a crystal will interact with the constituents of the matter, i.e., atomic nuclei, atomic electrons and, in the case of a conducting crystal, also conduction electrons. The fate of a projectile in a crystal is governed by the slowing down process from the initial energy to rest. For different projectiles such as electrons, protons and heavy ions, neutrons and γ-quanta, the defect arrangement is different due to the specific mechanism of energy loss of these particles.

In metals only the nuclear collision effect contributes to defect production. If enough energy is transferred to an atom by the primary projectile it will be displaced from the equilibrium position forming an interstitial-vacancy pair (Frenkel pair). The resulting Frenkel pairs may annihilate spontaneously if the distance between the two defects is close. Only if the interstitial is outside a certain recombination volume, which includes approximately 100 atoms, the Frenkel pair remains stable and can be annealed by thermal activated migration of the defect. In metals, the displacement energy E_d, which is the energy for stable Frenkel pair production, is between 20 and 50 eV and depends rather strongly on the collision direction with respect to the crystal lattice [4.8,9]. For instance, in copper, the displacement energy averaged over all crystal directions is $E_d(Cu) = 29$ eV [4.9]. In many cases the primary interstitial will further eject secondary atoms from their sites in a cascade of atomic collisions if the primary recoil energy $\bar{E}$ exceeds the displacement energy. The average number N_d of displaced atoms per primary knock-on was roughly estimated by KINCHIN and PEASE [4.10]:

$$N_d = \bar{E}/2\,E_d \quad , \quad (\bar{E} > 2\,E_d) \quad . \tag{4.1}$$

At higher defect concentrations, besides the simple Frenkel defects, a variety of defect configurations, such as divacancies, diinterstitials and clusters with diameters up to some 100 Å become possible through point defect agglomeration.

Irradiation with electrons. Electrons displace atoms by Coulomb interaction with the nuclei. In order to transfer the necessary displacement energy for a primary knock-on, electrons with relatively high energies are needed. For copper with a displacement energy $E_d \approx 30$ eV, a minimum electron energy of 0.55 MeV is required (see Fig.4.1). On the basis of unscreened Coulomb potentials, the cross section

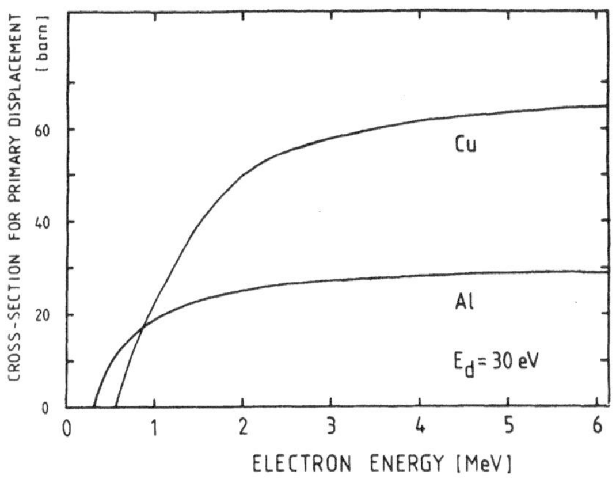

Fig.4.1. Cross section for primary displacement of atoms by electrons as a function of electron energy for aluminum and copper [4.12]

for primary knock-ons can be calculated for relativistic projectiles [4.11,12]. Numerical results are shown in Fig.4.1 for aluminum and copper. In many cases the transferred energy in a head-on collision is not more than a few times E_d (for 3 MeV electrons E_{max} = 405 eV in copper) so that the total number of displaced atoms is not much greater than the number of primary knock-ons. Electrons are therefore well suited for the production of isolated Frenkel pairs with a well-defined concentration.

Irradiation with neutrons. Defect production by neutron irradiation is accomplished mostly through hard-sphere type collisions. Neutrons with energies above a few hundred eV are able to transfer enough energy to a host atom in order to displace it. For fast neutrons ($E_n \approx 10$ MeV), which are most conveniently produced by fission, the primary recoil energy greatly exceeds the displacement energy and therefore initiates a collision cascade.

Part of the originally produced vacancies and interstitials may recombine spontaneously in the collision cascade region. Interstitials are created at some distance from the point of original impact because of a series of replacement collisions. Vacancies do not propagate in this way and are therefore left in the neighbourhood of the point of original impact. Production of a variety of vacancy clusters is possible in the collision cascade core. Although the relative local density of defects around a primary knock-on is rather high ($\sim 10^{-2}$), the relative concentration of displaced atoms averaged over a sample is only 10^{-20} - 10^{-21} because of the long range of neutrons in materials [4.13] (Fig.4.2).

Neutron scattering experiments are often accompanied by nuclear reaction effects. Thermal neutrons ($E_n \approx 0.1$ eV) are likely to react via neutron capture, which is always exothermic, leading to de-excitation γ-rays. These γ-quanta transfer a recoil to the compound nuclei, usually amounting to less than a few 100 eV. With this recoil energy from neutron capture only few Frenkel pairs can be produced, a situation quite similar to electron irradiation.

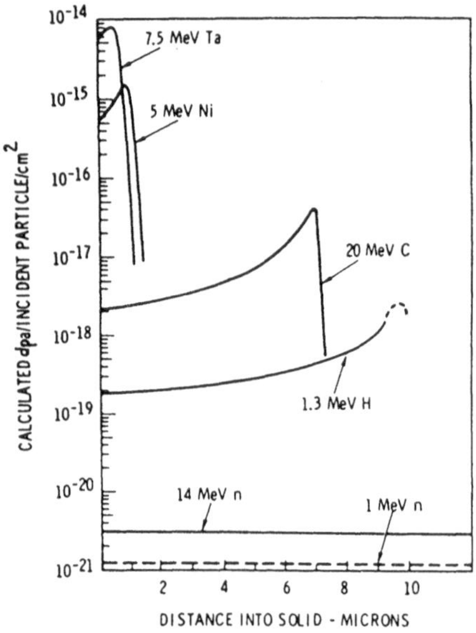

Fig.4.2. Displacement effectiveness for neutrons and different ions as a function of irradiation depth [4.13]

Irradiation with protons or heavy ions. Protons and heavy ions are, even at modest energies, able to displace atoms in a solid because of their mass. Since these projectiles interact with the host atoms by Coulomb forces, the cross section to produce a primary knock-on is much higher than for neutrons. Each primary knock-on produces a collision cascade, similar to the one induced for neutron irradiation. However, because of the increased displacement cross-section, the average distance between primary recoils is much smaller and for heavy ions the collision cascades may overlap because of the large stopping power. Mass and energy of the projectile determine the nature of the primary recoil spectrum.

Since the stopping power of charged heavy particles is rather high, the defects are concentrated in a thin layer of the irradiated material [4.14]. A typical damage depth profile together with the In-impurity distribution is shown in Fig. 4.3 for In in copper [4.15]. It is important to note that in this case the damage profile peaks closer to the surface than the impurity distribution. If the impurities act as probes for defect detection, as they do for microscopic methods, this may result in basically two fractions of impurities: the first one without defects in the close neighbourhood, the other one with high defect concentration around the impurities [4.16]. Figure 4.2 displays the different displacement effectiveness for different ions; neutrons are also included for comparison as a function of the irradiation depth.

Irradiation with γ-rays. γ-Quanta interact with solids by photo-electric effect, Compton effect or pair production. By these effects electrons with certain kinetic energies are released and are able to produce defects if their energy is higher

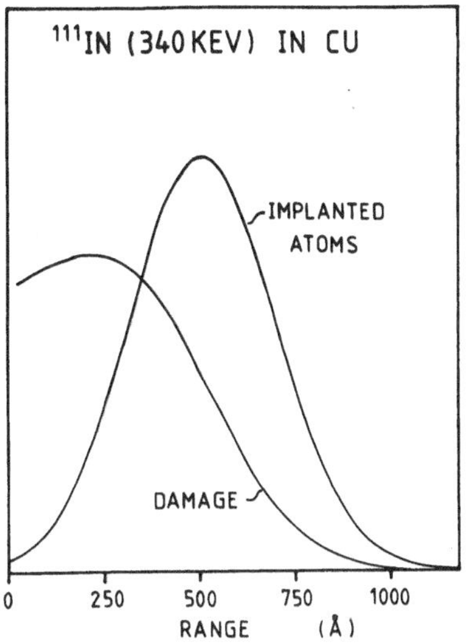

Fig.4.3. Damage depth profile and In-impurity distribution for 340 keV In-ions implanted into copper [4.15]

than the threshold energy for defect production. In copper irradiated with 1 MeV γ-rays, for instance, the Compton effect is most important, therefore the atoms can be displaced by recoil Compton electrons within the range of energy 0.5 MeV $\lesssim E_{Compt.electr.} \lesssim$ 0.8 MeV. For 0.7 MeV γ-rays, no defect could be created in copper because the maximum Compton-electron energy is below the displacement threshold energy.

b) *Defects in Thermal Equilibrium*

The thermal energy contained by a crystal is distributed over the lattice atoms as lattice vibrations. There is always a finite probability that sufficient vibrational energy is concentrated onto an atom to create a lattice defect. Only for a certain defect configuration a minimum of the appropriate thermodynamic energy function is reached and this governs the type of the most probable defect created. In order to introduce a defect in the lattice, the energy of formation E^F is needed; the concentration of these defects is given by thermodynamical considerations [4.11,17] as (k_B Boltzmann's constant)

$$C = C_0 \cdot \exp\left(-\frac{E^F}{k_B T}\right) \quad , \tag{4.2}$$

with the entropy factor $C_0 = \exp(\Delta S^F/k_B)$ (ΔS^F: formation entropy), which generally has values in the order of unity. If one assumes a defect with E^F = 1 eV and another with E^F = 5 eV, one finds at a temperature of 1000 K concentrations of 10^{-5} and 10^{-26}, respectively. This demonstrates that defects with E^F larger than a few eV can generally be neglected.

Table 4.1. Monovacancy formation energies E^F_{1v} obtained from quenching and positron annihilation experiments

Metal	Crystal structure	Monovacancy formation energy E^F_{1v} [eV] (Quenching)		(Positron annihilation)	
Al		0.66	[4.20]	0.66	[4.21]
Ni		1.60	[4.22]	-	
Cu		1.27	[4.23]	1.24	[4.24]
Ag	fcc	1.10	[4.25]	1.16	[4.26]
Pt		1.51	[4.27]	-	
Au		0.94	[4.28]	0.97	[4.28]
Pb		0.54	[4.29]	0.54	[4.30]
α-Fe		>1.55	[4.31]	1.5	[4.32]
Nb		>2.7	[4.33]	2.0	[4.32]
Mo	bcc	3.2	[4.34]	-	
Ta		-		2.2	[4.35]
W		3.9	[4.36]	3.5	[4.32]
Zn	hcp	-		0.53	[4.37]
Cd		0.42	[4.38]	0.47	[4.39]

Calculations of E^F for point defects in metals using the elastic continuum model, for instance [4.11], yield $E^F_{1v} \simeq 1$ eV for monovacancies and $E^F_{1i} = 4 - 5$ eV for interstitials depending on the interstitial configuration. This clearly shows that in thermal equilibrium, vacancies and interstitials are not produced at the same time, only vacancies originating from grain boundaries are existent in the metal even close to the melting point.

The formation energy of vacancies can be measured quite accurately through their concentration in thermal equilibrium. In positron annihilation experiments [4.18], for example, the average momentum density of the electrons which annihilate with positrons as a function of temperature is measured (s-shape curves) yielding the fraction of positrons trapped at vacancy sites. In quenching experiments, the equilibrium concentration of vacancies at a given temperature can be frozen in, if the sample is rapidly cooled down to temperatures, where the defects are immobile. For this a cooling rate of at least 10^4K/s is needed, otherwise a substantial part of the vacancies is lost during the cooling down process. The concentration of vacancy-type defects introduced can be determined from resistivity measurements because the resistivity increment $\Delta\rho$ is proportional to the defect concentration [4.19]:

$$\Delta\rho \propto \exp(-E^F/k_BT) \quad . \qquad (4.3)$$

Hence, variation of the sample temperature T before quenching allows the evaluation of the vacancy formation energy. Although only vacancies as defects have to be taken into account in quenching experiments, the conclusions are often hampered because of the vacancy losses during cooling and the introduction of impurity atoms from the cooling medium. In Table 4.1, experimental results for vacancy-formation energies obtained by quenching and positron annihilation are listed for several metals.

c) *Vacancy and Interstitial Configurations*

Vacancies, as they are created, are simply missing atoms on lattice positions. Interstitials on the other hand are atoms squeezed into the space between occupied lattice positions. Vacancies and, more important, interstitials exert a displacement field on their near neighbours, usually leading to a relaxation of the lattice around the defect. These point defect configurations, which may include more than single vacancies or interstitials, are of great interest. They exhibit a certain symmetry, which is one of the facts accessible to hyperfine methods.

Simple vacancy configurations are schematically shown in Fig.4.4 for the bcc and fcc lattices. The displayed configurations, being the most stable ones, are supported by theoretical considerations as well as experimental results [4.40]. Interstitial configurations are subject to strong relaxations; here the most probable configuration in the fcc lattice is the <100>-dumbbell and in the bcc lattice the <110>-dumbbell (Fig.4.5). Experimentally, these dumbbell configurations have been verified, for instance, by the diffuse scattering techniques for Al, Cu and Mo [4.41].

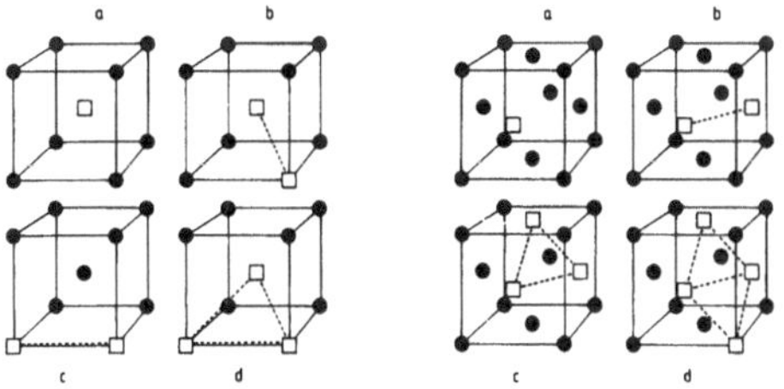

Fig.4.4. Simple vacancy configurations in cubic lattices: bcc (left side), fcc (right side) [4.40]

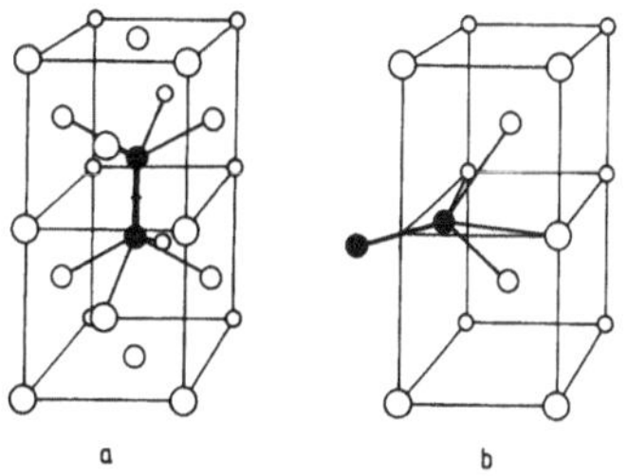

Fig.4.5. Most probable interstitial configurations in (a) fcc-lattice (<100>-dumbbell) and (b) bcc-lattice (<110>-dumbbell) [4.40]

d) *Migration of Defects*

The dynamical behaviour of lattice defects, besides their production, is an important aspect of defect studies. Defect mobility is responsible for the annealing of defects by their "anti"-defects and for the creation and growth of clusters.

In order to change the position of a defect, it has to overcome a potential barrier. The theoretical treatment of this problem is rather involved; most commonly it leads to a defect jump rate of the form [4.42]

$$\nu = \nu_0 \cdot \exp(-E^M/k_B T) \qquad (4.4)$$

Table 4.2. Monovacancy migration energies E_{1v}^{M} obtained from quenching experiments. For comparison, migration energies deduced from stage III after irradiation are listed

Metal	Crystal structure	Monovacancy migration energy E_{1v}^{M} [eV] (Quenching)		E_{III}^{M} [eV] (Irradiations)	
Al		0.65	[4.25]	0.59	[4.25]
Ni		1.3	[4.22]	1.03	[4.43]
Cu	fcc	0.72	[4.25]	0.70	[4.25]
Ag		-		0.66	[4.25]
Pt		1.42	[4.25]	1.43	[4.25]
Au		0.69-0.90	[4.43]	0.70-0.85	[4.43]
Pb		0.54	[4.29]	0.43	[4.44]
α-Fe		<1.0	[4.31]	0.5	[4.45]
Nb		<0.9	[4.33]	0.55	[4.46]
Mo	bcc	1.35	[4.34]	1.29	[4.25]
Ta		-		0.7	[4.47]
W		1.8	[4.36]	1.71	[4.48]
Zn	hcp	0.44	[4.49]	0.46	[4.50]
Cd		0.38	[4.38]	0.38	[4.51]

with ν_0 an attempt frequency, which is a typical phonon frequency in the order of 10^{13} Hz. E^M is the migration energy which represents the energy to pass the saddle point of the potential barrier.

Monovacancy migration energies can be derived from annealing experiments, for instance, after quenching. There are difficulties in the interpretation arising from simultaneous annealing of several mobil defects or due to defect-impurity interactions. Here, the self-diffusion process, in which the lattice atoms migrate via vacancies, is of great help in attaining reasonable self-consistency between the activation energy for self-diffusion $Q_{1v} = E_{1v}^{M} + E_{1v}^{F}$ and the separately determined terms. Results for migration energies of monovacancies obtained after quenching are collected in Table 4.2 for some metals. For comparison, migration energies deduced from resistivity recovery in stage III (see Fig.4.6) after irradiation are also listed.

Migration energies for self-interstitials in metals can only be studied via annealing of Frenkel pairs since self-interstitials alone cannot be introduced. Annealing of defects produced by irradiation at liquid helium temperatures proceeds through many annealing stages and makes the determination of the interstitial mobility somewhat difficult. A compilation of migration energies of interstitial defects, derived from annealing in stage I, is given in Table 4.3.

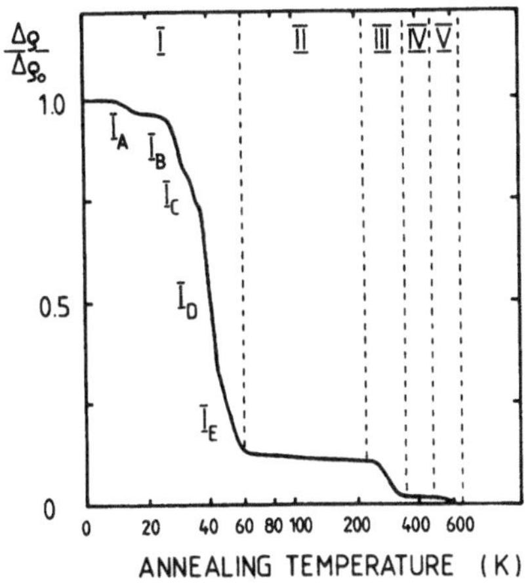

Fig.4.6. Isochronal resistivity recovery schematically shown for irradiated copper

Table 4.3. Self-interstitial migration energies E_i^M obtained from stage I in residual resistivity measurements

Metal	Crystal structure	Self-interstitial migration energy E_i^M [eV]	
Al		0.115	[4.52]
Ni		0.15	[4.52]
Cu		0.117	[4.52]
Ag	fcc	0.088	[4.53]
Pt		0.063	[4.52]
Au		-	
Pb		0.010	[4.52]
α-Fe		0.3	[4.52]
Nb	bcc	0.115	[4.52]
Mo		0.055	[4.54]
Ta		0.033	[4.52]
W		0.054	[4.52]
Zn	hcp	0.21	[4.55]
Cd		-	

e) *Interaction of Lattice Defects with Impurity Atoms*

In a lattice doped with defects, the presence of impurity atoms may change the dynamical behaviour of the defects considerably. Impurities, in general, represent a different potential compared to a regular lattice atom and are therefore able to trap or repell defects. In context of the microscopic methods described in this book, the binding between impurity atoms and defects needs special attention since most often the radioactive probe atoms are impurities.

The understanding of the interaction between defects and impurities is based on their electronic structure and/or the elastic interaction of the defect and impurity in a metal. Both approaches, the elasticity and electronic structure one, are related and show two aspects of the same problem.

In the elastic model for defect-impurity interactions, the binding energy is imagined by the stress field produced, for instance, by an over or undersized impurity acting onto the defect, which may also deviate in size from the host atoms [4.56]. In the electronic approach the excessive or deficient impurity charge is screened by conduction electrons and creates an oscillating electric potential which leads to an attraction or a repulsion of defects. The binding energy for defect and impurity, carrying excess charges Z_1 and Z_2 and being at a certain distance r, has the dependence [4.57,58]

$$E^B \propto - Z_1 \cdot Z_2 \cdot \frac{\cos(2k_F r)}{r^3} \tag{4.5}$$

(k_F Fermi vector).

This expression is obtained by using the solution for the electric potential at large distances. For distances in the order of a lattice constant, the above quoted asymptotic form is rather unaccurate; more precise expressions must be worked out [4.59].

For the calculation of binding energies the localization of the point defect and the impurity, which enters into the distance between the pair, are necessary to know. The first problem is the localization of the impurity in the pure metal. For many systems a substitutional position is accepted, however, there are exceptions, especially for lighter solutes. Here channeling experiments are of great help; considerable data for impurity atoms in different host materials have been collected [4.60].

Decorated with a defect, the impurity may relax away from its original position giving rise to an increase in binding energy. For impurity-interstitial pairs, mixed dumbbell configurations are theoretically and experimentally supported [4.61].

f) *Some Experimental Aspects in the Determination of Defect Properties*

In the preceeding paragraphs the fundamental properties characterizing defects have been outlined. The presence of defects in a lattice changes a number of basic physical properties like lattice and elastic constant, specific heat, electric resistivity and many others. From these property changes a considerable number of methods have developed. In Table 4.4 a collection of available methods for defect studies is presented including the probed physical quantity. The advantages inherent in each method are also quoted.

One of these methods, namely, the residual resistivity method, monitors the concentration of Frenkel pairs and is especially useful in characterizing the defect recovery for a given material in different temperature ranges. The method is based on the increase of specific resistivity due to scattering of conduction electrons at defects like vacancies, interstitials and their agglomerates. The sum of all contributions determines the total residual resistivity (Matthiessen's law).

Table 4.4. Comparison of methods for defect studies with respect to the probed physical quantity and their special sensitivity

Method	Probed Quantity	Special Sensitivity
Residual resistivity	Conduction electron scattering at defects	Concentrations of Frenkel pairs
Positron annihilation	Density and momentum of annihilation electrons	Concentration of vacancies
Elastic diffuse scattering	Atomic disorder by defects	Point defect structures
Inelastic relaxation	Elastic constants	Defect symmetry and reorientation
Transmission electron microscopy	Electron absorption	Direct observation of defect clusters
Channeling	Impurity atom displacement	Defect structures
Radioactive methods	Hyperfine interactions	Distinction between different types of defects

However, deviations of this additive ansatz are noticeable for large defect clusters or defect-impurity complexes.

The observation of defect annealing by thermal activation is usually done in an isochronal or isothermal annealing program. There, defects are first introduced into the sample at temperatures where defect migration can be neglected. This results in additional specific resistivity. In an isochronal annealing sequence, the sample is heated to a certain temperature for a fixed time interval, cooled down and the change of resistivity is then measured. This procedure is repeated with the sample successively heated to higher temperatures and the change of resistivity is observed as a function of the annealing temperature. In such an experiment most of the reduction of the resistivity happens in narrow temperature intervals, leading to annealing stages. A typical result, taken for copper after irradiation, is shown schematically in Fig.4.6. For an interpretation of the observed annealing stages, a detailed discussion of the migration mechanism for various types of defects would be needed. A basic problem arises because only the integral defect concentration is measured by residual resistivity. At an annealing stage the simultaneous annihilation of vacancies and interstitials is observed without knowing what defect part was mobil. This kind of ambiguity and others, of course, have led to two possible consistent explanations of the annealing stages: the one-interstitial model [4.62] and the two-interstitials model [4.63], which in the literature are presented with some controversy. Table 4.5 summarizes the most general features of these two models with respect to defect annealing at the different stages in pure metals. The discrepant predictions require the application of different methods to the same system in order to find the correct picture of the defect dynamics in metals.

Table 4.5. Comparison of the one and two-interstitials model with respect to defect annealing at different stages in pure metals

Stage	One-interstitial model	Two-interstitials model
I	Migration and agglomeration of interstitials	Migration of a metastable interstitial configuration (crowdion)
II	Rearrangement and growth of interstitial clusters	Conversion of crowdions to stage III interstitials (dumbbells)
III	Migration of monovacancies	Migration of dumbbells
IV	-	Migration of monovacancies
V	Dissolution of interstitial and vacancy clusters	Dissolution of interstitial and vacancy clusters

Here the methods listed in Table 4.4 are able to contribute to this problem, among them the nuclear methods which are the subject of this article.

4.1.2 Hyperfine Interaction Parameters

The study of defects on a microscopic basis relies on the change of the electromagnetic field in close vicinity of the defects. The change of the local electromagnetic field can be detected by radioactive probe nuclei, which are either defects themselves or are located near the defect. The information about the defects is transmitted to the probe nuclei by the interaction between defect-field and nuclear moments. This hyperfine interaction can be inferred from the properties of the radioactive decay.

Basically two different physical quantities are accessible experimentally: the strength and symmetry of the hyperfine interaction and the fraction of probe nuclei which are exposed to the defect field. The first quantity is important in characterizing the different types of defects, while the second one is relevant to the study of their mobility and annealing behaviour. In this chapter only a short description of the relevant hyperfine parameters is given—for the electric interactions and the magnetic and combined interaction—whereas further details can be found in the given references. Aspects of used experimental methods are described elsewhere [4.64].

a) *Electric Hyperfine Interaction*

Among the electric interactions only the isomer shift and the quadrupole interaction are important. The isomer shift corresponds to the lowest term in the multipole expansion of the electrostatic energy between nuclear charge and surrounding electronic charge. In the nonrelativistic approximation, it is the difference between the Coulomb interaction of a finite and a point nucleus with an s-electron density $|\psi(0)|^2$ at the nuclear site:

$$E_C = \frac{2}{3}\pi|\psi(0)|^2 \cdot Z \cdot e \cdot \langle r^2 \rangle \quad . \tag{4.6}$$

Z·e is the nuclear electric charge and $\langle r^2 \rangle$ the nuclear mean square radius. Another form of (4.6), modified to include relativistic effects, is given in [4.65]. The energy E_C leads only to a shift of the nuclear level and cannot directly be observed.

Only the Mössbauer effect is sensitive to these isomer shifts. There one observes the γ-transition of the isomeric state to the ground state which contains the difference between the isomer shifts of excited and ground state. This γ-radiation is brought into resonance through the Doppler effect, either by moving the source or absorber. If there is a change of the electron charge density at the probe nucleus site, because, for instance, a vacancy is nearby, the energy of the γ-radiation is then changed by the amount

$$\Delta S_C = \frac{2}{3}\pi Ze\left(\langle r_i^2 \rangle - \langle r_g^2 \rangle\right) \cdot \{|\psi(0)|_0^2 - |\psi(0)|_{defect}^2\} \quad , \tag{4.7}$$

where $\langle r_i^2 \rangle$, $\langle r_g^2 \rangle$ are the nuclear mean square radii of the isomeric and ground state, respectively, and therefore only nuclear properties. The change of the electron charge density due to the defect, described by the second bracket in (4.7), is the quantity which allows us to recognize different types of defects by their different isomer shifts.

The electric quadrupole interaction deals with the interaction of the nuclear quadrupole moment with the electric field gradient (efg) at the nuclear site. The efg is characteristic of the charge distribution around the nucleus; it is a tensor quantity which also contains the symmetry of the charge distribution. If one chooses the coordinate system along the principal axes of the efg-tensor (x, y, z are Cartesian coordinates) so that $V_{xy} = V_{xz} = V_{yz} = 0$ and $|V_{zz}| \geq |V_{yy}| \geq |V_{xx}|$, one gets the following hyperfine Hamiltonian:

$$\mathcal{H}_Q = \frac{eQV_{zz}}{4I(2I-1)}\left[3I_z^2 - I(I+1) + \frac{\eta}{2}(I_+^2 + I_-^2)\right] \quad , \tag{4.8}$$

where $I^{\pm} = I_x \pm iI_y$ are the angular momentum operators, I is the nuclear spin and Q the nuclear quadrupole moment. The parameter η contains the asymmetry of the efg-components: $\eta = (V_{xx} - V_{yy})/V_{zz}$ $(0 \leq \eta \leq 1)$. Because of the Poisson equation the efg tensor is fully described by only two parameters, V_{zz} and η, which reflect the strength and the symmetry of the charge distribution, respectively.

In general, numerical diagonalization is required to obtain the energy eigenvalues of $\mathcal{H}_Q$. For the special case of an axially symmetric interaction ($\eta = 0$) they can be derived analytically (z-axis parallel to the efg axis):

$$E_Q(M) = [3M^2 - I(I+1)]\hbar\omega_Q \tag{4.9}$$

with

$$\omega_Q = \frac{eQV_{zz}}{4I(2I-1)\hbar} \quad . \tag{4.10a}$$

This relation separates into pure nuclear quantities and the defect specific electric field gradient V_{zz}, which describes the change of the charge distribution caused by the defect.

The energy splitting leads to transition frequencies in the M-sublevel system given by (n = 0,1,2,...):

$$\omega_n = 6n\omega_Q \quad \text{with} \quad n = \frac{1}{2}\,|M^2 - M'^2| \quad \text{for halfinteger I}$$

$$\omega_n = 3n\omega_Q \quad \text{with} \quad n = |M^2 - M'^2| \quad \text{for integer I} \tag{4.10b}$$

(in the literature the fundamental frequency ω_1 is usually called ω_0). In the case of $\eta \neq 0$ the number of transition frequencies is unaltered for half-integer I, but for integer I the two-fold degeneracy is lifted and more transition frequencies become possible [4.66].

As an example the electric quadrupole splitting for a nuclear spin I = 5/2 is displayed in Fig.4.7 as a function of the asymmetry parameter .

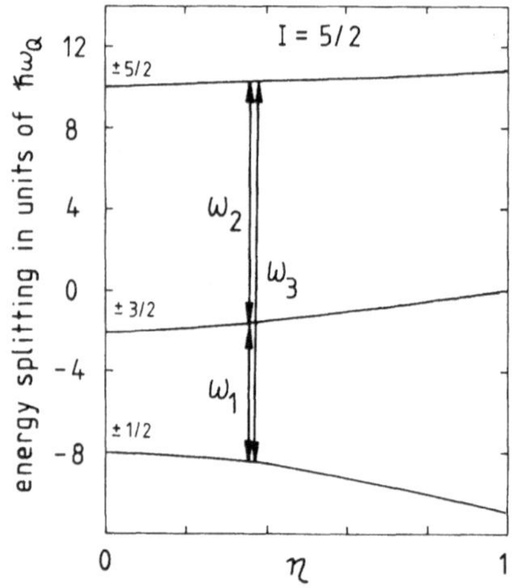

Fig.4.7. Electric quadrupole splitting for I = 5/2 as a function of the asymmetry parameter $\eta = (V_{xx} - V_{yy})/V_{zz}$

The most commonly used methods, which are based on the measurement of the hyperfine transition frequencies, are:

(i) nuclear resonance methods (e.g., nuclear quadrupole resonance NQR, nuclear acoustic resonance NAR);

(ii) low temperature method (e.g., nuclear orientation NO, nuclear magnetic resonance on oriented nuclei NMR/ON);

(iii) Mössbauer effect (ME);

(iv) perturbed angular correlations (PAC) or distributions (PAD).

Since most of the results which will be presented later were obtained by studying the electric quadrupole interaction and using the PAC or PAD method, some short remarks with respect to these methods shall be made here (for more details see Chap.5).

The PAC-method utilizes the isomeric nuclear state in the daughter nucleus of a radioactive probe atom which is populated as an intermediate state in a γ-γ-cascade. Through the detection of the first γ-quantum, the M-substates of the isomeric nuclear spin system are unequally populated with respect to the detection direction; therefore the second radiation, in general, is anisotropic. In the PAD-method a nuclear reaction directly produces the unequal M-substate population of the isomeric state. Besides this different population mechanism, all other descriptions are the same as for a γ-γ-cascade. Therefore, the further discussion is restricted to the PAC-method. The presence of a hyperfine interaction causes transitions within the aligned spin system which leads to a variation in time of the γ-γ-angular correlation. The angular correlation W(t) can be obtained in principle from the experimental coincidence counting rate of two detectors $N(t) = N_0 \exp(-t/\tau_N)\cdot W(t)$ by dividing it by the exponential decay function of the isomeric nuclear state with lifetime τ_N. Within the theory of perturbed angular correlations, the effect of hyperfine interactions is described by rather complex perturbation factors G(t) [4.67]; for our use, however, the result shall be highly simplified. The angular correlation can always be expressed in terms of a finite Fourier series

$$W(t) = 1 + \sum_{n=0} a_n \cdot \cos\omega_n t \quad , \qquad (4.11)$$

where the frequencies ω_n are the possible transition frequencies between the M-sublevels. The Fourier coefficients a_n are theoretically deducible and are tabulated even for complicated cases [4.68]. For the case where the electric field gradient has statistical orientation in the sample, the angular correlation can also be expressed in lowest order in terms of the perturbation factor $G_2(t)$:

$$W(t) = 1 + A_2 G_2(t) P_2(\cos\theta) \quad (G_2(0) = 1) \quad , \qquad (4.12)$$

where A_2 is the angular correlation coefficient, depending on spins and multipolarities of the γ-γ-cascade. The Legendre polynominal $P_2(\cos\theta)$ depends on the angle θ between the two detectors. If no electric field gradient is present at the probe nucleus, $G_2(t)$ is equal to 1 and the usual unperturbed γ-γ angular correlation is obtained. This simple case is illustrated in Fig.4.8a. The connection to the Fourier form, given above, is best seen for the example of randomly oriented, axial symmetric field gradients:

$$G_2(t) = \sum_n s_{2n} \cdot \cos\omega_n t \quad , \qquad (4.13)$$

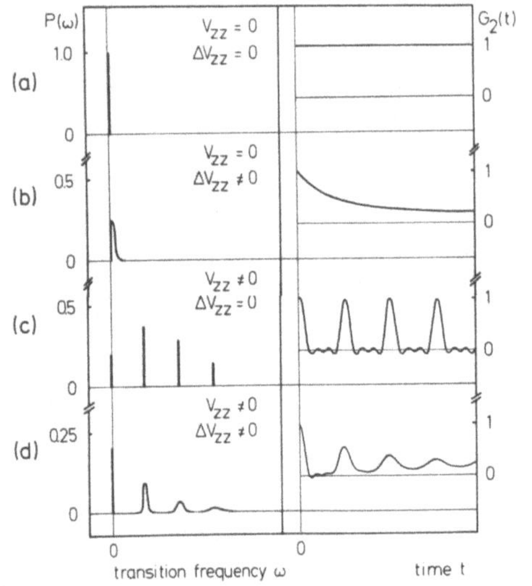

▲ Fig.4.8a-d. Transition frequencies with their weights (left side) and corresponding time modulation pattern $G_2(t)$ of the angular correlation (right side) for different cases of electric field gradients V_{zz} and electric field gradient distributions ΔV_{zz}

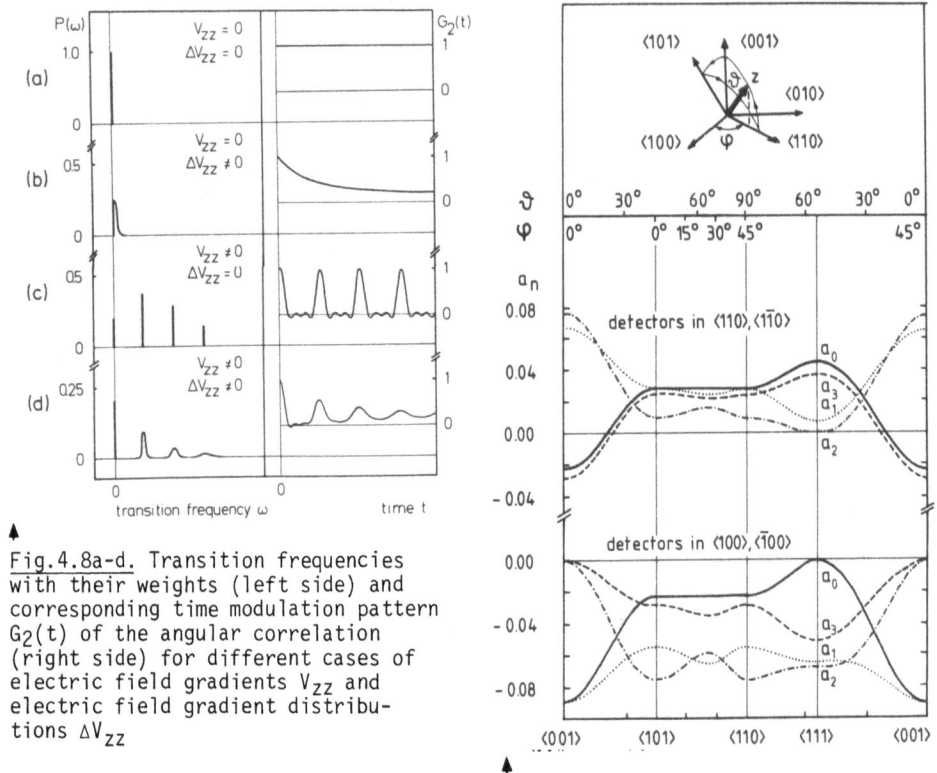

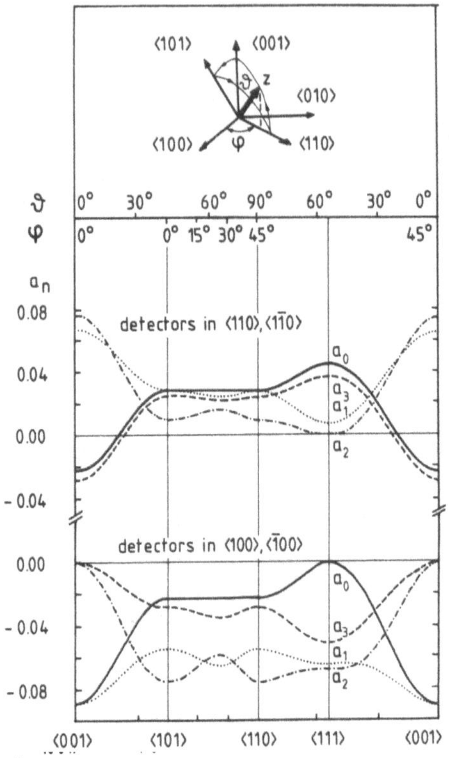

▲ Fig.4.9. Coefficients a_n [see (4.11)] for detectors pointing into a <110> and <1$\bar{1}$0> (top) and into a <100> and <$\bar{1}$00> (bottom) lattice direction as a function of the angles ϑ and φ defining the efg orientation z for the case of ^{111}In

where s_{2n} are tabulated parameters involving only vector coupling algebra [4.67]. In case I = 5/2, only three transition frequencies exist; this is illustrated for a unique efg in Fig.4.8c. On the left-hand side the transition frequencies with their weights are displayed, on the right the corresponding time modulation, described by $G_2(t)$, of the angular correlation is given. Figure 4.8 also depicts cases of efg distributions and their result onto the time modulation. In Fig. 4.8b the distribution with a certain width ΔV_{zz} centered around $V_{zz} = 0$ and in Fig.4.8d around a finite efg value are shown.

Another example is the case of a defined orientation of the (axial symmetric) efg with respect to the lattice of a single crystal sample; there the parameters a_n depend on the angles of the efg symmetry axis and the detector positions. This is demonstrated for the γ-γ-cascade of ^{111}In (I = 5/2) for two different detector geometries in Fig.4.9. One can clearly see changes in the parameters a_n and therefore also the sensitivity of the calculated spectra.

b) *Magnetic and Combined Hyperfine Interaction*

The magnetic hyperfine interaction is of interest for defect studies in the case of substances with internal magnetic fields. There the defects change the local magnetic field which can then be used, in full analogy to the isomer shift and electric field gradient, to characterize defects.

The applicable methods, in general, are the same ones as for electric quadrupole interactions, except that one now has to detect transitions within a nuclear spin system which is split by a magnetic field. The magnetic energy shift of an M-substate is given by

$$E_m(M) = -M\hbar\omega_L \tag{4.14}$$

with the Larmor frequency $\omega_L = -g\mu_N B_z/\hbar$ (g nuclear g-factor, μ_N nuclear magneton, B_z magnetic field along the z-axis). In the magnetic case the perturbed angular correlation behaves similar to the electric one. Again the angular correlation can be written as a finite Fourier series, with the only difference that the index n now has to fulfil the condition $n = |M - M'|$.

Defect studies in magnetic substances are, however, accompanied by a general complication. As described earlier, besides the influence of the defect onto the local magnetic field, additionally isomer shift and quadrupolar effects are expected. This makes it difficult to distinguish clearly between different defects in order to obtain more sound interpretations. Occasionally, for example, when using the PAD method, it may be advantageous to apply an external magnetic field for experimental reasons. More to this problem in Sect.4.2.2.

4.1.3 Nuclear Probes

The size of the electromagnetic moments and the lifetime of an excited nuclear state are the two properties which are most important so that radioactive nuclei can be used to observe a defect via the hyperfine interaction. In the case of source experiments (ME, PAC), in addition the properties of the parent (probe) activity, decay and lifetime have to be such that the desired method can be applied. All these requirements make it difficult, especially for source experiments, to find a large selection of probe isotopes. In in-beam experiments (PAD), the limitations are less serious because of the variety of available target materials and nuclear reactions. In general, the limited number of suitable probe-nuclei often turns out to be very impeding for systematic studies of defects.

A basic aspect in the use of microscopic methods is the time window in which the probes are sensitive to the defect fields. Obviously, the nuclei are only useful as probes as long as the isomeric state exists, which means the time window is governed by the nuclear lifetime. One consequence of this is that due to the short nuclear lifetime, the defect fields on a microscopic scale could be static,

Table 4.6. Probe nuclei for defect studies mentioned in this review. Nuclea which the hyperfine interaction is measured. (I: nuclear spin; $T_{1/2}$: half-lif pole moment; Q: electric quadrupole moment)

Probe/Daughter Nucleus	I	$T_{1/2}$ [ns]	E_{exc} [keV]	μ [μ_N]	Q [barn
$^{57}Co/^{57}Fe$	3/2	98	14	-0.155	+0.19
^{58}Co	4	10.2×10^3	53	+4.194	
^{67}Zn	9/2	333	605	-1.097	+0.61
^{67}Ge	9/2	70	734	-0.948	
^{69}Ge	9/2	2.8×10^3	398	-1.001	
^{71}Ge	5/2	84	175	+1.018	
^{73}As	9/2	5.6×10^3	428	+5.234	
^{99}Mo	5/2	$15\ \times10^3$	98	-0.775	
$^{100}Pd/^{100}Rh$	2	214	75	+4.324	0.076
^{107}Cd	11/2	70	845	-1.041	-0.94
^{109}Cd	11/2	10.6×10^3	463	-1.096	-0.92
$^{111}Ag/^{111}Cd$	5/2	84	247	-0.7656	+0.83
$^{111}In/^{111}Cd$					
^{111}Cd					
^{113}Sn	11/2	89	731	-1.300	-0.46
^{114}Sn	7	765	3088	-0.567	0.32
^{116}Sn	5	370	2367	-0.376	0.26
^{119}Sn	3/2	18	24	+0.633	0.094

^{112}Sb	8	536	796	+2.20	0.5
^{115}Sb	19/2	161	2796	-2.75	
^{120}Sb	3	247	88	+2.584	
^{115}Te	11/2	7.5×10^3	279	-0.963	
$^{125}Sb/^{125}Te$	3/2	1.5	36	+0.604	-0.31
$^{125}I/^{125}Te$					
$^{127}Te/^{127}I$	7/2	2	58	+2.54	-0.71
$^{129}Te/^{129}I$	5/2	17	28	+2.801	-0.68
^{131}I	7/2	8d	-	+2.742	-0.40
$^{129}Cs/^{129}Xe$	3/2	1	40	+0.58	-0.41
^{129}Xe					
$^{131}I/^{131}Xe$	1/2	0.5	80		
^{131}Xe					
$^{133}Xe/^{133}Cs$	5/2	6	81	+3.51	
$^{161}Tb/^{161}Dy$	5/2	29	26	+.0.595	+2.45
$^{169}Er/^{169}Tm$	3/2	4	8	+0.54	-1.2
$^{181}Hf/^{181}Ta$	5/2	11	482	+3.24	+2.51
^{208}Po	8	380	1530	+7.29	0.64
^{209}Po	17/2	98	1473	+7.62	0.39
^{210}Po	8	112	1557	+7.28	0.57

whereas in experiments on macroscopic properties within the measure time the fields may very well vary.

In Table 4.6, nuclei which are suitable for defect studies and their properties are presented. Only those nuclei which are mentioned in this review are included.

4.2 Hyperfine Investigations of Defects

4.2.1 Experiments with Radioactive Sources

The study of defects by hyperfine interaction methods has been mainly performed in cubic metals and especially in fcc metals. Two reasons are decisive: (i) the interpretation of the results is much simplified by the fact that only probe atoms marked with a defect exhibit a perturbation, while probe atoms on a regular substitutional cubic site are unperturbed. In a noncubic metal efg's are present even without defects. (ii) Most extensive data from classical methods exist about fcc and bcc metals, classification schemes as annealing stages are well established and general similarities are already observed within these classes.

In comparing existing data with classical results such as, for instance, residual resistivity measurements, the nuclear methods were able to prove their capability in extracting additional information from the microscopic viewpoint; they are about to show their complementary character and at the same time demonstrate considerable progress.

Several reviews have been published recently. VOGL [4.69] and NIESEN [4.70]have discussed general aspects of the study of point defects by hyperfine interaction methods. PLEITER [4.71] concentrated on the investigation of vacancy trapping in fcc metals studied by perturbed angular correlation. Introductions into the relevant hyperfine interaction theory and experimental techniques as well as selected examples of defect study are given by RECKNAGEL and WICHERT [4.72,73].

Definite systematic conclusions about unique defect interactions in different metals are a point of current discussion. DEICHER and WICHERT [4.74] have discussed the behaviour of trapped monovacancies in fcc metals, while PLEITER and HOHENEMSER [4.75] proposed a classification of vacancy defects according to the observed quadrupole interaction frequencies. More detailed and specific information, however, has to be accumulated in order to get a consistent picture of the various types of defects detected by hyperfine interaction methods.

In this chapter a survey of Mössbauer (Table 4.7) and perturbed angular correlation experiments (Table 4.8) will be given in which defects are detected by radioactive probe atoms. It will, in general, be restricted to the discussion of well-defined interactions, i.e., all experimental results are omitted where only a rather unspecified interaction is observed. These interactions are usually attributed to a broad hyperfine field distribution caused either by defects, which are not in the immediate neighbourhood of the impurity atoms, or by other influences such as grain boundaries or surface effects.

In general, the probe atoms experience different perturbations so that in the evaluation of perturbed angular correlation (PAC) experiments, the perturbation factor $G_2(t)$ in (4.13) has to be replaced by a generalized perturbation factor. For quadrupole interactions in a polycrystalline sample, assuming $\eta = 0$, it is given by

$$G_2(t) = \sum_n s_{2n}\left[f_0 + \sum_i f_i\ e^{-n\sigma_i t}\cos(\omega_n^i t)\right] \quad ; \qquad (4.15)$$

for magnetic interactions it reads

$$G_2(t) = f_0 + \sum_i f_i\ e^{-\sigma_i t}\cos(\omega_L^i\ t) \quad , \qquad (4.16)$$

where f_0 denotes the unperturbed fraction of nuclei, f_i is the fraction associated with unique interaction frequencies ω_n^i or ω_L^i and σ_i allows for a frequency distribution. The latter parameter corresponds to an increase of line width in the Fourier transforms.

The quadrupole interaction strength is generally expressed by the quantity ν_Q which is independent of the nuclear spin and is related to the fundamental quadrupole frequency ω_0 defined in Sect.4.1.2a,

$$\nu_Q = \frac{eQV_{zz}}{h} = \begin{cases} \dfrac{I(2I-1)}{3\pi}\ \omega_0 & \text{for halfinteger } I \\ \dfrac{2I(2I-1)}{3\pi}\ \omega_0 & \text{for integer } I \quad . \end{cases} \qquad (4.17)$$

$\omega_0 = 2\pi/T_0$, where T_0 is the observable period in an actual time spectrum (Fig.4.8c).

A measurement of the quadrupole interaction in a single crystal allows the orientation of the efg tensor to be determined. The observation of the γ-rays at various angles with respect to the crystal yields the orientation of the symmetry axis of the efg for $\eta = 0$ with respect to the crystallographic axis <hkl> [4.76]. The coefficients a_n depend drastically on this orientation, as demonstrated in Fig.4.9, whereas the transition frequencies are not affected.

Thus, the relevant information obtained from PAC investigations of defects in metals can be summarized as follows:

- ν_Q^i, ω_L^i denote the specific defect probe atom interaction, from which the field quantities V_{zz} and B can be derived;
- f_0, f_i denote the fractions of nuclei experiencing a certain interaction, and
- <hkl> and η determine the direction and the symmetry of the efg.

Correspondingly, the information from Mössbauer effect investigations can be summarized as:

Table 4.7. Hyperfine interaction parameters of defect configurations obtained fro nuclei in metals. Listed are the isomer shift ΔS relative to the unperturbed subs fine field MHF, the quadrupole interaction QI and the range of annealing temperat corresponding annealing stages, as known from residual resistivity measurements, to the interpretations of the authors

Host (lattice structure)	Probe	#	Relative ΔS [mm/s]	MHF [KG]	QI [mm/s]	ΔT [K]
Al(fcc)	$^{57}Co/^{57}Fe$	1	+0.42	-	0,17	40-200
		2	+0.42	-	0,17	40-200
		3	+0.32	-	-	230-350
	$^{119}Sn/^{119}Sn$	1	+0.6	-	-	300-500[b]
		2	-0.7	-	-	300-500[b]
	$^{169}Er/^{169}Tm$	1	~0	-	110-83	100-300
		2	~0	-	87	500-900[b]
Fe(bcc)	$^{125}I/^{125}Te$	1	~0.4	540(30)		[a]300
		2	~-2.0(5)	190(100)		[a]300
	$^{129}Te/^{129}I$	1	-0.3(1)	100(40		[a] 12-470[b]
		2	-0.6(2)	460(40)		[a] 12-970
	$^{133}Xe/^{133}Cs$	1	+0.3(1)	135(15)		[a] 7-575[b]
		2	+0.9(1'	<25		[a] 7-575[b]
	$^{131}I/^{131}Xe$	1	0	1130(70)		4-115[b]
		2	0	270(5)		4-115[b]
	$^{129}Cs/^{129}Xe$	1	0	1100(100)		
		2	0	600(125)		
Ni(fcc)	$^{129}Te/^{129}I$	1	-0.22(8)	271(7)		[a]~270
		2	-0.88(7)	160(9)		[a] 270-1050[b]
Mo(bcc)	$^{57}Co/^{57}Fe$	1	+0.48(1)	-	1.82(1)	[a] 100-214
		2	+0.45(1)	-	1.14(1)	[a] 100-460

Mo (bcc)		3	+0.45(1)	-	1.48(1)	214-370
		4	+0.47(1)	-	0.81(1)	460-500
		5	+0.47(1)	-	0.39(1)	460-500
	$^{133}Xe/^{133}Cs$	2	+0.36(7)	-	-2.6(4)	400-500
		3	+0.91(5)	-	3.6(4)	400-1000
		4	+1.21(6)	-	<0.8	400-1300[b]
Ag (fcc)	$^{57}Co/^{57}Fe$	1	+0.32	-	-	30-150
W (bcc)	$^{133}Xe/^{133}Cs$	2	+0.50(8)	-	-2.9(4)	500-600
		3	+1.10(7)	-	4.7(4)	550-1300[b]
		4	+1.5(8)	-	<1.0	550-1300[b]
Pt (fcc)	$^{133}Xe/^{133}Cs$	2	-0.92(8)	-	-6.2(4)	[a] 100-450
		3	-0.74(6)	-	5.0(4)	[a] 100-450
		4	-0.65(2)	-	-	[a] 100-850
		5	-0.30(2)	-	-	250-1300[b]

[a] Irradiation temperature.

[b] Highest temperature of annealing program

Table 4.8a. Hyperfine interaction parameters of defect configurations with ^{111}In/
turbed angular correlation (PAC) experiments. $\nu_Q = eQV_{zz}/h$ is the quadrupole inte
the asymmetry parameter of the efg; <hkl> is the efg orientation produced by the
lattice. ΔT is the range of annealing temperatures where defects are observed. Th
known form residual resistivity measurements, and trapped defect configurations r

Host (lattice structure)	#	ν_Q [MHz]	η	<hkl>	ΔT [K]	Stage
Al (fcc)	1	16.6(5)	0.54(3)	-	[a]55-160	I
	2	37(1)	<0.2	-	160-360	I
	3	133(5)	0	-	160-260	III
	4	66(3)	0.41(5)	-	200-240	III
Ni (fcc)	2	ω_L=39.9(3) [c]	-		300-600	III
	3	54(3) $\omega_L \sim 40$ [c]	<0.2	<111>	280-820	V
Cu (fcc)	1	116(1)	0	<110>	200-300	III
	2	181(1)	0	not <110>	180-320	III
	3	52(1)	0	<111>	350-750	V
	4	55(2)	0.48	-	350-750	V
	5	46.8(5)	0	<100>	40- 70	I
	6	19(1)	1	-	40- 70	I
Zn (hcp)	1	70.0(5)	<0.1	-	140-220	III
Nb (bcc)	1	88(3)	0	<111>	200-350	III
	2	105(2)	0.65	-	200-350	III
	3	180(5)	0	-	270-400	III
Mo (bcc)	1	125(1)	0	<111>	450-600	III
	2	155(1)	1	-	450-600	III
	3	98(1)	0	<100>	450-600	III
	4	~700	-	-	450-1500	V
	5	36(1)	1	-	300-1500	V

Pd (fcc)	1	86.5(15)	0	<110>	<430[d]	III
Ag (fcc)	1	82(2)	0	<110>	210-260	III
	2	36(2)	0	-	300-500[b]	V
Cd (hcp)	1	103(3)	1	-	110-140	III
	2	4-22	0	-	110-150	III
	3	0	-	-	[a] 77-300	(V)
W (bcc)	1	141(2)	0	<111>	300-800	III
	(2?)	318(4)	0	<110>	530-750	III
	2	181(5)	1	-	300-800	III
	3	263(5)	0	-	750-1250	(V)
Pt (fcc)	1	103(2)	<0.1	-	[a] 10-480	III
	2	210(2)	<0.1	-	270-400	III
	3	61(8)	0	-	270-400	III
Au (fcc)	1	102(1)	0.45(3)	-	190-240	III
	2	101(1)	0	not <110>	190-240	III
	3	91(1)	0	<110>	210-300	III
	4	40.0(5)	0	<111>	250-600	V
Pb (fcc)	1	28	1	-	<180	II

[a] Irradiation temperature.
[b] Highest temperature of annealing programs.
[c] Magnetic interaction: Larmor frequency ω_L.
[d] Though the daughter nucleus is ^{111}Cd the probe atom is ^{111}Ag instead of ^{111}In.

Table 4.8b. Hyperfine interaction parameters of defect configurations with [1] experiments

Host	#	ν_Q [MHz]	η	<hkl>	ΔT [K]	Stage
Nb (bcc)	1	42(2)	0	-	[a]30-100	I
	2	33(2)	1	-	[a]30-100	I

[a]Irradiation temperature

Table 4.8c. Hyperfine interaction parameters of defect configurations with [1] experiments

Host	#	ν_Q [MHz]	η	<hkl>	ΔT [K]	Stage
Pt (fcc)	1	668(5)	-	-	600-850	V
	2	333(8)	-	-	680-850	V

- ΔS, ΔE_Q, ΔE_m denote the defect-probe atom hyperfine interactions, from which the respective quantities $|\phi(0)|^2_{defect}$, V_{zz}, and B can be derived;
- f_0, f_i denote the fractions of nuclei experiencing a certain interaction. These fractions can be extracted from the areas $F_i = f_i \cdot (DWF)_i = f_i \cdot \exp(-k^2\langle x^2\rangle)^i_T$, thus taking into account the specific temperature influence onto each probe-defect complex by an appropriate temperature-dependent mean square vibration amplitude $\langle x^2\rangle_T$ (k is the wave number of the γ-ray; DWF: Debye-Waller factor).

The index for the defect-impurity configurations formed in each host metal (4.15,16) is usually given by the different authors in the order of their detection. This labeling is retained to avoid confusion with the original literature.

a) *Diamagnetic fcc Metals*

Up to now investigations have been carried out on the following elements: Al, Cu, Pd, Ag, Pt, Au and Pb.

Aluminum

Defect studies in Al have been performed with several nuclear probes by Mössbauer effect and perturbed angular correlation. In particular, the detailed ME investigations by VOGL and MANSEL [4.77-81] beginning in 1973 had an important impact on the whole research in this field.

Mössbauer Effect: In a series of experiments the influence of irradiation dose with fast neutrons and electrons and of isochronal annealing on the trapping of defects at Co impurities was studied. The trapping is noticeable by the appearance of a new Mössbauer line shifted with respect to the line originating from substitutional Co impurities and by a broadening of this line due to an unresolved quadrupole splitting. The fractional area of the line is a direct measure of the percentage of impurities having trapped defects.

In Fig.4.10 the fraction F of interstitial-decorated Co atoms is shown for different neutron doses as a function of annealing temperature and, for comparison, also the fractional $\Delta\rho/\Delta\rho_0$ of the radiation induced resistivity for the same specimen [4.80]. The irradiation was performed with fast neutrons at 4.6 K. It reveals that most of the trapping is correlated with a simultaneous steep decrease of the resistivity in the high temperature part of stage I (I_E), which is usually attributed to the free migration of interstitials. Together with the deduced hyperfine interaction, VOGL and MANSEL concluded that the Co impurity atom and the interstitial have formed a complex with noncubic symmetry at the impurity atom site.

Further trapping occurs in stage II where thermally instable interstitial agglomerates dissolve. Only at an annealing temperature of about 220 K (around stage III) do the trapped interstitials disappear due to annihilation by migrating vacancies.

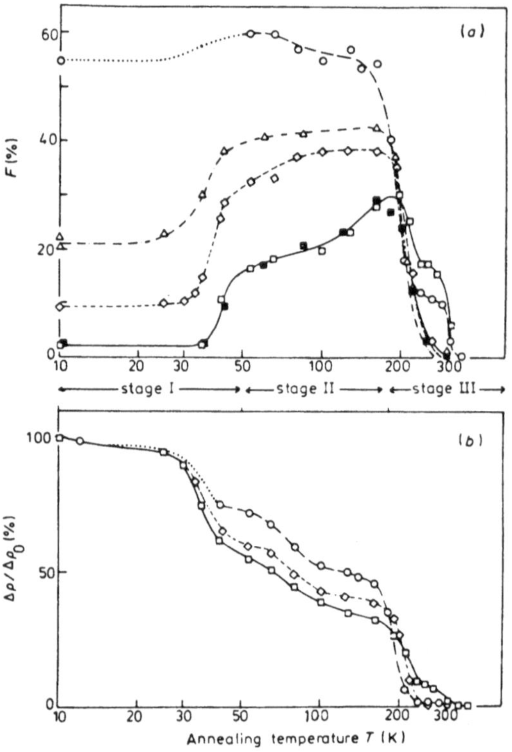

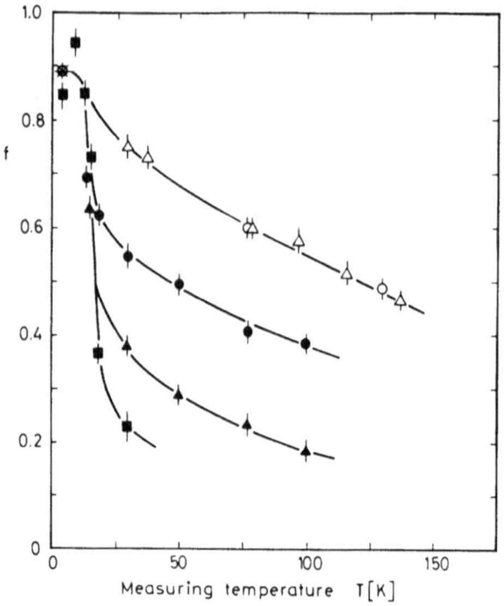

▲ Fig.4.11

◄ Fig.4.10

Fig.4.10. (a) Fraction of ^{57}Co atoms with trapped interstitials, (b) recovery of the radiation induced resistivity of Mössbauer samples as a function of isochronal annealing temperature and neutron irradiation doses: (■) (□) 0.2; (◇) 0.7; (△) 2.3; (O) 7.3 $\times 10^{18}$/cm^2 [4.80]

Fig.4.11. Temperature dependence of the Debye-Waller factor f of ^{57}Co impurities in aluminum with trapped interstitials. Lower three curves: after 2.8 MeV electron irradiation at about 100 K; irradiation doses in 10^{18}/cm^2: (●) 5.6; (▲) 1.37; (■) 0.95. Top curve: after fast neutron irradiation at 4.6 K and annealing at 160 K; irradiation doses in 10^{18}/cm^2: (△) 2.3; (O) 7.3 [4.81]

The temperature data at 10 K (below stage I) show that trapping had already occurred during the low temperature irradiation. This fraction increases with increasing irradiation dose. To decide whether the complexes formed after neutron irradiation contain one or more interstitials, MANSEL et al. produced isolated Frenkel pairs in the dilute ^{57}CoAl samples [4.81] by irradiation with 2.8 MeV electrons at 100 K, at temperature well above stage I, where only interstitials are mobile. They can only survive the recombination with vacancies if they are trapped at Co impurities. Assuming a Poisson distribution of the produced defects, the authors calculated that about 90% of the decorated Co impurities have trapped single interstitials after irradiation with a dose of 0.95×10^{18} e$^-$/cm^2. For higher doses the fraction of interstitial clusters at the Co atoms is increased. Isomer shift and quadrupole splitting are not sensitive to the degree of clustering; the Debye-

Waller factor, however, exhibits a pronounced dependence when measured as a function of temperature as shown in Fig.4.11 [4.81]. The most rapid decrease of the DWF takes place for the lowest irradiation dose, while for higher electron doses or increased multiple interstitial trapping, the reduction is less dramatic. The comparison with fast neutron irradiation displays an even higher fraction of interstitial clustering at the Co impurities. From the deduced fractions F_1 of trapped single interstitials and $F_{>1}$ of more than one trapped interstitial, one can see that already below T_A = 33 K nearly all Co impurities have trapped more than one interstitial during irradiation.

The observation of the drastic temperature dependence of the DWF around 20 K was attributed to a restricted diffusion of the ^{57}Co impurity atom in an octahedral "cage" [4.78]. These cage jumps are hindered when more than one interstitial atom is trapped. Most recent experiments have shown that the originally suggested mixed dumbbell configuration of ^{57}Co-interstitial (<100> cage) has to be replaced by a <111> cage, where the single interstitial atom adds to a trigonal complex around the Co atom [4.82,82a].

Beside the annihilation process of trapped interstitials by vacancies mobilized in stage III as mentioned before, direct evidence for the mobility of vacancies by trapping at ^{57}Co in aluminum was given by SASSA et al. [4.83,84]. They quenched diffused ^{57}Co$\underline{\text{Al}}$ from temperatures below the melting point to 233 K by a typical rate of 5×10^4 K/s. The outcome of an isochronal annealing process is shown in the Mössbauer spectra in Fig.4.12.

The new line which is shifted with respect to the substitutional line and slightly broadened (10%) is interpreted as resulting from trapped vacancies. This shift (+ 0.33 mm/s) corresponds to a larger electron density at the nuclear site compared to the solid solution peak, but it is slightly smaller than for trapped interstitials (+ 0.43 mm/s). The vacancy peak almost disappeared at 353 K due to detrapping. A comparison of this source experiment with earlier absorber experiments [4.84,85] where the vacancies are trapped at ^{57}Fe atoms exhibits only different detrapping temperatures which are attributed to different binding energies between vacancies and Co or Fe probes, respectively. In addition it should be mentioned that SASSA et al. [4.85a] and VERBIEST and PATTYN [4.85b] recently performed supplementary experiments in the ^{57}Co$\underline{\text{Al}}$ system.

Further Mössbauer experiments in aluminum are carried out with ^{119}Sn as probe atoms where the defects are either produced by quenching [4.86] or by direct implantation of the radioactive isotope [4.87]. Though in both experiments new Mössbauer lines are observed with fairly constant isomer shifts (ΔS = 1.6 - 1.9 mm/s and ΔS = 30 mm/s), the interpretation is by no means clear. According to NYLANDSTEDT LARSEN and WEYER [4.87], the line with the lower isomer shift depends on the concentration of other impurities in aluminum and is stable at temperatures

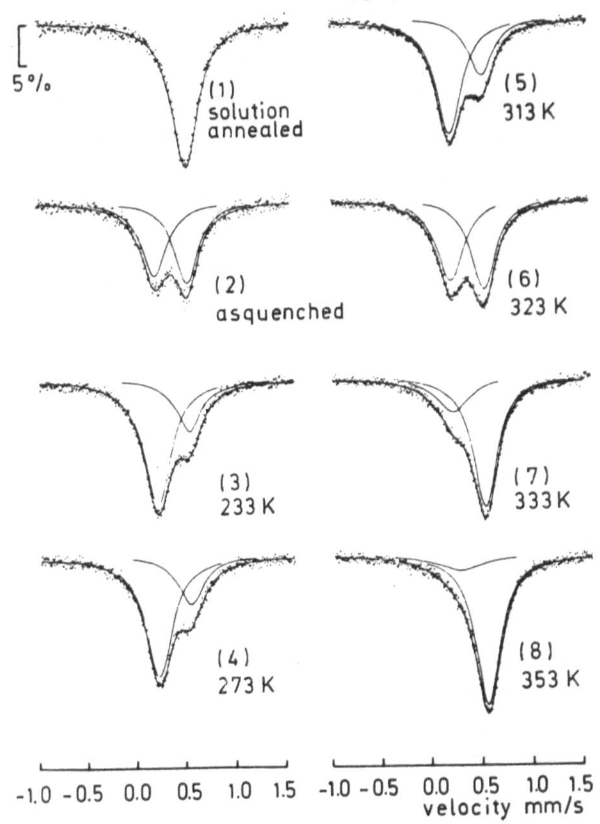

Fig.4.12. Mössbauer spectra of ^{57}Co in aluminium after quenching for different isochronal annealing temperatures. Measuring temperature is 77 K [4.83]

well above stage III. They therefore exclude the attribution of this line to be a monovacancy complex as suggested by KATO et al. [4.86]. The line with the higher isomer shift is interpreted by these authors as being due to vacancy clusters, while NYLANDSTEDT LARSEN and WEYER favour an interstitial cluster or an interstitial position of the tin atom.

A Mössbauer effect study on ^{169}Er atoms implanted at 7 K into Al was performed by WIT et al. [4.88]. They observed a strong decrease of the substitutional fraction between 100 and 300 K in favour of a new fraction showing a large quadrupole splitting. During annealing the average value of this doublet splitting ΔE_Q decreased from 110 mm/s to 83 mm/s, indicating that this new site is composed of different configurations most probably due to trapped vacancies. At higher annealing temperatures another site was found with a doublet splitting of ΔE_Q = 87 mm/s.

Perturbed Angular Correlation: PAC experiments to study defects in aluminum were performed with radioactive ^{111}In [4.89-91] and ^{111}Ag [4.92] nuclei. In both cases the quadrupole interaction of the excited state of ^{111}Cd with the electric field gradients caused by defects was measured and used to monitor the formation and dissolution of impurity-defect complexes by the usual isochronal annealing procedure.

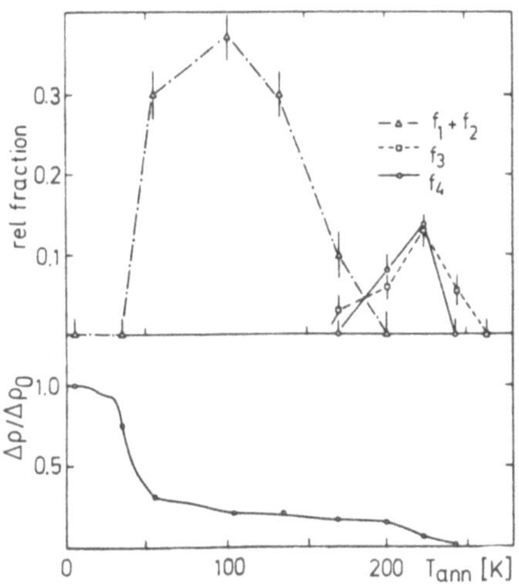

Fig.4.14. Fractions of ^{111}In atoms with trapped defects in Al and recovery of radiation induced resistivity (lower part) as a function of annealing temperature [4.90]

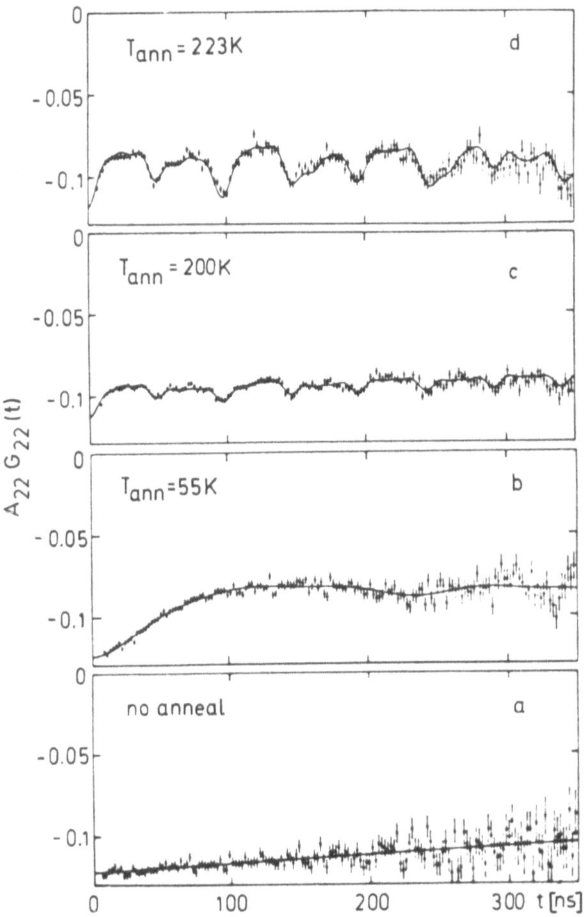

Fig.4.13. PAC spectra of electron irradiated ^{111}InAl for different annealing temperatures [4.90]

In all experiments high purity polycrystalline aluminium foils were chemically doped by diffusion with radioactive probe atoms, the concentration of which is typically much less than 1 ppm. The diffusion process already establishes an initial state of cubic symmetry around the probe atoms, which one expects if the atom is situated on a regular lattice site surrounded by an unperturbed aluminum lattice. Introduction of defects was performed by quenching [4.89] and electron irradiation [4.90,91]. Several well-defined interactions were observed, characterized by the quadrupole interaction frequency ν_Q and the asymmetry parameter η (Table 4.8). Since it is generally accepted that only vacancy-like defects are produced when metals are quenched, the trapped defect characterized by ν_{Q3} = 133 MHz was attributed by RINNEBERG and HAAS [4.89] to a di- or monovacancy, with some preference for the In-divacancy complex because of mobility arguments. The same interaction was observed after electron irradiation in stage III shown in Fig.4.13c,d and as fraction f_3 in Fig.4.14. This experiment [4.90] which included a simultaneous residual resistivity measurement (lower part of Fig.4.14) already revealed the other interactions observed so far in Al: an additional interaction (ν_{Q4} = 66 MHz; η = 0.41;) in stage III, attributed to a small vacancy cluster, and interactions caused by trapped interstitials in the temperature region of stage I and II. These latter ones, already indicated in Fig.4.13b and shown as a sum of fractions $f_1 + f_2$ in Fig.4.14 but not yet clearly resolved, were further investigated by BUTT et al. [4.91]. They assigned the interaction ν_{Q1} = 16.6 MHz, η = 0.54 as being due to a single interstitial, where the asymmetry parameter indicates that the interstitial and the In impurity do not form a mixed dumbbell. The mono-interstitial character of this configuration was confirmed in an experiment where 40 ppm stable In was alloyed into the Al sample in addition to the radioactive probe atoms, thus effectively suppressing a multiple interstitial trapping. The observed mono-interstitial fraction f_1 = 8.3% is in excellent agreement with the value expected from the estimated ratio of defect to In-impurity concentration of 0.1 in this experiment. The frequency ν_{Q2} = 37 MHz observed in Al well above stage I but below stage III is most probably due to interstitial clusters.

Recently, PAC experiments have also been performed after laser irradiation of ^{111}In-implanted aluminum [4.92a].

Copper

Perturbed Angular Correlation: In all radiation damage studies, copper, in many respects, has been considered as a prototype of fcc metals. Thus, also detailed investigations were performed using the PAC technique with ^{111}In as probe atoms [4.76,93-98]. Up to now six well-characterized defect-impurity configurations have been found (Table 4.8).

Polycrystalline as well as single-crystal copper samples were doped with ^{111}In either by diffusion or implantation. Defects were produced by quenching, electron-, proton- and heavy-ion irradiation and the investigations cover the whole temperature range of annealing stages I - V (10 - 800 K).

Two interstitial-impurity complexes [4.76] were observed between 40 and 70 K, the lower temperature corresponding to stage I_E, the annealing stage of freely mobile interstitials. In Fig.4.15 (upper part) the fractions of probe atoms with trapped interstitials are shown as a function of annealing temperature following proton irradiation. Both fractions ν_{Q5} and ν_{Q6} appear simultaneously and, as proved by further experiments, dissolved at least partly by detrapping [4.99]. The lower part of this figure exhibits a damage rate measurement at constant irradiation temperature T_{irr} = 52 K with increasing H^+ dose, clearly showing a saturation effect as observed in many defect studies. From investigations in polycrystalline samples, information can be gained only about the interaction frequency ν_Q and the asymmetry parameter η. To determine the orientation of the efg, single-crystal measurements have to be performed. The results of such an experiment are shown in Fig.4.16. The Fourier transforms of measured time spectra after H^+ irradiation at

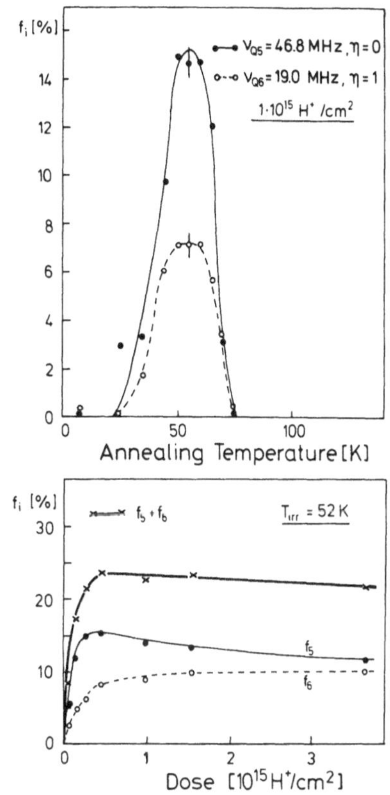

◄ Fig.4.15. Fractions of ^{111}In atoms with trapped interstitials in Cu as a function of annealing temperature (upper part). Dependence of the fractions from H^+-dosis at constant irradiation temperature of 52 K (lower part)

Fig.4.16. Fourier transforms of time spectra of ^{111}In-interstitial configurations in Cu single crystals after H^+-irradiation at 52 K. The spectra are recorded under different detector-crystal orientations and different angles between the detectors (left side). Comparison of the measured amplitude coefficients a_n with calculated s_n [corresponding to a_n of (4.11)] assuming the three efg-orientations <100>, <110> and <111> (right side) [4.76]
▼

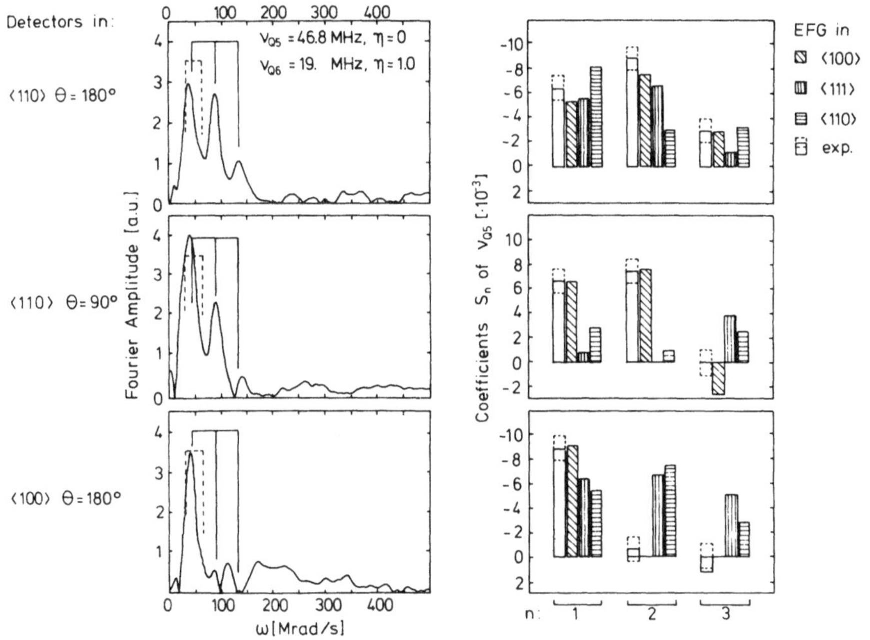

52 K are recorded under different detector-crystal orientations and different angles θ between the detectors (left side). It is evident that the amplitudes a_n [see (4.11)] of the interaction ν_{Q5} drastically depend on the detector geometry. On the right side of Fig.4.16 the experimental values are compared with calculated a_n assuming efg orientations in <100>, <111> and <110>directions. Only the <100> orientation of the efg agrees with the experiment which means that this interstitial-impurity configuration possesses axial symmetry along a <100> crystallographic direction; possible explanations are a mixed dumbbell in <100> or a copper dumbbell pointing into <100> on a next-nearest-neighbour site to the substitutional impurity [4.76]. This finding, however, is in contrast to theoretical considerations which predict an associated copper dumbbell next to an oversized impurity like the In-atom in a copper lattice.

Around stage III two further configurations are found which are characterized by ν_{Q1} = 116 MHz, η = 0, <110> orientation, and ν_{Q2} = 181 MHz, η = 0. Both configurations are observed after quenching, electron and proton irradiation [4.93,94]. In Fig.4.17 typical experimental time spectra are shown while in Fig.4.18, the fractions f_1 and f_2 of the two configurations are depicted as functions of annealing temperature (in Fig.4.18 the notation B is used instead of f). The low and high dose electron irradiations correspond to Frenkel pair concentrations at 4.2 K of 80 ppm and 565 ppm or, a resistivity increase of 16 and 113 nΩcm, respectively. Since both defect configurations are also observed after quenching, they have to be vacancy like. In principle, the mobility of only one defect type in stage III would—via multiple trapping—be sufficient to create the two configurations. But as a comparison of the two irradiation experiments shows, fraction f_1 cannot be populated via f_2 (see low-dose electrons) and, furthermore, f_2 is formed prior to f_1—as can be clearly seen from the H^+ irradiation. Thus, two defect types must be mobile. Electron irradiation produces preferably point defects and defect complexes mainly emerge from defect agglomeration at higher doses. Therefore, the configuration ν_{Q1} is attributed to a trapped monovacancy which is supported by the most likely orientation in <110> direction, while configuration ν_{Q2} is due to a divacancy or a small vacancy complex which is mobile in this stage.

These findings are supported by further experiments [4.98] and the obtained results are shown in Fig.4.19. Taking the same electron-irradiated pure copper samples, defect ν_{Q1} is predominantly formed in the low dose case (upper part) while in the high dose case (middle part), both fractions are about the same. This means that defect 1 should be the simpler one of them. A third Cu sample was also irradiated with a high dose (120 nΩcm), but this time the sample was additionally alloyed with 20 ppm inactive In, thus providing a drastic increase in trapping centers. The reduction of fraction 2 and the increase of fraction 1 with respect to the middle diagram means that above 230 K the small vacancy cluster (defect 2) is also formed via the defect 1, the monovacancy, by multiple trapping of vacancies at the In probe

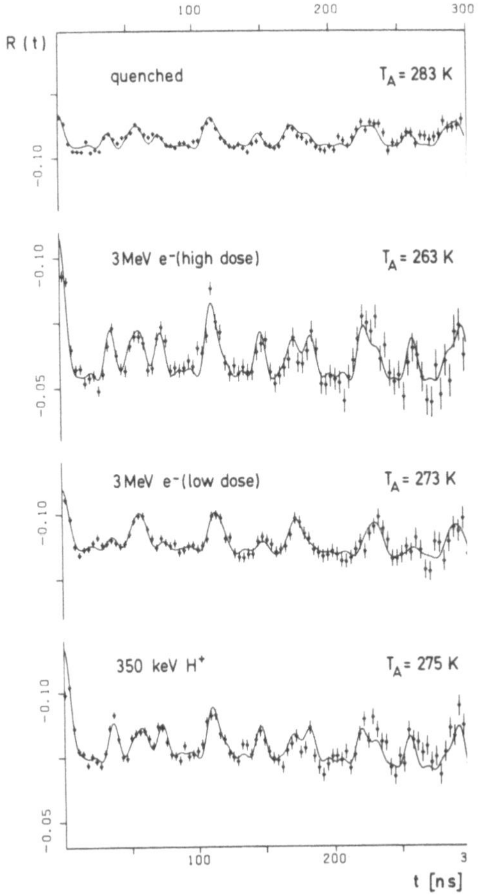

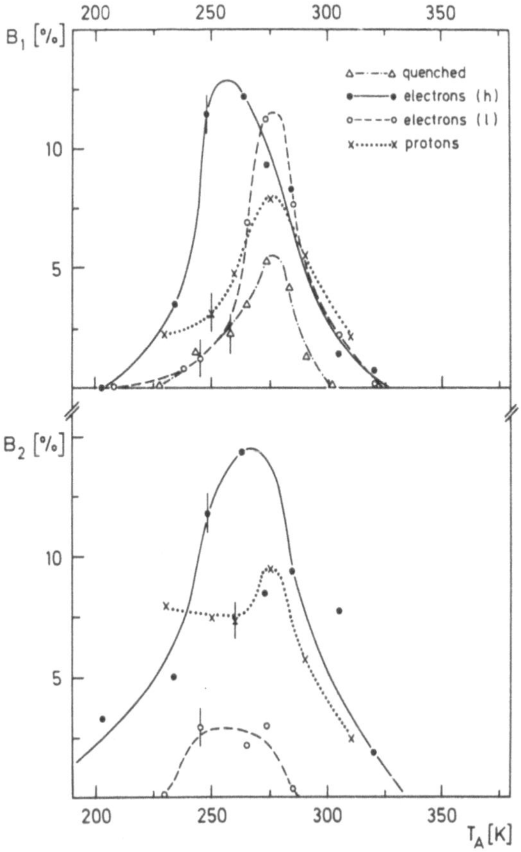

Fig.4.17

Fig.4.18

Fig.4.17. PAC spectra of ^{111}In in copper as observed after damaging by various procedures and annealing at temperatures T_A in the region of stage III [4.94]

Fig.4.18. Fractions of ^{111}In atoms in copper with trapped defects of type ν_{Q1} (upper part) and ν_{Q2} (lower part) as a function of annealing temperature. (h) and (ℓ) denote irradiation with high and low doses of electrons, respectively [4.94]

(probably a second one) so that ν_{Q2} describes a trapped divacancy. The "crossover" in the lower diagram around 230 K confirms the result of the proton irradiation, that the divacancy is already formed in the lattice and is more mobile than a monovacancy.

The two configurations 3 and 4 in copper are most probably vacancy clusters [4.93,95,97]. They appear around stage V. The trapped defect 3 with ν_{Q3} = 52 MHz, η = 0 and <111> orientation was observed after electron, proton and heavy-ion irradiation while defect 4 (ν_{Q4} = 55 MHz; η = 0.48) was only recognized after heavy-ion irradiation at a high dose [4.93]. Such vacancy clusters in fcc metals are known

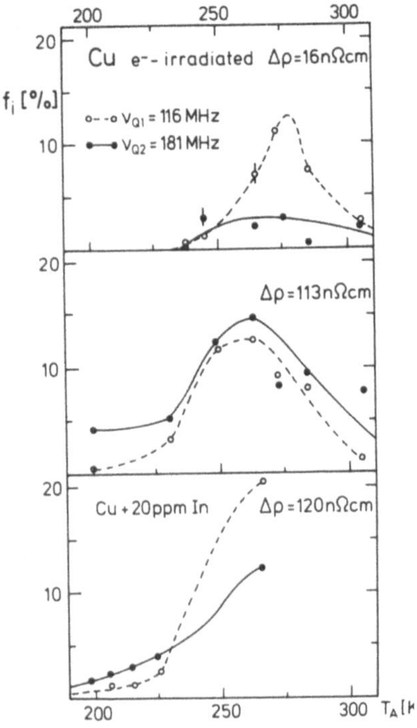

Fig.4.19

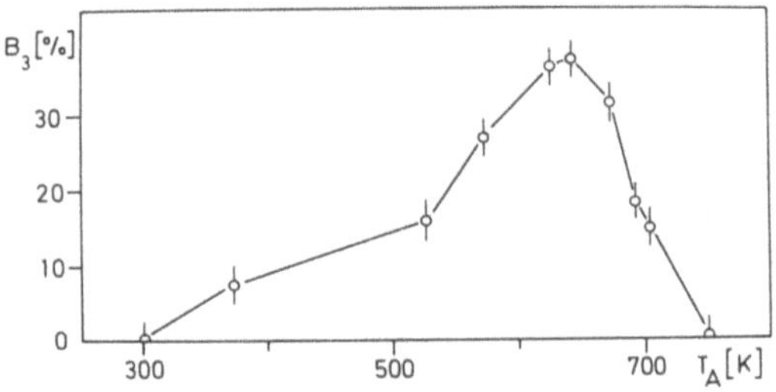

Fig.4.20

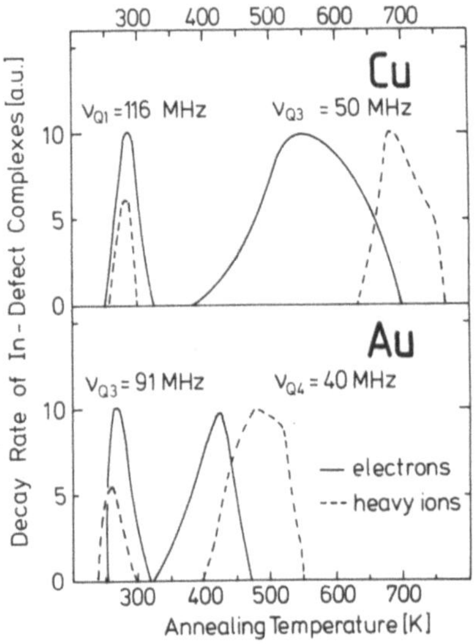

Fig.4.21

Fig.4.19. Fractions of ^{111}In atoms with trapped defects after electron irradiation with a low dose (16 nΩcm) and a high dose (113 nΩcm) in pure copper (<10^{-3} ppm ^{111}In) in comparison to a high dose irradiation (120 nΩcm) in Cu alloyed with 20 ppm inactive In [4.98]

Fig.4.20. Fraction of ^{111}In probe atoms with trapped vacancy clusters of type ν_{Q3} as a function of annealing temperature T_A [4.93]

Fig.4.21. Influence of damaging conditions on the thermal stability of the ^{111}In-vacancy cluster configuration in Cu(ν_{Q3}) and Au(ν_{Q4}). In contrast, the stability of the ^{111}In-monovacancy configuration (ν_{Q1} in Cu, ν_{Q3} in Au) is unchanged [4.72]

to form planar loops on a {111} plane [4.100]. Thus, the normal stacking sequence ABCAB of those planes is altered by a missing plane yielding a new sequence AB.AB, the stacking sequence of hcp-metals. A probe atom trapped in one of these stacking faults will experience a well-defined efg with an axial symmetry pointing into the crystallographic <111> direction. In Fig.4.20 the annealing behaviour of defect 3 is shown following correlated damage of the copper sample, i.e., the damage produced by the implanted radioactive ^{111}In ions themselves. At T_A = 750 K the sample is nearly completely annealed. In contrast to all other defect configurations discussed up to now, the annealing curve shown in Fig.4.20 depends on the type of irradiating particles. After electron irradiation the configurations decay at a lower annealing temperature than after heavy ion irradiation [4.97] which is explained by the different size of these loops and connected with their different thermal stability. In the upper part of Fig.4.21, the normalized decay rate of the fraction f is displayed, i.e., the high temperature part of f (T_A) has been differentiated with respect to the annealing temperature. The maxima of the decay rate of configuration ν_{Q3} differ by about 100 K between electron and heavy-ion irradiation, while — for comparison — the corresponding decay rate for the monovacancy configuration appears at the same annealing temperature. Because of the short-range sensitivity of the nuclear probe, however, no influence of cluster size on the efg is recognized, i.e., the immediate surroundings of the probe atom do not change with cluster size.

Palladium

Perturbed Angular Correlation: Studies were performed by BUTT et al. [4.101] with ^{111}Cd atoms produced after thermal neutron capture on enriched ^{110}Pd samples:

$$^{110}\mathrm{Pd}(n,\gamma)^{111}\mathrm{Pd} \xrightarrow[5.5h]{\beta^-} {}^{111}\mathrm{Ag} \xrightarrow[7.5d]{\beta^-} {}^{111}\mathrm{Cd} \quad .$$

Due to the capture process, the Pd atoms suffer a mean recoil energy $\bar{E}_R$ = 110 eV, enough to create up to three Frenkel pairs. Thus, some of the recoil atoms will make a replacement collision with one of their nearest neighbours, leaving a vacancy at the original position nearby the radioactive probe which in this case is ^{111}Ag. Therefore, the observed unique coupling constant ν_{Q1} = 86.5 MHz measured at the ^{111}Cd probe most probably reflects the interaction with a monovacancy. The complex dissociates between 235 K and 430 K, corresponding to the annealing stage III in Pd which is expected to be around 350 K. The same interaction frequency was observed after the ^{108}Pd $(\alpha,n)^{111}$Cd nuclear reaction (Sect.4.2.2a).

Silver

Mössbauer Effect: VOGL et al. [4.102] observed a similar behaviour for ^{57}Co in Ag after electron and fast neutron irradiation of the diluted alloys as in the case of aluminum. The large increase in the isomer shift of +0.32 mm/s is explained by interstitial trapping at the impurity atom forming a mixed dumbbell with a distance less than the nearest neighbour one. The drop in the Debye-Waller factor around 45 K is also explained by the activation of restricted cage jumps.

Perturbed Angular Correlation: Since the pioneering work of BEHAR and STEFFEN [4.103], several authors have investigated the problem of radiation damage by PAC methods in Ag [4.98,104-109]. The situation is not as clear as in other fcc metals because only one defect configuration trapped at ^{111}In in Ag could be uniquely identified. BUTT et al. [4.109] report ν_{Q1} = 80(2) MHz, η = 0 after e^--irradiation in the same annealing temperature regime as DEICHER et al. [4.98] who found ν_{Q1} = 82(2) MHz, η = 0 and a <110> orientation after proton irradiation of an unusually high dose as compared to Cu and Au. In both cases the absolute fraction of trapped defects is about a factor of 3-4 smaller than in comparable fcc metals, which might be a hint of a much smaller trapping radius and/or binding energy of In in Ag. Although additional identification of the vacancy character by quenching has been unsuccessful up to now, other similarities with Cu and Au suggest this configuration to be also due to a trapped monovacancy [4.74].

WODNIECKI et al. [4.108] have seen a trapped defect at ^{111}In nuclei recoil-implanted into Ag following the reaction ^{109}Ag(α,n)^{111}In and after proton irradiation [4.108a]. This defect is directly trapped at the implantation temperatures of 4.2 K and 77 K [4.108] and after thermal activation [4.108a] as in the experiments mentioned above, and the authors conclude from the measured mean coupling constant of ν_Q = 82 MHz that they have observed the identical configuration, a configuration already found in earlier experiments by THOMÉ and BERNAS [4.107] who investigated the damage dependence on α-particle dose rates. The same authors report the observation of a configuration with ν_{Q2} = 36.1 MHz after annealing above 300 K, i.e., after the monovacancy defect is annealed. It is probably due to vacancy clusters.

Platinum

Mössbauer Effect: MAGENDANS et al. [4.110] have reported a Mössbauer effect study of ^{161}Tb implanted into Pt at 5 K. Upon annealing they observed two nonsubstitutional fractions appearing around 250 K and 450 K which they attribute to the trapping of di- and monovacancies, respectively. VERBIEST et al. [4.110a] have studied platinum after implantation of ^{133}Xe at 100 K and 300 K. Vacancy trapping at the Xe impurities occurs around 300 K and 400 K. They observed four different defect sites which are ascribed to Xe atoms decorated by one, two, three, and four or more vacancies. This agrees with the PAC results of ^{111}In in Pt, where also in this temperature region the activation of defects was observed.

Perturbed Angular Correlation: Defects in platinum were studied by PAC with two different radioactive probes: ^{111}In and ^{181}Hf. In the case of In in Pt, the results of MÜLLER et al. [4.111] and PLEITER et al. [4.112] agree quite well. Both groups implanted the radioactive probes into high purity foils and investigated the defects produced by their own damage cascade (correlated damage). In spite of different implantation temperatures (293 K [4.111]; 10 K [4.112]), a configuration with ν_{Q1} = 103 MHz, $\eta \leq 0.1$ already appeared after implantation and remained stable up to 480 K. A second defect type ν_{Q2} = 210 MHz, $\eta \leq 0.1$ appeared above 270 K and dissolved around 400 K (in the experiment of MÜLLER et al., partly in favour of fraction 1). A third configuration with ν_{Q3} = 61 MHz, η = 0 is mentioned only by PLEIER et al. Both groups suggest that defect 1 is due to a monovacancy and defect 2 to a divacancy or small vacancy cluster. Both configurations appeared in a temperature region where a recovery stage was found by residual resistivity measurements, though the nature of this stage is still the subject of a controversy [4.113,114]. Radiation damage in Pt was also investigated by MÜLLER et al. [4.115,116] by implantation of ^{181}Hf ions at room temperature. By PAC measurements on the daughter nucleus ^{181}Ta, a well-defined defect was detected after annealing between 600 K and 850 K, which the authors assume to be due to a vacancy loop. However, this configuration was only observed after correlated damage and not after post-irradiation of an annealed sample with Pt ions, so that trapping of other impurities at the Hf probes cannot be completely ruled out as an explanation.

Gold

Perturbed Angular Correlation: The results of defect investigations in gold are very similar to those in copper—as far as stage III and V are involved—except that three instead of two defect configurations are observed around stage III [4.97,98,117]. Also, in this case results have only been obtained for ^{111}In as probe atoms. Figure 4.22 presents a collection of all data showing the influence of annealing temperature and mass of irradiating particles on the relative occurrence of the different defect types. Quenching data are also shown. Defects 1 and 2, characterized by ν_{Q1} = 102 MHz, η = 0.45 and ν_{Q2} = 101 MHz, η = 0, respectively, are formed predominantly after proton and heavy-ion irradiation. Their simultaneous occurrence during isochronal annealing as well as their nearly constant relative formation probability is a strong hint that the two configurations are produced by only one defect type trapped at two different positions at the probe atom, just giving rise to two different efg.

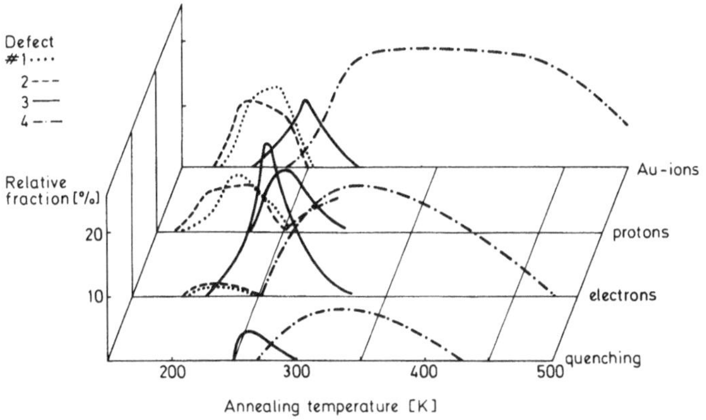

Fig.4.22. Influence of various damage procedures on the relative occurrence of different types of defects trapped at ^{111}In in gold as a function of annealing temperature. The different defects are explained in the text [4.117]

Configuration 3 could be uniquely identified as being a trapped monovacancy. A comparison of the time spectra after quenching and e^--irradiation reveals the identical quadrupole interaction and proves the vacancy character. An orientation measurement in a single crystal approved the expected <110> direction for a monovacancy. Therefore defect 1 and 2 are ascribed to be small defect clusters, probably divacancies.

A fourth defect type observed at annealing temperatures above 250 K shows similar behaviour to the one in copper which is due to a trapped vacancy cluster. Figure 4.22 clearly demonstrates also in this case the different thermal stabilities of clusters of different size as produced under different irradiation conditions. In the lower part of Fig.4.21 a comparison of the decay rates similar to Cu is shown [4.72,73].

As far as stage I is concerned, gold behaves quite differently to Al and Cu. In residual resistivity measurements no annealing stage I has ever been observed, leav-

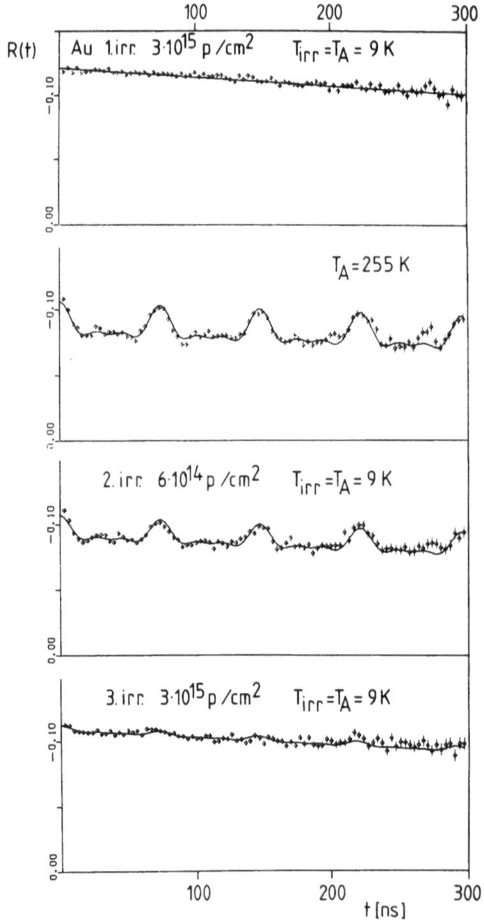

Fig.4.23a-d. Defect-antidefect reaction in Au. From above: (a) defect production at 9 K by protons; (b) trapping of monovacancies by ^{111}In atoms at T_A = 255 K; (c) and (d) post irradiation by protons at 9 K. Mobile interstitials annihilate the trapped vacancies [4.72]

ing it an open question as to whether interstitials are mobile at low temperatures or not. Since the investigation of defects by their hyperfine interaction with probe atoms is based on the trapping of the migrating defect at the probe atoms, PAC experiments could answer this question. But no unique hyperfine interaction due to trapping has yet been observed at ^{111}In, as can be seen from the first spectrum in Fig.4.23. However, via an indirect method, the defect-antidefect reaction, the mobility of interstitials has been proved down to 9 K [4.72,73]. The results are shown in the lower three spectra of Fig.4.23. Gold samples were doped with "In-vacancy" probes of defect type 3 by annealing proton irradiated specimens at 255 K. Subsequently, these probes were post-irradiated with protons of two different doses at 9 K. The amplitude of the modulation pattern characteristic of the trapped monovacancy gradually disappears. Since neither vacancies are mobile at these low temperatures nor is detrapping of vacancies from In possible, one has to assume that mobile interstitials annihilate the trapped vacancies.

Lead

Perturbed Angular Correlation: BUTT et al. [4.118] performed PAC studies on lead samples doped with ^{111}In after electron irradiation at 7 K, a temperature above stage I. Even at low irradiation doses, a high fraction of probe atoms experience a perturbation which is interpreted as trapping of interstitials and interstitial clusters in different configurations. Annealing between 120 K and 180 K (stage III) partially restores the cubic symmetry around the In atoms, but also leads to a well-defined configuration with ν_{Q1} = 28 MHz and η = 1 which is probably due to an In-interstitial complex. At the end of stage III all probe atoms have a cubic surrounding.

b) *Diamagnetic bcc Metals*

Niobium

Perturbed Angular Correlation: SIELEMANN et al. [4.119] have studied radiation-induced defects in high-purity niobium by using the PAC technique on the probe atoms ^{100}Rh and ^{111}Cd. In contrast to most other experiments discussed before, the respective parent nuclei ^{100}Pd and ^{111}In have been produced by heavy-ion reactions thus leading to a homogenous doping of very low concentration ($<10^{-2}$ ppm) over a thickness of about 20 μm. The applied nuclear reactions and β-decay schemes are:

$$^{93}\mathrm{Nb}(^{12}\mathrm{C},5n)^{100}\mathrm{Ag} \xrightarrow[2.3\mathrm{m}]{\beta^+,\mathrm{EC}} {}^{100}\mathrm{Pd} \xrightarrow[3.6\mathrm{d}]{\mathrm{EC}} {}^{100}\mathrm{Rh}$$

and

$$^{93}\mathrm{Nb}(^{22}\mathrm{Ne},4n)^{111}\mathrm{Sb} \xrightarrow[75\mathrm{s}]{\beta^+,\mathrm{EC}} {}^{111}\mathrm{Sn} \xrightarrow[35\mathrm{m}]{\beta^+,\mathrm{EC}} {}^{111}\mathrm{In} \xrightarrow[2.8\mathrm{d}]{\mathrm{EC}} {}^{111}\mathrm{Cd} \quad .$$

Isochronal annealing revealed that the two different probes trap defects in different temperature ranges.

The small Pd probes trap two defects during irradiation at 30 K with about the same probability, characterized by ν_{Q1} = 42 MHz; η = 0 and ν_{Q2} = 33 MHz; η = 1. Detrapping occurs around 100 K. In contrast to these findings, a well-defined modulation pattern of ^{111}In probes is only observed between 200 K and 250 K, resulting in defect configurations with ν_{Q1} = 87 MHz; η = 0 and ν_{Q2} = 105 MHz; η = 0.65. Above 270 K a third defect appears with ν_{Q3} = 177 MHz, η = 0.

The authors suggest trapping of interstitials at the Pd probes in agreement with the high mobility of these defect types above 4 K. The trapping of defects at the ^{111}In probes correlates to a recovery stage observed around 300 K. The ^{111}In results are confirmed by the investigations of VIANDEN and WINAND [4.120] who implanted ^{111}In with 80 keV into Nb at room temperature, leading to an estimated impurity atom concentration of 600 ppm in a depth of 170 Å. They have also seen the defect configurations 1 and 3 up to annealing temperatures of 350 K and 400 K, respectively. In all three experiments the trapped defects originated from their own recoil cascade, i.e., correlated damage is observed. Recently, SIELEMANN et al. [4.119a] performed additional experiments. They observed the formation of ν_{Q1} and ν_{Q2} around 250 K after e^--irradiation and determined the <111> orientation of the efg belonging to defect 1. They concluded that single vacancies migrate freely at 250 K.

Molybdenum

Mössbauer Effect: MANSEL et al. [4.121] have observed the trapping of interstitials at ^{57}Co impurities in molybdenum by Mössbauer spectroscopy. They irradiated diffused ^{57}CoMo samples at 100 K with 3 MeV electrons. Since interstitials in Mo are mobile at this temperature and vacancies are not, the interstitials can survive a recombination with vacancies only if they are trapped at ^{57}Co impurity atoms. In such a case new Mössbauer lines appear in addition to the unperturbed single line. In the first spectrum of Fig.4.24 the contribution of this unperturbed line is subtracted.

These results are confirmed by annealing experiments after fast neutron irradiation at 4.6 K. Figures 4.24b-e show Mössbauer spectra with subtracted substitutional lines after isochronal annealing measured a 4.2 K. A numerical analysis revealed five quadrupole doublets with central isomer shifts of about ΔS = +0.46 mm/s (lines 1 to 5 in Table 4.7). The contribution of the different defect lines are indicated in the figures. At T_A = 124 K configuration 1 is most pronounced, while at T_A = 214 K, it has disappeared and line 3 dominates due to the reorientation of configuration 1 into 3. At higher temperatures, line 3 also disappears and after T_A = 460 – stage III of Mo – configurations 4 and 5 are observed. Defect line 2, which is most pronounced after electron irradiation, anneals out in stage III as do line 4 and 5 at somewhat higher temperatures. Above stage III new lines occur as shown in Fig.4.24f,

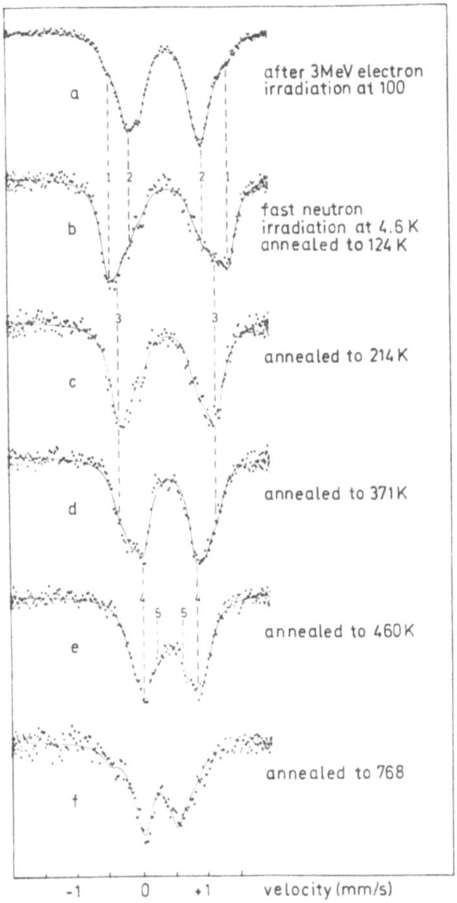

Fig.4.24. Mössbauer spectra of ^{57}Co in molybdenum after electron and neutron irradiation at different annealing temperatures T_A. The numbers refer to the different ^{57}Co-defect configurations described in the text. The substitutional Mössbauer line is substracted [4.121]

characterized by a small central shift. These new lines anneal only in stage V at about 1300 K. The five configurations tabulated in Table 4.7 are attributed to trapped mono, di- or triinterstitials, probably with different orientations around the ^{57}Co probe atom.

Lattice damage induced by implantation of radioactive ^{133}Xe in Mo has been studied by Mössbauer spectroscopy using the 81 keV transition in its daughter nucleus ^{133}Cs. Four overlapping single Lorentzian lines were observed by REINTSEMA et al. [4.122,123], the intensities of them varying with annealing temperature, while the isomer shifts and the linewidths remain unchanged. Since the Xe atom is large compared to the host atom, the authors assume only interactions with vacancies. The isomer shift is then taken as an indication for the space which is available for the impurity atom in the host lattice. The more the Xe(Cs) atom is compressed, the higher the electron density at its nucleus and therefore the more negative the isomer shift. Thus the most negative isomer shift δ_1 is attributed to the substitutional site (δ_1), while δ_2 to δ_4 are attributed to one or more vacancies trapped at Xe

Table 4.9. Isomer shifts (in mm/s) of the four sites of ^{133}Xe$\underline{Mo}$ and ^{133}Xe$\underline{W}$, taken at 4.5 K with respect to a CsCl absorber [4.127]

	δ_1	δ_2	δ_3	δ_4
^{133}Xe$\underline{Mo}$	-1.21(3)	-0.57(7)	-0.02(5)	0.58(7)
^{133}Xe$\underline{W}$	-1.59(4)	-0.63(8)	-0.01(6)	0.72(8)

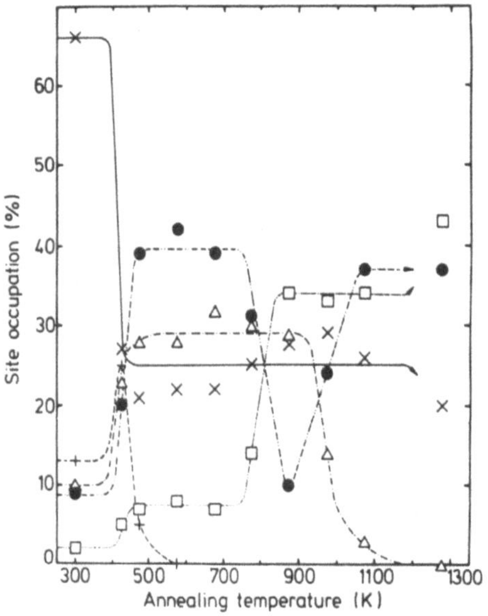

Fig.4.25. Site occupation of ^{133}Xe implanted into molybdenum: x site 1 (substitutional); □ site 2 (monovacancy); ● site 3 (divacancy); Δ site 4 ($\geq$ trivacancy) [4.123]

atoms (Table 4.9). A sharp decrease of the substitutional fraction near 430 K correlated with an increase for fraction 2 (Fig.4.25) further supports the trapping of vacancies, since they are becoming mobile at this temperature (stage III). A broadening of the line due to quadrupole interaction could not be observed because of the small quadrupole moment of the nuclear state involved.

Perturbed Angular Correlation. Defect studies in poly- and monocrystalline Mo samples were performed by WEIDINGER et al. [4.124,125] with the PAC technique using ^{111}In as probe atom. The radioactive ions were implanted into the specimen at room temperature, well below annealing stage III. The measurements were either carried out on the correlated damage cascade (ion irradiation) or on annealed samples after electron and proton irradiation, respectively. Three different defect configurations were identified around stage III, characterized by (i) ν_{Q1} = 125 MHz, η = 0, symmetry axis <111>; (ii) ν_{Q2} = 155 MHz, η = 1 and (iii) ν_{Q3} = 98 MHz, η = 0, <100>. Figure 4.26 shows three representative Fourier transforms of angular correlation spectra from which the parameters have been deduced. The annealing behaviour of the trapped fractions for different irradiation conditions are displayed in Fig.4.27 which also includes a differentiated resistivity recovery curve of stage III [4.125]. The defect

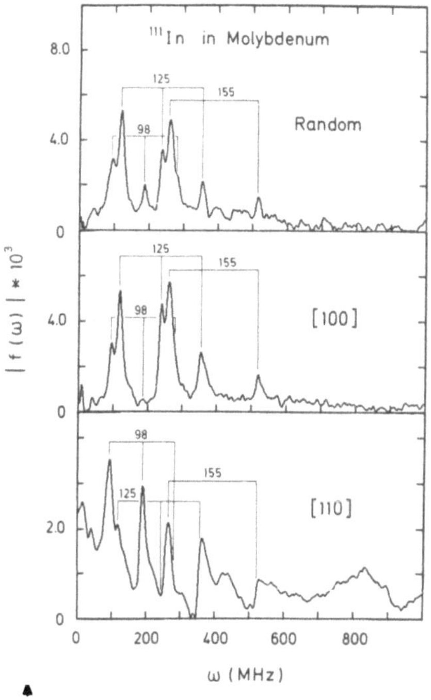

Fig.4.26. Fourier transforms of PAC spectra of ^{111}In implanted into molybdenum after annealing at 470 K. The spectra measured at a randomized crystal and at a crystal with two distinct orientations with respect to the γ-detectors show a drastic difference in the intensity of the different defect lines from which the orientation of the respective efg's are determined [4.124]

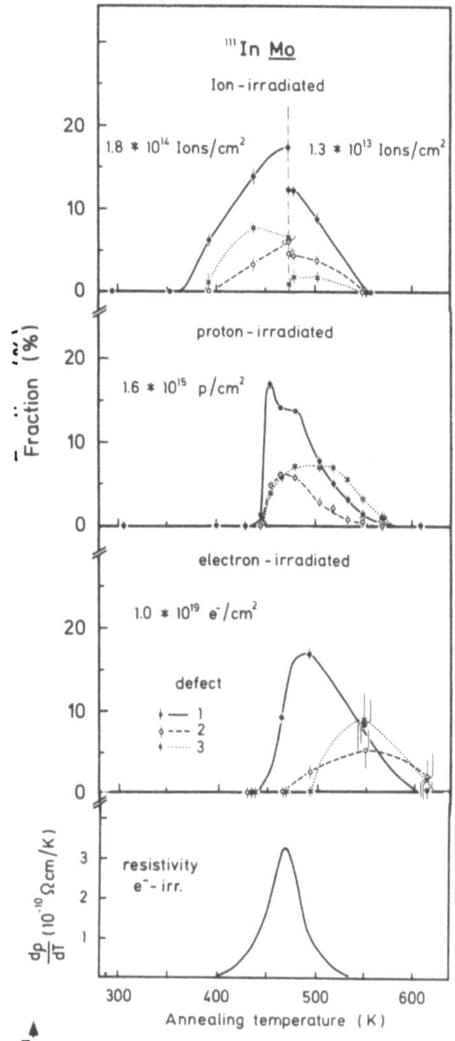

Fig.4.27. Fractions of ^{111}In atoms in molybdenum with different trapped defects for various irradiation procedures as functions of annealing temperature. Ion-irradiated corresponds to damage produced by the ^{111}In implantation for two different doses (correlated damage). For comparison, a differential resistivity recovery curve for an e$^-$ irradiated sample is shown in the lower part of the figure [4.125]

configurations are attributed to the trapping of a mono-, di- and trivacancy at the In atom, respectively. The geometrical arrangement in the bcc lattice is shown in Fig.4.4.

Obvious differences in the trapping process can be deduced from Fig.4.27. While in the ion-irradiated sample the trapping starts around 370 K due to the nearby vacancies of the correlated damage cascade, the trapping after proton and electron irradiation is strongly correlated with the mobility of vacancies in stage III. The appearance of the more complicated defects 2 and 3 at lower annealing temperatures after proton irradiation compared to electron irradiation reflects the higher vacancy concentration in the former case and at the same time, is a hint that the di- and trivacancy configurations are most probably formed by multiple trapping. Assuming that the decrease at high temperatures is due to detrapping, one has to conclude that configuration 3 is the most stable one.

Configurations 4 and 5 (Table 4.8) are only observed after ion irradiation and are stable up to temperatures above 1500 K. The exact annealing temperature depends on the sample preparation and/or the implantation dose. They are suggested as being vacancy clusters, though an exact proof is still missing.

Tungsten

Mössbauer Effect: Similar behaviour for ^{133}Xe implanted in W to that in Mo is reported by REINTSEMA et al. [4.122,123] (Table 4.9), the only difference is connected to the higher temperature of stage III (570 K).

Perturbed Angular Correlation: An investigation as performed by VIANDEN et al. [4.120] in Nb was carried out in tungsten by PÜTZ et al. [4.126]. They identified two defects trapped at the ^{111}In probe atom: Defect 1 with ν_{Q1} = 141 MHz, η = 0 and <111> orientation between 550 K and 800 K, which they attributed to a trapped monovacancy since it appears around stage III [4.127] and exhibits the expected orientation within a bcc lattice such as the defect configuration 1 seen in Mo. Defect 2 with ν_{Q2} = 318 MHz; η = 0 and <110> orientation was only observed in a single crystal experiment. Figure 4.28 shows a time spectrum in a single crystal taken for an annealing temperature of 650 K.

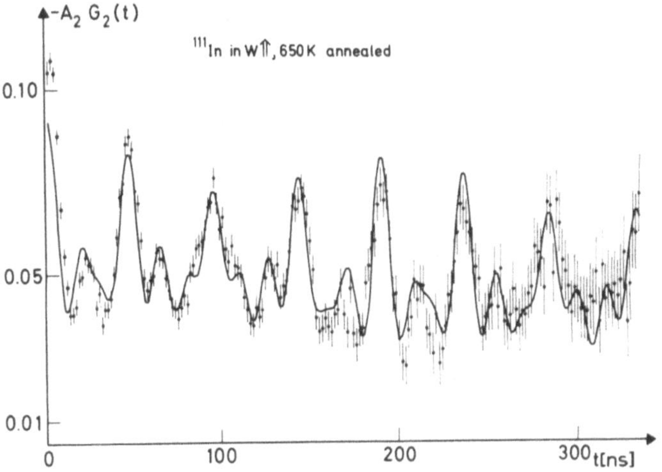

Fig.4.28. PAC spectrum for ^{111}In implanted into tungsten single crystal annealed at 650 K. The solid line represents a fit to ^{111}In-defect configurations 1 and 2 where the efg orientations are in <111> and <110> directions, respectively [4.126]

In a more recent publication, PÜTZ et al. [4.126a] extended their studies on vacancy trapping at ^{111}In in tungsten. They observed two new configurations characterized by ν_{Q2} = 181 MHz, η = 1 and ν_{Q3} = 263 MHz, η = 0. Defect 2 is connected with stage III, whereas defect 3 occurs in the temperature range above stage III. The absence of the defect with ν_{Q2} = 318 MHz, η = 0 quoted earlier in [4.126] is not mentioned in the new publication. Probably both frequencies attributed to ν_{Q2} (318 and 181 MHz) describe the same defect, with the discrepancy being due to a different η-analysis.

c) *Ferromagnetic Cubic Metals*

In ferromagnets, a radioactive probe atom interacts with the internal magnetic field of the lattice and—usually, in addition—an external magnetic field. Thus, in the case of an unperturbed lattice, the PAC spectrum normally exhibits a sinusoidal modulation of frequency $2\omega_L$. If a defect is trapped, the Larmor frequency may change and, if a quadrupole interaction is also present, a combined interaction occurs. In Mössbauer experiments an additional isomeric shift may exist.

Iron

Mössbauer Effect: REINTSEMA et al. [4.128-134] performed Mössbauer effect measurements with the heavy 5(s,p) elements Te, I, Xe and Cs implanted at different temperatures into polycrystalline iron foils. The analysis of the spectra, examples of which are shown in Fig.4.29, revealed three to four different sites, classified by their magnetic hyperfine field values. A high field site (h) is interpreted as corresponding to substitutional atoms with an unperturbed surrounding; an intermediate field site (i) with the association of one vacancy, a low field site (l) with two or three vacancies and in the ^{133}Xe case a fourth site (z), though invisible because its recoilless fraction is close to zero, is attributed to at least three vacancies. Besides different field values, these findings are deduced from less negative isomer shifts for the vacancy-associated sites with respect to the substitutional site due to a lower electron density at the probe atoms (Table 4.7). Smaller recoilless fractions are also observed.

From these results, from the site occupation behaviour at different implantation temperatures and annealing sequences, the authors conclude that vacancies become mobile around 500 K. At this temperature an increase of the low field site is correlated with a decrease of the other fractions. A comparison of the hyperfine magnetic field for the different impurity atoms shows striking systematic behaviour for all sites (Fig.4.30). For ^{161}Tb in iron, only one vacancy-associated site was observed [4.135].

Nuclear Quadrupole Resonance: In the first successful nuclear magnetic resonance experiment on oriented nuclei (NMR/ON), a resonance was observed for a vacancy as-

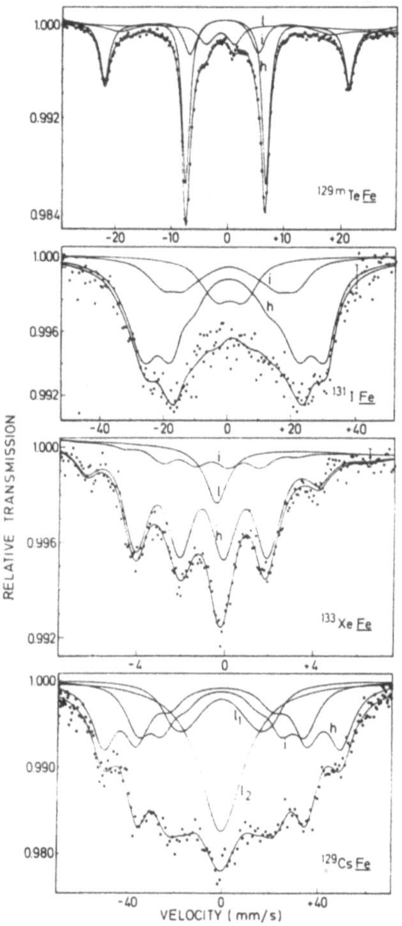

Fig.4.29. Mössbauer spectra of implanted sources of ^{129m}Te, ^{131}I, ^{133}Xe and ^{129}Cs in iron. In each case the spectra are decomposed into at least three components: (h) high field (substitutional-) site; (i) intermediate field site; (l) low field site [4.134]

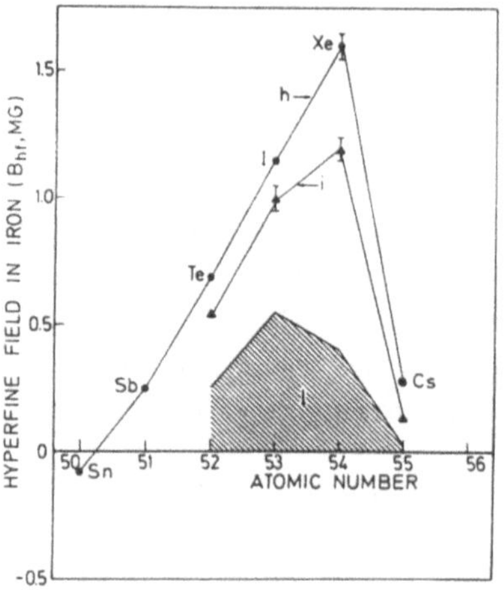

Fig.4.30. Magnetic hyperfine fields of the 5 (s,p) elements and Cs in iron. (h) high field (substitutional-) site; (i) intermediate field site; (l) low field site. The shaded area indicates that there may be more than one component for each case [4.134]

sociated site at ^{131}I in iron. VISSER et al. [4.136] implanted ^{131}I, presumably at 300 K, in single crystals mounted with their crystalline axis <100>, <110> and <111> parallel to the magnetic polarizing field B = 0.15 T, which determines the magnetic hyperfine interaction axis. Beside the main substitutional resonance at 683.8 MHz (Fig.4.31), they observed additional weak resonances displaced from the Zeeman frequency by the quadrupole interaction frequency $(eQV_{zz}/4\hbar)\cdot(3\cos^2\theta-1)/2$, where θ is the angle between the magnetization axis and the <111> axis of V_{zz}, the direction expected for an impurity-monovacancy configuration in a bcc lattice. For B applied along the <100> axis, the term $(3\cos^2\theta-1)$ is zero for all <100>-<111> angles so that there is no quadrupole splitting. The observed single line is due to the reduced magnetic hyperfine field. For B along the <100> direction a symmetric splitting of this line is expected, while for B along <111>, an asymmetric splitting with $eQV_{zz}/4\hbar$ and $-eQV_{zz}/12\hbar$ should occur. This is clearly proven by the experimental results as shown in Fig.4.31 though the appearance of a third line in the B parallel to the <111> case at $+eQV_{zz}/12\hbar$ is still unexplained.

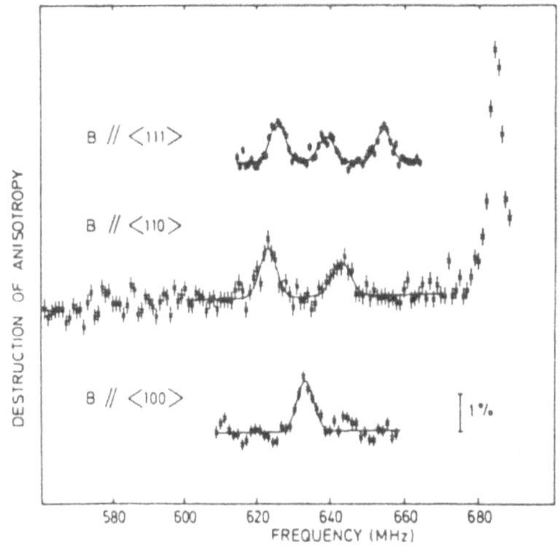

Fig.4.31. NMR/ON spectra of a vacancy-associated site on ^{131}I implanted in single-crystal iron observed with the host magnetized along <111>, <110> and <100> directions [4.136]

Nickel

Mössbauer Effect: As discussed in the case of iron, Mössbauer experiments were also performed in Nickel using ^{129m}Te as implanted radioactive probes [4.131,132]. The spectra could be fitted with three components (Table 4.7), a substitutional one with high magnetic field (B_{hf} = 295 KG at 4.2 K), an intermediate component, also displaying a sharp hyperfine field (B_{hf} = 266 KG at 4.2 K) which is attributed to the trapping of one vacancy since this configuration disappears after annealing at 470 K, and a low field site ($B_{hf} \sim$ 175 KG) which exhibits a broader distribution and is probably due to several trapped vacancies. In contrast to other Mössbauer measurements, the vacancy-associated intermediate site for $^{129}I\underline{Ni}$ as well as for $^{129}I\underline{Fe}$ displays an isomer shift of higher s-electron density than for the substitutional site, a result not yet understood.

Perturbed Angular Correlation: Several PAC experiments were carried out to study radiation damage in nickel with ^{111}In as probe atoms [4.137-142]. HOHENEMSER et al. [4.138] implanted ^{111}In at liquid He temperature and observed the influence of correlated damage in an isochronal annealing sequence. Besides the unperturbed fraction f_1 they identified two perturbed fractions above T_A = 300 K as displayed in Fig.4.32. A typical time differential PAC spectrum taken after annealing to 400 K, together with its Fourier-transform is shown in Fig.4.33. Two components with sharp frequencies $2\omega_{L1}$ and $2\omega_{L2}$ can be seen: the higher frequency ω_{L1} corresponds to the unperturbed fraction, while the lower one is assumed to arise from the defect-associated site 2. Since no quadrupole contribution leading to a combined interaction is observed for this defect, the authors assume a cubic surrounding which might be due to a tetra-vacancy around the probe atom relaxed to the center of the tetrahedron. Between T_A = 500 K and 800 K, a third configuration appears which exhibits about the same magnetic interaction as found for fraction 2 but, in addition, shows

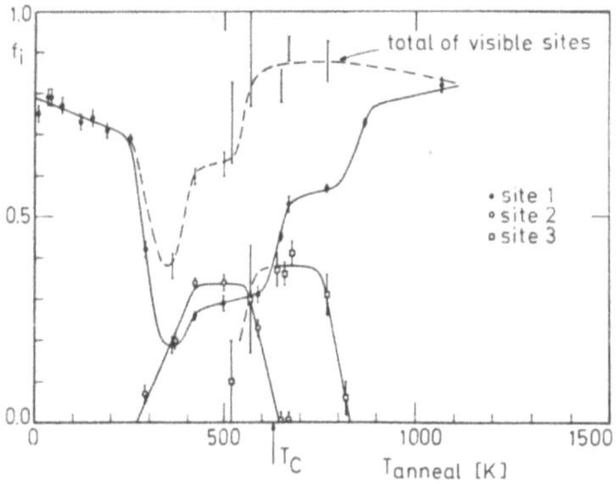

Fig.4.32. Annealing behaviour of different fractions of ^{111}In implanted into nickel. Site 1 refers to the substitutional fraction, site 2 and 3 to vacancy-associated ^{111}In sites [4.138]

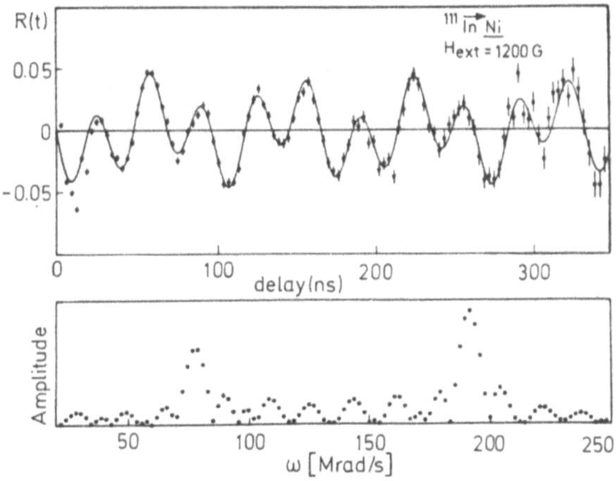

Fig.4.33. Magnetic interaction at ^{111}In implanted in Ni after annealing at 400 K. The Fourier transform (lower part) shows two components with sharp frequencies $2\omega_{L1}$ and $2\omega_{L2}$. The sample was magnetized perpendicular to the detector plane [4.138]

a rather well-defined quadrupole interaction (ν_{Q3} = 56 MHz) with axial symmetry, pointing in the <111> direction. This configuration is interpreted as resulting from three vacancies around a relaxed probe atom. It is argued that the relaxation in both cases originates from the larger size of the In atom compared to the Ni host atoms. Concerning the occurrence of fraction 3, new PAC measurements show that this configuration is already formed at 280 K [4.142].

d) *Hcp Metals*

In contrast to diamagnetic cubic metals, a noncubic metal already exhibits quadrupole interaction if a probe atom occupies a substitutional lattice site with an unperturbed surrounding due to the intrinsic efg.

Cadmium

A defect-free situation is displayed in the last spectrum (T_A = 290 K) of Fig.4.34 for an annealed ^{111}In doped cadmium sample, together with the Fourier transform on the right side of this figure, which shows the characteristic three line pattern of an axial symmetric efg. WITTHUHN et al. [4.143] irradiated such a doped specimen with 10 MeV protons at 77 K and carried out an isochronal annealing program. At 100 K (first spectrum), mainly the intrinsic signal is observed, whereas between 110 K and 140 K (stage III of Cd), an additional signal is visible caused by a trapped defect with ν_{Q1} = 103 MHz and η = 1 (second spectrum for T_A = 125 K). This defect was also seen after quenching, thus proving its vacancy character.

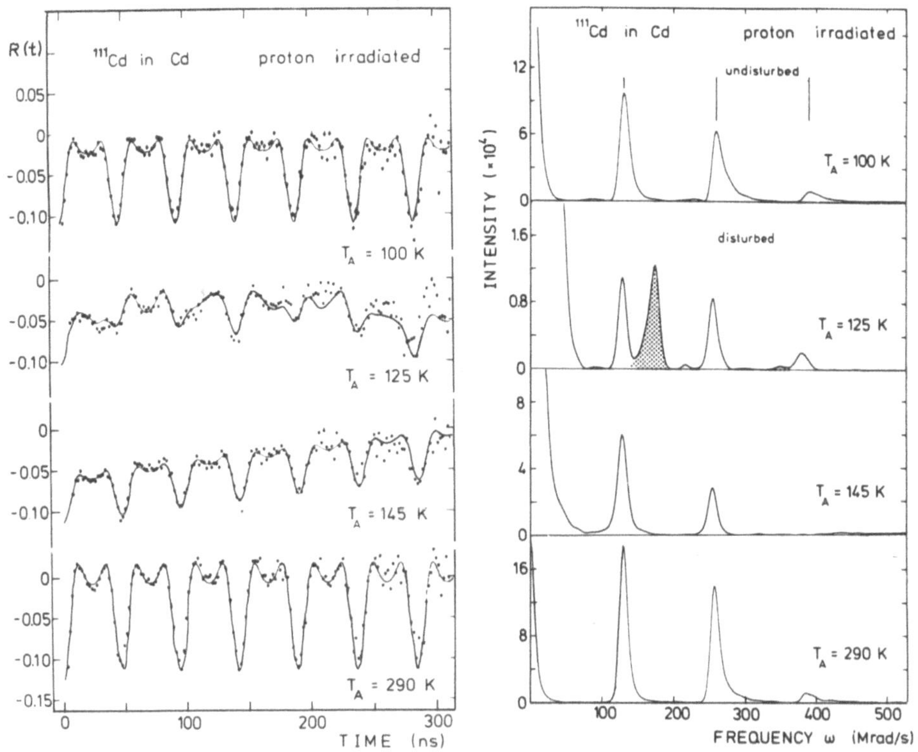

Fig.4.34. PAC spectra of ^{111}In atoms in Cd after proton irradiation, annealed at different temperatures (left side) with the corresponding Fourier transforms (right side). The undisturbed pattern due to the hexagonal structure and the disturbed one (dotted area) are indicated [4.143]

An interesting feature of this experiment is the observation of another defect with vanishing efg, to be deduced from the offset of the three upper spectra in comparison to the lowest one in Fig.4.34. This configuration, which removes the characteristic efg of the hcp lattice, must have cubic symmetry. It can be explained by analogy to the influence of clusters in fcc metals (see defects 3 and 4 in Cu and Au, respectively) - by a change of the hcp stacking sequence ABAB to ABCAB, thus generating a local fcc lattice with vanishing efg and therefore ν_{Q3} = 0 MHz.

Zinc

An experiment similar to the one in Cadmium was carried out in Zinc by SEEBÖCK et al. [4.144,144a] who observed a trapped defect at ^{111}In between 140 K and 220 K, characterized by ν_{Q1} = 70.0 MHz and $\eta \leq 0.1$. Figure 4.35 shows the recovery behaviour of proton irradiated Zn. The appearance of the defect configuration is obviously correlated to a reduction of unperturbed lattice sites. The same interaction parameters were observed in a quenching experiment. On the basis of symmetry considerations and taking into account the ionic radii of In and Zn, the authors favour a monovacancy or divacancy-In complex as an explanation for this defect.

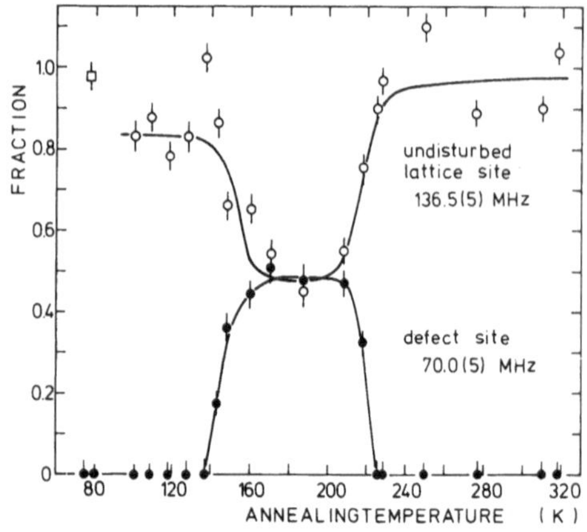

Fig.4.35. Annealing behaviour of proton irradiated Zn doped with ^{111}In. The vacancy-associated defect site anticorrelates with the separately determined undisturbed fraction [4.144]

e) *Summary*

In the preceeding review on defect investigations by nuclear methods using probe atoms from radioactive sources, a considerable amount of detailed information collected in recent years has been presented. The very nature of these methods to determine physical quantities on a microscopic scale has been proved to be successful in providing information complementary to other techniques and by these means in contributing to the solution of sometimes controversial or open problems. Thus, e.g., the activation of monovacancies in copper and gold around stage III could be unequivocally proved, supporting the one-interstitial-model (Table 4.5).

The success of the nuclear methods stems mainly from their high resolution and sensitivity which allows them to distinguish between different defect configurations and by attributing well-defined parameters for their characterization to recognize them under different experimental conditions. Independent of the defect production process, they can be detected as soon as a positive binding energy exists between defect and impurity atom or - if not - via annihilation in a defect-antidefect process, as has been shown in a few examples.

Since up to now the hyperfine interaction between an impurity and a defect has not been able to be calculated theoretically, the characterization and discrimination of different defects has had to be carried out completely by experimental means. Though in most cases the reliable information is only drawn from the more complicated systems where impurities are a part of the investigated 'pure' metals, the dilution of impurities is usually so high, that the probe atoms can be considered as independent observers. Due to the localized sensitivity given on the one hand by a limited trapping radius, and on the other hand by the short-sightedness of the interaction, the nuclear methods are able to directly detect the mobility of point-defects, a procedure which is scarcely possible with other methods. Together with the detailed determination of the interaction parameters, the method allows the microscopic classification of point defects within a lattice.

The application of general assumptions to explain the nature of defects, however, has to be carefully examined in each case if definite statements are to be deduced. For example, it was widely accepted from energetic arguments that an oversized impurity favourably traps vacancies, and not interstitials, while an undersized one behaves the other way around. But there are several cases like the oversized In atom in Cu or the undersized Co atom in Al, where both kinds of defects are trapped. Therefore, a characterization of a defect only by geometrical arguments has to be confirmed by additional independent proofs. The same holds for the influence of defects upon the isomeric shift. Though it is believed that a nearby vacancy reduces the electron density at the impurity atom resulting in a more positive shift, there are contradictory results such as, for example, for ^{129}I in Ni and Fe, where the assumed association of vacancies exhibits a negative shift compared to the unperturbed substitutional site.

Throughout the review a discussion in context with thermodynamic quantities has been omitted, in spite of the fact that in many papers the "microscopic" results are compared with values deduced from conventional techniques. In his review, NIESEN [4.70] has tabulated vacancy migration enthalpies obtained from hyperfine interaction experiments together with stage III migration enthalpies from resitivity measurements. Though a clear and not surprising correlation exists, most of the hyperfine data yield lower values. This may be roughly explained by the fact that the trapping impurity influences the activation enthalpy, which might be especially important if only a small number of jumps are involved. But it is exactly this feature which needs further detailed investigation. More careful annealing experiments involving wide concentration ranges of well-characterized defects have to be carried out and the influence of the probe atom on the surrounding potentials has to be studied.

As far as the binding energies between impurities and defects for a given host are concerned, the very nature of the disappearance of hyperfine interaction has to be determined. Several competing processes may cause it such as, e.g., annihilation by antidefects, trapping of additional defects, etc., but only detrapping allows a definite conclusion about those energies. There are only a few cases where this is proved by experiments. Such cases are, e.g., the detrapping of interstitials in Cu or the vacancy detrapping in some fcc metals, where the vacancies were produced by low dose electron irradiation. One of the difficulties of drawing unambiguous conclusions stems from the fact that the disappearance of a certain interaction correlated to a simultaneous increase of the unperturbed fraction still incorporates the ambiguity of a cubic defect configuration around the impurity atom, at least in cases where only quadrupole interactions are observed in a cubic lattice.

There have already been a few attempts to set up systematics. Definite conclusions would be much easier, of course, if hyperfine fields could be calculated exactly. DEICHER and WICHERT [4.74] have proposed a set of required properties to characterize trapped monovacancies at In impurities in fcc metals. For all configurations so far selected, they compared the quadrupole interaction frequencies after proper normalization of the different lattice constants. Though this procedure is a rather rough one, it shows that other effects like lattice/impurity relaxation or charge screening have to be taken into account to completely understand the vacancy impurity complex. Localization studies by channeling may reveal information about the relaxation of an impurity in the presence of defects [4.60].

Systematics on more complicated configurations starting from divacancies to larger clusters are not yet evident, mainly because of the lack of sufficient definite data. Similarities for trapped clusters in Cu and Au have already been mentioned in the preceding discussion, but corresponding configurations in other fcc host metals have still to be identified.

The largest amount of PAC results exists for the In/Cd impurity and they are specific in as much as this probe is concerned. To estimate the influence of the special impurity on defect studies, however, other probes have to be used to a larger extent.

This has been done in the Mössbauer effect studies in Iron for 4 subsequent impurities of 5(s,p) elements and DE WAARD [4.134] has pointed out the obvious systematic behaviour of magnetic hyperfine fields of the vacancy-associated sites in analogy to the dependence of the substitutional site fields.

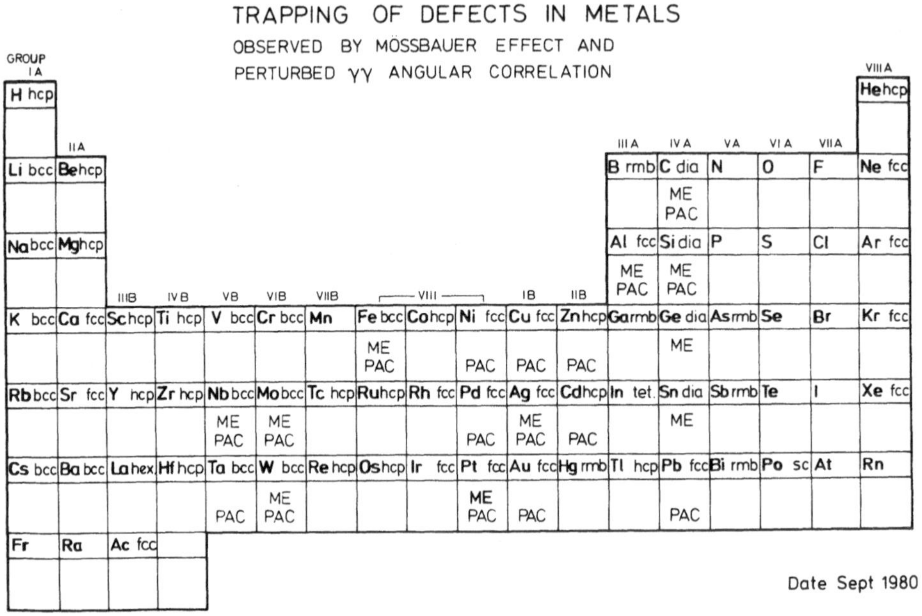

Fig.4.36. Collection of host elements where defects are investigated by Mössbauer effect (ME) or perturbed angular correlation (PAC) [4.73]

In conclusion, it can be summarized that the present status of defect studies by hyperfine interaction techniques has proved to be a powerful tool in the microscopic understanding of that field. Many host metals have already been attacked (Fig.4.36) though the scope of their investigation still varies widely. Different defect production techniques have been applied. Their specific influence on the results, however, needs further consideration. A comparison of them profits from the individual recognition of each kind of defect. This also allows the application to other damaging procedures such as, e.g., cold working [4.145a,b]. More refined techniques such as defect-antidefect reactions, quantitative rate measurements, impurity/defect concentration dependences and combination with channeling [4.145c,d], only elaborated in a few pioneering experiments up to now, have to be applied to other cases. Probably more of those techniques will come up in the future. Finally, the extension of defect studies beyond pure metals should be mentioned as it has already been carried out on some alloys [4.3,4] or systems where, in addition to the probe atoms, gases are introduced [4.145].

4.2.2 In-Beam Experiments

The discussion of the results in the previous chapter had demonstrated that the main advantages of PAC over other techniques (i.e., residual resistivity), normally used in the study of radiation damage, are its sensitivity to specific defects and the ability to selectively pursue the behaviour of different defect types. However, the small amount of probe isotopes suitable for PAC can become a drawback if the results become dependent on the binding of the defects with the probe atom. The method of the perturbed angular distribution after nuclear reactions provides an alternative, since a much larger selection of isomeric nuclear states can be reached by nuclear reactions. However, such in-beam experiments are very often much more involved and, in addition, the defects are always produced by the recoil of the probe nuclei so that only correlated radiation damage can be studied. In this section, an attempt will be made to review this in-beam part of radiation damage research and to convey the simple models which have been developed to understand the observed effects.

In an in-beam experiment, the radioactive probe nuclei are produced within the bursts of a pulsed particle beam; during the beam-off period the de-excitation γ-rays from the nuclear state of interest are detected. Therefore the beam-off period, which is adjusted to the nuclear life time, is the time window through which the defect situation is observed. After the nuclear reaction has taken place, the recoiling impurity atom dissipates its energy through atomic-collision cascades and finally this atom comes to rest in a limited volume of high lattice disorder. During this energy spike, which equilibrates between 10^{-12} and 10^{-10} s, migrating and annealing of defects occurs [4.146]. If the implanted impurity presents an attraction to the lattice defects, a certain fraction of the excited nuclei may end up with a nearest-neighbour defect after the equilibration of the collision spike.

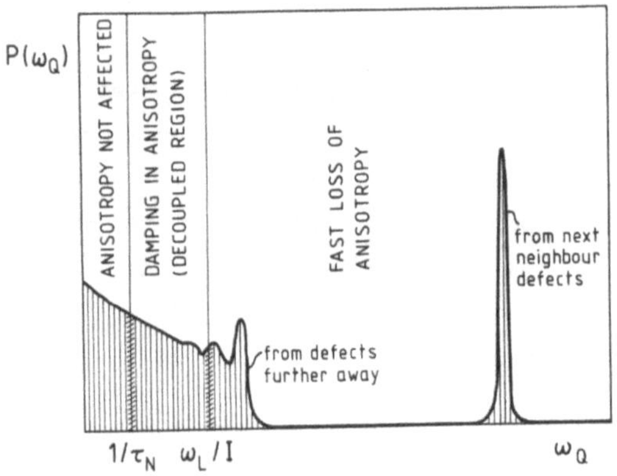

Fig.4.37. Schematic representation of the electric field gradient distribution, characterized by the quadrupole frequency ω_Q, at a probe. The effect of an external magnetic field, indicated by the Lamor frequency ω_L, is also illustrated

This results in a typical distribution of the electric field gradients at probe sites, which is illustrated in Fig.4.37. For weak electric field gradients or small ω_Q [see (4.2.10)] from defects at some distance where $\omega_Q \ll 1/\tau_N$, no influence on the γ-anisotropy during the nuclear lifetime τ_N is observed. On the other hand, strong electric field gradients and distributions of them, originating, for instance, from next-neighbour defects, may cause changes of the γ-anisotropy within the time (Δt) between anisotropy production and the start of anisotropy observation, resulting in an integral loss of γ-anisotropy. At lower temperatures the described defect situation remains static within the nuclear lifetime, whereas at higher temperature defects which are not trapped, may migrate away from the probe impurity, or the trapped defects start to diffuse around the impurity and finally overcome the binding and escape. Therefore, the fraction of probe nuclei exposed to perturbing electric field gradients during their lifetime diminishes with increasing temperatures. This effect has been known for quite a while; satisfying explanations were reported first for the system ^{69}Ge and ^{67}Zn probes in copper [4.147] and zinc [4.148].

For a quantitative understanding of the temperature-dependent anisotropy loss, it is necessary to know the exact electric field gradient distribution and the dynamical behaviour of all involved defects. If one restricts the discussion to a unique defect gradient originating from a certain trapped defect, the theoretical treatment becomes rather simple. Furthermore, as long as the binding energy is assumed to be small compared to the migration energy of the free defect, the rotational diffusion of the defect around the impurity can be neglected. With these simplifications and the assumption of many detrapping events during the interval Δt before the anisotropy observation starts, the time-dependent perturbation factor $G_2(t)$ [see (4.12,13] can be expressed in terms of a temperature-dependent, time integral factor [4.147]:

$$G_2(T) = (1 - f) + f \sum_n s_{2n}\{1 + [(\omega_n/\nu_0)\,\exp(E^S/k_BT)]^2\}^{-1} \qquad (4.18)$$

with f describing the fraction of nuclei with trapped defects at low temperatures. The attempt frequency ν_0 and the activation energy $E^S = E^M + E^B$ for detrapping are the parameters which determine the detrapping frequency

$$\nu = \nu_0 \exp(-E^S/k_BT) \quad . \tag{4.19}$$

So far only one defect type has been assumed which was expressed by a unique efg, but efg distributions can be incorporated [4.149]. The model can also be extended by including the defect's rotational migration around the impurity before the release [4.148,150]. However, the present accuracy of the experiments makes it difficult to justify these refinements. In the following section a survey of the experimental in-beam data will be given; it is certainly not a complete one since many of the data are unpublished and therefore not always accessible to the authors.

a) *Cubic Metals*

Most of the experimental in-beam work was done on cubic metals, especially the fcc metals. In a cubic lattice only those probe impurities which are associated with defects have been exposed to electric field gradients. The fraction of strongly perturbed nuclei leads to an anisotropy loss, for the unperturbed nuclei a time-independent γ-anisotropy is expected. Time-independent anisotropies are rather difficult to determine in in-beam experiments (in contrast to PAC experiments with a four-detector geometry), therefore an external magnetic field is applied in order to observe the so far unperturbed fraction of nuclei via their magnetic interaction.

This magnetic interaction is now competing with the (electric) defect interactions (combined interaction); the influence of the external field onto the observed anisotropy loss is illustrated in Fig.4.37. A magnetic interaction, characterized by the Larmor precession ω_L, decouples the defect interaction up to the limit given approximately by the condition [4.151]

$$\omega_L > I \cdot \omega_Q \quad . \tag{4.20}$$

Usually, only those defects with stronger electric quadrupole interactions now contribute to a fast anisotropy loss. This demonstrates the dependence of the experimentally observed fraction of decoupled nuclei on the strength of the external magnetic field.

For a fixed external magnetic field, Fig.4.38 shows a typical experimental series of γ-anisotropy modulations for the system of ^{208}Po nuclei in lead as a function of the measuring temperature [4.149,152]. The plotted quantity R(t) is obtained from the coincidence spectra of two γ-detectors with the beam pulse. The detectors are positioned symmetrically to the beam axis and at 90° with respect to each other; thus within our used notation we obtain $R(t) = a_2(T) \cos 2\omega_L t$ with $a_2(T) \approx (3/4)A_2^{eff} = (3/4)A_2 \cdot G_2(T)$. As one can see from the data, the modulation amplitude $a_2(T)$ increases drastically with temperature; at lower temperatures a damping of the modu-

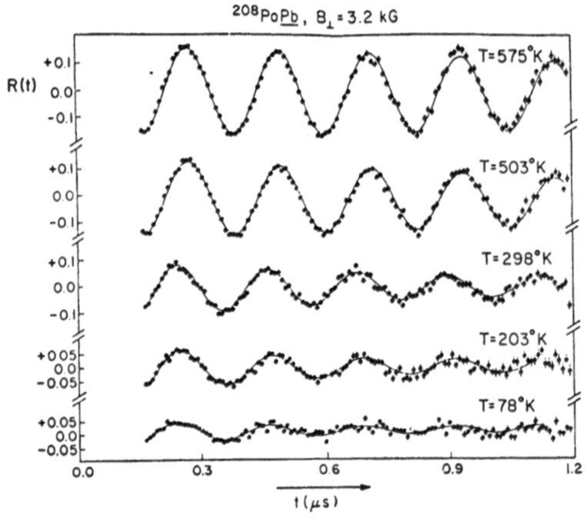

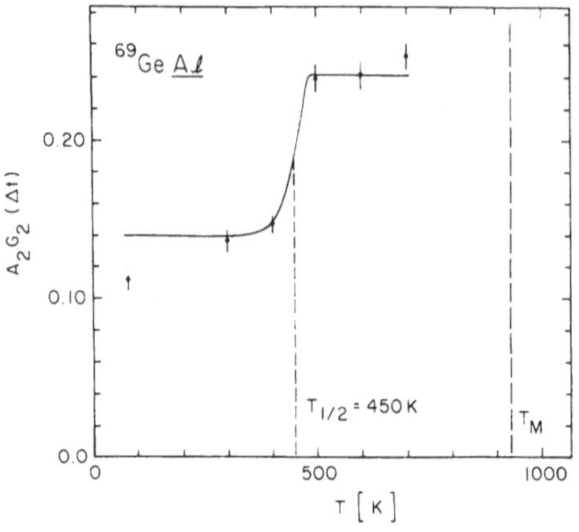

Fig.4.38

Fig.4.39

Fig.4.38. Spin rotation spectra, caused by a magnetic hyperfine interaction, for ^{208}Po (8^+) in Pb for different temperatures [4.152]

Fig.4.39. Temperature dependence of the γ-anisotropy observed in an external magnetic field. The solid line is a fit to data using (4.18) [4.153]

lation can be seen as an additional effect, originating from efg-distributions which are decoupled in the external magnetic field [4.151]. It can be clearly observed that the γ-anisotropy produced by the pulsed beam is strongly attenuated within very short times (<150 ns for ^{208}Po<u>Pb</u>). If one could measure closer to the time-zero point, which is not easily feasible because of the finite beam burst and γ-background, it would be possible to follow the anisotropy loss caused by undecoupled efg's. Such experiments have not been successful so far.

Aluminium

For aluminium, one in-beam experiment is known for ^{69}Ge impurities [4.153]. Figure 4.39 shows the quantity $A_2G_2(T)$ which corresponds to the unperturbed fraction of ^{69}Ge nuclei oscillating in the external magnetic field. At 450 K a reduction of the anisotropy loss is observed; the solid line is a fit to the data points using (4.18). From this an activation energy $E^S \simeq 0.4$ eV is deduced; this value is too low for the detrapping of a monovacancy (see Table 4.2) and might possibly belong to a small vacancy cluster.

Copper

^{67,69}Ge and ^{67}Zn impurities have been simultaneously produced in copper by a nuclear reaction in an external magnetic field [4.147]. Figure 4.40 shows the anisotropy as a function of temperature normalized to the liquid value. In the case of Ge nuclei, a strong reduction (~50%) of the anisotropy is observed at temperatures below

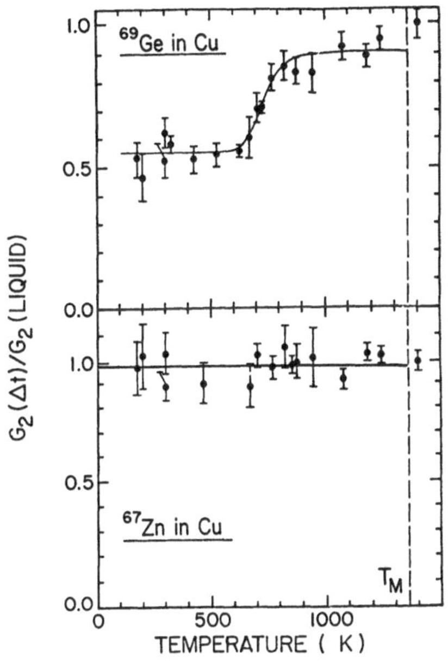

Fig.4.40. Temperature dependence of the γ-anisotropy observed in an external magnetic field for ^{69}Ge and ^{67}Zn probes in copper metal [4.147]

600 K; at higher temperatures the perturbing defects in the vicinity of the Ge probes are annealed within the nuclear lifetime. According to a fit of (4.18) to the data, an activation energy of E^S = 0.64(14) eV is yielded. Assuming a binding energy of the defect to the Ge probe of 0.1-0.2 eV, a defect migration energy of ~0.5 eV is obtained, also hinting at the migration of a divacancy or a small vacancy cluster. ^{73}As in copper [4.154] shows very similar anisotropy behaviour to ^{69}Ge; in the temperature region between 700 K and 900 K the free anisotropy is restored, whereas below 700 K only 20% of the ^{73}As nuclei can be observed through the precession effected by the external field.

A very interesting aspect is revealed by the ^{67}Zn probes in copper. In contrast to Ge Cu, they are not affected by defects at all temperatures. Decoupling because of different electromagnetic moments can be excluded so that different trapping behaviour between Ge and Zn probes within the nuclear production and start of the anisotropy observation (typically Δt = 10 ns in this experiment) should be the reason. Zn in copper does not attract defects whereas 50% of the Ge probes end up with trapped defects.

Palladium

In palladium, ^{111}Cd probe nuclei have been produced with the reaction ^{108}Pd(α,n) ^{111}Cd [4.155,156]. In this case no external magnetic field was applied. A distribution of electric field gradients around zero value was observed with two recovery stages around 150 K and 1400 K (Fig.4.41). These two stages are attributed to free interstitial migration [$E_i^M \simeq 0.10(2)$ eV] and free vacancy migration [$E_v^M \simeq 1.55(15)$ eV]. In addition, a unique electric field gradient was observed at about 4% of the

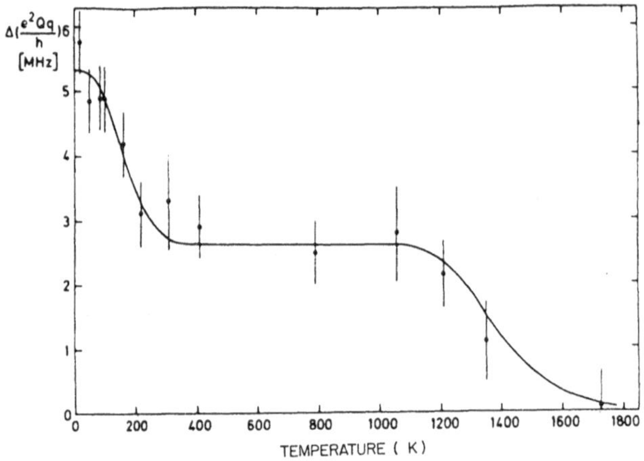

Fig.4.41. Temperature dependence of the measured width of the quadrupole frequency distribution of ^{111}Cd nuclei in Pd [4.156]

Cd probes up to 400 K due to athermic trapping of vacancies. The interaction frequency ν_Q = 87 MHz is the same as the one observed in the PAC experiment for the Cd-monovacancy configuration (see Sect.4.2.1a); above 400 K target temperature the unique efg vanishes. The probe atom in both experiments is the same, while the trapping atom is different: Ag in the PAC and Cd in the PAD case.

Silver

In silver, several in-beam experiments have been reported [4.157,158]. ^{107,109}Cd and ^{113}Sn probe nuclei have been used to study the correlated damage and the experiments were done with external magnetic field. A characteristic feature of the results is the only gradual annealing of the anisotropy, in contrast to the cases of aluminium, copper and palladium (Fig.4.42). This can originate from a broad distribution of electric field gradients which would be due to a negligible or missing defect-probe binding energy (see Fig.4.37; the next neighbour peak would be missing). Inserting a wide efg distribution into (4.18) leads to a more gradual annealing stage. In Fig.4.42, the effect of decoupling of electric field gradients by the external magnetic field can also be seen for ^{109}Cd at 0.5 KG and 35 KG. The annealing behaviour for ^{107}Cd at 35 KG in silver should be the same as for ^{109}Cd at 35 KG, since their quadrupole moments are very similar. However, the very different life times of the isotopes [$\tau_N(^{107}$Cd) = 110 ns, $\tau_N(^{109}$Cd) = 15.4 μs] require other pulsed beam conditions. Therefore the time before the anisotropy is observed is much larger for ^{109}Cd (Δt = 6 μs) and, consequently, weaker electric field gradients contribute to the anisotropy loss as compared to the ^{107}Cd case.

Lead

A great variety of experiments has been carried out on the system Po in lead [4.149,152a,159]. In particular, the influence of decoupling in the external magnetic field was demonstrated experimentally and theoretically formulated. As dis-

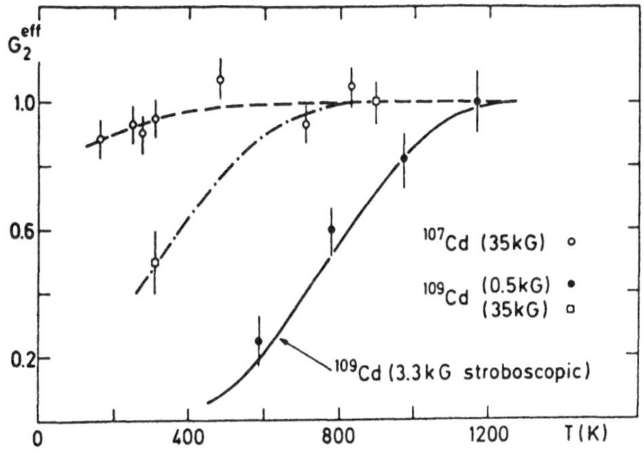

Fig.4.42. Temperature dependence of the attenuation of the initial spin-rotation amplitudes for ^{107}Cd and ^{109}Cd in Ag. The dashed curves are guides to the eye [4.157]

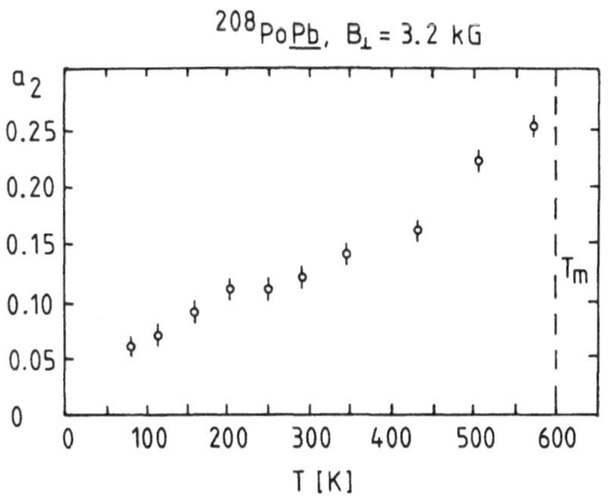

Fig.4.43. Temperature dependence of the γ-anisotropy observed in an external magnetic field for ^{208}Po nuclei in Pb [4.149]

cussed before, the decoupled electric field gradients manifest themselves as a damping, the nondecoupled electric field gradients as a fast anisotropy loss. As in the silver case, the anisotropy parameter a_2 shows no pronounced annealing stages, although complete annealing can be obtained close to the melting point. As an example, the results for ^{208}Po in lead are shown in Fig.4.43; consistent results are obtained with ^{209}Po and ^{210}Po isotopes. Experiments with high magnetic fields in the ^{210}Po Pb case show a somewhat more defined annealing temperature for the remaining nondecoupled ^{210}Po probes [4.159]. From this an estimate of the activation energy yields $E^S \approx 0.35(10)$ eV.

Other Cubic Metals

There are several more in-beam experiments exhibiting the above discussed anisotropy loss; complete temperature dependencies, however, have not been measured for these systems. A systematic study of ^{112}Sb in Al, Cu, Au, Pb, Ta and ^{69}Ge in Al, Ag, Cu, Au, Pb, at room temperature [4.153] shows the anisotropy loss as a function of the different host metals. It is remarkable that in gold a high fraction of the implanted probes are strongly perturbed which makes gold a problematic catcher for other hyperfine studies. In molybdenum, hyperfine experiments have been carried out with ^{99}Mo nuclei [4.160] showing complete anisotropy loss below 1300 K. ^{58}Co nuclei have been implanted in iron [4.161]; up to 1100 K no anisotropy was found, but around 1170 K anisotropy returned. This effect coincides with the α-γ-phase transition in iron so that other than radiation damage effects might be responsible.

b) *Noncubic Metals*

In-beam radiation damage studies in noncubic metals are reported for Zn, Cd, In [4.164] and Sn; only some of them, however, have been extensively investigated. In noncubic metals the radioactive probe nuclei are exposed to an electric field gradient due to the crystal structure. Radiation induced field gradients are superimposed. In these experiments, mainly the fraction of probe nuclei which interacts only with the pure crystal field gradient was inspected as a function of temperature. Usually the field gradients from radiation damage in the noncubic metals are also relatively high, hence fast loss of anisotropy is observed. Typical PAD-spectra are shown in Fig.4.44 for the case of ^{120}Sb nuclei in tin at two different temperatures [4.162]. The modulation amplitude is given by $R(t) = A_2^{eff}\, G_2^{latt}(t)$, where $G_2^{latt}(t)$ is the usual perturbation factor for an axial symmetric field gradient in a polycrystalline probe [see (4.13)] and $A_2^{eff} = A_2G_2(T)$ containing the damage associated perturbation factor $G_2(T)$ [see (4.18)]. At the higher temperature the effective anisotropy is slightly higher, indicating annealing of defect fields. The change of the modulation frequencies with temperature is a well-known effect and is discussed in Chap.5.

Zinc

For the hexagonal Zn, in-beam experiments with ^{67,69,71}Ge and ^{67}Zn probes are reported [4.148,163-165]. A large fraction of probe atoms experiences strong radiation damage fields at lower temperatures; at around 400 K annealing occurs. Figure 4.45 displays the fast anisotropy loss for ^{67}Ge and ^{71}Ge in a zinc single crystal [4.165]. Noteworthy is the high anisotropy at around 100 K, dropping at temperatures up to 400 K. This effect could indicate that at these low temperatures the defect is not mobile enough to be trapped. Another interesting result, which is not yet understood, can be observed from these accurate data. The fraction of ^{71}Ge perturbed by defects is smaller than for ^{67}Ge; a similar deviation is suggested for ^{113}Sn and ^{116}Sn in Cd [4.166]. From a fit to the Ge<u>Zn</u> data using a diffusion model including rotational migration of the defect around the impurity, a binding energy of $E^B \approx 0.07$ eV can be deduced with the assumption of $E^M = 0.50$ eV [4.148,165]. In the case of ^{67}Zn probes, no anisotropy loss is observed; this is attributed to the "host-character" so that no defects can be trapped in the vicinity of the probe atom.

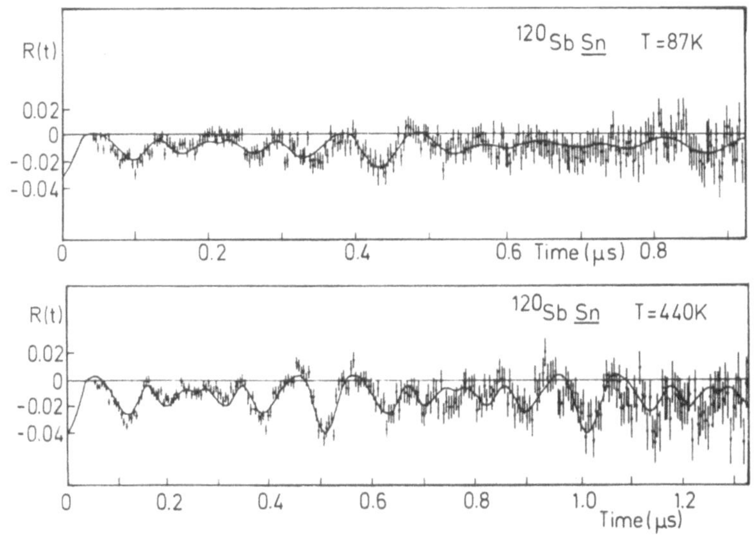

Fig.4.44. Spin rotation spectra, caused by an electric hyperfine interaction, for ^{120}Sb in Sn for two temperature values [4.162]

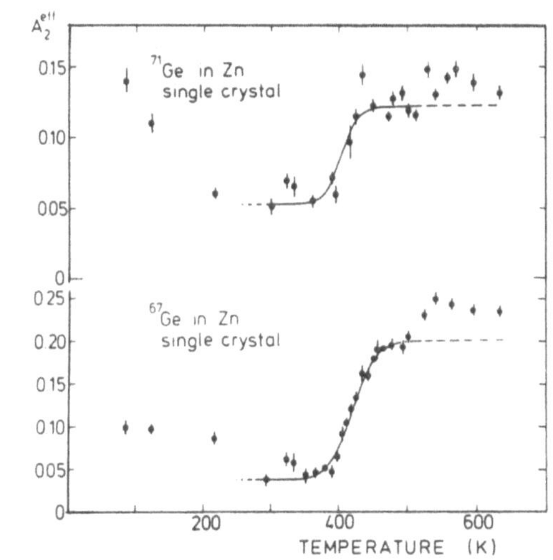

Fig.4.45. Modulation amplitudes A_2^{eff} observed as a function of the target temperature for ^{71}Ge and ^{67}Ge in a Zn single crystal [4.165]

Cadmium

One of the very first in-beam radiation damage studies was performed on the system ^{111}Cd in cadmium [4.167]. Up to now, experiments on ^{67}ZnCd, ^{67}GeCd [4.164], ^{113,114,116}SnCd [4.166,166a], and ^{115}SbCd [4.168] have been added.

In the pure system ^{111}CdCd, no fast anisotropy loss was observed [4.167,169], probably due to the host character of the probes. However, a large fraction of ^{111}Cd probes experience field gradient distributions which anneal around 400 K. In the impurity systems a pronounced annealing stage is observed, around 360 K for ^{67}ZnCd and ^{67}GeCd, around 450 K for ^{113,114,116}SnCd and ^{115}SbCd. The annealing stage of ^{115}SbCd is rather peculiar; below 460 K all ^{115}Sb nuclei are perturbed;

Table 4.10. Results obtained from perturbed angular distribution (PAD) experiment of the fast anisotropy loss: T_{Anneal}: temperature at which half of the fast aniso external magnetic field B_{ext}) and ω_Q (due to the internal efg of the host lattice which the loss is observed. Δt: time between anisotropy production and start of a values obtained by using $\omega_Q > \pi/6 \cdot I \cdot \Delta t$

Host	Probe (spin I)	Characterization of fast anisotropy loss	T_{Anneal} [K]	ω_L from B_{ext} [Mrad/s]	ω_Q from host lattice[a] [Mr (temp. [K])
Al	$^{69}Ge(9/2^+)$	steep	450	11.6	
Cu	$^{67}Ge(9/2^+)$	steep	730	9.56	
	$^{69}Ge(9/2^+)$	steep	730	10.2	
	$^{67}Zn(9/2^+)$	flat	-	11.1	
	$^{73}As(9/2^+)$	steep	700	2.79	
Zn	$^{67}Ge(9/2^+)$	steep	420		4.6(RT)
	$^{69}Ge(9/2^+)$	steep	430		3.8(374)
	$^{71}Ge(5/2^+)$	steep	400		2.7(633)
	$^{67}Zn(9/2^+)$	flat	-		2.0(300)
Pd	$^{111}Cd(5/2^+)$	flat[b]	-		
Ag	$^{107}Cd(11/2^-)$	flat		31.7	
	$^{109}Cd(11/2^-)$	gradual	800	0.477-30.5	
	$^{113}Sn(11/2^-)$	gradual	500	25.8	

Cd	$^{67}Ge(9/2^+)$	gradual	350		2.5(38
	$^{67}Zn(9/2^+)$	gradual	350		1.7(31
	$^{111}Cd(5/2^+)$	gradual	460		19.6(29
	$^{113}Sn(11/2^-)$	steep	450		3.0(30
	$^{114}Sn(7^-)$	steep	450		1.2(48
	$^{116}Sn(5)$	steep	450		1.9(48
	$^{115}Sb(19/2)$	steep	450	14.3	0.06(5
Sn	$^{67}Ge(9/2^+)$	gradual	300		1.7(44
	$^{120}Sb(3^+)$	steep	300		4.5(RT
	$^{115}Te(11/2^-)$	steep	280		0.19(0
Pb	$^{208}Po(8^+)$	gradual	300	13.2	
	$^{209}Po(17/2^-)$	gradual	300	13.9	
	$^{210}Po(8^+)$	gradual	300	57.5	

[a] R. Vianden, Hyperfine Interactions *4*, 956 (1978).

[b] Annealing stages of resolved interaction frequencies have been observed at 160, 4 characterized by the annealing of a unique quadrupole frequency of ω_Q = 13.7(3) M

that means all ^{115}Sb probes have trapped defects and within a very small temperature interval, the full anisotropy is gained back. These data require an activation energy of $E^S = 0.79(2)$ eV, which under the assumption of a trapped monovacancy yields a rather high binding energy of about $E^B \approx 0.4$ eV [4.168].

Tin

In tin, in-beam data are known for ^{67}Ge<u>Sn</u> [4.164], ^{115}Te<u>Sn</u> [4.170] and ^{120}Sb<u>Sn</u> [4.162,171]. The system ^{120}Sb<u>Sn</u> has already been presented (Fig.4.44) as an example; the anisotropy annealing is shown in Fig.4.46 which exhibits a moderate annealing at around 300 K. Within the model of [4.148] which considers rotational diffusion of the defect before the break-up, an activation energy of $E^S = 0.29(3)$ eV is obtained. The ^{115}Te<u>Sn</u> experiments show a rather pronounced annealing stage also at 280 K; however, even close to the melting point at 505 K only about half of the anisotropy can be observed. For the system ^{67}Ge<u>Sn</u>, a broad annealing stage again centered around 300 K was found.

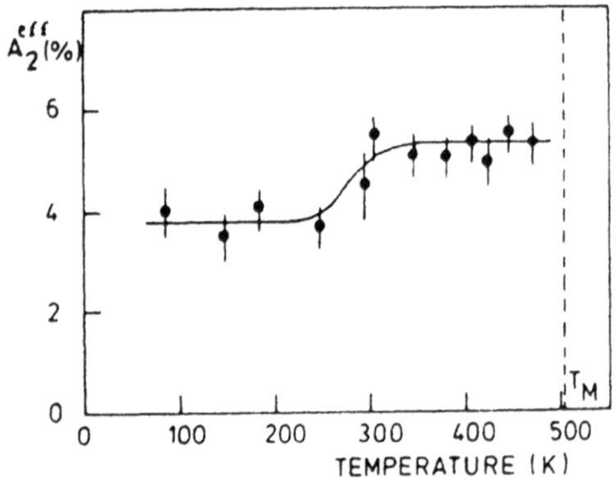

Fig.4.46. Temperature dependence of the spin rotation amplitude of the electric quadrupole interaction for ^{120}Sb in Sn [4.162]

c) *Summary of Nuclear Reaction Experiments*

Most of the in-beam experiments are accompanied by the phenomenon of an unresolved loss of γ-anisotropy and only in a few cases have the quadrupole frequencies originating from defects been resolved. In Table 4.10 the most relevant experimental results about this fast anisotropy loss are listed for all the known systems, in particular a lower limit of the defect interaction frequencies is extracted from the unobserved time Δt of the γ-anisotropy using $\Delta t > 0.25 \cdot 2\pi/\bar{\omega}$. Here $\bar{\omega}$ is a mean value representing all the transition frequencies for a given nuclear spin I (4.13) and is given in good approximation by $\bar{\omega} = 3 \cdot I \cdot \omega_Q$ [4.152a]. The obtained frequency limits are quite consistent with the data known from PAC experiments. In the pure systems Zn<u>Zn</u> and Cd<u>Cd</u>, no or only little integral loss of anisotropy is observed, hinting to the fact that the impurity character of probes in the different hosts is obviously playing a major role in most of all the other systems. Basically,

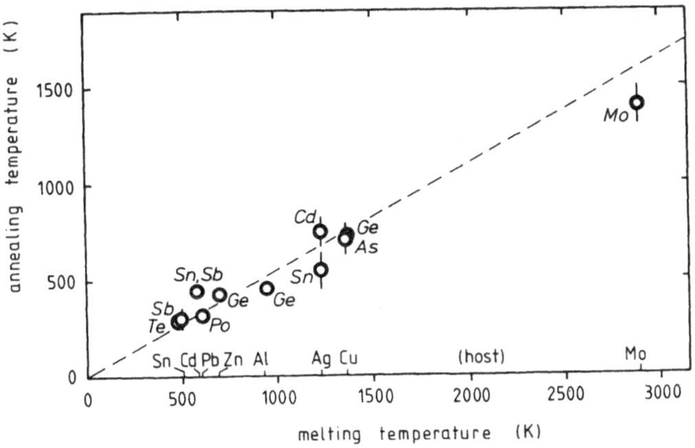

Fig.4.47. Correlation between annealing temperature and melting point for various host metals

two categories of annealing behaviour of the fast anisotropy loss are recognizable: annealing in a narrow temperature interval (step-like annealing curves) or the annealing over a wide temperature range (gradual annealing curves). An interpretation of the first group with pronounced annealing stages is based on trapped defects at the probe sites with rather definite electric field gradients. The break up of this binding as a function of temperature sets the probe free from the perturbation fields and full anisotropy can be observed. It was pointed out that the observed annealing temperatures (temperature at which half of the anisotropy loss is annealed) show a correlation to the melting point [4.172]: $T_{anneal} \propto 0.6\ T_m$ (Fig.4.47). This observation has a practical application, namely, for hyperfine experiments where an unreduced γ-anisotropy is desired during the nuclear lifetime. From the defect point of view, this correlation indicates the connection to stage III in resistivity experiments, after proper scaling of the annealing time [4.5]. The results obtained for the activation energy E^S are in fair agreement with the migration energies for monovacancies, considering small corrections due to the binding energy. However, besides these more general conclusions, detailed results on specific point defects, like the monovacancy, are not available yet. There is only one experiment, the ^{111}Cd<u>Pd</u> case, where a unique field gradient was observed and attributed to a trapped vacancy. The detailed information, however, is not hampered principally, but is restricted by the experimental circumstances so far. For example, if one could succeed in resolving the unique electric field gradient for the trapped defects, the problem of the binding of a whole variety of probe impurities to the defect could be attacked. By this, the influence of site and valence of the impurities onto the trapping process could be elucidated systematically. The other group of data reveal either no influence of radiation damage on the probes or a broad distribution of field gradients, possibly from more than one type of defect with different migration energies, resulting in a gradual annealing up to the melting point. These cases indicate missing defect-impurity trapping, so that one is only dealing with a statistical array of defects in the vicinity of the probes. Here experiments with decoupling of the defect field as a function of an external magnetic field, as was tried for the PoPb system [4.152], are helpful in determining the electric field gradient distributions and by that attain more information on the defect properties.

Acknowledgement. We would like to thank Drs. H. Bernas, E. Dafni, M. Deicher, O. Echt, P. Heubes, M.H. Rafailovich, and A. Weidinger for valuable comments and discussions. The support of the Bundesminister für Forschung und Technologie is gratefully acknowledged.

References

4.1 E. Karlsson, R. Wäppling (eds.): Proc. of Intern. Conf. on Hyperfine Interactions Studied in Nuclear Reactions and Decay, Uppsala, Sweden (1974), Physica Scripta *11* (1975)

4.2 B.I. Deutsch, H. de Waard, R. Coussement, M. Rots, L. Vanneste (eds.): Proc. of Intern. Meeting on Hyperfine Interactions, Leuven, Belgium (1975), Hyperfine Interact. *2* (1976)

4.3 R.S. Raghavan, D.E. Murnick (eds.): Proc. of IV. Intern. Conf. on Hyperfine Interactions, Madison, USA (1977), Hyperfine Interact. *4* (1978)

4.4 G. Kaindl, H. Haas (eds.): Proc. of V. Intern. Conf. on Hyperfine Interactions, Berlin, Germany (1980), Hyperfine Interact. *9*, *10* (1981)

4.5 A. Seeger, D. Schumacher, W. Schilling, J. Diehl (eds.): Vacancies and Interstitials in Metals, Proc. of Intern. Conf., Jülich, Germany (1968) (North-Holland, Amsterdam 1970)

4.6 N.L. Peterson, R.W. Siegel (eds.): Proc. of Intern. Conf. on the Properties of Atomic Defects in Metals, Argonne, Illinois, USA (1976), J. Nucl. Mater. *69*, *70* (1978)

4.7 M.T. Robinson, F.W. Young (eds.): Fundamental Aspects of Radiation Damage in Metals, Proc. of Intern. Conf., Gatlinburg, USA (1975), US-ERDA-CONF 751006 (1976)

4.8 P. Lucasson: In [Ref.4.7, p.42]

4.9 T.E. Mitchell, G. Das, E.A. Kenik: In [Ref.4.7, p.73]

4.10 G.H. Kinchin, R.S. Pease: Repts. Prog. Phys. *18*, 1 (1955)

4.11 M.W. Thompson: *Defects and Radiation Damage in Metals*, Cambridge Monographs on Physics, ed. by A. Herzenberg, J.M. Ziman (Cambridge University Press, Cambridge 1969)

4.12 O.S. Oen: USAEC unclassified report (1965) ORNL - 3813

4.13 D.I.R. Norris: Radiat. Eff. *14*, 1 (1972)

4.14 K.B. Winterbon: Radiat. Eff. *13*, 215 (1972)

4.15 J.P. Biersack, L.G. Haggmark: Nucl. Instrum. Methods *174*, 257 (1980)

4.16 H. Bernas, M.O. Ruault, B. Jouffrey: Phys. Rev. Lett. *27*, 859 (1971)

4.17 A.C. Damask, G.J. Dienes: *Point Defects in Metals* (Gordon Breach, New York 1971)

4.18 R.N. West: In *Positrons in Solids*, Topics in Current Physics, Vol.12, ed. by P. Hautojärvi (Springer, Berlin, Heidelberg, New York 1979)

4.19 J.E. Bauerle, J.S. Koehler: Phys. Rev. *107*, 1493 (1957)

4.20 A.S. Berger, S.T. Ockers, M.K. Chason, R.W. Siegel: J. Nucl. Mater. *69*, *70*, 734 (1978)

4.21 M.J. Fluss, L.C. Smedskjaer, M.K. Chason, D.G. Legnini, R.W. Siegel: J. Nucl. Mater. *69*, *70*, 586 (1978)

4.22 W. Wycisk, M. Feller-Kniepmeier: J. Nucl. Mater. *69*, *70* , 616 (1978)

4.23 R.R. Bourassa, B. Lengeler: J. Phys. F *6*, 1405 (1976)

4.24 J.L. Campbell, C.W. Schulte, R.R. Gingerich: J. Nucl. Mater. *69*, *70* , 609 (1978)

4.25 R.W. Balluffi: J. Nucl. Mat. *69*, *70* , 240 (1978)

4.26 W. Triftshäuser, J.D. Mc Gervey: Appl. Phys. *6*, 177 (1975)

4.27 J.J. Jackson: In *Lattice Defects in Quenched Metals*, ed. by R.M.J. Cotterill, M. Doyama, J.J. Jackson, M. Meshii (Academic, New York 1965) p.467
4.28 R.W. Siegel: J. Nucl. Mater. *69, 70* , 117 (1978)
4.29 W.C. Knodle, J. Koehler: J. Nucl. Mater. *69, 70* , 620 (1978)
4.30 W. Triftshäuser: Phys. Rev. B *12*, 4635 (1975)
4.31 H. Wever, W. Seith: Phys. Stat. Sol. (a) *28*, 187 (1975)
4.32 K. Maier, H. Metz, D. Herlach, H.E. Schaefer: J. Nucl. Mater. *69, 70* , 589 (1978)
4.33 I.A. Schwirtlich, H. Schultz: Phil. Mag. A *42*, 613 (1980)
4.34 I.A. Schwirtlich, H. Schultz: Phil. Mag. A *42*, 601 (1980)
4.35 K. Maier, H. Metz, D. Herlach, H.E. Schaefer, A. Seeger: Phys. Rev. Lett. *39*, 484 (1977)
4.36 K.D. Rasch, R.W. Siegel, H. Schultz: Phil. Mag. A *41*, 91 (1980)
4.37 A. Seeger: J. Phys. F *3*, 248 (1973)
4.38 J.P. Simon, D. Vostry, J. Hillairet, V. Levy: Phil. Mag. *31*, 145 (1975)
4.39 K.P. Singh, R.N. West: J. Phys. F *6*, L267 (1976)
4.40 P.H. Dederichs, C. Lehmann, H.R. Schober, A. Scholz, R. Zeller: J. Nucl. Mater. *69, 70* , 176 (1978)
4.41 P. Ehrhart: J. Nucl. Mater. *69, 70* , 200 (1978)
4.42 C.P. Flynn: *Point Defects and Diffusion* (Clarendon Press, Oxford 1972)
4.43 G. Antesberger, K. Sonnenberg, P. Wienhold: J. Nucl. Mater. *69, 70* , 660 (1978)
4.44 H. Schroeder, R. Lennartz, U. Dedek: In [Ref.4.7, p.411]
4.45 W. Decker, J. Diehl, A. Dunlop, W. Frank, H. Kronmüller, W. Mensch, H.E. Schaefer, B. Schwendemann, A. Seeger, H.P. Stark, F. Walz, M. Weller: Phys. Stat. Sol. (a) *52*, 239 (1979)
4.46 K. Faber, H. Schultz: Radiat. Eff. *31*, 157 (1977)
4.47 K. Faber, J. Schweikhardt, H. Schultz: Scripta Metall. *8*, 713 (1974)
4.48 J. Cornelis, L. Stals, P. de Meesters, J. Roggen, J. Nihoul: J. Nucl. Mater. *69, 70* , 704 (1978)
4.49 J.P. Simon, P. Vostry, J. Hillairet, P. Vajda: Phys. Stat. Sol. (b) *64*, 277 (1974)
4.50 J. Roggen, J. Cornelis, J. Nihoul, L. Stals: Phil. Mag. *35*, 1205 (1977)
4.51 J.P. Simon, C. Minier: Radiat. Eff. *13*, 137 (1972)
4.52 F.W. Young: J. Nucl. Mater. *69, 70* , 310 (1978)
4.53 R. Rizk, P. Vajda, A. Maury, A. Lucasson, P. Lucasson: J. Appl. Phys. *48*, 481 (1977)
4.54 S. Okuda: In [Ref.4.7, p.361]
4.55 D. Schumacher: In [Ref.4.5, p.889]
4.56 J.D. Eshelby: In *Solid State Physics*, Vol.3, ed. by F. Seitz, D. Turnbull (Academic, New York 1956)
4.57 J. Friedel: Phil. Mag. *43*, 153 (1952)
4.58 A. Blandin: J. Phys. Rad. *22*, 507 (1961)
4.59 C.A. Sholl: Proc. Phys. Soc. *91*, 130 (1967)
4.60 S.T. Picraux: In *New Uses of Ion Accelerators*, ed. by J.F. Ziegler (Plenum, New York 1975) p.229
4.61 H. Wollenberger: J. Nucl. Mater. *69, 70* , 362 (1978)
4.62 W. Schilling, P. Ehrhart, K. Sonnenberg: In [Ref.4.7, p.470]
4.63 A. Seeger: In [Ref.4.7, p.493]
4.64 H.H. Rinneberg: At. Energy Rev.*17* , 477 (1979)
4.65 D.A. Shirley: Rev. Mod. Phys. *36*, 339 (1964)
4.66 E.N. Kaufmann, R.J. Vianden: Rev. Mod. Phys. *51*, 161 (1979)
4.67 H. Frauenfelder, R.M. Steffen: In *Alpha-, Beta-, and Gammay-Ray Spectroscopy*, Vol.2, ed. by K. Siegbahn (North-Holland, Amsterdam 1965)
4.68 E. Dafni, R. Bienstock, M.H. Rafailovich, G.D. Sprouse: Atomic Data and Nuclear Data Tables *23*, 315 (1979)
4.69 G. Vogel: Hyperfine Interactions *2*, 151 (1976), and in: *Hyperfine Interactions*, Proc. of XVII Winter School, Bielsko-Biala, Poland, ed. by R. Kulessa, K. Krōlas (1979) p.127
4.70 L. Niesen: Hyperfine Interact. *10*, 619 (1981)

4.71 F. Pleiter: In *Nuclear Physics Methods in Materials Research*, ed. by K. Bethge, H. Baumann, H. Jex, F. Rauch (Vieweg, Braunschweig 1980) p.174
4.72 E. Recknagel, Th. Wichert: Nucl. Instrum. Methods *182/183*, 439 (1981)
4.73 Th. Wichert, E. Recknagel: In *Nuclear and Electron Resonance Spectroscopies Applied to Materials Science*, ed. by E.N. Kaufmann, G.K. Shenoy (Elsevier, New York 1981) p.211
4.74 M. Deicher, Th. Wichert: Abstracts of the V. Intern. Conf. on Hyperfine Interactions, Berlin, Germany (1980)
4.75 F. Pleiter, C. Hohenemser: Phys. Rev. B *25*, 106 (1982)
4.76 M. Deicher, O. Echt, E. Recknagel, Th. Wichert: In *Nuclear and Electron Resonance Spectroscopies Applied to Material Science*, ed. by E.N. Kaufmann, G.K. Shenoy (Elsevier, New York 1981) p.435
4.77 G. Vogl, W. Mansel, W. Vogl: J. Phys. F *4*, 2321 (1974)
4.78 G. Vogl, W. Mansel, P.H. Dederichs: Phys. Rev. Lett. *36*, 1497 (1976)
4.79 W. Mansel, G. Vogl: J. Physique *37*, C6-19 (1976)
4.80 W. Mansel, G. Vogl: J. Phys. F*7*, 253 (1977)
4.81 W. Mansel, H. Meyer, G. Vogl: Radiat. Eff. *35*, 69 (1978)
4.82 W. Petry, G. Vogl, W. Mansel: Phys. Rev. Lett. *45*, 1862 (198o)
4.82a W. Petry, G. Vogl: Z. Physik B *45*, 207 (1982)
4.83 K. Sassa, H. Goto, Y. Ishida, M. Kato: Scr. Metall. *11*, 1029 (1977)
4.84 K. Sassa, H. Goto, Y. Ishida: J. Physique *40*, C2-559 (1979)
4.85 C. Janot, H. Gilbert: Phil. Mag. *27*, 545 (1972)
4.85a K. Sassa, S. Umeyama, Y. Ishida, H. Yoshida: J. Phys. F *11*, L163 (1981)
4.85b E. Verbiest, H. Pattyn: Phys. Rev. B *25*, 5097 (1982)
4.86 M. Kato, Y. Ishida, K. Sassa, S. Umeyama, M. Mori: J. Physique *35*, C6-309 (1974)
4.87 A. Nylandsted Larsen, E. Weyer: J. Phys. F *9*, 27 (1979)
4.88 H.P. Wit, N. Teekens, L. Niesen, S.A. Drentje: Hyperfine Interact. *4*, 674 (1978)
4.89 H. Rinneberg, H. Haas: Hyperfine Interact. *4*, 678 (1978)
4.90 H. Rinneberg, W. Semmler, G. Antesberger: Phys. Lett. *66*A, 57 (1978)
4.91 R. Butt, W. Semmler, R. Keitel: Phys. Lett. *80*A, 29 (1980)
4.92 H. Rinneberg, H. Haas: Annual Report, Hahn-Meitner-Institut, Berlin (1976), p.84
4.92a F. Pleiter, K.G. Prasad: Phys. Lett. *84* A, 345 (1981)
4.93 O. Echt, E. Recknagel, A. Weidinger, Th. Wichert: Z. Physik B *32*, 59 (1978)
4.94 Th. Wichert, M. Deicher, O. Echt, E. Recknagel: Phys. Rev. Lett. *41*, 1659 (1978)
4.95 O. Echt, E. Recknagel, A. Weidinger, Th. Wichert: Hyperfine Interact. *4*, 706 (1978)
4.96 O. Echt, E. Recknagel, G. Schätz, A. Weidinger, Th. Wichert: Hyperfine Interact. *4*, 585 (1978)
4.97 M. Deicher, O. Echt, E. Recknagel, Th. Wichert: Hyperfine Interact. *10*, 667 (1981)
4.98 M. Deicher, E. Recknagel, Th. Wichert: Hyperfine Interact. *10*, 675 (1981)
4.99 M. Deicher: Dissertation, Konstanz (1983)
4.100 M. Wilkins: In [Ref.4.7, p.98]
4.101 R. Butt, H. Haas, T. Butz, W. Mansel, A. Vasquez: Phys. Lett. *64*A, 309 (1977)
4.102 G. Vogl, W. Mansel, W. Petry, V. Gröger: Hyperfine Interact. *4*, 681 (1978)
4.103 M. Behar, R.M. Steffen: Phys. Rev. C *7*, 788 (1973)
4.104 L. Thomé, H. Bernas: Phys. Rev. Lett. *36*, 1055 (1976)
4.105 H. Bernas, L. Thomé, F. Sage: Hyperfine Interact. *2*, 348 (1976)
4.106 L. Thomé, H. Bernas: Hpyerfine Interact. *4*, 702 (1978)
4.107 L. Thomé, H. Bernas: Hyperfine Interact. *5*, 361 (1978)
4.108 P. Wodniecki, K. Królas, M. Marszalek, L. Thomé: Hyperfine Interact. *10*, 721 (1981)
4.108a P. Wodniecki, B. Wodniecka, Pham Quoc Hung, A. Fazeli, K. Krōlas, L. Thome: Acta Phys. Pol. A *60*, 817 (1981)
4.109 R. Butt, R. Keitel, G. Vogl: Annual Report, Hahn-Meitner-Institut, Berlin (1979) p.68

4.110 F.C. Magendans, J.P. Kikkert, L. Niesen: Abstracts of the V. Intern. Conf. on Hyperfine Interactions, Berlin (1980)
4.110a E. Verbiest, H. Pattyn, J. Odeurs: Nucl. Instrum. Methods *182/183*, 515 (1981)
4.111 H.G. Müller, K. Kusch: Hyperfine Interact. *4*, 697 (1978)
4.112 F. Pleiter, W.Z. Venema, A.R. Arends: Hyperfine Interact. *4*, 693 (1978)
4.113 K. Sonnenberg, W. Schilling, K. Mika, K. Dettmann: Radiat. Eff. *16*, 65 (1972)
4.114 W. Frank, A. Seeger: Radiat. Eff. *31*, 7 (1976)
4.115 H.G. Müller, U. Pütz, K. Krusch, K. Freitag: Z. Physik B *35*, 327 (1979)
4.116 H.G. Müller, K. Krien, U. Pütz, F. Reuschenbach, R. Trzcinski, K. Freitag: Z. Physik B *32*, 315 (1979)
4.117 M. Deicher, E. Recknagel, Th. Wichert: Radiat. Eff. *54*, 155 (1981)
4.118 R. Butt, H. Rinneberg, W. Semmler: Annual Report, Hahn-Meitner-Institut, Berlin (1977) p.59
4.119 R. Sielemann, H. Metzner, S. Klaumünzer, R. Butt, H. Haas, G. Vogl: Hyperfine Interact. *10*, 701 (1981)
4.119a R. Sielemann, H. Metzner, R. Butt, S. Klaumünzer, H. Haas, G. Vogl: Phys. Rev. B *25*, 5555 (1982)
4.120 R. Vianden, P.M.J. Winand: Hyperfine Interact. *10*, 713 (1981)
4.121 W. Mansel, J. Marangos, G. Vogl: Hyperfine Interact. *10*, 687 (1981)
4.122 S.R. Reintsema, J. Odeurs, E. Verbiest, H. Pattyn, R. Coussement: Hyperfine Interact. *4*, 685 (1978)
4.123 S.R. Reintsema, J. Odeurs, E. Verbiest, H. Pattyn: J. Phys. F *9*, 1511 (1979)
4.124 A. Weidinger, R. Wessner, Th. Wichert, E. Recknagel: Phys. Lett. *72*A, 369 (1979)
4.125 A. Weidinger, R. Wessner, E. Recknagel, Th. Wichert: Nucl. Instrum. Methods *182/183*, 509 (1981)
4.126 U. Pütz, H.J. Rudolph, R. Vianden: Hyperfine Interact. *10*, 709 (1981)
4.126a U. Pütz, A. Hoffmann, H.J. Rudolf, R. Vianden: Z. Phys. B *46*, 107 (1982)
4.127 J. Cornelis, S. Stals, P. De Meester, J. Roggen, J. Nihoul: J. Nucl. Mater. *69*, *70* , 704 (1978)
4.128 H. de Waard, R.L. Cohen, S.R. Reintsema, S.A. Drentje: Phys. Rev. B *10*, 3760 (1974)
4.129 S.R. Reintsema, S.A. Drentje, H. de Waard: Radiat. Eff. *24*, 145 (1975)
4.130 S. Bukshpan, W. Hilbrands, S.R. Reintsema, H. de Waard: Hyperfine Interact. *2*, 356 (1976)
4.131 S.R. Reintsema, H. de Waard, S.A. Drentje: Hyperfine Interact. *2*, 367 (1976)
4.132 S.R. Reintsema: Mössbauer Studies of Implantation Damage in Iron and Nickel, Thesis, Groningen (1976)
4.133 S.R. Reintsema, S.A. Drentje, H. de Waard: Hyperfine Interact. *5*, 167 (1978)
4.134 H. de Waard: In *Hyperfine Interactions*, Proc. of XVII Winter School, Bielsko-Biala, Poland, ed. by R. Kulessa, K. Królas (1979) p.189
4.135 H.P. Wit, L. Niesen, H. de Waard: Hyperfine Interact. *5*, 233 (1978)
4.136 D. Visser, L. Niesen, H. Postma, H. de Waard: Phys. Rev. Lett. *41*, 882 (1978)
4.137 C. Hohenemser, A.R. Arends, H. de Waard: Phys. Rev. B *11*, 4522 (1975)
4.138 C. Hohenemser, A.R. Arends, H. de Waard, H.G. Devare, F. Pleiter, S.A. Drentje: Hyperfine Interact. *3*, 297 (1977)
4.139 F. Pleiter: Hyperfine Interact. *5*, 109 (1977)
4.140 F. Pleiter, A.R. Arends, H.G. Devare: Hyperfine Interact. *3*, 87 (1977)
4.141 R.M. Suter, M. Haoni, C. Hohenemser: Hyperfine Interact. *4*, 711 (1978)
4.142 A.R. Arends, H. Hasper, C. Hohenemser, J.G. Mullen, G. van Opbroek, F. Pleiter: Hyperfine Interact. *10*, 659 (1981)
4.143 W. Witthuhn, A. Weidinger, W. Sandner, H. Metzner, W. Klinger, R. Böhm: Z. Physik B *33*, 155 (1979)
4.144 R. Seeböck, F. Simonato, R. Keitel, W. Witthuhn: Hyperfine Interact. *10*, 695 (1981)
4.144a F. Simonato, W. Engel, S. Hoth, R. Keitel, R. Seeböck, W. Witthuhn: Phys. Lett. *84* A, 393 (1981)
4.145 H. de Waard, F. Pleiter, L. Niesen, D.W. Hafemeister: Hyperfine Interact. *10*, 643 (1981)
4.145a H.G. Müller: In *Nuclear Physics Methods in Material Research*, ed. by K. Bethge, H. Baumann, H. Jex, F. Rauch (Vieweg, Braunschweig 1980) p.418

4.145b G. Scott Collins, G.P. Stern, C. Hohenemser: Phys. Lett. *84* A, 289 (1981)
4.145c G. Lindner, K. Bendel, M. Deicher, R. Minde, E. Recknagel, Th. Wichert: Nucl. Instrum. Methods *194*, 193 (1982)
4.145d Th. Wichert, G. Lindner, M. Deicher, E. Recknagel: Phys. Rev. B *24*, 7467 (1981)
4.146 P. Sigmund: Appl. Phys. Lett. *25*, 169 (1974)
4.147 G. Schatz, M.H. Rafailovich, W.A. Little, G.D. Sprouse: Phys. Rev. Lett. *35*, 1086 (1975)
4.148 G. Hempel, H. Ingwersen, W. Klinger, W. Loeffler, W. Sandner, W. Witthuhn: Phys. Lett. *55*A, 51 (1975)
4.149 E. Dafni, M.H. Rafailovich, W.A. Little, G.D. Sprouse: Hyperfine Interact. *4*, 743 (1978)
4.150 F. Bosch, H. Spehl: Internal Report, University of Freiburg (1973), unpublished
4.151 E. Dafni, G.D. Sprouse: Hyperfine Interact. *4*, 777 (1978)
4.152 E. Dafni, M.H. Rafailovich, W.A. Little, G.D. Sprouse: Phys. Rev. C *23*, 90 (1981)
4.152a E. Dafni: PhD Thesis, Stony Brook (1978)
4.153 M.H. Rafailovich: PhD Thesis, Stony Brook (1980)
4.154 M.H. Rafailovich, G. Schatz, W.A. Little, G.D. Sprouse: Bull. Am. Phys. *21*, 76 (1976)
4.155 H. Bertschat, H. Haas, F. Pleiter, E. Recknagel, E. Schlodder, B. Spellmeyer: Phys. Rev. B *12*, 1 (1975)
4.156 H. Bertschat, O. Echt, H. Haas, E. Ivanov, F. Pleiter, E. Recknagel, E. Schlodder, B. Spellmeyer: Hyperfine Interact. *2*, 339 (1976)
4.157 W. Semmler, F. Abildskow, K.B. Nielsen, J. Sindholt, G. Weyer: Hyperfine Interact. *4*, 755 (1978)
4.158 R. Böhm, J. Christiansen, P. Heubes, R. Keitel, W. Klinger, W. Sandner, W. Witthuhn: Hyperfine Interact. *4*, 758 (1978)
4.159 P. Heubes, R. Keitel, W. Klinger, W. Loeffler, W. Sandner, W. Witthuhn, R. Langkau, H. Krause: Hyperfine Interact. *4*, 746 (1978)
4.160 T.V. Ragland, R.J. Mitchel, R.P. Scharenberg: Phys. Rev. C *18*, 2494 (1978)
4.161 M. Becker, H. Bertschat, H.-E. Mahnke, E. Recknagel, R. Sielemann, B. Spellmeyer, Th. Wichert: Proc. Intern. Conf. on Hyperfine Interactions in Excited Nuclei, Rehovot, Israel, Vol.4, ed. by G. Goldring, R. Kalish (Gordon and Breach, London 1970) p.1192
4.162 M. Ionescu-Bujor, A. Iordachescu, E.A. Ivanov, D. Plostinaru, S. Vajda: Hyperfine Interact. *7*, 241 (1979)
4.163 W. Bartsch, W. Leitz, W. Semmler, R. Sielemann, Th. Wichert: Phys. Lett. *54*A, 66 (1975)
4.164 W. Bartsch, B. Lamp, W. Leitz, H.-E. Mahnke, W. Semmler, R. Sielemann, Th. Wichert: Annual Report, Hahn-Meitner-Institut, Berlin (1975) p.92
4.165 R. Böhm, J. Christiansen, W. Klinger, R. Keitel, W. Loeffler, W. Sandner, W. Witthuhn: Hyperfine Interact. *4*, 763 (1978)
4.166 O. Echt, H. Haas, E. Recknagel, E. Schlodder: Annual Report, Hahn-Meitner-Institut, Berlin (1975) p.99
4.166a M. Ionescu-Bujor, A. Iordachescu, E.A. Ivanov, D. Plostinaru, G. Pascovici: Hpyerfine Interact. *11*, 171 (1981)
4.167 J. Bleck, R. Butt, H. Haas, W. Ribbe, W. Zeitz: Phys. Rev. Lett. *29*, 1371 (1972)
4.168 R.E. Shroy, G. Schatz, D.B. Fossan: Hyperfine Interact. *4*, 738 (1978)
4.169 M. Menningen, H. Haas, H.H. Bertschat, R. Butt, H. Grawe, R. Keitel, R. Sielemann, W.-D. Zeitz: Phys. Lett. *77* A, 455 (1980)
4.170 S. Vajda, E.A. Ivanov, A. Iordachescu, D. Plostinaru: Rev. Roum. Phys. *22*, 1121 (1977)
4.171 R. Böhm, J. Christiansen, W. Engel, S. Hoth, R. Keitel, W. Klinger, H. Metzner, W. Sandner, R. Seeböck, W. Witthuhn: Annual Report, University of Erlangen-Nürnberg (1977/78) p.54
4.172 G. Schatz, P. Heubes: Hyperfine Interact. *4*, 751 (1978)

5. Electric Quadrupole Interaction in Noncubic Metals

W. Witthuhn and W. Engel

With 39 Figures

The knowledge of the charge distribution is of basic importance for the detailed understanding of the properties of a solid. In noncubic solids this charge distribution creates an electric field gradient (EFG). The quadrupole hyperfine interaction (QI) between the nuclear quadrupole moment Q and the EFG at the site of the nucleus results in an energy-splitting of the nuclear levels. The investigation of this QI in noncubic metals and alloys can be carried out by a variety of experimental techniques. The application of the perturbed angular correlation methods has been especially fruitful in this field. Its advantages, essential for QI studies in metals, are the ease with which temperature and pressure dependencies can be measured and the applicability to pure metals as well as to systems with extremely diluted impurity probe atoms. Thus, most experimental data have been obtained by this technique during the last few years.

The first experiments on the EFG in metals were reported more than two decades ago [5.1-3]. The interest in this field increased drastically after the pioneering work of RAGHAVAN [5.4] who demonstrated the perturbed angular correlation technique to be extremely powerful in this field. The experimental data collected within the following years revealed two remarkable systematic trends: the first is related to the temperature dependence of the quadrupole interaction. It was discovered at Erlangen [5.5,6] that in most cases this temperature variation follows fairly accurately a simple $T^{3/2}$ relation. This behaviour is found in pure metals as well as in dilute binary alloys and in many cases it holds over the entire temperature range up to the melting point.

The second important systematics is the correlation between the ionic and the electronic field gradients [5.7,8]. This empirical result was highly unexpected in terms of the conventional ansatz of the EFG based on two independent sources, one from the lattice and the other from the conduction electrons. The "universal correlation" rather implies—besides the dependence on the geometrical array of the lattice atoms—a strong relation of the electronic gradient to the nonsphericity of the probe ion cores, which is described by the Sternheimer antishielding factor.

Stimulated by these recent experimental trends emerging from the large body of available experimental data, significant progress in the theoretical approaches can be recognized. But we are still not able to predict reliable quantitative numbers from the theory for a given experimental situation. Calculations have been carried out only for a few pure metals, whereas the theory of the EFG at impurity probe atoms in highly diluted alloys—most experimental data fall into this category—is at the very beginning. Here, certainly, a challenge for future theoretical work exists.

Both theoretical and experimental aspects of the quadrupole interaction in metals have been reviewed by several authors. ROWLAND [5.9], DRAIN [5.10] and BARNES [5.11] dealt with the nuclear resonance techniques and its applications to pure systems. Theoretical calculations based on first principle approaches are summarized by DAS [5.12]. The articles by RAGHAVAN [5.13], NISHIYAMA and RIEGEL [5.14] and KAUFMANN [5.15] demonstrate the growing experimental information and theoretical interest. The most recent and comprehensive review is given by KAUFMANN and VIANDEN [5.16]. The rapid progress in this field can also be deduced from the numerous contributions to the last conferences on hyperfine interactions [5.17-21].

After a short summary of the quadrupole interaction theory in Sect.5.1 the experimental methods will be discussed in Sect.5.2. The systematics mentioned above will be discussed in more detail in Sect.5.3, together with experimental trends of the pressure dependence and the influence of different impurities and data in alloys. In Sect.5.4 the current status of the theory of the EFG in metals will be reviewed with applications to the experimental data, especially to the recent systematic trends. In the conclusion we try to trace the directions of future experimental and theoretical activities.

5.1 Electric Quadrupole Hyperfine Interaction

The nuclear electric quadrupole interaction has been treated in detail by several authors [5.22-24]. In this section we give only the basic ideas and formulas without paying attention to a rigorous derivation.

The Hamiltonian describing the electric quadrupole interaction of a localized charge distribution $\rho_N(\underline{r})$, for example a nucleus in an external potential $V(\underline{r})$, is given by

$$\hat{H}_Q = \frac{1}{6} \sum_{\alpha,\beta=1}^{3} e\hat{Q}_{\alpha\beta} \frac{\partial^2 V}{\partial x_\alpha \partial x_\beta}(\underline{r}_0) \quad , \tag{5.1}$$

where $\underline{r}_0$ denotes the radius vector to the centre of the charge and $x_1 = x$, $x_2 = y$, $x_3 = z$ are the cartesian coordinates.

The operators

$$\hat{Q}_{\alpha\beta} = \frac{3Q}{2I(2I-1)} (\hat{I}_\alpha \hat{I}_\beta - \frac{2}{3} \hat{\underline{I}}^2 \delta_{\alpha\beta}) \tag{5.2}$$

form the symmetric and traceless quadrupole moment tensor operator. Here $\hat{\underline{I}}$ is the vector operator of the nuclear spin and Q is the expectation value of Q_{zz} in the state with maximum I_z eigenvalue I:

$$Q = \frac{1}{e} \int \rho_N(\underline{r})(3z^2 - r^2)d^3r \quad . \tag{5.3}$$

The second partial spatial derivatives in (5.1)

$$\frac{\partial^2 V}{\partial x_\alpha \partial x_\beta}(\underline{r}) = V_{\alpha\beta}(\underline{r}) \tag{5.4}$$

are the cartesian components of the symmetric tensor of the electric field gradient. The expression (5.1) gives the correct second-order contribution to the electric interaction energy only if the EFG tensor has a vanishing trace. In this case the spherical components of the EFG tensor $\phi_2^{(m)}$ are often used:

$$\begin{aligned}
\phi_2^{(0)} &= \frac{1}{4}\sqrt{\frac{5}{\pi}}\, V_{zz} \quad , \\
\phi_2^{(\pm 1)} &= \pm \frac{1}{2}\sqrt{\frac{5}{6\pi}}\, (V_{xz} \pm iV_{yz}) \quad , \\
\phi_2^{(\pm 2)} &= \frac{1}{4}\sqrt{\frac{5}{6\pi}}\, (V_{xx} - V_{yy} \pm 2iV_{xy}) \quad .
\end{aligned} \tag{5.5}$$

With the quadrupole moment tensor defined in an analogous way, the quadrupole interaction term has the form

$$H_Q = \frac{4}{5}\pi \sum_m (-1)^m \hat{Q}_2^{(m)} \phi_2^{(-m)} \quad . \tag{5.6}$$

As a consequence of the Laplace equation the trace of the EFG tensor can vanish only if the density of the charges $\rho(\underline{r})$ producing the external potential is zero at $\underline{r}_0$. If, however, $\rho(\underline{r}_0)$ does not vanish, there is an additional contribution to the interaction energy

$$W_Q = \frac{1}{6}\Delta V(\underline{r}_0) \int d^3r \cdot r^2 \rho_N(\underline{r}) \tag{5.7}$$

and the trace of the total EFG tensor is nonzero. In this case the EFG tensor can be split into two contributions:

$$V_{\alpha\beta} = V_{\alpha\beta}^{(1)} + V_{\alpha\beta}^{(2)} \quad . \tag{5.8}$$

The first contribution $V_{\alpha\beta}^{(1)}$ originates from the charge density at $\underline{r}_0$. It is proportional to the unit tensor and has a nonvanishing trace

$$\sum_\alpha V_{\alpha\alpha}^{(1)}(\underline{r}_0) = \frac{1}{\varepsilon_0}\rho(\underline{r}_0) \quad . \tag{5.9}$$

The second contribution $V_{\alpha\beta}^{(2)}$ is generated by the charges outside $\underline{r}_0$ and therefore its trace is zero:

$$\sum_{\alpha} V_{\alpha\alpha}^{(2)}(\underline{r}_0) = 0 \quad . \tag{5.10}$$

In its system of principle axes, $V_{\alpha\beta}^{(2)}$ has two independent components and is usually described by its zz-component $V_{zz}^{(2)}$ and the asymmetry parameter

$$\eta = \frac{V_{xx}^{(2)} - V_{yy}^{(2)}}{V_{zz}^{(2)}} \quad . \tag{5.11}$$

Conventionally, the principle axes are chosen such that

$$|V_{xx}^{(2)}| \leq |V_{yy}^{(2)}| \leq |V_{zz}^{(2)} \quad . \tag{5.12}$$

According to this convention η is a positive number not greater than one:

$$0 \leq \eta \leq 1 \quad .$$

In the following sections the index (2) in $V_{\alpha\beta}^{(2)}$ is omitted where the meaning is obvious from the context. The principal component V_{zz} is denoted very often by eq in the literature.

In an arbitrarily chosen coordinate system the nonvanishing matrix elements of H_Q in the angular momentum representation have the form

$$\begin{aligned} \langle IM|H_Q|IM\rangle &= \frac{eQ}{4I(2I-1)}[3M^2 - I(I+1)](2V_{zz}^{(2)} - V_{xx}^{(2)} - V_{yy}^{(2)}) \\ \langle IM|H_Q|IM \pm 1\rangle &= \frac{eQ}{4I(2I-1)}(2M \pm 1)[(I \pm M + 1)(I \mp M)]^{\frac{1}{2}} \cdot (V_{xz}^{(2)} \pm iV_{yz}^{(2)}) \\ \langle IM|H_Q|IM \pm 2\rangle &= \frac{eQ}{4I(2I-1)}[(I \pm M + 2)(I \pm M + 1)(I \mp M - 1)(I \mp M)]^{\frac{1}{2}} \\ &\quad \cdot (V_{xx}^{(2)} - V_{yy}^{(2)} \pm 2iV_{xy}^{(2)}) \quad . \end{aligned} \tag{5.13}$$

Here it is assumed that the EFG is independent of coordinates. Otherwise one has to take the average of the EFG components over the motion of the nucleus.

Depending on the symmetry properties of the external potential, the quadrupole splitting is different. This can easily be seen if one uses the system of principle axes.

i) At a regular lattice site in a metal with cubic symmetric lattice, only an EFG $V_{\alpha\beta}^{(1)}$ proportional to the unit tensor can be present. The contribution $V_{\alpha\beta}^{(2)}$ must vanish: as the crystal axes are equivalent its diagonal elements are equal. According to definition (5.10), this EFG has a vanishing trace. Therefore all components are zero.

From (5.13) it can be seen that at a regular lattice site in a cubic metal, no quadrupole splitting can be observed, but only a shift of the nuclear energy levels according to (5.7) occurs.

ii) In a crystal with axial symmetry, the contribution $V^{(2)}_{\alpha\beta}$ is, in general, nonzero and two of its diagonal components are equal: $V^{(2)}_{xx} = V^{(2)}_{yy} = - V^{(2)}_{zz}/2$. In this case all off-diagonal elements of H_Q vanish and the energy eigenvalues are given by

$$E_M = \frac{eQ}{4I(2I - 1)} [3M^2 - I(I + 1)] \cdot V^{(2)}_{zz} \quad . \tag{5.14}$$

By means of the quadrupole frequency ω_Q defined as

$$\omega_Q = \frac{eQ}{4I(2I - 1)\hbar} V^{(2)}_{zz} \quad , \tag{5.15}$$

the difference between adjacent energy levels can be written as

$$E_{M+1} - E_M = 3(2M + 1)\hbar\omega_Q \quad . \tag{5.16}$$

That is, the quadrupole splitting is not equidistant, the spacings have integer ratios and the levels are twofold degenerate because states with the same magnetic quantum number but of opposite sign possess the same energy.

iii) In the case of nonaxial symmetry, H_Q is nondiagonal in the angular momentum representation. Its energy eigenvalues depend not only on $V^{(2)}_{zz}$ but also on the asymmetry parameter η. For integer nuclear spin the degeneracy of the ±M states is removed but not for half-integer spin. The dependence of the eigenvalues on the asymmetry parameter in the case I = 1 and I = 3/2 is shown in Fig.5.1.

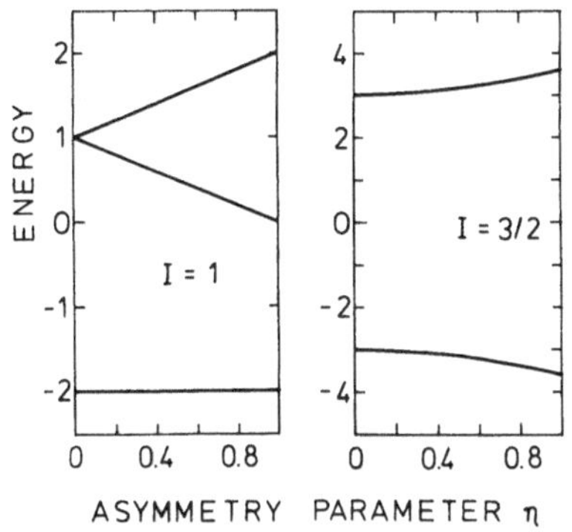

Fig.5.1. The variation of the substate energies with the asymmetry parameter η for spin I = 1 and I = 3/2. The energy eigenvalues are given in units of $eQV_{zz}/4I(2I - 1)$

5.2 Experimental Methods

The measurements of electric field gradients in solids are all based on the detection of the quadrupole hyperfine interaction. A variety of experimental techniques – some of them with sophisticated ramifications – have been developed for this purpose. They can, in general, give information on any hyperfine interaction and most of them have in fact been applied in the past to magnetic dipole interaction.

Here we concentrate on the various aspects of the methods related to the QI in metals and—in view of the excellent and detailed reviews published on each technique—summarize only the characteristic features of each and its limitations and range of application.

The complete information on the QI requires the measurement of

i) at least one energy-difference of the nuclear level split into unequally spaced M-sublevels yielding the magnitude of the quadrupole coupling constant $\nu_Q = eQV_{zz}/h$ and

ii) the sublevel ordering, which is determined by the sign of ν_Q. The sign of the EFG, however, still remains undetermined unless the sign of the nuclear quadrupole moment Q is known.

The experimental techniques can be divided into two categories:

a) the "Energy Methods". Here the energies of the split level are directly involved in the measuring process. The time dependence of the M-States is irrelevant for these methods.

b) the "Precession Methods", where the time dependency of the M-State eigenfunctions has to be considered explicitly. Here the information on the QI is obtained from interference terms in the coherent time evolution of the M-state eigenvectors. In a semiclassical picture, this corresponds to a precession of the nuclear spins around the field direction. Experimentally this precession is detected by resonance methods or by its influence on the emitted radiation pattern.

5.2.1 Energy Methods

In this class of methods we have to discuss the specific heat experiments, the nuclear orientation techniques and the Mössbauer effect.

a) *Specific Heat Measurements*

The nuclear level splitting due to the QI is typically of the order of $\Delta E_Q \approx 10^{-7}$ eV. This corresponds to temperatures in the order of a few mK. Therefore, the direct measurement of this energy requires (and is restricted to) very low temperatures of the sample.

In a metal the specific heat C_p is given by

$$C_p = AT^3 + \gamma T + C_N \quad , \tag{5.17}$$

where the first term represents the lattice vibrations and the second one describes the influence of the conduction electrons. Both contributions are frozen out at mK

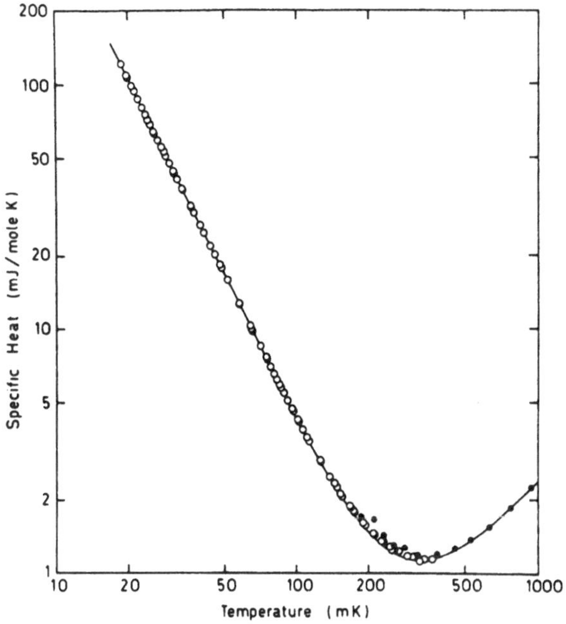

◄ Fig.5.2

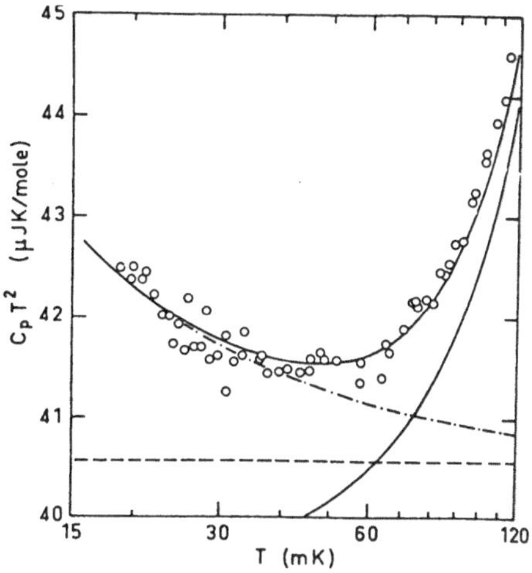

▲ Fig.5.3

Fig.5.2. The specific heat of Re metal at low temperatures measured by GREGERS-HANSEN et al. [5.26] (open circles) and SMITH and KEESOM [5.27] (full circles). The solid lines gives the fit of (5.17) with (5.18) to the data

Fig.5.3. The low temperature region of the data given in Fig.5.2 here plotted as C_pT^2 versus T. The solid line fitting the experimental data corresponds to a negative quadrupole interaction, the lower solid line to a positive one with the same magnitude. The dashed line represents the first term, the dashed-dotted curve the first two terms of the quadrupole heat-capacity (5.19) with a negative interaction [5.26]

temperatures. Here the nuclear (or hyperfine) specific heat C_N dominates. It can be expanded in a power series with inverse powers of the temperature T (for the exact expression of the "Schottky-anomaly" of the specific heat and for more details see [5.25]):

$$C_N = c_2T^{-2} + c_3T^{-3} + \dots \quad . \tag{5.18}$$

In nonmagnetic noncubic metals the QI determines this nuclear specific heat. The first two leading terms are given in this case by [5.26]

$$C_N \equiv C_Q = a(I)\left(\frac{e^2qQ}{k_B}\right)^2 T^{-2} + b(I)\left(\frac{e^2qQ}{k_B}\right)^3 T^{-3} \quad , \tag{5.19}$$

where a and b are constants depending on the nuclear spin I and k_B the Boltzmann constant. At very low temperatures (Figs.5.2,3) the second term in (5.19) becomes significant and due to the odd power in e^2qQ, the sign of the QI can be extracted. As an example the results of an experiment on Re metal [5.26] are shown in Figs. 5.2,3.

Measurements of C_Q can be made, in principle, on every metal and alloy provided a nuclear ground state with spin $I \geq 1$ exists. However, only a few experiments have been reported. The disadvantages of the method are serious: the experiments are possible only at temperatures below 1 K, therefore the temperature dependence of the QI cannot be investigated in principle and the study of the pressure dependence is nearly impossible. Secondly, only abundant isotopes can be measured since gram quantities are required for a determination of C_Q and experiments on samples with natural isotope composition cannot be interpreted unambiguously.

b) *Nuclear Orientation*

As in the specific heat measurements, the nuclear orientation techniques (NO, NMR/ON, NQR/ON) require low temperatures where k_BT is small compared to the splitting by the QI. The nuclear orientation is detected by observing the spatial anisotropy of the radiation emitted from the radioactive nuclei embedded in single crystals. The Boltzmann distribution forces an enhanced population $P(|M|)$ of the energetically lowest $|M|$ sublevel. The resulting nuclear alignment[1] $[P(M) = P(-M) \neq p(M')]$ is then reflected by the anisotropy of the nuclear radiation. The structure of the radiation pattern depends on the value of $|M|$. This feature allows the extraction of the sign of the QI because the most strongly populated level is the lowest one, which can be the $|M| = I$ or the $|M| = 0$, 1/2 level depending on the sign of ν_Q.

The angular distribution of the radiation emitted into a solid angle $d\Omega$ at an angle θ with respect to the alignment axis (an axially symmetric QI is assumed) can be expanded conveniently in Legendre polynomials [5.28]:

$$W(\theta,T) = \frac{d\Omega}{4\pi} \sum_{\substack{k \\ \text{even}}} B_k(\nu_Q,T) U_k(X_1) F_k(X_2) P_k(\cos\theta) \quad , \tag{5.20}$$

where the orientation parameters $B_k(\nu_Q,T)$ contain the information on the QI:

$$B_k(\nu_Q,T) = \left(\sqrt{2I+1} / \sum_M e^{-E_M/k_BT}\right) \sum_M (-1)^{I+M} \sqrt{2k+1} \times \begin{pmatrix} I & I & k \\ -M & M & 0 \end{pmatrix} e^{-E_M/k_BT} \quad , \tag{5.21}$$

with the energy E_M given in (5.14). The quantities $U_k(X_1)$ and $F_k(X_2)$ are decay-scheme parameters depending on the characteristics of the nuclear decay, i.e., on spins, multipole orders and mixing ratios δ of the radiations involved. They have been tabulated for many cases [5.29].

---

1 The term "alignment" (as well as "polarization") are special cases of nuclear "orientations" which denote, in general, an unequal population of the M-sublevels.

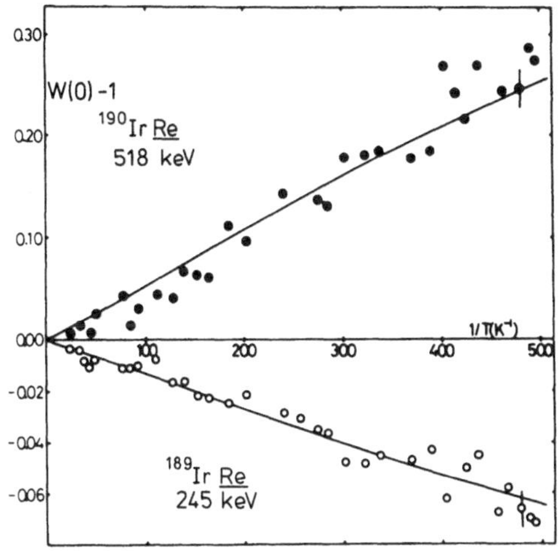

Fig.5.4. Pure electric quadrupole NO. The figure shows the anisotropy W(0)-1 of the γ-radiations emitted in the isomeric transitions in ^{189}Ir and ^{190}Ir. The nuclei are aligned by the QI at low temperatures in a rhenium single crystal. The data show the characteristic linear dependence on 1/T [5.30]

In the "high-temperature" region ($k_BT > h\nu_Q$), the orientation parameter can be expanded in powers of $h\nu_Q/kT$. The results of a recent experiment by MURRAY et al. [5.30] on Ir isotopes in Re single crystals are shown in Fig.5.4. The "anisotropy" W(0) - 1 is plotted over the reciprocal temperature, with W(0) being the count rate at the angle $\theta = 0^\circ$ normalized to that measured from the unoriented ensemble. It clearly shows the characteristic linear dependence on T^{-1}.

The NO method can be applied to all radioactive isotopes with spin $I \geq 1$ and a half-life of at least several hours. Pure metals, alloys and—due to the detection of single atoms by their decay-radiation—extremely diluted systems can be investigated. Low temperatures, however, and single crystals are required. This may be the reason why only a few NO experiments on QI have been performed up to now. But recent improvements in both cooling power and minimum temperature will certainly widen the accessible field in the next future.

Instead of observing the decay of the anisotropy during the warming-up process of the sample, one can hold it at constant temperature and destroy the alignment by a NMR/NQR experiment. By these resonances on oriented nuclei (NMR/ON or NQR/ON), the QI at impurity sites in metals can also be detected because of the high signal-to-noise ratio of the radioactive decay. The methods allow the separate investigation of nonequivalent lattice sites, since distinct field gradients correspond to different resonance frequencies. By means of the conventional NO method only an average EFG can be measured.

The pure NQR/ON has yet to be observed; NMR/ON has been applied successfully to ferromagnetic metals. Contributions to the level splitting due to QI well below 1% of the total interaction strength can be measured. Figure 5.5 gives the results obtained by VISSER et al. [5.31] for ^{131}I ions implanted into iron single crystals. Besides the main resonance, subresonances are observed which depend on the relative angle between the crystalline axis and the polarizing magnetic field B (for details

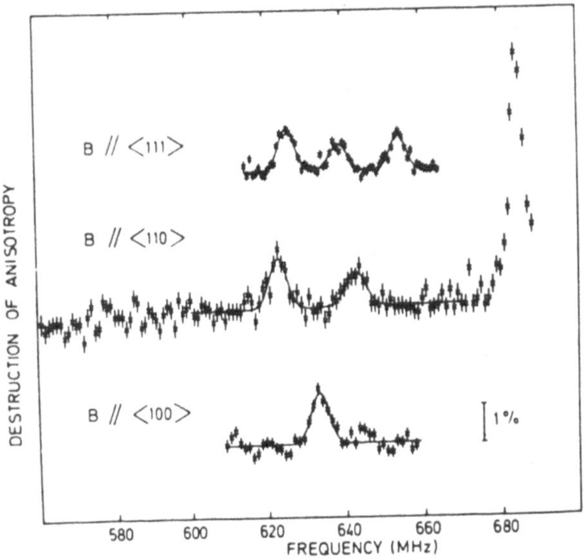

Fig.5.5. NMR/ON resonances of nonsubstitutional I-ions implanted into iron single crystals for three different orientations of the applied magnetic field with respect to the crystal axes. The subresonances are split by the quadrupole interaction [5.31]

see [5.31]). The data clearly demonstrate the high sensitivity and resolution of the NMR/ON techniques.

c) *Mössbauer Effect*

During the last twenty years the Mössbauer effect (ME) [5.32] has been applied to a large number of problems in different fields of physics, chemistry and biology. Several review articles have been published dealing with general aspects of the ME (e.g., [5.33]) or focussing on special topics.

The extremely high energy-resolution of the ME is based on the recoilless emission and absorption of γ-radiation in solids. The amount of the energy transferred to the recoiling atom is of the order of 10^{-2} eV and therefore small compared to the γ-energy. It is, however, still very large compared to the natural line width of the emitted γ-rays (the "standard Mössbauer-nucleus" ^{57}Fe has a line-width of 4.6×10^{-9} eV). The probability for recoilless emission (and absorption) is finite because the lattice is a quantized system and cannot be excited in an arbitrary fashion. The Debye-Waller factor, i.e., the fraction f of events without lattice excitation, is given in the Debye-approximation at low temperatures by

$$f = \exp\left[-\frac{E_R}{k_B\theta_D}\left(\frac{3}{2}+\frac{\pi^2T^2}{\theta_D^2}\right)\right] \quad \text{for} \quad T \ll \theta_D \quad , \tag{5.22}$$

with E_R being the recoil energy, θ_D the Debye-temperature and k_B the Boltzmann constant. Due to this factor the ME requires, in general, low temperatures and can be observed at room temperatures with sufficient efficiency only in favourable cases [such as for the 14.4 keV level of ^{57}Fe with $f(300\ K)\approx0.9$].

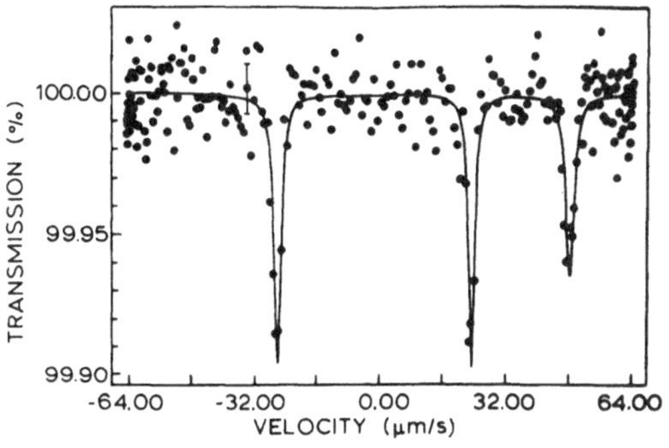

Fig.5.6. Mössbauer spectrum of ^{67}Zn in a polycrystalline enriched Zn metal absorber. A ^{67}GaCu source was used; the measurements were carried out at 4.2 K [5.34]

The level splitting due to the QI is in a great many cases comparable to or larger than the available energy resolution. Thus, this splitting can be measured directly by a relative motion of source and absorber resulting in a shift of the γ-ray energy due to the Doppler shift (relative velocities in the order of mm/s are required). In general the excited level as well as the ground state are split leading to rather complex Mössbauer spectra. Therefore, in practice, unsplit ("single line") sources or absorbers are used. The spectra yield information not only on the level splitting but also on the level ordering determined by the sign of the QI and in many cases on the ratio of the excited to the ground state moments. An example of a high resolution Mössbauer spectrum is shown in Fig.5.6 for ^{67}Zn in a powdered Zn metal absorber. In special cases (as in the mostly used Mössbauer nucleus ^{57}Fe with spins $I = 3/2$ and $I = 1/2$ for the excited and the ground state), ambiguities concerning the sign of the QI are present which can be avoided by using single crystals or external magnetic fields.

The application of the ME to QI studies in metals is limited because (i) the Mössbauer transition has to be a low-energy ground state transition, (ii) some of the frequently used Mössbauer nuclei have short meanlifes (corresponding to rather large national line widths) and/or small quadrupole moments. This leads to unresolved QI-patterns, where the interpretation is complicated and the accuracy only limited; (iii) the temperature dependence of the ME itself via the Debye-Waller factor seriously limits the investigation of the temperature dependence of the EFG in the high temperature region.

5.2.2 Precession Methods

In the following sections we discuss the precession methods. This category comprises the classical radiofrequency methods (NMR,NQR) and the perturbed angular correlation techniques. In the past both methods have been applied to numerous problems. The latter, however, has proven to be extremely powerful for QI studies in metals and the overwhelming part of the experimental data are obtained by this method. Therefore we give some aspects in more detail.

a) *Nuclear Resonance Methods*

Since the first successful demonstration by BLOCH et al. [5.35] and PURCELL et al. [5.36], the nuclear magnetic resonance (NMR) has become an extremely valuable tool for the investigation of magnetic and electric fields in bulk material. The field of applications ranges from physics to chemistry and biology. Reviews of nuclear magnetic resonance with emphasis on the metallic state were given by ROWLAND [5.9] and DRAIN [5.10]. In noncubic metals the quadrupole splitting of the nuclear energy levels can be observed directly without an applied external magnetic field. In pure metallic systems this nuclear quadrupole resonance NQR has been observed first in Ga metal by KNIGHT et al. [5.2]; a review was given by BARNES [5.11].

The basic principle of the nuclear resonance technique is the detection of the absorption of radio-frequency power in the case of resonance between the externally applied rf-field frequency and the frequency corresponding to the level splitting. Because of the smallness of the effect, the method requires a macroscopic number ($\gtrsim 10^{20}$) of nuclei contributing to the absorption process and therefore can be applied only to nuclear ground states of stable and relatively abundant nuclear species.

In metallic systems the skin effect seriously limits the general applicability of the NMR/NQR methods. To reduce this restriction one usually has to work with finely divided powder samples. But even then the signal-to-noise ratio is rather small, primarily due to crystalline imperfections producing inhomogenous broadening of the resonance signal. Therefore the NQR has yet to be observed in a number of noncubic metals. However, in favourable cases, like In, Ga and Sb metal, the obtainable precision of the method (the relative experimental error of the resonance frequency is of the order of 10^{-3} to 10^{-5}) cannot be reached, in general, by any other technique. Here also the variation of important solid state parameters such as temperature and pressure do not add difficulties on the experimental side in principle.

In contrast to the pure magnetic interaction where only one resonance frequency is observed, in pure NQR several transition frequencies occur which are given by

$$\nu_{|M|\to|M+1|} = \frac{3e^2qQ}{4I(2I-1)h}\,(2|M|+1) \quad . \tag{5.23}$$

A typical example is shown in Fig.5.7 for In metal. The figure shows the $|M| = 5/2 \to 7/2$ resonance lines for both isotopes ^{113}In and ^{115}In. The shift of the resonance lines corresponds to the slightly different quadrupole moments of both isotopes; the different intensities reflect the different isotope abundance.

The pure quadrupole resonance has been observed only in a few metals up to now. In less favourable cases it is very often appropriate to apply a substantial magnetic field and observe the quadrupole effects such as perturbations on the magnetic resonance. The single-line spectrum is split into 2I separate lines. In first-order perturbation, the transition frequencies for a single crystal are given by

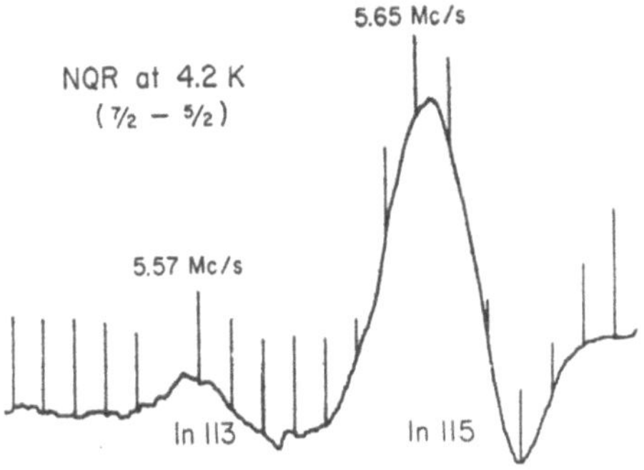

Fig.5.7. NQR resonance signal of the (5/2 → 7/2) transitions of ^{113}In and ^{115}In in In metal at 4.2 K [5.37]

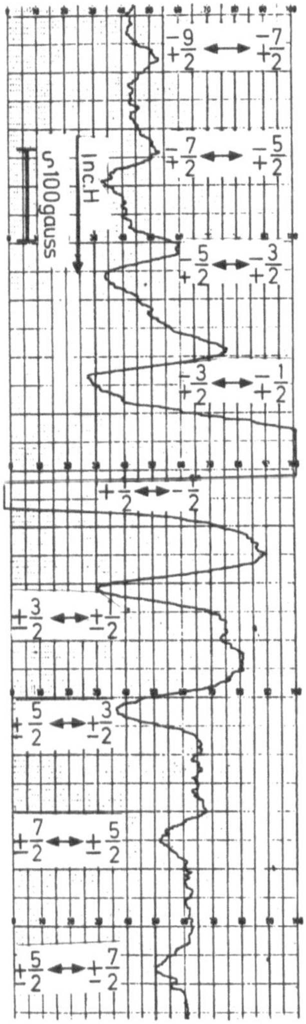

Fig.5.8. NMR of ^{99}Tc in technetium metal. Besides the central resonance, eight satellite resonances due to the quadrupole interaction are observed [5.38]

$$\nu_{M,M+1} = \nu_L + \frac{3e^2qQ}{4I(2I-1)h}\left(3\cos^2\theta - 1 + \eta\sin^2\theta\cos 2\varphi\right)\left(M+\frac{1}{2}\right) \quad . \tag{5.24}$$

Here, $\nu_L = gH/h$ denotes the Larmor frequency and θ and φ describe the direction of the magnetic field with respect to the principal axes system of the EFG. A typical pattern observed for ^{99}Tc in Tc metal is shown in Fig.5.8 [5.38]. Here the central resonance at about 10 Mc/s is seen together with eight satellite resonance lines.

If the quadrupole and magnetic interaction energies are comparable, first-order perturbation must not be applied. In these cases the second-order perturbation can give reliable results on the QI. For axial symmetry of the QI the shift of the central line is given by

$$\Delta\nu_Q = \frac{\nu_Q^2}{16\nu_L}\left[I(I+1) - \frac{3}{4}\right] \sin^2\theta(1 - 9\cos^2\theta) \quad . \tag{5.25}$$

By varying the magnetic field strength, the quadrupole effect can be separated from other competing shifts like anisotropic Knight shift (for details see [5.10]).

The major problem inherent in NMR and NQR experiments — the limited rf-skin depth — is avoided by the nuclear acoustic resonance method (NAR). Here the nuclear spins absorb energy from a sound wave propagating within the single crystal sample. The oscillating EFG created by the sound wave couples to the quadrupole moment of the nucleus. The mechanism of this coupling is still not completely understood. The only experiment reported so far was carried out on Re metal by BUTTET and BAILY [5.39]. This method certainly offers new experimental access to the QI studies in metals. It may, however, be limited to those cases where large single crystals can be grown and where the quadrupole interaction strength is rather large.

b) *Perturbed Angular Correlation (Distribution) Methods*

Most experimental data on the QI in metals have been extracted from *T*ime-*D*ifferential-*P*erturbed-*A*ngular-*C*orrelation (*D*istribution) measurements, conventionally abbreviated as the TDPAC (TDPAD) method, or shorter as the PAC (PAD) method. These techniques offer significant advantages in metals since they are highly sensitive due to the detection of nuclear radiations and therefore extremely diluted systems can be studied. Limitations due to the skin effect do not exist as in classical NMR/NQR and the measurements are not restricted to low temperatures in the mK range, as required for specific heat and nuclear orientation experiments. Thus, the temperature range from mK to temperatures beyond the melting points of the metals or alloys can be covered. In this respect the methods are superior to the Mössbauer effect where the signal depends on a Debye-Waller factor and decreases with increasing temperatures.

The isomeric γ-radiation need not be a ground state transition as in ME. Therefore a large number of suitable nuclei is available; in practice, an appropriate isomeric state can be found in nearly all elements. The selection of long-living states (10^{-7} - 10^{-4} s) increases the energy-resolution (QI-splittings of the order of 10^{-9} eV can be detected). In favourable cases the accuracy of the QI measurements approaches that of the nuclear resonance methods.

Principle of Method. The observation of the QI is based on the detection of an anisotropic radiation pattern and its change in time caused by this interaction. The basic requirement of an anisotropic emission probability is a nonrandom nuclear spin orientation, i.e., the M-sublevels must not be populated equally. Each γ-transition between certain states M (of the isomeric state $|I,M\rangle$) and M_f (of the final state $|I_f,M_f\rangle$) has a characteristic angular distribution with respect to the quantization axis. This distribution depends on the multipolarity L of the radiation and

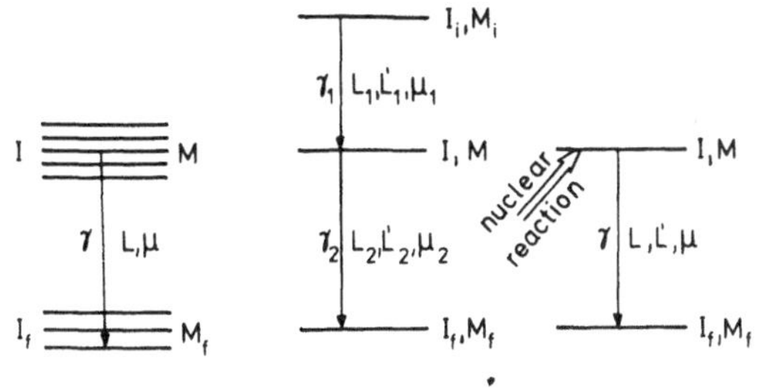

Fig.5.9. Nuclear levels and γ transitions involved in PAC/PAD experiments

on $(M - M_f)$. The observed γ-radiation is a superposition of these $M \rightarrow M_f$ transitions and is therefore emitted anisotropically; an equal population of the M-states would result in an isotropic radiation intensity.

The nonrandom spin orientation is provided by the detection of a preceding radiation populating the isomeric level (TDPAC method) or by populating the isomeric state via a nuclear reaction (TDPAD method) (Fig.5.9). In both methods the population process defines an axis. The QI causes a change of the population of the M-substates. This time evolution can be interpreted classically as a precession of the nuclear spins around the axis of the QI. Due to this precession the angular distribution of the decay γ-radiation becomes time dependent. The observation of this characteristic time dependence yields the information on the QI.

It should be emphasized that even in case of no QI, the angular distribution of the decay radiation is generally anisotropic. If the intermediate level is split by a QI, no effect of this interaction is observable if the symmetry axis of the QI and the axis defined by the populating process coincide.

The Perturbed Angular Correlation. The probability $W(\underline{k}_1,\underline{k}_2,t)$ of emission of the γ-radiation γ_1 into the direction $\underline{k}_1$ and of γ_2 into the direction $\underline{k}_2$ is given by [5.40]

$$W(\underline{k}_1,\underline{k}_2,t) = \sum_{k_1N_1} \sum_{k_2N_2} A_{k_1}(1)A_{k_2}(2)[(2k_1+1)(2k_2+1)]^{-\frac{1}{2}} \cdot G_{k_1k_2}^{N_1N_2}(t)Y_{k_1}^{*N_1}(\theta_1,\phi_1)\,Y_{k_2}^{N_2}(\theta_2,\phi_2) \quad . \tag{5.26}$$

The information on the hyperfine interaction is contained in the time-dependent perturbation factor $G_{k_1k_2}^{N_1N_2}$. The spherical harmonics Y_k represent the angular part of the correlation function with θ_i and ϕ_i being the angles of the radiation k_i with respect to the quantization axis. The angular correlation coefficients A_k are determined by the spins of the involved states, the multipole orders L,L' of the radiations and their mixing ratios δ. These coefficients are tabulated (e.g., in [5.40]).

In the case of a static quadrupole interaction with axial symmetry the perturbation factor has the form

$$G^{N_1N_2}_{k_1k_2}(t) = [(2k_1 + 1)(2k_2 + 1)]^{\frac{1}{2}} \sum_{MM'} \begin{pmatrix} I & I & k_1 \\ M' & -M & N_1 \end{pmatrix} \begin{pmatrix} I & I & k_2 \\ M' & -M & N_2 \end{pmatrix}$$

$$\times \exp\left[-\frac{i}{\hbar}(E_M - E_{M'})t\right] \quad . \tag{5.27}$$

Many experiments are carried out on polycrystalline samples (powder sources). The perturbed angular correlation function applicable for these cases is obtained by averaging over all directions of the QI symmetry axes:

$$W(\theta,t) = \sum_{k \text{ even}} A_k(1)A_k(2)G_{kk}(t)P_k(\cos\theta) \quad . \tag{5.28}$$

Here, θ denotes the angle between the directions $\underline{k}_1$ and $\underline{k}_2$ and P_k the Legendre polynomials. The perturbation factor then is given by

$$G_{kk}(t) = \sum_n s_{kn} \cdot \cos(n\omega_0 t) \tag{5.29}$$

with

$$s_{kn} = \sum_{MM'}{}' \begin{pmatrix} I & I & k \\ M' & -M & M-M' \end{pmatrix}^2 \quad .$$

The basic frequency ω_0 is related to the quadrupole frequency ω_Q defined in (5.15) by

$$\omega_0 = \begin{cases} 6\omega_Q & \text{for I half-integer} \\ 3\omega_Q & \text{for I integer} \quad . \end{cases} \tag{5.30}$$

Equations (5.28) with (5.29,30) are the basic expressions which allow the parameter-free calculation of the influence of a static axially symmetric EFG on the angular correlation. They are related directly to the experimentally observed spectra.

Usually the PAC spectra are taken with a multidetector arrangement. In a typical four-detector system one has, in principle, 12 coincidence spectra; in practice, however, two detectors are used as start-detectors (A and B) and two serve as stop-detectors (C and D). The coincidence rates are given by

$$N_{ij}(\theta,t) = N_0\varepsilon_i\varepsilon_j \, e^{-t/\tau} \cdot W(\theta,t) \quad , \tag{5.31}$$

where the exponential factor describes the decay of the isomeric state and $W(\theta,t)$ denotes the angular correlation function defined in (5.28).

The different detector efficiencies ε_i (i = A,B,C,D) can be cancelled in the case k_{max} = 2 by forming the ratio

$$R(t) = \frac{2}{3}\left[\left(\frac{(AC)(BD)}{(AD)(BC)}\right)^{\frac{1}{2}} - 1\right]$$

$$= \frac{2}{3}\left(\frac{W(180^\circ)}{W(90^\circ)} - 1\right) \approx A_{22}G_{22}(t) \quad . \tag{5.32}$$

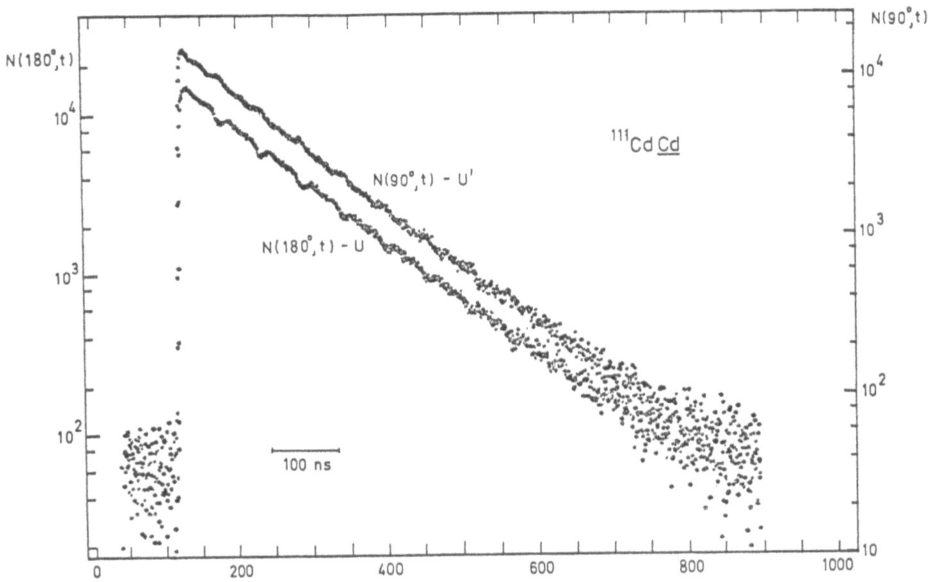

Fig.5.10. Background corrected coincidence rates according to (5.31) for two detector pairs encompassing angles of 90° and 180°, respectively. The data show the exponential decay of the isomeric level of ^{111}Cd (E_x = 247 keV, I = 5/2, $T_{\frac{1}{2}}$ = 84 ns) modulated by the perturbation factor due to the quadrupole interaction in polycrystalline Cd metal

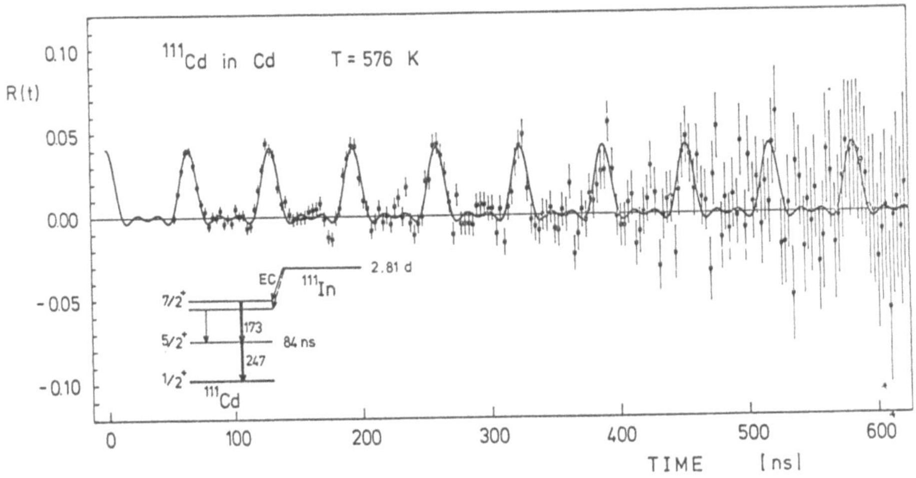

Fig.5.11. Typical TDPAC-spectrum as calculated according to (5.32) from the data given in Fig.5.10. The solid line represents a fit of the theoretical expression, see (5.29), for a spin I = 5/2 in a polycrystalline sample

Typical primary intensity spectra at two angles are shown in Fig.5.10. The ratio corresponding to (5.32) formed from these spectra is given in Fig.5.11. A typical spectrum for in-beam experiments (TDPAD method) is shown in Fig.5.12 for a single crystalline lattice (Ge in Zn [5.41]).

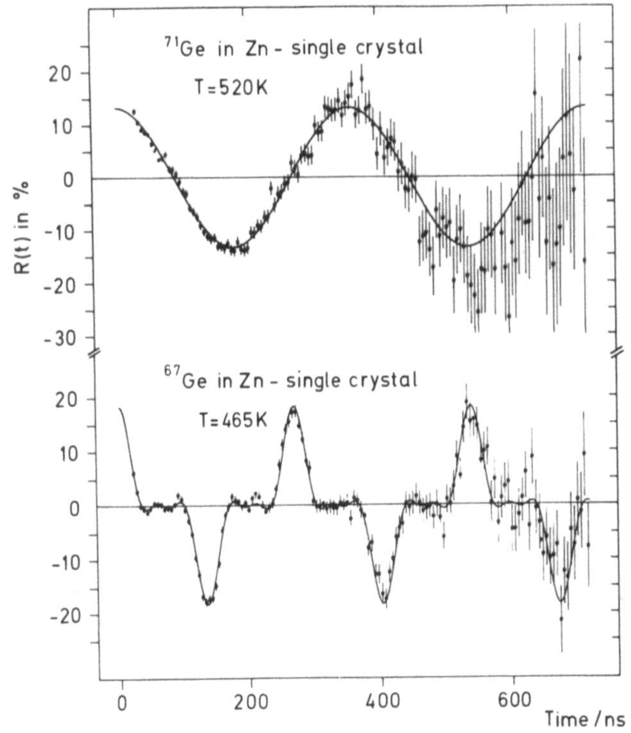

Fig.5.12. In-beam modulation patterns (TDPAD-method) of the quadrupole interaction of ^{67}Ge and ^{71}Ge in Zn. The isomeric states were populated via the nuclear reactions ^{64,68}Zn(α,n)^{67,71}Ge on a Zn single crystal. The c-axis of the single crystal was oriented perpendicular to the α-beam direction and the detector plane. The solid lines are fits of the theoretical function for spin I = 9/2 (^{67}Ge) and I = 5/2 (^{71}Ge) to the data [5.41]

Measurements of the Sign by PAC Methods. In standard perturbed angular correlation measurements, the intermediate states are aligned and therefore do not yield the sign of the QI. In a classical sense this means that no net precession of the nuclear spins can be observed. The detection of the sign requires that the nuclei are polarized. This can be realized in PAC experiments by (i) the detection of the γ-ray polarization, i.e., one selects only the nuclear spins precessing in one direction or (ii) by an unequal population of the +M and -M states. This question is treated in more detail in Chap.6, see also [5.42].

The first method has not yet been applied to the measurement of the sign. The second technique can be realized experimentally by detecting a β-γ-directional correlation. Because of parity nonconservation in nuclear β-decay, the detection of the emission direction of the β-particles selects an ensemble of polarized nuclei. Therefore, one has to observe only the directional correlation of the β- and γ-radiations.

The theory of the β-γ-perturbed angular correlation was first reported by HARRIS [5.43] for allowed β-decay. Here we want to demonstrate the method on the first-forbidden β-decay, since most experiments used the first-forbidden decay of ^{111}Ag $\rightarrow$ ^{111}Cd.

The β-γ-angular correlation function can be obtained from (5.26) by inserting the phase factor $(-1)^{k_1+k_2}$. It was omitted in (5.26) since in the case of a γ-γ-cas-

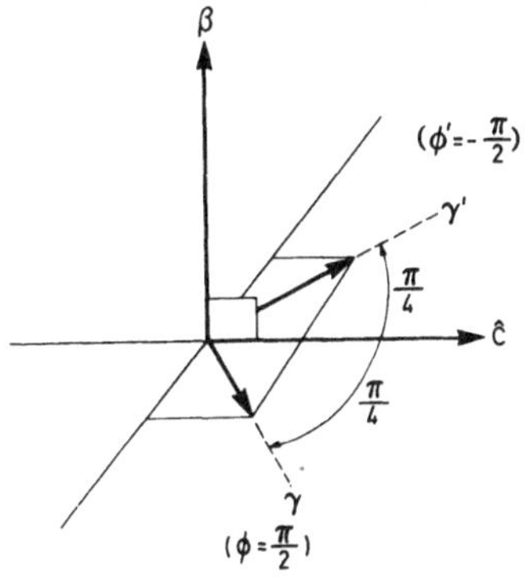

Fig.5.13. Geometry used in β-γ-perturbed angular correlation experiments. The γ-detectors are at the positions γ and γ'; ĉ denotes the c-axis of the single crystal

cade, $k_1 + k_2$ is always even. Assuming an axially-symmetric field gradient, one obtains for the perturbation factor [5.44]

$$G^{NN}_{k_1k_2}(t) = \sum_n s^{k_1k_2}_{nN} \begin{cases} \cos n\omega_0 t & k_1+k_2 \text{ even} \\ (-i)\sin n\omega_0 t & k_1+k_2 \text{ odd} \end{cases} . \tag{5.33}$$

The terms with odd values of $(k_1 + k_2)$ correspond to nonzero polarization of the nuclei and therefore allow the determination of the sign of ω_0.

For the unique first-forbidden β-decay of ^{111}Ag, the angular correlation W(±,t) has been calculated explicitly by RAGHAVAN et al. [5.42] for the geometry shown in Fig.5.13 ($\phi = \pm\frac{1}{2}\pi$):

$$W(\pm,t) = 1 \mp 0.18\,\frac{v}{c}\sin 2\omega_0 t \mp 0.08\,\frac{v}{c}\sin 3\omega_0 t - 0.10\cos 3\omega_0 t \quad . \tag{5.34}$$

Experimentally one observes the coincidence rates at angles $\phi = \pm\frac{1}{2}\pi$:

$$N(\pm,t) = N_0\, e^{-t/\tau} W(\pm,t) \tag{5.35}$$

and conventionally forms the ratio

$$A(t) = \frac{2[N(-) - N(+)]}{[N(-) + N(+)]} \quad . \tag{5.36}$$

Inserting (5.34) one obtains

$$A(t) \approx a \sin 2\omega_0 t + b \sin 3\omega_0 t \quad , \tag{5.37}$$

where a and b are positive parameters. Typical spectra of ^{111}Cd in various host metals are shown in Fig.5.14. From the phases of the functions A(t), the signs of the QI can be deduced to be positive for Cd<u>Sn</u> and Cd<u>Ti</u> and to be negative for Cd<u>Be</u> (for details see [5.8,42]).

Besides the general applicable PAC and PAD methods, some ramifications have been found to be practicable in a few systems. The NMR/PAC version was first experimentally demonstrated by MATTHIAS et al. [5.45]. This method has later been

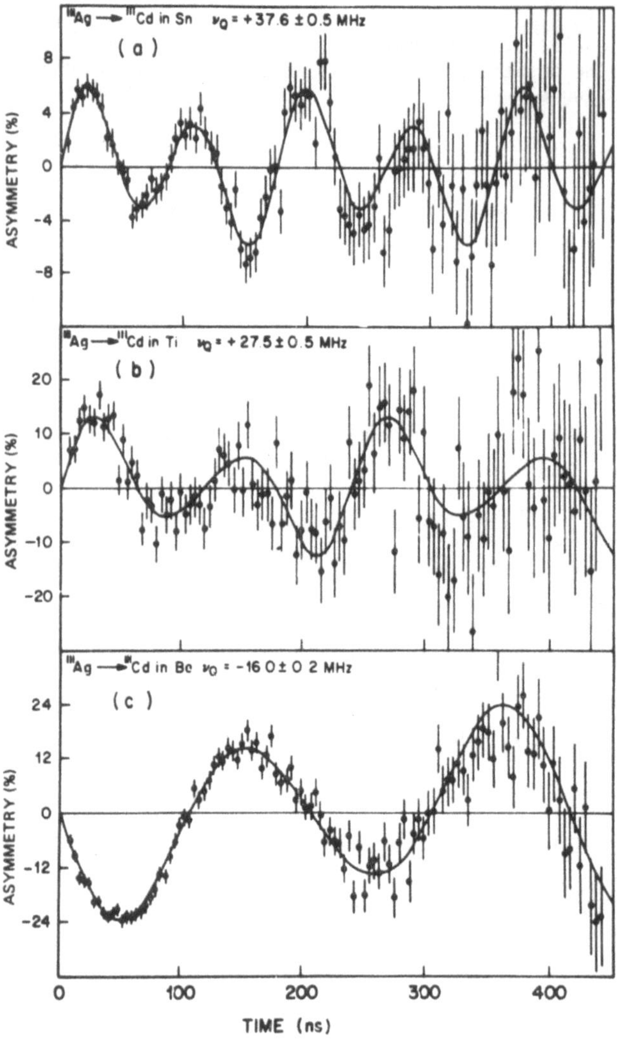

Fig.5.14. Typical β-γ-angular correlation spectra for ^{111}Cd in single crystals of Sn, Ti, and Be. The solid lines are fits of the asymmetry A(t) defined by (5.37) to the data [5.8]

used on polarized β-instable light nuclei, where the β-ray asymmetry is destroyed by a NMR-experiment (Chap.6).

In magnetic hyperfine interaction measurements on isomeric states produced with a pulsed ion beam, a stroboscopic method has been developed by CHRISTIANSEN et al. [5.46,47]. Here the resonance between the fixed pulsing frequency of the particle accelerator and the Larmor frequency is observed, the latter being varied by tuning the external magnetic field, i.e., the Larmor precession frequency. This principle has been applied by SCHATZ et al. [5.48] to the QI of Ge in Zn (Fig.5.15). Since the internal electric field gradient cannot be varied, the accelerator pulsing frequency has to be tuned in this technique. Because of the technical problems involved this method has not been applied to other systems so far.

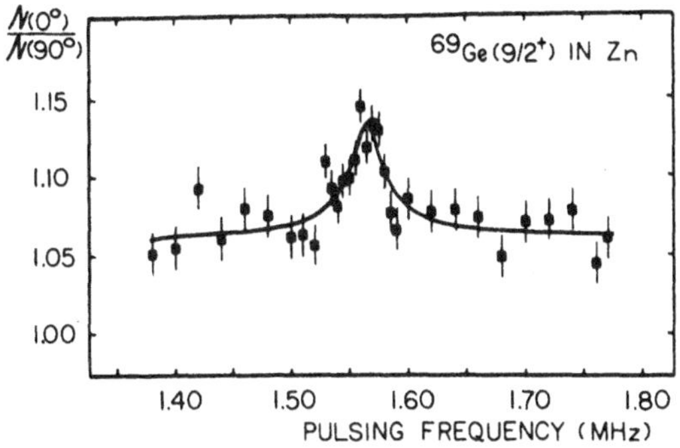

Fig.5.15. Stroboscopic resonance (QSOPAD) of the quadrupole interaction of ^{69}Ge (spin I = 9/2, $T_{1/2}$ = 2.9 μs) in polycrystalline Zn [5.48]

5.3 Experimental Data and Systematic Trends

The nuclear quadrupole interaction has been investigated so far in nearly all noncubic metals (except for some metals of the rare earths and actinides). Experiments have not been carried out in the pure metals only; the quadrupole coupling constants have also been determined in about 400 "impurity systems". Here the field gradient is measured at a probe nucleus which is a chemical impurity in the host lattice. Most data have been obtained in highly diluted systems, i.e., the concentration of the probe atoms is far below the concentrations that can be produced by standard alloying techniques. Only very few experiments have been reported on binary alloys and intermetallic compounds.

The comparison of the experimental results with theory requires more information than only the magnitude of the interaction energy. In this respect the sign of the EFG is of great importance. It cannot be obtained, however, by all experimental techniques. Furthermore, the dependence of the EFG on lattice parameters, on temperature and pressure, as well as on the concentration of the constituents in binary alloys can yield very valuable information and can critically test the theoretical models. In view of these requirements, most of the available data provide only incomplete information. The need for complete information on a given system has been fully recognized in the last few years and has stimulated recent experimental activity into that direction.

It is far beyond the scope of this article to discuss all the experimental data in detail. We rather intend to review the systematic trends of the EFG emerging from the data and to relate them to the present understanding of the sources of the EFG in noncubic metals. For the purpose of completeness we have listed the experiments on quadrupole interaction in noncubic metals reported so far in the appendix.

The early experimental observations did not reveal clear systematic trends of the EFG. The results were compared individually to theoretical or phenomenological models and revealed, in general, unsatisfying agreement. In the mid-seventies, however, empirical systematics were recognized. These principal trends can be classified into three categories:

i) the "universal correlation" reported first by RAGHAVAN et al. [5.7,8];
ii) the $T^{3/2}$ relation describing the temperature dependence of the EFG [5.5,6];
iii) the correlation between the impurity valence and the EFG at the impurity site [5.49,50].

Besides these, further systematic trends can be recognized from recent data on pressure dependence, binary alloys, transition metals, rare-earth metals and distorted cubic metals. All of them cannot be discussed in this article; a recent summary was given by KAUFMANN [5.15] and References therein.

The calculation of the electric field gradient in metals will be discussed in Sect.5.4; here we quote only the commonly used ansatz. Conventionally one uses the parametrization

$$V_{zz} = (1 - \gamma_\infty)V_{zz}^{ext} + (1 - R)V_{zz}^{local} \tag{5.38}$$

of the EFG at the probe nucleus in a solid. Here the term V_{zz}^{ext} denotes the gradient due to the charges of ions and electrons of the lattice outside the atom or ion of interest. The Sternheimer factor γ_∞ describes the distortion of the probe ion shell by the external EFG. This distortion results in a large enhancement ("antishielding") of the external field gradient (Sect.5.4). The second term of (5.38) represents the gradient due to sources within the electron shell of the probe ion such as nonspherical unfilled orbitals, e.g., open 4f electron shells.

The separation of the different sources of the EFG in (5.38) can be justified for ionic solids where the electron shells do not overlap. However, the application to metals of (5.38) without modification—as done by many authors—raises serious questions, since the charge density of the conduction electrons of the metal cannot be separated in a simple way into an external and local fraction. Disregarding these difficulties, one usually writes for noncubic metals

$$\begin{aligned} V_{zz} &= (1 - \gamma_\infty)V_{zz}^{latt} + (1 - R)\hat{V}_{zz}^{el} \\ &= V_{zz}^{ion} + V_{zz}^{el} \quad . \end{aligned} \tag{5.39}$$

Here, the first term describes the EFG due to the ions and the second one represents the EFG due to the conduction electrons. This ansatz must be regarded as a rough approximation: the EFG due to the lattice ions, the localized shell electrons and the conducting electrons can not be separated because these field sources interact. The second open question is the use of the Sternheimer factors γ_∞ which are defined for free ions only. A more detailed discussion of these questions is given in Sect. 5.4.

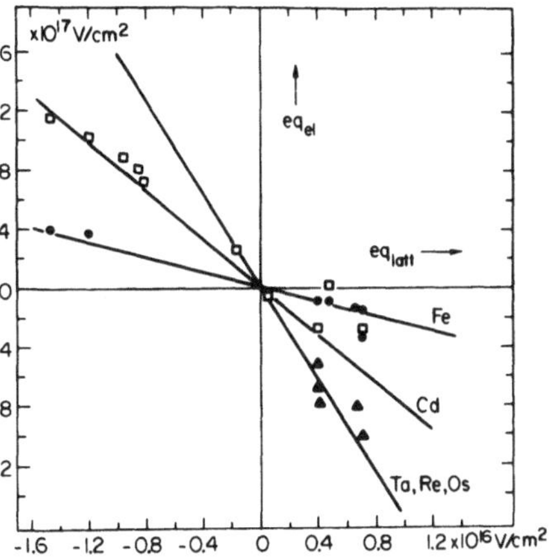

Fig.5.16. The electronic gradient vs the lattice gradient for Fe, Cd, and Ta, Re, Os impurities in different host lattices. The electronic gradient (eq_{el}) was calculated from the experimental EFG by subtracting the theoretically calculated lattice gradient (eq_{latt}). The slopes of the straight lines are roughly in the ratio of the Sternheimer factors γ_∞ for Fe, Cd, and Ta [5.13]

5.3.1 The Universal Correlation

A remarkable and unexpected correlation between the ionic and the electronic field gradient defined by (5.39) was discovered by RAGHAVAN et al. [5.7,8,13]. They calculated the lattice gradient V_{zz}^{latt} for about 25 coupling constants, where the sign of the EFG was determined. The correlation between the lattice gradient and the electronic gradient, the latter obtained from the difference $(V_{zz}^{exp} - V_{zz}^{ion})$, is shown in Fig.5.16. Two features are obvious:

i) the electronic term (denoted by eq_{el} in Fig.5.16) depends linearly on the lattice gradient (eq_{latt}). This result can be understood qualitatively, since it seems reasonable that the charge distribution of the conduction electrons reflects the structure and the symmetry of the lattice.
ii) the slopes of the straight lines in Fig.5.16 are rouhgly in the ratio of the Sternheimer factors of the probe ions. This indicates a proportionality of the electronic gradient to the ionic gradient, i.e., the distorted probe ion shell significantly influences the electronic contribution.

The so-called universal correlation is shown in Fig.5.17. The filled circles represent the data on which the correlation is based. The open circles give experimental data where only the magnitude but not the sign of the EFG was measured. Here the data were plotted with a choice of sign such that the best agreement with the correlation was achieved (for details see [5.8]). Omitted from the figure are data on rare-earth atoms with open 4f shells where the major contribution to the EFG arises from the 4f electrons (these have been treated by PELZEL [5.51] and more recently by ERNST et al. [5.52]).

The correlation can be represented by the empirical expression [5.7]

$$V_{zz} \approx + K(1 - \gamma_\infty)V_{zz}^{latt} \quad , \tag{5.40}$$

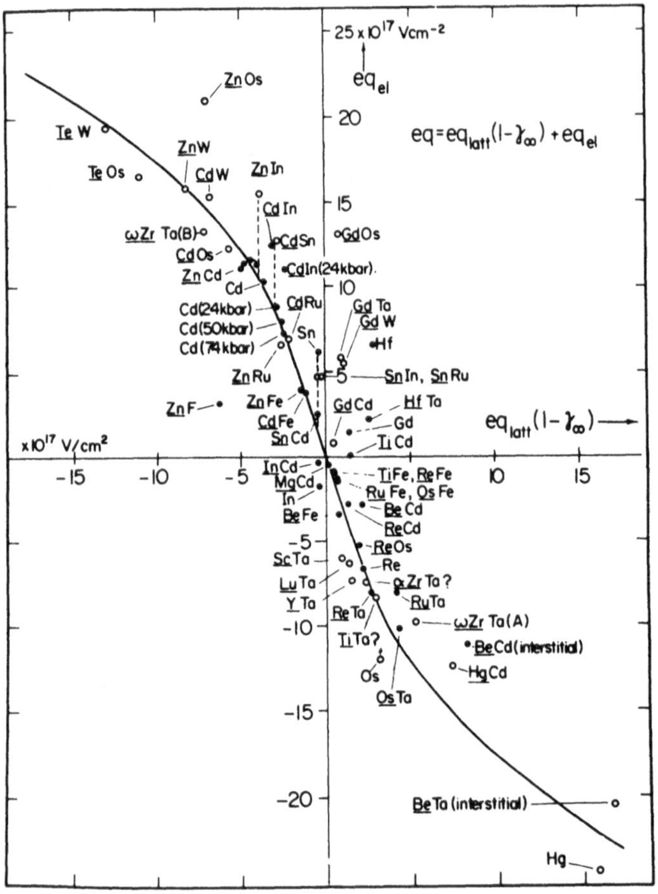

Fig.5.17. Universal correlation of the ionic and the electronic field gradient in noncubic metals. Filled circles indicate data for which the sign was measured, open circles are data without measurement of the sign. The location is predicted in these cases. The solid line does not represent a theoretical fit, it rather illustrates the proportionality [5.7]

the constant K being of the order of $K \approx -3$. The approximative character of this expression can be seen from Fig.5.17. Several points lie off the main curve. These discrepancies were related to valence electron effects, to possibly nonregular lattice sites of the probe ions (as for F in Zn), to deviations of the real ion charges from the assumed ones (in the calculations the nominal integer valence of the host elements were taken), or to host lattice relaxtions in the impurity systems. However, in all cases these explanations were based on qualitative arguments rather than on detailed theoretical calculations. Some data were found in the first or third quadrant, indicating an unexplained breakdown of the general sign-reversal behaviour.

The correlation was extended recently by ERNST et al. [5.52]. They found an additional positive branch of the correlation with a constant $K \approx +2$. Figure 5.18 shows this extended correlation for pure systems. ERNST et al. [5.52] propose the

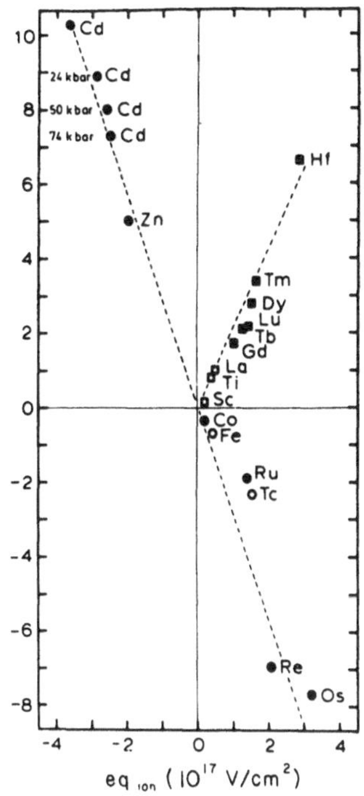

Fig.5.18. The extended universal correlation for pure systems [5.52]

original correlation (K ≈ -3) to be valid for all group IIb, VIIb, and VIIIb metals and the positive branch (K ≈ +2) for all group IIIb and IVb hexagonal metals. Recent data on impurity systems [5.53] support this revised correlation. This new positive correlation has not yet been explained theoretically.

The basic feature of the proportionality of the electronic EFG to the Sternheimer factor of the probe ion given in the original negative correlation was reproduced recently [5.54-56] by first principle calculations, where the antishielding effects on conduction electrons were included explicitly for cadmium and zinc. The pseudopotential approach applied by LODGE [5.57] also corroborates the experimental results. All these calculations are carried out for pure metals only. The theoretical interpretation of the impurity systems still remains completely unsolved.

5.3.2 The Temperature Dependence of the EFG

The first measurements of the quadrupole interaction in metals revealed a temperature dependence of the EFG considerably stronger than that observed in ionic crystals. Figure 5.19 gives as an example the experimental results for ^{115}In in In metal obtained by HEWITT and TAYLOR [5.37]. The interaction frequency decreases continuously by a factor of two from 4 K to the melting point.

In the most simple approach, the measured temperature dependencies were related to thermal lattice expansions by calculating the lattice gradients V_{zz}^{latt}, see (5.39), using the known temperature dependencies of the lattice constants. Only

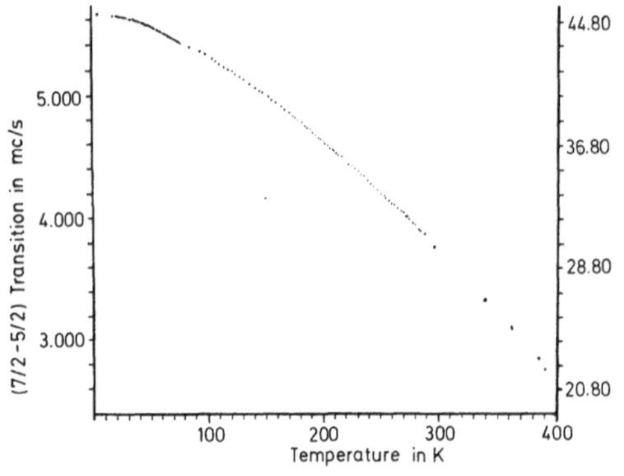

Fig.5.19

Fig.5.20

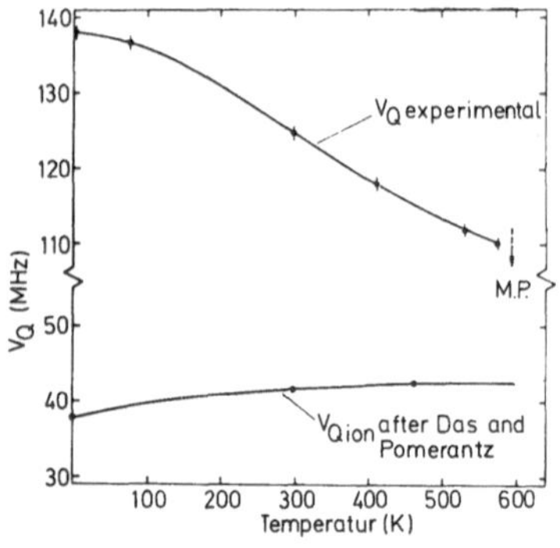

Fig.5.19. The temperature dependence of the EFG at In in In metal. The data demonstrate the high accuracy of the NQR method and the relatively strong variation of the EFG with temperature [5.37]

Fig.5.20. Experimental and calculated temperature dependence of the EFG of Cd in Cd metal. The lower curve represents the ionic contribution obtained from lattice sum calculations using the tabulated temperature dependence of the lattice constants. This anomalous variation of the EFG was observed first by RAGHAVAN and RAGHAVAN [5.4]

poor agreement was obtained: the predicted variation of the EFG with temperature was, in general, too weak and was in many cases in the opposite direction to the experimental results as demonstrated in Fig.5.20. In view of these discrepancies, a substantial temperature dependence of the local conduction electron contribution was invoked by several authors. Within a model proposed by WATSON et al. [5.58], the temperature dependence was explained as being due to a thermally induced repopulation of conduction electron states at the Fermi level. However, the general applicability of this model was questioned first by measurements of the temperature dependence of Sc, Cd, and Ta impurities in Ti metal [5.59,60]. The observed variation which is different for all three impurities implies a substantial probe dependence of EFG(T), not included in the picture based on the Fermi surface electrons.

A highly unexpected discovery was reported by HEUBES et al. [5.5] and discussed in more detail by CHRISTIANSEN et al. [5.6]. They noticed a remarkably simple and universal relationship between field gradient and temperature (Fig.5.21):

$$V_{zz}(T) = V_{zz}(0)(1 - BT^{3/2}) \quad . \tag{5.41}$$

Besides the trivial normalization factor $V_{zz}(0)$, the temperature dependence of the EFG is described in many cases over the entire temperature range up to the melting point by only one free parameter B. This empirical temperature rule was found to be valied for pure noncubic metals, for most impurity systems and for binary alloys and intermetallic compounds. The exceptions from relation (5.41) known so far are most rare-earth systems, some systems including transition metals and a few cases where the probe ions may not occupy well-defined lattice positions.

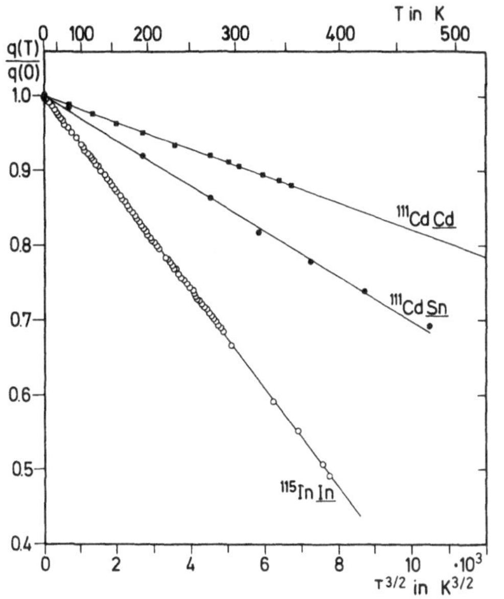

Fig.5.21. The empirical $T^{3/2}$ temperature dependence of the EFG in metals shown for Cd, In, and Sn. The data are normalized at T = 0 K. The filled symbols give the results of PAC-experiments [5.6], the open circles are the NQR data [5.37] shown originally in Fig.5.19

The parameter B represents the strength of the temperature variation. It is positive in nearly all cases, the values ranging from 1×10^{-5} $K^{-3/2}$ to about 8×10^{-5} $K^{-3/2}$ [5.6]. Small deviations from this relation were observed for several systems at low temperatures. Here the experimental field gradients are always smaller than the values predicted from an extrapolated $T^{3/2}$ curve. In some dilute alloys the coefficient B is determined essentially by the host lattice (such as Cd and In). In other cases (such as Zn), there are different groups of probe nuclei with distinct temperature parameters B. In a third class (such as Ti or Sn), the coefficient B depends strongly on the individual impurity.

The most surprising feature of (5.41) is its simplicity. So far no theoretical model reproducing the analytical form of this relation has been reported. A physically attractive idea explaining the observed $T^{3/2}$ power law was discussed at Erlangen [5.6,61,62]. It is based on the interaction of neighbouring atomic quadrupoles. Elementary excitations in this array of quadrupoles ("quadrons") results in a $T^{3/2}$-temperature dependence of the EFG in analogy to the $T^{3/2}$ law of the magnetization of ferromagnetic solids derived by BLOCH [5.63]. The observed inverse proportionality between the temperature parameter B and the absolute QI strength [Ref. 5.6, Fig.10] could be explained naturally within this quadron picture. New data, however, question this correlation [5.64]. Further implications and objections are discussed briefly in Sect.5.4. At present the thermal lattice vibrations are regarded as being essentially responsible for the temperature dependence of the EFG in metals. For ionic crystals and molecular solids, this has long been known [5.65,66]. The application to metals was first reported by QUITMANN et al. [5.67]; more detailed investigations were given by JENA [5.68], NISHIYAMA et al. [5.69], THOMPSON et al.

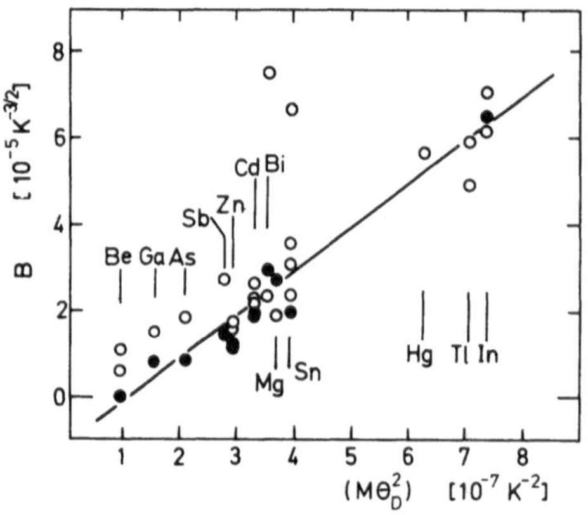

Fig.5.22. The strength of the temperature dependence of the EFG represented by the slope parameter B versus the "spring constant" of the host lattice. The filled circles give the pure metals, the open symbols denote impurity systems. The references of the experimental data are given in the appendix

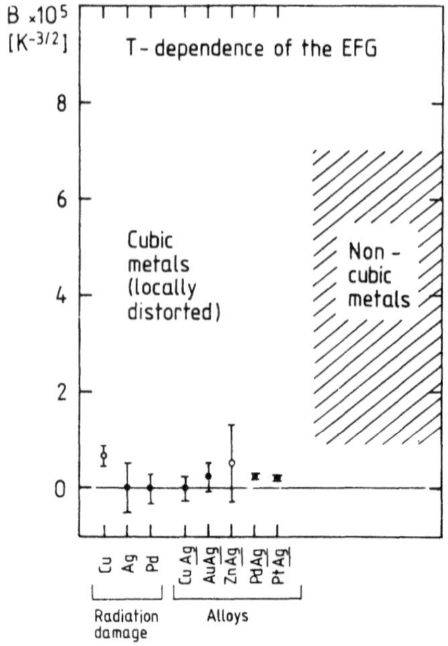

Fig.5.23. Temperature dependence of the EFG in locally distorted cubic metals and alloys. The range of the values of the temperature parameter B, see (5.41), observed in noncubic metals is indicated by the hatched area [5.73]

[5.70] and recently by ENGEL et al. [5.71]. These approaches are discussed in Sect.5.4.

The phonon picture is supported by the experimental correlation between the temperature parameter B and the host Debye temperature [5.64,72] as shown in Fig.5.22. The general proportionality between B and the effective spring constant of the lattice is obvious. The deviations have been explained so far only qualitatively by invoking local vibrational modes. In this context the results obtained in distorted cubic lattices are remarkable (Chap.4). Here the EFG at the impurity site is essentially temperature independent. This feature, demonstrated in Fig.5.23, has not yet been explained theoretically.

5.3.3 The Pressure Dependence of the EFG

The measurement of the variation of the EFG with pressure provides the most direct information on its volume dependence. The comparison of data obtained from experiments on temperature and pressure dependencies allows explicitly the isolation of temperature effects on V_{zz} not correlated with the variation of the lattice parameters with temperature, as first demonstrated by RAGHAVAN et al. [5.74] (Fig.5.24).

At constant temperature the variation of the EFG with hydrostatic pressure can be written [5.75]

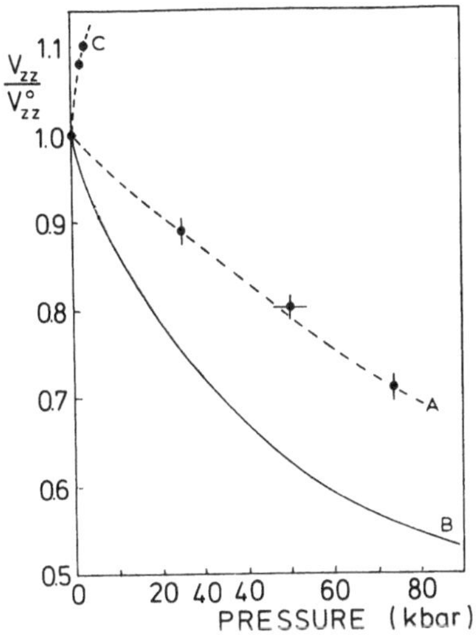

Fig.5.24. The pressure dependence of the EFG at ^{111}Cd in Cd metal. The experimental data are given by curve A, the predictions from the point charge model (see text) are given by curve B. The comparison with data obtained from temperature variation measurements (curve C) plotted at the pressure corresponding to the same change in axial ratio of the lattice yields an opposite response of the EFG to the change of the lattice constants by temperature and pressure, respectively [5.74]

$$\left(\frac{\partial \ln V_{zz}}{\partial P}\right) = \left(\frac{\partial \ln V_{zz}}{\partial \ln V}\right)_{T,c/a} \left(\frac{\partial \ln V}{\partial P}\right)_T + \left(\frac{\partial \ln V_{zz}}{\partial \ln \frac{c}{a}}\right)_{T,V} \left(\frac{\partial \ln \frac{c}{a}}{\partial P}\right)_T \quad . \tag{5.42}$$

The subscripts T, V, and c/a denote constant temperature, volume, and c/a ratio, respectively. An analogous equation can be written for unaxial pressure. Then by applying both hydrostatic as well as unaxial pressure to a single crystalline sample, the two unknown quantities $(\partial \ln V_{zz}/\partial \ln V)_{T,c/a}$ and $(\partial \ln V_{zz}/\partial \ln c/a)_{T,V}$ in (5.42) can be extracted from the experiment, if the changes of the volume and the c/a ratio with pressure are known.

Most experiments, however, were carried out under hydrostatic pressure only. Here the first factor of the second term in (5.42) is not at hand. This difficulty was overcome by BUTZ and KALVIUS [5.75], who assumed for close-packed metals the approximation

$$V_{zz} \sim \left[\frac{c}{a} - \left(\frac{c}{a}\right)_0\right] \tag{5.43}$$

at constant volume and temperature. The ratios $(c/a)_0 = 1.633$ for hcp metals and $(c/a)_0 = 1$ for tcp metals were used. The volume coefficients obtained by this procedure from experimental data are shown in Fig.5.25. They deviate systematically from the value -1 predicted from the point charge model. The response of the EFG to a congruent compression of the lattice is evidently much stronger and implies that the conduction electron charge density does not simply scale with volume. These discrepancies were related qualitatively to band structure variations caused by the increase of the average charge density [5.74,76]. Within the approach of NISHIYAMA and RIEGEL [5.77] using a dynamic Hartree screened Coulomb potential, very good agreement is obtained for In, Cd, and Zn, whereas the data on Sn are reproduced only approximately. The striking disagreement observed in the case of Tl [5.78] indicates that this model is not applicable generally. Some details are given in Sect.5.4.

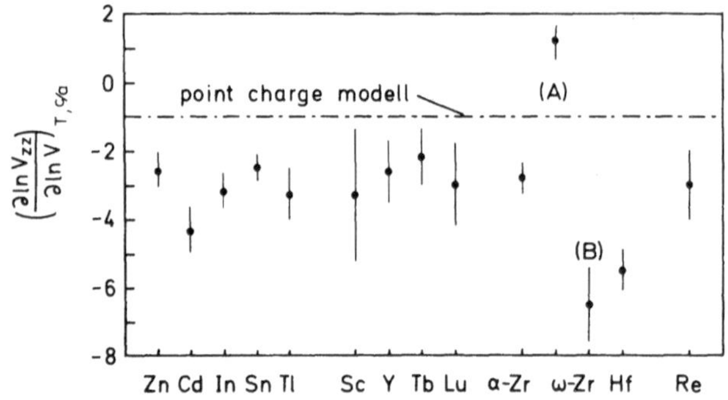

Fig.5.25. The volume coefficients representing the response of the EFG to congruent compression for different metals [5.75]

5.3.4 Impurity Valence Effects of the EFG

Most measurements of the EFG were carried out in highly diluted impurity systems. At present no generally applicable theory exists for this class of experiments. Therefore, the predictions of unknown field gradients can be based solely on systematic correlations between EFG and parameters like probe ion valence or impurity size, or on the respective difference of these quantities to that of the host lattice.

The influence of impurities on the electric field gradient was first systematically investigated by LEITZ et al. [5.49]. For nontransition metals, they compared the magnitude of the EFG at the site of the impurity with that of the pure host within a charge screening picture. Here the effective impurity charge is given to first order by the difference between impurity and host valence $\Delta Z = Z_{imp} - Z_{host}$. This charge difference is partially screened by the conduction electrons; these will redistribute themselves near the impurity and can be regarded in some sense as valence electrons. The results obtained by LEITZ et al. are given in Fig.5.26. The valence differences were corrected for the size effect, i.e., the differences in atomic volumes of host and impurity. The apparent correlation demonstrated by the figure supports the basic ideas of the charge screening picture.

This model was modified slightly by COLLINS [5.50] who found a more striking correlation between the conduction electron gradient V_{zz}^{el} and the probe valence Z itself (see Fig.5.27). In this picture, the impurity with valence Z is surrounded by Z con-

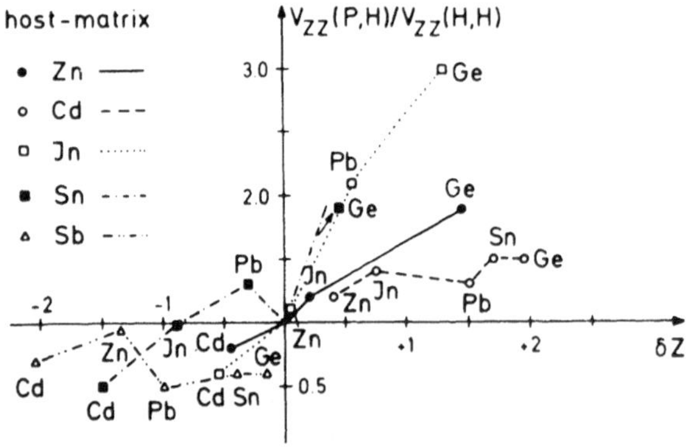

Fig.5.26. The ratios of the EFG at impurity sites and that of the host metals versus the effective size-corrected valence difference. The lines are drawn to guide the eye only [5.49]

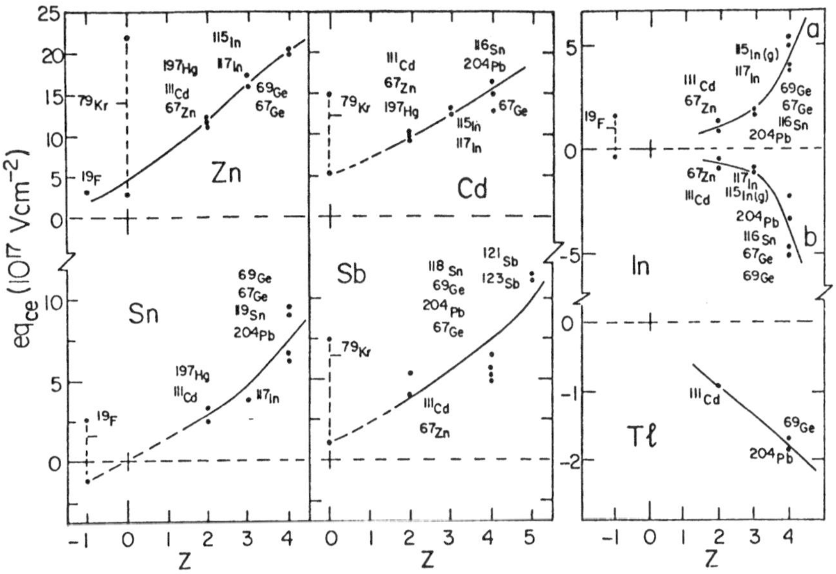

Fig.5.27. The conduction electron EFG versus the probe valence Z. The two alternative sets of data in the case of In metal are given for the two different possibilities of the relative signs of V_{zz}^{el} and V_{zz}^{ion} [5.50]

duction electrons resulting in neutral pseudo-atoms [5.79]. The data bear out the expected correlation between V_{zz}^{el} and Z, which is linear in rough approximation. The size effect turned out to be small and did not show any systematic trend responsible for the observed scattering of the isovalent probe atoms.

5.4 The EFG in Metals

In the following, a survey is given over the approaches taken so far for the calculation of electric field gradients in metals. The basic ideas rather than the mathematical details are presented.

According to the conception that a metal is composed of ions and conduction electrons, the EFG in metals is a combination of an ionic and an electronic contribution. Compared to the computation of the electronic EFG, the calculation of the ionic EFG is less extensive and difficult. In order to calculate the electronic EFG one has to know the true charge density distribution of the conduction electrons.

But the exact wave functions of the conduction electrons cannot be calculated because of their mutual interaction. Therefore, self-consistent methods have to be applied in order to find good approximations for the real wave functions or charge densities. Two classes of models for the calculation of the electronic contribution to the EFG have been developed.

i) Within the wave function approach, the electronic EFG is directly computed from the electronic wave functions. The total potential experienced by an electron is assumed to be independent of the coordinates of the other electrons. Then it is possible to construct one-electron wave functions of Bloch-type $\Psi_{nk}(\underline{r})$ and so the electronic charge density $\rho_e(\underline{r})$ has the general form

$$\rho_e(\underline{r}) = 2e \sum_n \int |\Psi_{n\underline{k}}(\underline{r})|^2 \frac{d^3k}{(2\pi)^3} \quad . \tag{5.44}$$

Here the sume extends over all occupied bands and the integral over all occupied regions of $\underline{k}$-space. The electronic EFG then is given by

$$V^e_{\alpha\beta} = \int \rho_e(\underline{r}) \frac{3x_\alpha x_\beta - \underline{r}^2\delta_{\alpha\beta}}{r^5} d^3r \quad . \tag{5.45}$$

ii) In the second approach, screening potentials are used. By means of the quantum-mechanical Liouville equation, the linear density response of a homogeneous electron gas to the ionic potential $V^c(\underline{q})$ is determined [5.79,80]. This charge density gives rise to the screening potential

$$V^{scr}(\underline{r}) = \int \frac{1 - \varepsilon(\underline{q})}{\varepsilon(\underline{q})} V^c(\underline{q}) \frac{d^3k}{(2\pi)^3} \tag{5.46}$$

whose second derivative yields the electronic EFG. In some cases the Hartree dielectric function

$$\varepsilon_H(\underline{q}) = 1 + \frac{4\pi e^2}{\underline{q}^2} \sum_{\underline{k}\leq k_F} \frac{f_0(\underline{k}) - f_0(\underline{k} + \underline{q})}{E(\underline{k} + \underline{q}) - E(\underline{k})} \tag{5.47}$$

is used. Here k_F is the Fermi momentum, $E(\underline{k})$ the electron energy and $f_0(\underline{k})$ is the occupation probability of the state $|\underline{k}\rangle$.

Improved versions of the form

$$\varepsilon(\underline{q}) = 1 + \frac{\varepsilon_H(\underline{q}) - 1}{1 - [\varepsilon_H(\underline{q}) - 1]\cdot G(\underline{q})} \tag{5.48}$$

including exchange and correlation have been used so far only by LODGE and SHOLL [5.81].

Before going into the details of these models, we discuss calculations on Sternheimer antishielding and summarize the results of lattice sum calculations.

5.4.1 Antishielding

Although the experimentally determined quadrupole splitting of the nuclear energy levels is caused by the EFG due to the charges surrounding the probe ion, the QI is not completely determined by this external EFG, $V^{ext}_{\alpha\beta}$. The observed interaction is influenced and often amplified by the electron shell of the probe ion. This effect is called Sternheimer antishielding [5.82-91] and can be viewed in two equivalent ways [5.88,92].

In the first approach it is assumed that the nuclear electric quadrupole moment (EQM) induces a quadrupole moment Q^{ion} in its own ion shell. This ionic EQM is proportional to the nuclear EQM: $Q^{ion} = -\gamma_\infty Q$ [5.92,93]. The proportionality factor γ_∞ is called the Sternheimer factor; values of Sternheimer factors obtained from relativistic Hartree-Fock-Slater calculations were tabulated by FEIOCK and JOHNSON [5.94]. In this picture the total interaction energy is composed of the interaction energy between the external EFG and both the nuclear and the ionic EQM and can therefore be written as

$$E_Q^{ext} = (1 - \gamma_\infty)\,\frac{1}{6}\sum_{\alpha,\beta} Q_{\alpha\beta}V_{\alpha\beta}^{ext} \quad . \tag{5.49a}$$

According to the second conception, the initially spherical probe ion shell is deformed by the external EFG and therefore produces an additional EFG at the nuclear site. This induced EFG is given by $(-\gamma_\infty \cdot V_{\alpha\beta}^{ext})$ and, consequently, the total interaction energy is the same as before.

If the field sources are partly inside the ion shell, as in the case of unfilled electron shells, the original EFG, $V_{\alpha\beta}^{loc}$, is much less amplified in general. The associated antishielding factor [5.82,86,95-103] is usually labelled with R and the total interaction energy takes the form

$$E_Q^{loc} = (1 - R)\,\frac{1}{6}\sum_{\alpha,\beta} Q_{\alpha\beta}V_{\alpha\beta}^{loc} \quad . \tag{5.49b}$$

In a metal, the total EFG is composed of an external EFG as well as a local EFG due to conduction electrons and possibly present unfilled electron shells. Therefore, expressions (5.49a,b) were combined to give the conventional parametrization of the EFG at a probe nucleus in a metal:

$$V_{\alpha\beta}^{total} = (1 - \gamma_\infty)V_{\alpha\beta}^{ext} + (1 - R)V_{\alpha\beta}^{loc} \quad . \tag{5.50}$$

This equation has been widely used in the past. But the interpretation of the first term as the EFG solely due to the ions and the second one as the electronic EFG turned out to give unsatisfactory results. Therefore, a closer analysis of the antishielding of the electronic contribution was needed.

The basic problem of antishielding is to take into account all contributions to the potential energy of the nucleus which depend on the relative orientation of the nuclear EQM and the external EFG, where the meaning of "external" has to be clarified. If "external" is interpreted as "external to the probe nucleus", there clearly would be only one contribution. In this case, however, the *real* distribution of *all* charges surrounding the nucleus had to be known and antishielding really does not exist at all. Antishielding arises only if "external" is interpreted as "due to charges other than core electrons" and if the nucleus and the ion core are considered to be initially spherically symmetric.

When the source charges are completely outside the probe ion core, the factor $(-\gamma_\infty)$ is involved, as mentioned above. When the density of the field sources $\rho(\underline{r})$ penetrates the ion shell, as is the case with conduction electrons, the radially dependent Sternheimer antishielding factor

$$\gamma(r) = -\frac{1}{Q}\left[\int_0^r dr' Q_i(r') + r^5 \int_r^\infty dr' Q_i(r')\cdot(r')^{-5}\right] \quad , \tag{5.51}$$

has to be taken [5.89,93], rather than the atomic antishielding factor R. Here, $Q_i(r')$ is the induced quadrupole moment density due to the nuclear moment and r represents the radial distance between the nucleus and a point-like field source. The antishielded EFG then has the general form

$$V_{\alpha\beta} = \int d^3r[1 - \gamma(r)] \frac{3x_\alpha x_\beta - \underline{r}^2\delta_{\alpha\beta}}{r^5} \rho(\underline{r}) \quad . \tag{5.52}$$

DAS and coworkers [5.104,105] have carried out detailed investigations of antishielding effects in metals. They first calculated the core wave functions perturbed by the nuclear EQM and then the Coulomb interaction energy between these perturbed core electrons and the conduction electrons, retaining only terms linear in Q. Since the orthogonal plane wave functions which were used to describe the conduction electrons are a superposition of plane-wave and core-state wave functions, individual antishielding factors arise for the plane-wave and core components. In the case of Cd and Zn, the values of the antishielding factor for the core components were found to be of the order of magnitude of the atomic antishielding factor, but for the plane-wave components, values of $\sim 0.6\ \gamma_\infty$ were obtained. The latter result is due to the fact that the plane wave component only partially penetrates the core and leads to drastically improved theoretical values for the EFGs in these metals.

A different approach was chosen by KAUFMANN and VIANDEN [5.106] in order to compute the interaction energy between the deformed ion core and conduction electrons. They determined the linear density response of a free electron gas to the total field of the distorted core assuming that the core radius can be neglected. From the quadrupole moment for this screening charge, they obtained an explicit expression for the radially dependent antishielding factor:

$$\gamma(r') = \gamma_\infty - \frac{2(1 - \gamma_\infty)}{3\pi} (r')^3 \int_0^\infty dk\cdot k^2\left(\frac{1}{\varepsilon(k)} - 1\right) j_2(kr') \quad , \tag{5.53}$$

where $j_2(kr')$ is the spherical Bessel function of second order. By means of this antishielding factor they calculated the total EFG at the nucleus following the approach of NISHIYAMA et al. [5.69]. They found an expression analogous to (5.93) in Sect.5.5.5c, but now with the enhancement factor α being determined. Values for

α are typically ~2 which indicates that the EFG is substantially enhanced due to the nonsphericity of the screening charges around the probe ion.

The most detailed investigation was carried out by LODGE [5.107]. He showed that in addition to the contributions to the QI discussed above, there are at least three further terms, $\Delta E_Q(i)$. They arise from the interaction (i) of a quadrupole moment density of the conduction electrons induced by the nuclear EQM, $Q^{el.ind.}_{(r)}(r)$, with the lattice field gradient, V_{zz}^{latt}, (ii) of $Q^{el.ind.}(r)$ antishielded by the core with a field gradient density of the conduction electrons induced by the lattice EFG, $V_{zz}^{el.ind.}(r)$, and (iii) of $Q^{el.ind.}(r)$ with $V_{zz}^{el.ind.}(r)$ antishielded by the ion core. These additional terms can be written as

$$\Delta E_Q(1) = -V_{zz}^{latt} \int_0^\infty dr\, Q^{el.ind.}(r)\cdot\beta(r) \quad , \tag{5.54}$$

$$\begin{aligned}\Delta E_Q(2) + \Delta E_Q(3) = &- \int_0^\infty dr\, Q^{el.ind.}(r)\left[r^{-5}\int_0^r dr' V_{zz}^{el.ind.}(r')\cdot\beta(r')\cdot(r')^5 \right.\\ &\left. + \int_r^\infty dr'\, V_{zz}^{el.ind.}(r')\cdot\gamma(r')\right]\\ &- \int_0^\infty dr\, Q^{el.ind.}(r)\left[r^{-5}\gamma(r)\int_0^r dr'\, V_{zz}^{el.ind.}(r')\cdot(r')^5\right.\\ &\left. + \beta(r)\int_r^\infty dr'\, V_{zz}^{el.ind.}(r')\right]\\ &- \int_0^\infty dr\, Q^{el.ind.}(r)\left[r^{-5}\gamma_\infty\int_0^r dr'\, V_{zz}^{el.ind.}(r')(r')^5\right.\\ &\left. + \gamma_\infty\int_r^\infty dr'\, V_{zz}^{el.ind.}(r')\right] \quad ,\end{aligned} \tag{5.55}$$

where the last term is a correction for counting both $\Delta E_Q(2)$ and $\Delta E_Q(3)$ twice.

In the expressions above, a new radially dependent antishielding factor $\beta(r)$ appears which can be regarded as the dual of $\gamma(r)$. According to its definition

$$\beta(r) = -\frac{1}{V_{zz}^{latt}}\left[r^{-5}\int_0^r dr'\, V_{zz}^{core}(r')\cdot(r')^5 + \int_r^\infty dr'\, V_{zz}^{core}(r')\right] \quad , \tag{5.56}$$

it describes the amplification of the QI due to the distortion of the ion core by the lattice field gradient. Here $V_{zz}^{core}(r)$ is the induced EFG density.

As pointed out before, the problems with antishielding arise because of the artificial distinction of two groups of electrons, the core and the conduction electrons. This procedure works well when the distortion of the ion core does not lead to a noticable redistribution of the field sources. This surely comes true in the case of lattice EFG. But as KAUFMANN and VIANDEN [5.106] have shown, a perturbed core causes an appreciable change in the conduction electron density. Therefore it may be questioned to what extent, e.g., the results of DAS and coworkers [5.56] would be modified if such a redistribution of the electron gas was taken into account. In the theory of LODGE [5.107], such effects are principally included but it has not been applied to a specific case up to now.

5.4.2 The Lattice Sum

The simplest way to think of a metal is as follows: point ion charges $Z\cdot e$ are situated at lattice sites $\underline{r}_i$ within a uniformly distributed electron gas. The value of Z is usually assumed to be the nominal valence of the metal in question and the electron density is chosen such as to guarantee the electrical neutrality of the crystal. Because of its uniform distribution, the electron gas cannot cause any nuclear quadrupole splitting. Therefore, the calculation of the EFG at a probe ion ($i = 0$) reduces to a sum over the individual contributions of all point ion charges

$$V_{\alpha\beta}^{\mathrm{latt}} = \sum_{i \neq 0} \int Z\cdot e\cdot \delta(\underline{r} - \underline{r}_i) \frac{3x_\alpha x_\beta - \underline{r}^2 \delta_{\alpha\beta}}{r^5} d^3r \quad . \tag{5.57}$$

The direct lattice sum (5.57), however, converges only slowly. Therefore other rapidly converging summation procedures were developed using transformation to reciprocal lattice space [5.108-111].

For hexagonal close-packed (hcp) lattices, DAS and POMERANTZ [5.112] showed that V_{zz}^{latt} is a simple function of the lattice constants a and c within the range shown in Fig.5.28:

$$V_{zz}^{\mathrm{latt}} = \frac{e}{a^3}\cdot \left[0.0065 - 4.3584 \left(\frac{c}{a} - \sqrt{\frac{8}{3}}\right)\right] \quad . \tag{5.58}$$

The dependence of the EFG on the axial ratio over a wide range was given by DE WETTE [5.113] for hcp and tetragonal close-packed (tcp) lattices and is shown in Fig.5.29.

Two zero crossings of the tcp curve are due to cubic symmetry whereas the third crossing is accidental. The zero crossing of the hcp curve is also accidental and according to (5.58) does not occur at the ideal axial ratio $c/a = \sqrt{8/3}$. Although the contribution from the next-nearest neighbours vanishes at the ideal ratio, one cannot conclude that the contribution from the remainder of the lattice is also zero because the symmetry of the total lattice remains the same for all axial ratios.

In the work of BODENSTEDT and PERSCHEID [5.114], not only the ions were regarded as point charges, but also the conduction electrons were assumed to be concentrated at the centres of the faces of Wigner-Seitz polyhedra and their contribution to the total EFG was therefore computed by point-charge lattice sum methods. The relative amount of charge centered between the basal planes of a hexagonal crystal and within these planes was determined such that the electrostatic and elastic forces form a stable crystal at the actual axial ratio. The EFG for Zn derived from this model

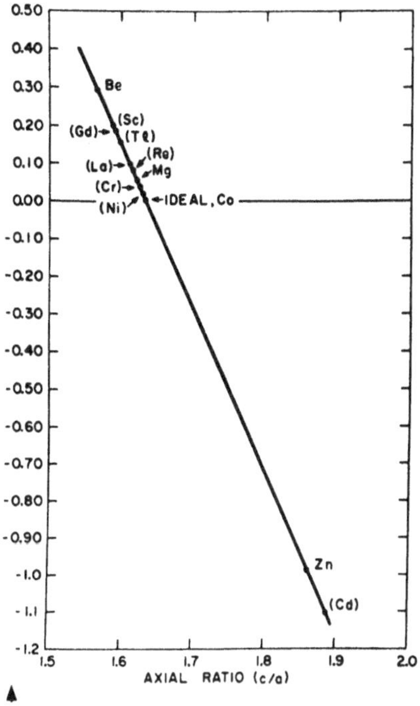

Fig.5.28. Ionic contribution to the EFG in hexagonal close-packed metals vs axial ratio, c/a. The EFG is expressed in the dimensionless form $a^3V_{zz}^{ion}/Z$ [5.112]

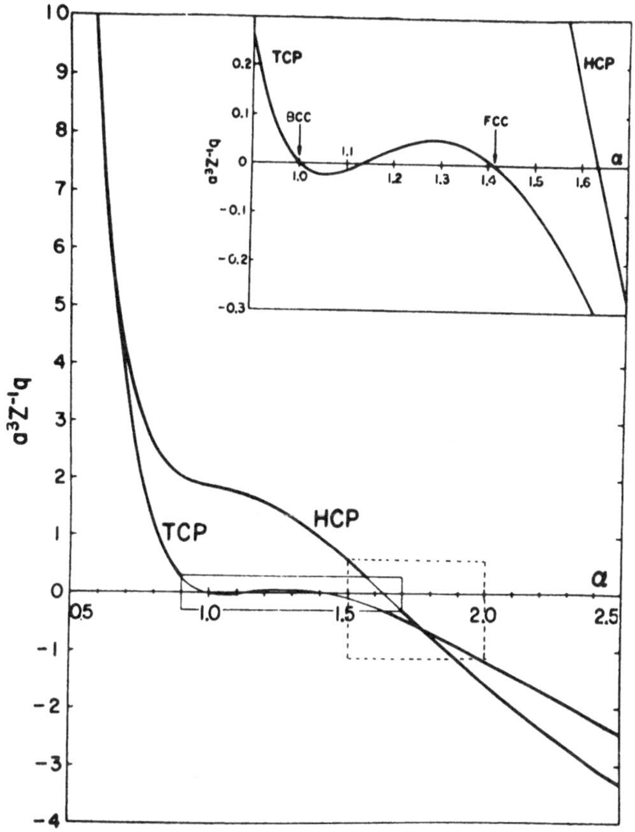

Fig.5.29. Ionic contribution to the EFG in hexagonal (hcp) and tetragonal (tcp) close-packed metals as a function of axial ratio. The inset shows the enlarged central section [5.113]

agreed with the experimentally determined one both in sign and magnitude. In spite of its simplicity, this model seems to reflect the essential features of the EFG in metals. The assumptions made about the conduction electrons are, however, far from reality, because they are not localized at distinct positions in the lattice.

5.4.3 Wave Function Approaches

As pointed out at the beginning of this section, it is impossible to calculate the exact wave functions of the conduction electrons. Therefore, one has to find good approximations for the real wave function. Bandstructure calculations have shown that there is a variety of proper wave functions each having a certain range of applicability. Extreme cases are the simple metals like aluminium in which the conduction electrons seem to be nearly free and the semimetals where tight binding wave functions can be applied successfully.

A common feature of all these wave functions is that they are of the Bloch type because of the translational symmetry of the lattice: the wave function translated by a lattice vector $\underline{n}$ can differ from the untranslated one only by a phase factor dependent on the translation vector

$$\Psi(\underline{r} + \underline{n}) = \lambda(\underline{n}) \cdot \Psi(\underline{r}) \quad . \tag{5.59}$$

Applying successive translations yields $\lambda(\underline{n} + \underline{n}') = \lambda(\underline{n}) \cdot \lambda(\underline{n}')$ and therefore the wave functions have the general form

$$\Psi_{\underline{k}}(\underline{r}) = \langle \underline{r} | \underline{k} \rangle_{\text{Bloch}} = \exp(i\underline{k}\underline{r}) \cdot u(\underline{r}) \quad , \tag{5.60}$$

where the function $u(\underline{r})$ is invariant under transformation

$$u(\underline{r}) = u(\underline{r} + \underline{n}) \quad . \tag{5.61}$$

a) *Wannier Functions*

One way of constructing wave functions which reflect the translational symmetry of the lattice is to construct a linear combination of wave functions $W(\underline{r} - \underline{n})$ which are attributed to a single atom

$$\Psi(\underline{r}) = \sum_{\underline{n}} \alpha_n W(\underline{r} - \underline{n}) \quad . \tag{5.62}$$

The functions $W(\underline{r} - \underline{n})$ are the so-called Wannier functions which have to be orthogonal.

This kind of wave function has been used by HYGH and DAS [5.93] in order to calculate the EFG for the semimetal antimony. The Wannier functions were constructed from orthogonalized atomic orbitals, i.e., from orthogonal linear combinations of atomic wave functions.

Although only rough estimates for the Sternheimer antishielding factors R and γ_∞ were made, the calculated magnitude of the total EFG was close to the experimental value determined by HEWITT and WILLIAMS [5.115].

b) *Augmented Plane Waves*

In at least two cases, augmented plane waves (APW) were used. An APW function is obtained if one assumes that the potential experienced by a conduction electron is spherically symmetric inside the so-called APW spheres each centered at an ionic site and constant in the space between. Therefore, an APW function is a plane wave outside the spheres and a linear combination of solutions to the Schrödinger equation inside the sphere. DAS and RAY [5.116] used this type of wave function to calculate the EFG in dysprosium. They found that the conduction electron contribution is opposite in sign to the lattice EFG.

Similar calculations for tin done by COLLINS and BENCZER-KOLLER [5.117] also resulted in a sign reversal. Experimental and theoretical values differ, however, by a factor ~2.5. This discrepancy is said to be due in part to various approximations in the calculation of the electronic contribution and in part to the uncertainty in the value of the nuclear quadrupole moment.

c) *Orthogonalized Plane Waves*

A widely used type of wave function is the orthogonalized plane wave (OPW) which is constructed in the following way: the periodic function $u(\underline{r})$ in (5.60) can be expanded in plane waves

$$u(\underline{r}) = \sum_{\underline{g}} C_{\underline{g}} \exp(i\underline{g}\underline{r}) = \sum_{\underline{g}} C_{\underline{g}} \langle \underline{r}|\underline{g}\rangle \quad , \tag{5.63}$$

where $\underline{g}$ is a reciprocal lattice vector. A Bloch state then has the form

$$|\underline{k}\rangle_{\text{Bloch}} = \sum_{\underline{g}} C_{\underline{g}} |\underline{k} + \underline{g}\rangle \quad . \tag{5.64}$$

Because of the rapid oscillations of the real wave function inside the ion core, the sum (5.64) has to be carried out over a wide range of $\underline{g}$-vectors. In order to obtain rapid convergence to each component of the sum (5.64), a linear combination of core electron states $|\alpha,i\rangle$ of each ion i in the crystal is added:

$$|\underline{k} + \underline{g}\rangle_{\text{OPW}} = |\underline{k} + \underline{g}\rangle - \sum_{\alpha,i} \mu_{\underline{g}\alpha} |\alpha,i\rangle \quad . \tag{5.65}$$

Since the states $|\alpha,i\rangle$ and $|\underline{k} + \underline{g}\rangle_{\text{OPW}}$ are eigenstates of the same Hamiltonian they have to be orthogonal. From this requirement the coefficients $\mu_{\underline{g}\alpha}$ are determined as

$$\mu_{\underline{g}\alpha} = \langle \alpha|\underline{k} + \underline{g}\rangle \quad . \tag{5.66}$$

The states $|\underline{k} + \underline{g}\rangle_{\text{OPW}}$ defined in this way are single orthogonalized plane wave states. A Bloch state can then be obtained by summing over only a few OPW states:

$$|\underline{k}\rangle_{\text{Bloch}} = \sum_{\underline{g}} \lambda_{\underline{g}} |\underline{k} + \underline{g}\rangle_{\text{OPW}} \quad . \tag{5.67}$$

The coefficients $\lambda_{\underline{g}}$ are determined when the potential experienced by the conduction electrons is known, and the core wave functions can reasonably be taken from free ion calculations. As a third prerequisite for the determination of the EFG, the shape and location of the Fermi surface which defines the limits of integration in (5.44) has to be known.

These conditions were met in the case of beryllium [5.118] and magnesium [5.119] and good agreement of theory with experiment could be achieved. For beryllium the electronic contribution was found to be negative and 25% of the lattice EFG, whereas for magnesium, its sign was determined as being positive and its magnitude by a factor ~2.5 greater than the lattice field gradient.

In the case of zinc and cadmium where no real potentials were available, use was made of pseudopotentials. A pseudopotential is constructed on the base of OPW functions such that it absorbs the core state contributions:

$$W = V + \sum_{\alpha} (E - E_\alpha)|\alpha\rangle\langle\alpha| \quad . \tag{5.68}$$

Here, V is the real potential, E_α are the core state energies and E is the energy of the conduction electron. The solutions of the redefined Schrödinger equation, which are the pseudo wave functions, can be represented by a sum over only a few plane wave states:

$$|\underline{k}\rangle_{Ps} = \sum_{\underline{g}} \lambda_{\underline{g}} |\underline{k} + \underline{g}\rangle \quad , \qquad (5.69)$$

where the coefficients λ_g are the same as in (5.67).

After calculating the λ_g by means of the pseudopotential, the real wave function is obtained by replacing the plane waves with the OPW functions.

Using pseudopotentials for zinc and cadmium which were appropriate only for Fermi surface electrons, DAS and coworkers [5.12,120] showed that the negative lattice contribution is overcompensated by a large positive electronic contribution. Although the sign prediction turned out to be right, the calculated magnitude deviated nearly a factor of 3 in the case of cadmium and a factor ~2 in the case of zinc from the experimental values.

Recalculations [5.54,104] showed that the $\ell = 2$ component of the plane wave part of the conduction electron density is relatively small inside the core region (Fig.5.30). This led to a drastically increased antishielding factor and so a quantitative agreement with the experiments could be obtained.

In later work, THOMPSON et al. [5.56] made use of a first principles pseudopotential valid for the total Fermi volume. For cadmium they showed that the various contributions to the total EFG, except in a few instances, do not differ substantially from their earlier results obtained with an empirical pseudopotential. The net electronic contribution differs by less than 5% indicating a good overall reliability of the procedure followed so far.

d) *Fermi Surface Electrons*

As pointed out above, one has to know the charge density for all electron states within the Fermi volume in order to calculate the electric field gradient. However, because of the observation that the QI at vanadium sites in V_3X type intermetallics scales with the electronic density of states at the Fermi surface (Fig.5.31), it may be asked whether Fermi surface electrons play a dominant role.

The first ones to investigate this question were WATSON et al. [5.58]. They used a tight-binding APW model. The charges outside the APW sphere were considered to be the source of the lattice field gradient. This field gradient perturbs the wave functions inside the APW sphere. First, the charge densities are spatially distorted analogously to the Sternheimer antishielding and second, it causes a redistribution of the occupied states. The latter effect was assumed to dominate. After several simplifying assumptions, WATSON et al. arrived at the following expression for this particular contribution to the EFG:

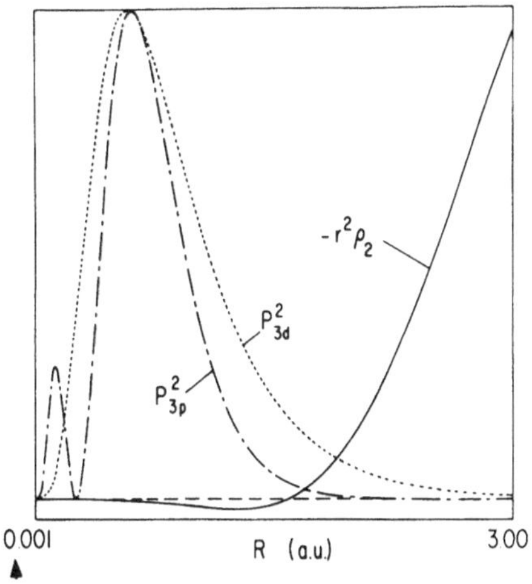

Fig.5.30. The radial density distributions of the $\ell = 2$ component of the conduction electrons and the 3p, 3d core electrons in Zn as a function of distance to the nucleus. A major part of the conduction electron density is external to the ion core [5.54]

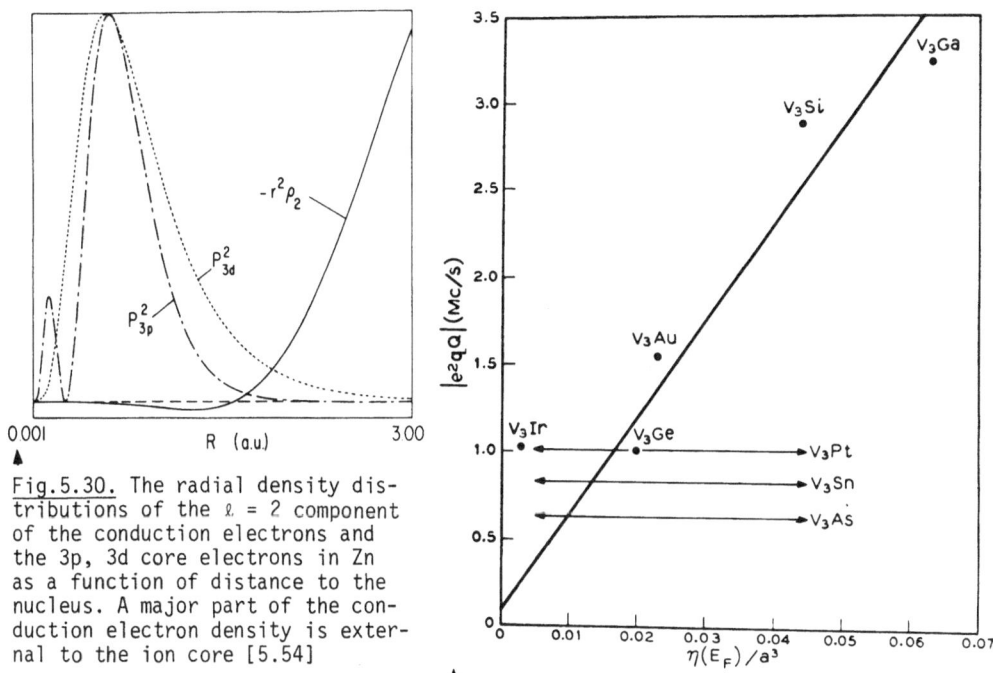

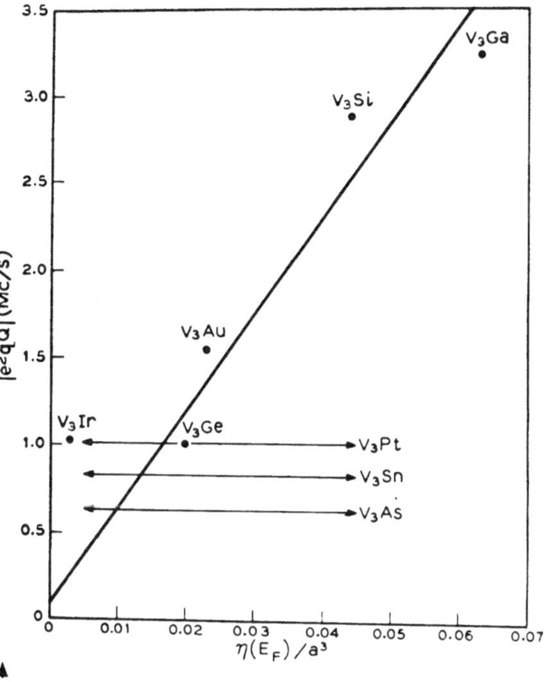

Fig.5.31. The magnitude of the quadrupole coupling constant for V^{51} for various V_3X (X = Si, Ge, Ga, etc.) compounds as a function of the density of states at the Fermi level, $\eta(E_F)$, normalized to lattice cell volume. In the cases of X = Pt, Sn and As, values of $\eta(E_F)$ had not been obtained experimentally. The correlation between the EFGs and $\eta(E_F)$ suggest that the Fermi surface electrons may dominantly determine the EFG in a metal [5.58]

$$V_{zz}^{FS} \approx -2e^2 V_{ZZ}^{latt} \cdot \eta(E_F) \cdot \langle r^2 \rangle \cdot \langle r^{-3} \rangle [P_2(\cos\theta)\rangle]^2 \quad . \tag{5.70a}$$

Here the expectation values, indicated by angular brackets, have to be taken at the Fermi surface and $\eta(E_F)$ is the density of states at the Fermi level. Obviously V_{zz}^{FS} is proportional to V_{zz}^{latt} and the overall proportionality constant is positive, indicating an antishielding effect.

Using experimental values for $\eta(E_F)$ and inserting the expectation values from atomic orbital calculations, the ratio $V_{zz}^{FS}/V_{zz}^{latt}$ was found to be in a region from $\approx$-10 to -650 for p and d-band metals. Thus, a general sign reversal of the total EFG relative to the lattice contribution is expected. However, for the p-band metal indium, e.g., the total EFG has the same sign as the lattice field [5.122] and also no sign reversals were observed in Lu, Hf and Ti [5.8,123-125].

Further objections to the results obtained by WATSON et al. were made by KOLK [5.121] who scrutinized this model. He came to the conclusion that the contribution V_{zz}^{FS} given in (5.70a) was overestimated. KOLK also estimated the contributions to the EFG arising from the perturbation of the conduction electron wave functions and the core electron wave functions. He found also that none of these terms could be responsible for a sign reversal since either they have the wrong sign or their order of magnitude is too small.

A refined model was proposed by PIECUCH and JANOT [5.26]. They considered the metal anisotropy as a weak perturbation in the electronic Hamiltonian and used a linear response theory of a tight-binding model. As a result the local EFG contains two contributions. The first is due to the crystal field anisotropy and has the form derived by WATSON (5.70a) [5.58]; the second arises from Coulomb transfer integrals between nearest neighbours.

In the static approximation, the new term is proportional to the density of d-electron states at the Fermi level $n_d(E_F)$ and explicitly proportional to the lattice field gradient. PIECUCH and JANOT assumed the lattice EFG to be due only to the nearest neighbours and obtained for the local EFG in hexagonal metals the expression

$$V_{zz}^{loc} = \frac{8}{35} f(\tfrac{c}{a}) \cdot e \cdot \langle r^{-3} \rangle \frac{n_d(E_F)}{1 - \frac{1}{10} \cdot (U - 2J) \cdot n_d(E_F)} \times \left(\frac{Ze^2 \langle r^2 \rangle}{a^3} - \frac{3}{4} |dd\sigma| \right) \quad . \tag{5.70b}$$

Here $|dd\sigma|$ denotes a transfer integral, U and J are the Coulomb and exchange interaction between atomic states, respectively, and f is a function of the axial ratio c/a:

$$f(\tfrac{c}{a}) = \left(\frac{c}{a} - \sqrt{\frac{8}{3}}\right) \Big/ \sqrt{\frac{8}{3}} \quad . \tag{5.71}$$

Rough estimates of the EFG were made for several transition metals. The agreement with experiment was, however, rather poor [5.127]. In two cases, Re and Hf, the signs can be compared with experiment, but only for Hf it agrees.

A different approach was used by D'ONOFRIO and IRALDI [5.128] to calculate the EFG in the magnetically ordered rare-earth metals Tb, Dy, Ho and Tm. For these metals the experimentally observed EFG is somewhat smaller than the EFG due to the ordered 4f-shell electrons. They could explain this deviation by assuming that the conduction electron states at the Fermi level are redistributed through the interaction with the 4f-electrons. The signs of the calculated field gradients agree in all cases with the experimentally determined ones and good agreement for the magnitudes in the cases of Tb and Dy was found.

5.4.4 Potential Approaches

In the potential approach, the explicit use of electron wave functions is avoided. The conduction electron contribution to the EFG is calculated from screening potentials.

Probably the first to use this method was SHOLL [5.129]. For the electron-ion potential, the bare Coulomb potential of point ions V^C was chosen. The total EFG, therefore, has the form

$$V_{zz}(\underline{r}_0) = -\frac{1}{3}\sum_{i\neq 0}\int \frac{V^C(q)}{\varepsilon(q)}\, \underline{q}^2 P_2(\cos\theta_q)\, \exp[i\underline{q}(\underline{r}_0 - \underline{r}_i)]\, \frac{d^3\underline{q}}{(2\pi)^3} \quad . \tag{5.72}$$

Here, $P_2(\cos\theta_q)$ is the second-order Legendre polynomial. For the Hartree dielectric function ε, the explicit formula (5.79,80)

$$\varepsilon(q) = 1 + \frac{2m_e\cdot e^2 k_F}{\pi\hbar^2 q^2}\left(1 + \frac{4k_F^2 - q^2}{4k_F q}\cdot \ln\left|\frac{2k_F + q}{2k_F - q}\right|\right) \tag{5.73}$$

was used, where m_e is the electronic mass. Because $\varepsilon(q)$ is a slowly varying function in general, the main contribution to the screened potential at large distances $r \gg k_F^{-1}$ arises from the logarithmic singularity in $\varepsilon(q)$ at $q = 2\,k_F$. Therefore the screened potential V^{scd} has the asymptotic form

$$V^{scd}_{(q)} \approx \frac{2m_e\cdot e^2 k_F^2 V^C(2k_F)}{\pi^2\hbar^2\varepsilon(2k_F)}\cdot\frac{\cos(2k_F r)}{(2k_F r)^3} \quad . \tag{5.74}$$

By means of the second radial derivative

$$V^{scd}_{rr}(r) = r\cdot\frac{d}{dr}\left[\frac{1}{r}\frac{d}{dr}V^{scd}(r)\right] \quad , \tag{5.75}$$

the spherical components to the total EFG at a probe nucleus can be written as

$$V_2^{(m)\,total} = \alpha(1 - \gamma_\infty)\sum_{i\neq 0}\frac{1}{3}\sqrt{\frac{4\pi}{5}}\, Y_2^{(m)}(\Omega_i) V^{scd}_{rr}(\underline{r}_i) \quad . \tag{5.76}$$

The additional factor α is introduced because the density of the field sources penetrate the probe ion core and therefore γ_∞ cannot be expected to be the correct antishielding factor. By comparing (5.76) with experiment, the effective amplification factor α was determined to ~3 for Ga and ~5 for Indium. With the knowledge of these factors, SHOLL [5.129] was able to successfully explain the experimental quadrupole relaxation times in the liquid metals.

KAUFMANN and VIANDEN [5.16] pointed out that the use of the second derivative of the asymptotic form of the screened potential is not strictly correct since differentiation should precede an asymptotic expansion. However, the leading term is unaffected and is given by

$$V^{scd}_{rr}(r) \approx -A(2k_F)^2\,\frac{\cos(2k_F r)}{(2k_F r)^3} \quad , \tag{5.77}$$

where A is the prefactor in (5.74).

Whereas SHOLL assumed the asymptotic form of the EFG to be applicable also for next-nearest neighbours, NISHIYAMA et al. [5.14,69,130] performed screened lattice sums using the exact screened Coulomb potential. A comparison of the second radial derivative of the exact potential and its asymptotic approximation is shown in Fig. 5.32. Although the deviations are obviously considerable at next-nearest neighbour positions (indicated by the arrows), the total EFG is insensitive to this point.

A more promising attempt was made by LODGE and SHOLL [5.81]. They first assumed the ions as having finite size, second they chose a nonlocal pseudopotential to describe the electron-ion interaction and third, use was made of dielectric functions including correlation and exchange.

The finite size of the ions is reflected by the utilized bare potential ω_0. Two such model potentials were used: (i) the Shaw potential, which is given by

$$\omega_0 = -\frac{Z}{\gamma} - \sum_{l=0}^{l_0} \theta[R_e(E) - r]\left(\frac{Z}{R_e(E)} - \frac{Z}{r}\right) |l\rangle\langle l| \quad . \tag{5.78}$$

Here, $\theta(r)$ is the step function

$$\theta(r) = \begin{cases} 1 & r > 0 \\ 0 & r < 0 \end{cases} ,$$

the $R_e(E)$ are radii within which the potential is modelled and l_0 is the maximum angular momentum quantum number for which core states exist; (ii) the Ashcroft potential, which is given by

$$\omega_0 = -\frac{Z}{r}\,\theta(r - R_c) \quad , \tag{5.79}$$

where R_c is a measure of the ion core radius.

Although pseudopotential theory gives the true charge density outside the ion cores, it gives only the total charge within the cores, which is called the depletion hole charge. Therefore, within this model it is not possible to calculate the total EFG in a crystal but only the EFG due to the charges outside the probe

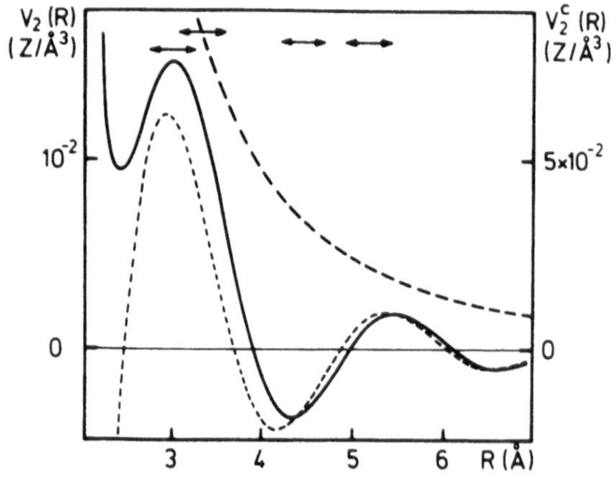

Fig.5.32. Comparison of the second radial derivative of the exact screened Coulomb potential (solid line), its asymptotic approximation (short dashed line) and the pure Coulomb potential $V_2^c(R)$ (long dashed line). The arrows indicate the vibration amplitudes of the first to fourth neighbouring ions at the melting point in Cd [5.14]

ion core. This EFG contains two contributions: one from the point ions V_{zz}^{ion}, whose charges are corrected by the depletion hole charge, and a second from the screening charges V_{zz}^{scr}, whose density is given by

$$\rho^{scr}(r) = \sum_{\underline{q}, i \neq 0} \frac{q^2}{4\pi} \left[\frac{1 - \varepsilon_H(q)}{\varepsilon(q)} V(q) + \frac{1 - \varepsilon(q)}{\varepsilon(q)} V_{dh}(q) + \frac{\varepsilon_H(q)}{\varepsilon(q)} g(q)\right] \cdot \exp[i\underline{q}(\underline{r} - \underline{r}_i)] \quad . \tag{5.80}$$

Here, $V(q)$ and $V_{dh}(q)$ are the Fourier transforms of the point-ion potential and the depletion hole potential, respectively; $g(q)$ is the form factor appropriate to screening of the nonlocal potential. For the dielectric function $\varepsilon(q)$ (5.48), the Hubbard-Sham form with

$$G_{HS}(q) = \frac{1}{2} \cdot \frac{q^2}{q^2 + k_F^2 + 4k_F/\pi a_0} \tag{5.81}$$

or the Shaw-Pynn form with

$$G_{SP}(q) = \frac{1}{2} \left[1 - \exp\left(\frac{-q^2}{\beta k_F^2}\right)\right] + \frac{\gamma q^2}{k_F^3} \exp\left(- \frac{\alpha}{\gamma} \cdot \frac{q^2}{k_F}\right) \tag{5.82}$$

was used, thus including exchange and correlation.

The calculations were carried out for the metals Be, Mg, Zn, Cd and In. A comparison between the Ashcroft potential and Shaw potential results for V_{zz}^{scr} shows good agreement for Mg, Zn and Cd but large deviations for Be and In. The calculations using the Shaw potential resulted in values for V_{zz}^{scr} opposite in sign to V_{zz}^{ion} and smaller in magnitude. The screening electron contribution is found to dominate the depletion hole contribution so that the total conduction electron EFG is always opposite in sign to V_{zz}^{ion}. The inclusion of correlation and exchange effects between conduction electrons turned out to have only a small effect on the screening term. Only in the case of Be could quantitative agreement with experiment be achieved, whereas the effective antishielding factor $|\alpha|$ ranged from $\sim$6 for Mg, Cd and In to $\sim$10 for Zn.

If there is no doubt about the validity of the method of calculation, one must conclude that for Mg, Cd, In and Zn there is a large contribution to the EFG arising from the conduction electrons inside the probe ion core. LODGE [5.131] has investigated this problem for the metals Be and Mg. He calculated the local contribution V_{zz}^{loc} on the base of orthogonal plane waves. For Mg, the inclusion of V_{zz}^{loc} substantially improved the agreement with experiment while for Be the result was less satisfactory. LODGE also considered the perturbation of the conduction electrons by the nuclear electric quadrupole moment. The effect was found to be of the order of 10% of the calculated total EFG and of opposite sign for Be and Mg. A recalculation of the EFG in Be, Zn, Cd and In by LODGE [5.132] using a dielectric function of the Singwi-Sjölander-Tosi-Land form

$$G_{SSTL}(q) = A\left[1 - \exp\left(- B \frac{q^2}{k_F^2}\right)\right] \quad , \tag{5.83}$$

yielded improved values for the magnitude of the EFG in these metals. They agree with the experimental values within a factor of two. The signs of the EFGs, however, agree only for Zn and Cd.

A possible explanation why the potential approach has not yet been completely successful in explaining the magnitudes and signs of the EFG was given by ENGEL et al. [5.71,133]. They pointed out that the screening potentials used so far represent only the lowest order part of the indirect interaction of the ions via the electron gas.

As a consequence of the elimination of the conduction electron coordinates in the Schrödinger equation of the total crystal, not only those long-ranged two-body interactions exist but also virtual short-ranged nonpairwise forces. They are due to scattering of the conduction electrons at more than two ions. But because the scattering ions need not be different, every higher-order force also contributes to every lower order interaction. From this it follows especially that the strength of the two-body interaction is corrected by the higher-order contributions.

The nonpairwise forces play an important role in the formation of the vibrational spectrum because they establish the stability of the crystal [5.134]. Therefore, one can conclude that these forces cannot be neglected in the calculation of electric field gradients in metals.

5.4.5 Temperature Dependence

From point-ion lattice sums it can be seen that thermal lattice expansion cannot account for the strong temperature dependence of quadrupole frequencies observed in most solids [5.4]. The first to explain the temperature dependence of the EFG in nonmetals in terms of thermal vibrations was BAYER [5.65] whose theory was later generalized by KUSHIDA et al. [5.66] and WANG [5.135]. A more detailed theory was developed by SCHEMPP and SILVA [5.136,137]. They discussed a model based on the diatomic linear chain. The proposed dominant T^4-dependence of the resonance frequency at low temperatures was in excellent agreement with the very accurate data of UTTON [5.138,139], as can be seen from Fig.5.33 . The thermal lattice vibrations are consequently the leading source of temperature variation in nonmetals. In metals, however, the situation may be completely different due to the presence of conduction electrons whose contribution to the total EFG turned out to overcompensate that of the metal ions. That a mechanism other than thermal lattice vibrations may play an important role is suggested by the empirical rule found by CHRISTIANSEN et al.

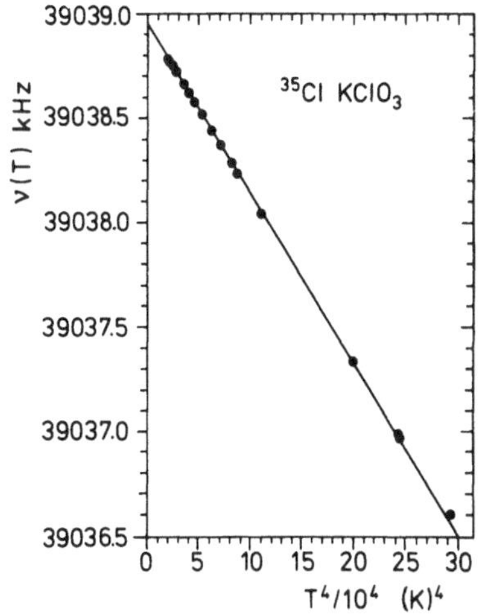

Fig.5.33. ^{35}Cl NQR-frequency ν in $KClO_3$, plotted against T^4 up to 23 K [5.137]

[5.6]. They observed that in many metals the temperature dependence of the EFG shows a $T^{3/2}$ behaviour:

$$V_{zz}(T) = V_{zz}(0)(1 - BT^{3/2}) \quad . \tag{5.84}$$

Because this relation is similar to Bloch's law describing the temperature dependence of the saturation magnetization in ferromagnetic materials, the authors proposed the existence of quasiparticles analogous to magnons. In this model the ionic quadrupole moments induced by the lattice EFG are assumed to be coupled via the conduction electrons. This coupling would then allow for collective excited modes called quadrons. At zero temperature the ionic quadrupole moments are aligned with the external EFG and the EFG at the site of the nucleus is maximum. At elevated temperatures when quadrones are excited, the alignment is reduced, resulting in a reduction of the EFG at the nucleus. Like magnons these quadrones should, however, contribute to specific heat. An estimate shows that their contribution has to be of the same order of magnitude as the magnon contribution. But there is no evidence from specific heat measurements that such excitations do exist. Therefore, the chance of finding a theory which yields an analytic $T^{3/2}$ law for the EFG in metals is decreased. Nevertheless, because of its simplicity, (5.84) remained a challenge to the theorists.

Within the approaches described in Sects.5.4.3,4 several authors have also proposed models for the temperature dependence of the EFG. Effects of Fermi-surface electrons as well as effects of thermal lattice vibrations have been discussed.

a) *Fermi Surface Electrons*

As a natural consequence of their theory, WATSON et al. [5.58] proposed that the temperature dependence of the EFG is mainly caused by thermal repopulation effects associated with the Fermi-Dirac distribution function f(E). They assumed that the

EFG due to a segment j of the Fermi surface $\langle 2r^{-3}P_2(\cos\theta)\rangle_j$ and the density of states $n(E)_j$, vary slowly with respect to $\partial f(E)/\partial E$ in an energy range $2k_BT$ around the Fermi energy E_F. Under these conditions they found that the temperature dependence of the local EFG can be written as

$$V_{zz}^{loc}(T) - V_{zz}^{loc}(0) = \sum_j \left[\Delta n(T)_j \langle 2r^{-3}P_2(\cos\theta)\rangle_j + \frac{\pi^2}{3}(k_BT)^2 n(E_F) \times \left(\frac{\partial\langle 2r^{-3}P_2(\cos\theta)\rangle_j}{\partial E}\right)_{E_F}\right] + O(T)^4 \quad . \qquad (5.85)$$

The first term represents thermal repopulation $\Delta n(T)_j$, into or out of the jth Fermi surface region, and the second term represents the change of the EFG due to thermal repopulation normal to the Fermi surface. The estimated temperature dependence of the EFG exceeded the experimentally observed one by an order of magnitude.

Within the more detailed approach of PIECUCH and JANOT [5.126,127], qualitative predictions of the temperature dependence in the pure 3d, 4d, and 5d metals were made. They neglected effects of thermal lattice expansion and assumed that the temperature dependence of the local EFG (5.70b) is mainly caused by that of the electronic density of d-states at the Fermi level $n_d(E_F)$, which is proportional to the magnetic susceptibility, χ_0. From the relative size of the local and lattice EFG and the known temperature variation of χ_0, they concluded that in most transition metals the total EFG decreases with increasing temperature. For Hf and Ru, an increase with temperature was predicted and for Y the total EFG should be nearly constant or weakly increasing. Unfortunately, experimental results for the pure metals are not available. But measurements at the probe ion ^{181}Ta in Hf [5.140] and Y [5.141] show a strong decrease with temperature. Because experiments on sp-metals show that the qualitative temperature behaviour of the EFG does not depend on the probe ion, one can conclude that these predictions will not agree with experiment.

b) *Wave Function Approach*

Within the wave function approach, the leading source of temperature dependence of the EFG is the perturbance of the electronic part by the phonons. Due to the thermal motion of the ions, the electron-ion potential is smeared out. As a consequence, the electronic wave function and the charge density, respectively, change with temperature resulting in a temperature dependence of the electronic EFG. Following the procedure of KASOWSKY [5.142], who successfully explained the temperature variation of the Knight shift in Cd, the influence of the phonons is taken into account via T-dependent pseudopotential matrix elements:

$$\langle \underline{k} + K|V_{Ps}|\underline{k}\rangle = \exp[-W(\underline{K},T)]\langle \underline{k} + K|V_{Ps}^0|\underline{k}\rangle \quad . \qquad (5.86)$$

Here, $W(\underline{k},T)$ is the Debye-Waller factor which can be expressed in the harmonic approximation by the mean square displacements $\langle u_{\underline{k}}^2\rangle_T$ of the ions:

$$W(\underline{K},T) = \frac{1}{2} \underline{K}^2 \langle u^2_{\underline{K}} \rangle_T \quad . \tag{5.87}$$

The temperature variation of the mean square displacements can be evaluated within the Debye model and is then given by

$$\langle u^2 \rangle_T = \frac{3\hbar^2}{Mk_B\theta_D} \left[\frac{1}{4} + \left(\frac{T}{\theta_0}\right)^2 \int_0^{\theta_D/T} \frac{zdz}{e^z - 1}\right] \quad . \tag{5.88}$$

Here, θ_D is the Debye temperature, k_B is Boltzmann's constant and M is the ion mass.

JENA [5.68] studied the temperature dependence of the electronic EFG in terms of an infinite-order perturbation theory and showed that the coefficients in (5.64) can be expressed by just the matrix elements (5.86). He expanded the exponential factor to first order in W and finally got for the electronic charge density the expression

$$\rho(\underline{r},T) = \rho_0(\underline{r}) - \tilde{\rho}_0(\underline{r})\phi(T/\theta_D) \quad . \tag{5.89}$$

Here, $\phi(T/\theta_D)$ is the quantity in brackets in (5.88). This Debye integral ϕ contains the entire T dependence of the electronic charge density and therefore completely determines the T dependence of the electronic EFG. As the ionic EFG is only weakly T-dependent, JENA argued that ϕ also describes the T dependence of the total EFG:

$$V_{zz}(T) = V_{zz}(0)[1 - \beta\phi(T/\theta_D)] \quad . \tag{5.90}$$

The function ϕ shows a T^2 behaviour at T values well below the Debye temperature and increases linearly with T at high temperatures and thus follows a behaviour similar to $T^{3/2}$ in an intermediate region. JENA pointed out that a better agreement with experiment might be achieved by taking into account anharmonic effects as well as realistic phonon frequency spectra.

In later work, JENA [5.143] calculated the electronic EFG in Mg at 295 K. The coefficients of linear combination of OPWs were obtained by solving numerically the Schrödinger equation of the conduction electrons at distinct points in the first Brillouin zone. As before, the temperature dependence was incorporated by using potential matrix elements of the form (5.86). By comparing the results with earlier calculations [5.119], he found that with increasing temperature, the s-component of the electron wave function increases at th expense of its p-content, resulting in a decrease of the electronic EFG. The lattice contribution of the EFG was calculated by using the formula of DAS and POMERANTZ [5.112] (5.58). The agreement between theory and experiment was good in spite of the approximations made in the calculation.

COLLINS and BENCZER-KOLLER [5.117] did a similar calculation for tin metal. The mean square displacements appearing in the Debye-Waller factor were obtained from experiment. They found a decrease of 40% between 4 K and 500 K for the calculated total EFG, whereas the experimental EFG decreased by 25% over the same range.

A more detailed investigation has been presented by THOMPSON et al. [5.70]. In addition to the influence of thermal lattice expansion on the lattice EFG, they took into account effects from anisotropic lattice vibrations. For this purpose a

Taylor expansion of the lattice EFG in the ionic displacements $\underline{u}^{(i)} = (u_\alpha^{(i)})$ was carried out. They considered the vibrations to be harmonic. Therefore, first-order contributions were neglected and in second order, only diagonal terms needed to be retained:

$$V_{zz}^{ion}(T) = V_{zz}^{ion}(0) + \frac{1}{2}\sum_{i,\alpha}\left[\frac{\partial^2 V_{zz}^{ion}}{\partial(R_\alpha^{(i)})^2}\right]_0 \cdot \langle(u_\alpha^{(i)})^2\rangle_T \quad . \tag{5.91}$$

Here, $V_{zz}^{ion}(0)$ is the EFG of the static lattice at 0 K and the $\underline{R}^{(i)}$ describe the positions of the ions. The zero subscript outside the brackets indicates that the derivatives have to be evaluated at the equilibrium positions of the ions.

In the isotropic approximation $\langle u_x^{(i)}\rangle_T = \langle u_y^{(i)}\rangle_T = \langle u_z^{(i)}\rangle_T$, the lattice EFG is temperature independent. For a localized isotropic charge density produces the same potential as a proper point charge at its center. Therefore, its EFG at distant points is independent of its dimension. For this reason only anisotropic vibrations can cause a temperature dependence of the lattice EFG. On the other hand, the electronic EFG is found to be temperature dependent mainly due to the isotropic part of lattice vibrations, whereas changes resulting from anisotropy and anharmonicity turned out to be less than 5% of the isotropic contribution.

c) *Potential Approach*

Within the potential approach the influence of thermal vibrations can be taken into account by substituting $\underline{r}_i = \underline{r}_i^{(0)} + \underline{u}_i$ in the summation over screened potentials (5.72), where $\underline{r}_i^{(0)}$ denotes the equilibrium position of the ith ion. The temperature dependence of each Fourier component of the EFG is then described by the temperature mean value of the exponential factor

$$\exp[i\underline{q}(\underline{r}_0^{(0)} - \underline{r}_i^{(0)})]\langle\exp[i\underline{q}(\underline{u}_0 - \underline{u}_i)]\rangle_T \quad . \tag{5.92}$$

As a consequence, each component has a different temperature dependence and therefore in the calculation of the T dependence of the total EFG, the integral over all Fourier components has to be evaluated directly. Additionally, the EFGs due to the screening charges inside and outside the probe-ion core have to be calculated separately because they undergo different antishielding.

As a first approximation, the thermal vibrations can be assumed as being harmonic, uncorrelated and isotropic and an effective antishielding factor $\gamma_{eff} = \alpha(1 - \gamma_\infty)$ may be introduced. If, additionally, use is made of the asymptotic approximation, this results in the expression

$$\langle V_{zz}\rangle_T = \alpha(1 - \gamma_\infty)V_{zz}^{scd}(T)\ \exp(-\tfrac{4}{3}\,k_F^2\langle u^2\rangle_T) \tag{5.93}$$

for the temperature dependence of the EFG at the site of a probe nucleus. Here, $V_{zz}^{scd}(T)$ is the sum of the screened potentials for the static lattice and depends on temperature via thermal lattice expansion. The influence of thermal vibration is in this approximation described by a simple exponential factor containing the Fermi momentum k_F.

The first to investigate the temperature dependence of the EFG within the potential approach were QUITMANN et al. [5.67]. They expanded the exponential factor (5.92) up to first order in $\langle \underline{u}^2 \rangle_T$. After carrying out summation and integration they got the result

$$\langle V_{zz} \rangle_T = \alpha(1 - \gamma_\infty)\left[V_{zz}^{scd}(T) - \frac{1}{2}\sum_\alpha \frac{\partial^2 V_{zz}^{scd}(T)}{\partial u_\alpha^2} \cdot \langle u_\alpha^2 \rangle_T\right] \quad . \tag{5.94}$$

This equation yields a net temperature dependence of the EFG even under the assumption of isotropic vibrations, because contrary to the case of lattice EFG discussed earlier (5.91), the probe nucleus is, in general, situated inside the screening charge cloud. The temperature dependence of $\langle u^2 \rangle_T$ was described in the Debye approximation (5.88) and the resulting function fitted to the experimental data for ^{115}In$\underline{\text{In}}$ [5.37]. The temperature dependence of the EFG was reproduced rather well but not perfectly. This is mainly due to the fact that the Debye model is applicable only in the low temperature region.

NISHIYAMA and RIEGEL [5.14] pointed out that the mean square displacements in a variety of metals follow a $T^{3/2}$ relation rather than a Debye integral with fixed Debye temperature. They calculated the temperature dependence of the EFG according to (5.94) [5.69] taking explicitly into account the variation of $V_{zz}^{scd}(T)$ due to thermal expansion and used experimental values for the Fermi momentum k_F. The T dependence of the mean square displacements was described by a $T^{3/2}$ relation fitted to experimental data. They obtained good agreement for the normalized T dependence of the EFG with experiment, as can be seen from Fig.5.34.

The validity of the approximations made in the models discussed so far has been investigated by LODGE [5.132]. He calculated the EFGs due to the screening charges outside and inside the probe-ion core separately and compared the influence of anisotropic and isotropic vibrations as well as the asymptotic approximation and the directly calculated EFG. He found significant effects of vibrational anisotropy though they tend to cancel in the total EFG. Comparison of the isotropic vibrational model and the asymptotically separated isotropic vibrational model revealed large discrepancies in the local EFG. These discrepancies, however, cancel when they are added to give the total EFG (Fig.5.35). LODGE pointed out that this cancellation appears to be fortuitous in the sense that it depends on the relative values of the antishielding factors. If these factors were made equal, as done by QUITMANN et al. [5.67] and NISHIYAMA et al. [5.69], the total EFG for Cd would show a 300% decrease over the experimentally investigated temperature range instead of the 30% decrease actually observed.

As can be seen from expression (5.91), the EFG at a probe nucleus depends on displacement differences rather than on displacements. The assumption of uncorrelated vibrations made in the models described above is therefore only approximate. Espe-

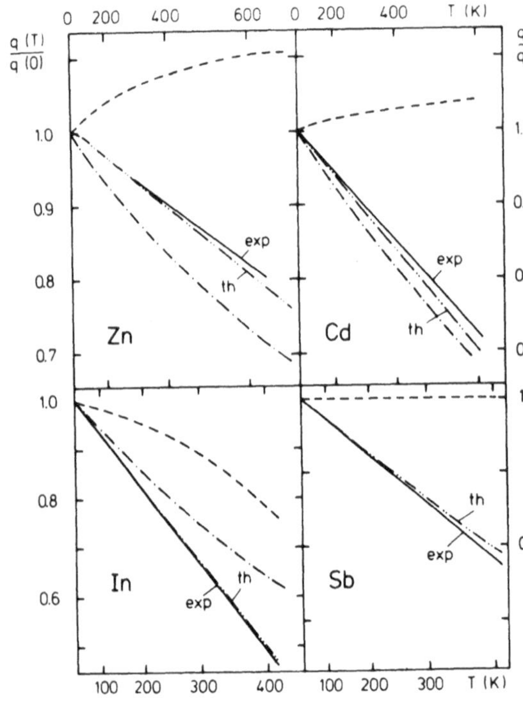

◄ Fig.5.34

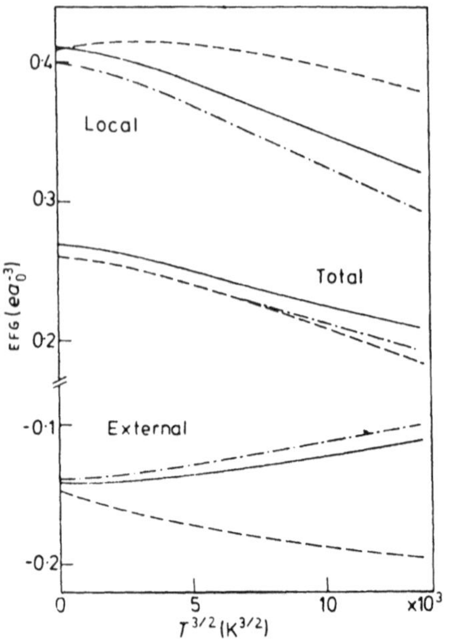

▲ Fig.5.35

Fig.5.34. Comparison of the theoretical temperature variation of the EFG according to (5.93) (dot-dot-dash line) with experiment (solid line) for Zn, Cd, In and Sb. The dashed line shows the temperature dependence of the screened lattice sum $V_{zz}^{scd}(T)$, and the dot-dash line shows the variation of the Debye-Waller factor with temperature. All curves are separately normalized to unity at 0 K [5.77]

Fig.5.35. Comparison of the temperature dependence of the local, external and total EFGs calculated within the anisotropic vibrational model (solid lines), the isotropic vibrational model (dashed lines) and the asymptotically separated isotropic vibrational model (dot-dash lines) [5.132]

cially in the low temperature region where displacement correlations play an important role, this approximation is expected to fail.

In the investigation of ENGEL et al. [5.71,133] these correlations were taken into account explicitly. They expanded the EFG at the nuclear site in a power series of displacement differencies up to second order. Neglecting influences of thermal lattice expansion and assuming the thermal vibrations to be harmonic in the low temperature region, they obtained the result

$$\nu_Q(T) = \frac{eQ}{3h} \langle 2\bar{V}_{zz} - \bar{V}_{xx} - \bar{V}_{yy} \rangle_T$$

$$= \frac{eQ}{h} \left\{ \bar{V}_{zz}^{harm} + \tilde{\phi}\left[\frac{1}{8} + \frac{1}{\omega_D^4} \int_0^{\omega_D} d\omega \cdot \omega^3 \bar{n}(\omega,T)\right]\right\} \quad ,$$

$$T \lesssim \theta_D/10 \qquad (5.95)$$

for the temperature variation of the quadrupole coupling constant ν_Q valid for pure nontransition metals and in the temperature region $T \lesssim \theta_D/10$. Here $\bar{V}_{zz}^{harm}$ denotes the temperature independent part of the EFG at the nucleus originating from the harmonic part of the potential, ω_D is the Debye frequency and $\bar{n}(\omega,T)$ is the mean occupation number. The quantity $\tilde{\phi}$ depends on the second derivatives of the EFG, on mean products of phonon polarization vectors and mean square sound velocity. Using literature values for the Debye frequency, the function (5.95) was fitted to the experimental data for ^{115}In<u>In</u> [5.37] and good agreement was obtained, as can be seen from Fig.5.36.

In the high temperature limit, influences of anharmonicity were also considered. For nontransition metals where anharmonicity is weak, the pseudo harmonic approximation [5.144] was applied. This means that the lifetime of the phonons are regarded as being infinite and only the phonon frequencies are temperature dependent. Taking into account only lowest-order terms of the self-energy of the phonons, they finally obtained the result that the linear T dependence of the quadrupole coupling constant in the harmonic approximation is corrected by a term quadratic in temperature:

$$\nu_Q(T) = \frac{eQ}{h}\left(\bar{V}_{zz}^{harm} + \bar{\phi}^{(1)}T + \bar{\phi}^{(2)}T^2\right) \quad , \qquad T > \theta_D \quad . \tag{5.96}$$

Here, $\phi^{(1)}$ depends on the second derivatives of the EFG and $\phi^{(2)}$ additionally on the temperature independent factor of the anharmonic corrections to the harmonic phonon frequencies.

The function (5.96) was fitted to the experimental data for ^{115}In<u>In</u> [5.37] and good agreement was obtained (Fig.5.37).

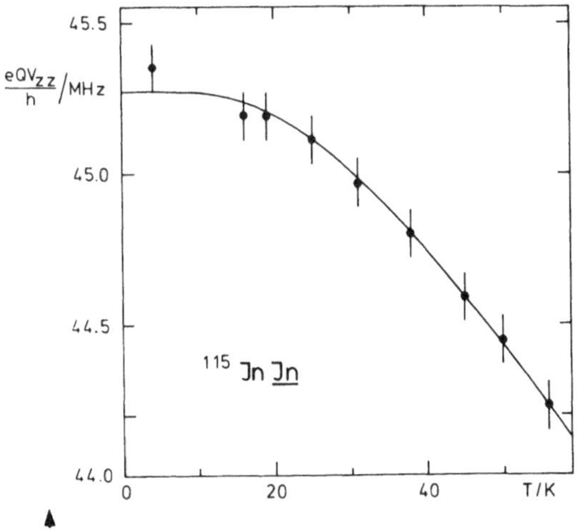

▲ Fig.5.36

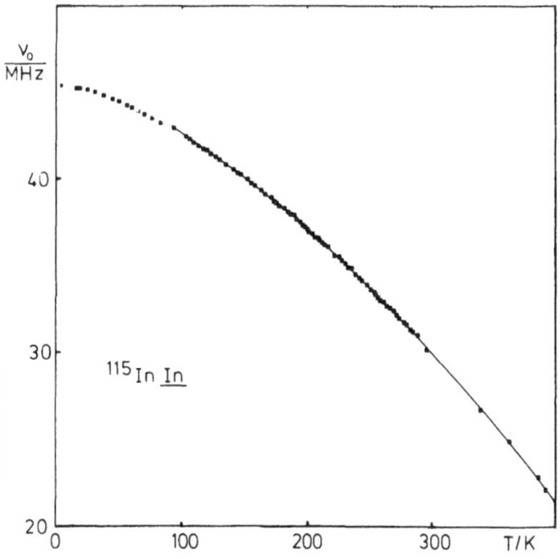

Fig.5.37 ►

Fig.5.36. Least-squares fit of the low temperature approximation (5.95) to the experimental data for ^{115}In<u>In</u> [5.37]. [5.71]

Fig.5.37. Least-squares fit of the high temperature approximation (5.96) to the experimental data for ^{115}In<u>In</u> [5.37]. [5.71]

5.4.6 Pressure and Concentration Dependence

The pressure dependence of the EFG was studied theoretically in two cases. MOHAPATRA et al. [5.120] utilized a pseudopotential approach in order to obtain energy bands and band wave functions for Cd. The calculations were performed with lattice parameters appropriate for atmospheric pressure and 50 kbar. The theoretical value of the ratio V_{zz} (1 bar)/V_{zz} (50 kbar), which was later corrected slightly to 1.28 [5.12], is in satisfactory agreement with the experimental value of 1.25 [5.74].

Whereas MOHAPATRA et al. considered the lattice constants to be the only parameters depending on pressure, NISHIYAMA and RIEGEL [5.77] started their analysis from expression (5.93) and, in addition, took the variation of the mean square displacements with pressure into account. From the definition of the Grüneisen constant $\gamma = -d \ln\theta_D/d \ln V$ with V representing the atomic volume and from the relations $\langle u^2\rangle_T \propto \theta_D^{-2}$ valid for $T \gtrsim \theta_D$ and $k_F \propto V^{-1/3}$, they obtained the expression

$$\frac{k_F^2(P_1)\langle u^2\rangle_T(P_1)}{k_F^2(P_2)\langle u^2\rangle_T(P_2)} = \left(\frac{V(P_1)}{V(P_2)}\right)^{2\gamma-2/3} .$$

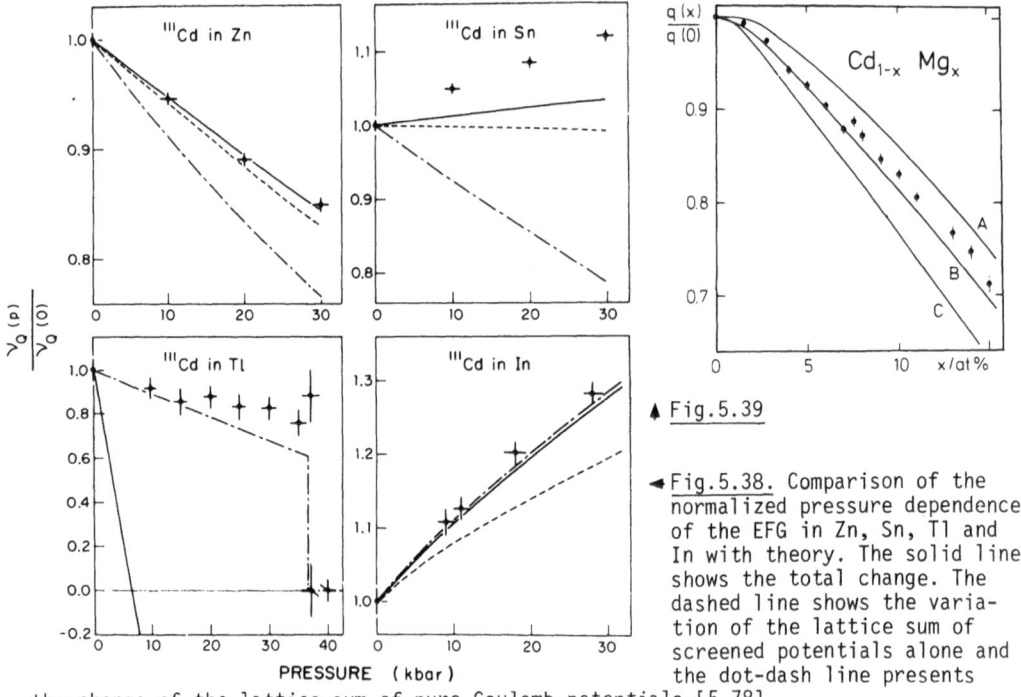

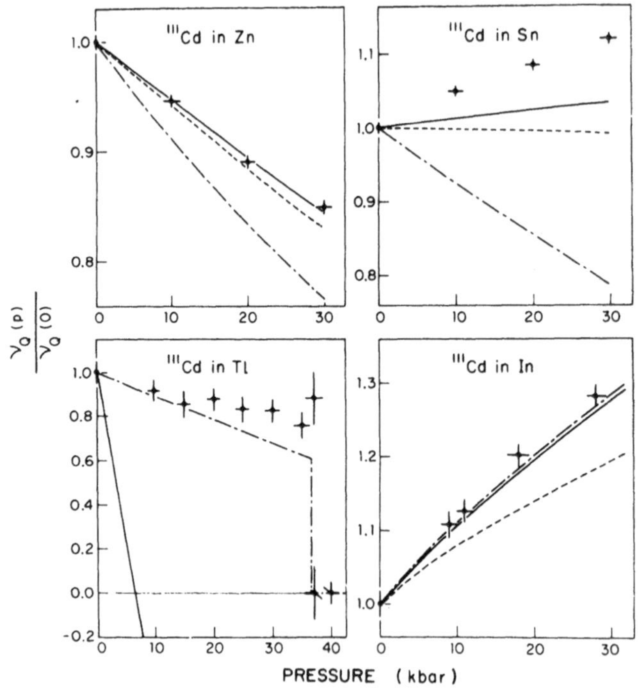

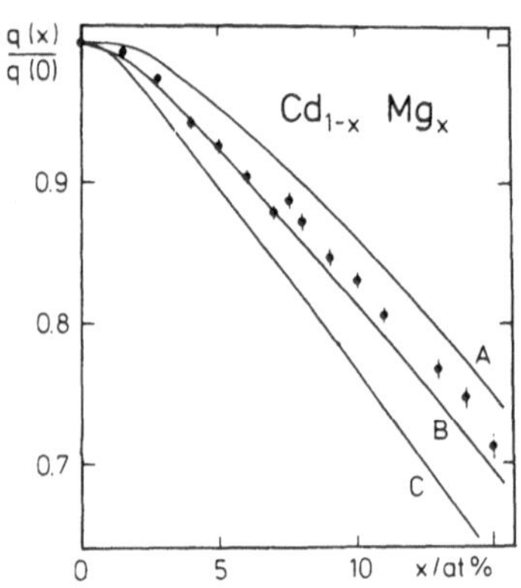

▲ Fig.5.39

◄ Fig.5.38. Comparison of the normalized pressure dependence of the EFG in Zn, Sn, Tl and In with theory. The solid line shows the total change. The dashed line shows the variation of the lattice sum of screened potentials alone and the dot-dash line presents the change of the lattice sum of pure Coulomb potentials [5.78]

Fig.5.39. Comparison of the normalized dependence of the EFG at Cd-sites in CdMg alloys [5.146] with theory. Curve A shows the total change, curve B the variation of the lattice sum of screened potentials alone and curve C the change of the lattice of pure Coulomb potentials [5.77]

With the help of this equation, the pressure dependence of the EFG could be expressed in terms of the P variation of the lattice constants alone.

Good agreement of the theoretical results with the experimental data for In [5.145], Cd [5.74] and Zn [5.78] could be obtained. It should be noted that the normalized pressure dependence of the EFG is reproduced without any adjustable parameter. In contrast to the variation of the EFG with temperature, where the Debye-Waller factor plays the dominant role, here the lattice sum of screened potentials is the more sensitive quantity. However, without considering the P-dependence of the Debye-Waller factor, quantitative agreement would not have been obtained. But as can be seen from Fig.5.38, the theoretical curves deviate from the data for Sn and Tl [5.78], indicating the limited range of applicability of the model.

Equation (5.93) was also applied to the dependence of the EFG on concentration x in $Cd_{1-x}Mg_x$ alloys, where the constituents are isoelectronic. NISHIYAMA and RIEGEL [5.77] assumed that the variation of the EFG with concentration is only due to the changes in the lattice constants and the Debye-Waller factor. The variation of k_F and V_{zz}^{scd} was again calculated from experimentally determined lattice constants. The change of the mean square displacements was estimated within the Debye approximation by means of the known variation of electrical conductivity and elastic constants.

The good agreement between theory and experiment [5.146] shown in Fig.5.39 demonstrates that the range of applicability of this model is not restricted to the temperature dependence of the EFG in metals.

5.5 Conclusion

During the last few years, the electric field gradient in noncubic metals has been investigated with an undiminished intensity. However, it can be recognized that we are now beyond the most exciting explorative phase bearing out unexpected systematics and first reliable theoretical pictures. The present activities are characteristic for the phase of consolidation: the basic features of the numerous experimental results can be regarded as explained by theoretical models. However, many details and some important more general aspects still remain unsolved and require substantial extensions and further developments of the theory. In particular, no theoretical models for the EFG at impurity sites and in disturbed cubic metals have been proposed so far.

On the experimental side, the amount of information collected so far is impressive. But there still exists a need for measurements of the sign of the EFG in a number of important systems. In many cases the accuracy of data on temperature and pressure dependencies has to be improved substantially in order to significantly test the theoretical approaches. The investigation of the EFG in alloys and intermetallic compounds has yielded first results (see, e.g., [5.147,148]), but is still at the very beginning. Here, in the future, new insights can be expected. Furthermore, systematic experiments on distorted cubic metals (Chap.4) or on cubic metals deformed by uniaxial strain can certainly widen the basic understanding of the EFG. Most recently, first investigations of the quadrupole interaction in semimetals and semiconducting alloys have been reported [5.149-151]. Here the conduction electron density can be varied in a well-known manner over many orders of magnitude. The results obtained so far are very promising and will certainly stimulate activities into that direction.

5.6 Appendix

Table of Experimental Data on Quadrupole Interaction in Noncubic Metals

The following table summarizes the experimental data on quadrupole interaction in noncubic metals published up to June 1981, including data from the tables in [5.152-155]. The host lattices are arranged according to the groups of the periodic system of the elements, the probe atoms (column 2) are listed in the order of increasing atomic number. For measurements on excited levels, the energy (in keV) of this level is given in brackets. The quadrupole coupling constants observed at room temperatures (T = 293 K ± 5 K) are given in column 3. Deviations from this temperature are indicated; the entries are also valid for the EFG data (column 4) and are not repeated there. The field gradients were calculated using the quadrupole moments tabulated by [5.155] or [5.432]. In all cases where the sign of the QI has been determined, it is given in column 3; a sign given for the EFG indicates a known sign of the nuclear quadrupole moment too. The entries (T), (P) or s.cry. in the last column indicate that temperature or pressure dependencies of ν_Q have been determined or that the experiments were carried out on single crystals.

The experimental methods are abbreviated:

ME	Mössbauer Effect
NMR	Nuclear Magnetic Resonance
NQR	Nuclear Quadrupole Resonance
β-NMR	Nuclear Magnetic Resonance on β-instable nuclei
NO	Nuclear Orientation at low Temperatures
SH	Specific Heat
TDPAC	Time Differential Perturbed Angular Correlation
e-TDPAC	Electron-γ TDPAC
TDPAD	Time Differential Perturbed Angular Distribution following nuclear reactions
IMPAC	Perturbed Angular Correlation following Coulomb Excitation and Recoil Implantation
QSOPAD	Stroboscopic Observation of the Perturbed Angular Distribution

Host	Probe Nucleus	Coupling-Constant eQV_{zz}/h [MHz]	EFG V_{zz} [10^{15} V/cm^2]	Method	Reference	Remarks
Be	^{9}Be	0.0564(3)	4.40(26	NMR	[5.1,157-160]	(T)
	^{12}B	0.0549(5)	13.3(13)	β-NMR	[5.161-163]	s.cry.
	^{12}N	0.237(10)	22.3(51)	β-NMR	[5.163-164]	s.cry.
	^{19}F (197)	6.30(-)	217(-)	TDPAD	[5.165]	(T)
	^{57}Fe (14)	- 13.1(3)	-282(8)	ME	[5.166-170]	(T),(P)
	^{99}Ru (90)	32.5(4)	584(130)	TDPAC	[5.171]	(T)
	^{100}Rh (75)	8.86(5)	480(13)	e-TDPAC	[5.172,173]	(T)
	^{111}Cd (247)	- 16.0(2)	- 79.7(10)	e-TDPAC	[5.8,174]	s.cry. EFG at subst. lattice site
		54.6(5)	272(30)	TDPAC NO	[5.173-175]	s.cry. EFG at interst. lattice site
	^{175}Hf	-173(20) (at mK)	-265(40)	NO	[5.176]	s.cry.
	^{178}Hf (93)	-202(27) (at 4.2K)	-432(60)	ME	[5.177]	
	^{181}Ta (482)	228(20)	376(25)	TDPAC e-TDPAC	[5.173,178]	(T),s.cry.
	^{193}Ir	108(1) (at 4.2K)	638(160)	ME	[5.189]	
	^{197}AU	31(12) (at 4.2K)	233(90)	ME	[5.371]	EFG at subst. site
		+278(8) (at 4.2K)	+2090(80)	ME	[5.371]	EFG at interst. site I
		+560(11) (at 4.2K)	+4211(160)	ME	[5.371]	EFG at interst. site II
	^{197}Hg (134)	54(1)	2760(200)	e-TDPAC	[5.179,376]	EFG at interst. site
	^{197}Hg (299)	-110(20 (at mK)	-309(80)	NO	[5.179]	
	^{199}Hg (158)	344(6)	1497(110)	e-TDPAC	[5.179]	EFG at two different lattice sites
		- 29.7(-)	-129(10)	e-TDPAC	[5.179]	
Mg	^{12}B	0.0465(5)	- 11.2(1)	β-NMR	[5.163,180, 181]	s.cry.
	^{13}B	0.139(2)	12.0(12)	β-NMR	[5.180]	s.cry.
	^{12}N	0.058(2)	5.5(9)	β-NMR	[5.163,164, 181]	s.cry.
	^{25}Mg	0.281(4)	5.3(2)	NMR	[5.182,183]	
	^{111}Cd (247)	- 7.5(2)	- 37(4)	TDPAC e-TDPAC	[5.184] [5.8,185]	(T),s.cry.

(Appendix cont.)

Host	Probe nucleus	Coupling-constant eQV_{zz} [MHz]	EFG V_{zz} [10^{15} V/cm^2]	Method	Reference	Remarks
Sc	^{44}Sc (68)	1.90(10)	37(4)	TDPAC	[5.186]	
	^{45}Sc	2.02(3)	40(4)	NMR	[5.187]	
	^{57}Fe (14)	18.56(22)	400(10)	ME	[5.188,189]	(T)
	^{99}Ru (90)	104.4(3)	1880(400)	TDPAC	[5.189,190]	(T)
	^{155}Gd	129(2) (at 4.2K)	336(34)	ME	[5.197]	
	^{178}Hf	-367(23) (at 4.2K)	+786(54)	ME	[5.191]	
	^{193}Ir	725(5)	4280(1100)	ME	[5.198]	(T)
	^{181}Ta (482)	312(12)	514(20)	TDPAC	[5.141,192]	(T),(P)
Y	^{45}Sc	5.20(-)	98(5)	NMR	[5.187]	
	^{57}Fe (14)	7.0(7)	151(15)	ME	[5.193,194]	
	^{99}Ru (90)	30.4(4)	547(115)	TDPAC	[5.190]	(T)
	^{100}Rh (75)	32.1(6)	1750(400)	TDPAC	[5.195]	
	^{111}Cd (247)	14.2(3)	71(8)	TDPAC	[5.196]	
	^{155}Gd	164(2) (at 4.2K)	426(45)	ME	[5.197]	
	^{181}Ta (482)	358.1(17)	590(30)	TDPAC	[5.141,192]	(T),(P)
	^{193}Ir	+358(4) (at 4.2K)	+2115(540)	ME	[5.191,198]	
	^{197}Au	+297(9) (at 4.2K)	+2233(80)	ME	[5.191]	
La	^{99}Ru (90)	27.8(6)	500(109)	TDPAC	[5.199]	
	^{133}La (536)	12.0(8)	142(15)	TDPAC	[5.200]	
	^{139}La	7700(2)	131.5(18)	NQR	[5.201]	EFG at two slightly different sites in α-La
		6798(4) (at 6K)	127.8(17)	NMR	[5.202,203]	

(Appendix cont.)

Host	Probe nucleus	Coupling-constant eQV_{zz}/h [MHz]	EFG V_{zz} [10^{15} V/cm^2]	Method	Reference	Remarks
Ti	^{44}Sc (68)	8.09(11)	159(16)	TDPAC	[5.59]	(T)
	^{47,49}Ti	7.7(-)* (at 1...4.2K)	118(5)*	NMR	[5.204]	
	^{53}Fe (14)	- 2.27(40)	- 48.9(15)	ME TDPAC	[5.205] [5.193,206]	
	^{99}Ru (90)	29(3) (at 4.2K)	521(120)	ME	[5.191]	
	^{111}Cd (247)	+ 27.8(2)	+139(17)	TDPAC e-TDPAC	[5.60] [5.8]	(T),s.cry.
	^{178}Hf (93)	-390(9) (at 4.2K)	+836(30)	ME	[5.177]	
	^{181}Ta (482)	341(3)	536(34)	TDPAC	[5.60,207, 208]	(T)
	^{193}Ir	111(2) (at 4.2K)	656(170)	ME	[5.198]	
	^{197}Au	64(14) (at 4.2K)	481(105)	ME	[5.191]	
	^{197}Hg (134)	13.8(9)	705(60)	e-TDPAC	[5.209]	

*Average value for both isotopes; the EFG was calculated using an average value of $\bar{Q}(^{47,49}Ti) = 0.27b$.

Host	Probe nucleus	Coupling-constant eQV_{zz}/h [MHz]	EFG V_{zz} [10^{15} V/cm^2]	Method	Reference	Remarks
Zr	^{57}Fe (14)	7.7(7)	166(17)	ME	[5.193]	α-Zr
	^{91}Zr	18.7(3) (at 4.2K)	368(37)	NMR	[5.210]	α-Zr
	^{92}Mo (2761)	16.0(12)	189(16)	TDPAD	[5.211]	α-Zr
	^{94}Mo (2953)	23.7(12)	188(12)	TDPAD	[5.211]	α-Zr
	^{99}Ru (90)	25(11) (at 4.2K)	450(200)	ME	[5.191]	
	^{111}Cd (247)	14.1(2)	70(8)	TDPAC	[5.15]	
	^{178}Hf (93)	-403(14) (at 4.2K)	+864	ME	[5.177]	
	^{181}Ta (482)	312(3)	514(31)	TDPAC	[5.207, 212-214]	α-Zr
	"	278(2)	458(27)	TDPAC	[5.215,216]	ω-Zr,(T), (P) lattice site A
	"	384(2)	633(38)	TDPAC	[5.213,215, 216]	ω-Zr, (T),(P) lattice site B

(Appendix cont.)

Host	Probe nucleus	Coupling-constant eQV_{zz}/h [MHz]	EFG V_{zz} [10^{15} V/cm^2]	Method	Reference	Remarks
Zr (cont.)	^{193}Ir	114(2) (at 4.2K)	674(175)	ME	[5.198]	
	^{197}Au	258(6) (at 4.2K)	1940(70)	ME	[5.191]	
Hf	^{57}Fe (14)	7.7(8)	166(18)	ME	[5.193]	
	^{99}Ru (90)	31(4) (at 4.2K)	557(75)	ME	[5.191]	
	^{111}Cd (247)	+ 27.2(5)	+136(16)	e-TDPAC	[5.217]	s.cry.
	^{175}Hf	+600(75) (at mK)	+920(150)	NO	[5.125]	s.cry.
	^{176}Hf (88)	-456(20) (at 4.2K)	+943(95	ME	[5.218,219]	
	^{178}Hf (93)	-446(3) (at 4.2K)	+956(28)	ME	[5.218-222]	s.cry.
	^{180}Hf (93)	-434(15) (at 4.2K)	+945(25)	ME	[5.218,222]	
	"	-353(80)	+768(175)	IMPAC	[5.223]	
	^{180}Hf (1142)	+996(60) (at mK)	+936(95)	NO	[5.125]	s.cry.
	^{181}Ta (6.2)	+443(24)	+470(70)	ME	[5.124]	
	^{181}Ta (482)	322(4)	530(32)	TDPAC	[5.76,140, 207,214. 224-229]	(T),(P)
	^{193}Ir	195(3) (at 4.2K)	1152(300)	ME	[5.198]	
	^{197}Au	222(37) (at 4.2K)	1670(280)	ME	[5.191]	
Mn	^{55}Mn	1.71(1)	17.7(9)	NMR	[5.230-232]	(T) EFG in β-Mn
Tc	^{57}Fe (14)	- 3.0(5)	- 65(11)	ME	[5.233]	
	^{99}Tc	5.716(-)	70(70)	NMR	[5.38]	
Re	^{57}Fe (14)	8.1(9) - 2.3(5)	174(20) - 50(11)	ME ME	[5.193] [5.205]	
	^{111}Cd (247)	- 29.2(6)	-145(18)	e-TDPAC	[5.8]	s.cry.

(Appendix cont.)

Host	Probe nucleus	Coupling-constant eQV_{zz}/h [MHz]	EFG V_{zz} [10^{15} V/cm^2]	Method	Reference	Remarks
Re (cont.)	^{177}Lu	neg.QI (at mK)	neg. EFG	NO	[5.234]	s.cry.
	^{181}Ta (6.2)	-520(5)	-551(50)	ME	[5.124,156]	s.cry.
	^{181}Ta (482)	378(2)	623(37)	TDPAC	[5.235,236]	(T),(P)
	^{182}Re	-223(7) (at mK)	-488(57)	NO	[5.237]	s.cry.
	^{183}Re	-257(10) (at mK)	-462(200)	NO	[5.238]	s.cry.
	^{184}Re	-386(25) (at mK)	-443(172)	NO	[5.238]	s.cry.
	^{185}Re	271(23) (at 4.2K)	467(97)	NAR	[5.39]	s.cry.
	^{186}Re	-117(8) (at mK)	-440(160)	NO	[5.238]	s.cry.
	^{187}Re	255.67(13) (at 4.2K)	481(109)	NAR	[5.39]	s.cry.
	^{187}Re (206)	376(6)	471(100)	TDPAC	[5.186]	
	^{185,187}Re	-246(2)* (at mK)	442(95)*	SH	[5.26,27, 239,240]	s.cry.
	^{188}Re	- 92(10) (at mK)	-475(180)	NO	[5.238]	s.cry.
	^{183}Os	-289(10) (at mK)	-383(15)	NO	[5.237]	
	^{186}Os (137)	+151(13) (at 4.2K)	-367(45)	ME	[5.241,242]	
	^{188}Os (155)	+108(20) (at 4.2K)	-288(54)	ME	[5.241]	
	^{186}Ir	+218(11) (at mK)	-355(23)	NO	[5.30,237]	
	^{188}Ir	-107.8(33) (at mK)	-354(20)	NO	[5.30]	
	^{189}Ir	- 85.9(36) (at mK)	-355(40)	NO	[5.30]	
	^{190}Ir	-247.7(74) (at mK)	-359(18)	NO	[5.30]	
	^{193}Ir	- 76(7) (at 4.2K)	-449(60)	ME	[5.191]	
	^{197}Au	- 43(17) (at 4.2K)	-323(128)	ME	[5.191]	
	^{197}Hg (299)	-110(15) (at mK)	-309(45)	NO	[5.243]	

*average for both isotopes

(Appendix cont.)

Host	Probe nucleus	Coupling-constant eQV_{zz}/h [MHz]	EFG V_{zz} [10^{15} V/cm^2]	Method	Reference	Remarks
Fe	^{57}Fe (14)	3.96(68) (at 48K)	85(15)	ME	[5.244,245]	EFG in ε-Fe
Ru	^{57}Fe (14)	- 3.32(10)	- 71.5(25)	ME	[5.205]	
	^{99}Ru (90)	2.7(5)	49(12)	TDPAC	[5.186]	
	^{100}Rh (75)	1.0(3)	54(16)	TDPAC	[5.186]	
	^{178}Hf (93)	220(23) (at 4.2K)	471(50)	ME	[5.177]	
	^{181}Ta	-378(10)	-401(30)	ME	[5.124]	
	^{193}Ir	49(5) (at 4.2K)	289(75)	ME	[5.246]	
	^{197}Au	50(20) (at 4.2K)	376(150)	ME	[5.191]	
Os	^{57}Fe (14)	- 3.5(1)	- 75(3)	ME	[5.205]	
	^{111}Cd (247)	65(-)	324(40)	TDPAC	[5.247]	s.cry.
	^{177}Lu	neg. QI (at mK)	(?)	NO	[5.234]	s.cry.
	^{178}Hf	317(18) (at 4.2K)	680(43)	ME	[5.177]	
	^{181}Ta (6.2)	-566(10)	-600(70)	ME	[5.124]	
	^{181}Ta (482)	389(2)	641(40)	TDPAC	[5.248]	
	^{186}Os (137)	+179(9) (at 4.2K)	435(51)	IPAC ME	[5.249] [5.242]	
	^{189}Os (70)	172(7) (at 4.2K)	1206(102)	ME	[5.251]	
	^{187}Ir (434)	258(12)	344(35)	TDPAC	[5.250]	
	^{193}Ir	54(2) (at 4.2K)	319(82)	ME	[5.191,252,253]	
Co	^{57}Fe (14)	- 1.49(12)	- 32(3)	ME	[5.254]	(T)
	^{59}CO	2.90(7) (at 4.2K)	30.0(20)	NMR	[5.255]	(T)
	^{111}Cd (247)	0.94(6)	4.7(6)	TDPAC	[5.256]	

(Appendix cont.)

Host	Probe nucleus	Coupling-constant eQV_{zz}/h [MHz]	EFG V_{zz} [10^{15} V/cm^2]	Method	Reference	Remarks
Co (cont.)	^{193}Ir	- 36(10) (at 4.2K)	-213(60)	ME	[5.191]	
	^{197}Au	- 14(6)	-105(45)	ME	[5.191]	
Zn	^{12}B	0.095(3) 0.266(12)	23(2) 64(6)	β-NMR	[5.164]	EFG at two different lattice sites
	^{19}F (198)	+ 8.4(10)	-289(48)	TDPAD	[5.165, 257-259]	(T), s.cry.
	^{57}Fe (14)	+ 12.2(7)	+263(15)	ME	[5.193,260, 261]	(T)
	^{65}Zn	pos. QI (at mK)	(?)	NO	[5.123]	
	^{67}Zn	+ 13.62(8) (at 4.2K)	376(38)	NQR SH ME	[5.263] [5.264] [5.265-267]	
	^{67}Zn (604)	45.5(2)	308(30)	TDPAD	[5.6,262]	(T)
	^{69}Zn (439)	- 44.0(44) (at mK)	308(31)	NO	[5.123,268]	s.cry.
	^{68}Ga (1230)	97(6)	557(34)	TDPAD	[5.269]	
	^{70}Ga (879)	47.4 (at 560K)	(500?)	TDPAD	[5.270]	(T)
	^{67}Ge (734)	107.5(8)	1590(600)	TDPAD	[5.271,272]	(T)
	^{69}Ge (398)	91.5(2) (at 633K)	1515(550)	TDPAD QSOPAD	[5.6,271] [5.48]	(T)
	^{71}Ge (175)	17(3) (at 633K)	1400(1000)	TDPAD	[5.271]	
	^{73}As (428)	20.2(4) (at 653K)	(?)	TDPAD	[5.273]	
	^{77}Br (130)	96(6)	(?)	TDPAC	[5.274]	
	^{79}Kr (147)	86.9(6)	877(65)	TDPAC	[5.250,275]	(T)
	^{83}Kr (9)	103(5)	926(50)	e-TDPAC	[5.250]	
	^{99}Ru (90)	21.0(10)	378(82)	TDPAC	[5.186]	
	^{100}Rh (75)	11.2(2)	609(150)	TDPAC e-TDPAC	[5.186] [5.276]	(T)
	^{111}Cd (247)	+133.1(9)	663(16)	TDPAC e-TDPAC	[5.186] [5.8,78, 277-279]	(T),(P)

(Appendix cont.)

Host	Probe nucleus	Coupling-constant eQV_{zz}/h [MHz]	EFG V_{zz} [10^{15} V/cm^2]	Method	Reference	Remarks
Zn (cont.)	^{111}Cd (396)	-139(15) (at mK)	676(72)	NO	[5.175]	s.cry.
	^{117}In (659)	181.7(15)	1192(60)	TDPAC	[5.280]	(T)
	^{129}In (28)	-413(15)	2512(200)	e-TDPAC	[5.281]	
	^{132}I (278)	510(40) (at 77K)	12940(1000)	e-TDPAC	[5.281]	(T)
	^{132}I (50)	390(110)	5760(1600)	IPAC	[5.281]	
	^{152}Sm (122)	66(5)	192(15)	IMPAC	[5.282,283]	s.cry.
	^{154}Sm (82)	57(3)	177(9)	IMPAC	[5.282,283]	s.cry.
	^{175}Hf	+211(10) (at mK)	+323(15)	NO	[5.176]	s.cry.
	^{181}Ta (482)	747(6)	1231(73)	TDPAC	[5.285]	(T)
	^{182}W (100)	-395(56)	+796(130)	IMPAC	[5.282-284]	s.cry.
	^{184}W (111)	-309(32)	+799(90)	IMPAC	[5.282-284]	s.cry.
	^{186}W (123)	-316(25)	+797(90)	IMPAC	[5.282-284]	s.cry.
	^{188}Os (155)	461(37)	1230(160)	IMPAC	[5.282,283]	s.cry.
	^{190}Os (187)	592(48)	1830(230)	IMPAC	[5.282,283]	s.cry.
	^{192}Os (206)	489(57)	1555(310)	IMPAC	[5.282,283]	s.cry.
	^{198}Au	+160(8) (at mK)	(?)	NO	[5.176]	s.cry.
	^{199}Au	+125(8) (at mK)	(?)	NO	[5.176]	s.cry.
	^{197}Hg (134)	33.1(6)	1690(125)	e-TDPAC	[5.209,286]	(T)
	^{197}Hg (299)	+480(20) (at mK)	1350(130)	NO	[5.243]	
	^{199}Hg (158)	+346(9)	1510(110)	e-TDPAC	[5.286]	

(Appendix cont.)

Host	Probe nucleus	Coupling-constant eQV_{zz}/h [MHz]	EFG V_{zz} [10^{15} V/cm^2]	Method	Reference	Remarks
Cd	^{57}Fe (14)	+ 13.1(5)	+282(13)	TDPAC ME	[5.287] [5.193]	
	^{67}Zn (604)	39.5(4) (at 310K)	268(30)	TDPAD	[5.72,288, 289]	(T)
	^{67}Ge (734)	58.7(5)	870(300)	TDPAD	[5.72,288, 290]	(T)
	^{79}Kr (147)	44.0(8)	(?)	TDPAC	[5.275]	(T)
	^{99}Ru (90)	25.4(6)	457(100)	TDPAC	[5.186]	
	^{100}Rh (75)	8.83(9)	480(120)	e-TDPAC	[5.291]	(T)
	^{105}Cd (?)	173(1) (at 483K)	612(60)	TDPAD	[5.293]	
	^{107}Cd (845)	139(1) (at 438K)	612(60)	TDPAD	[5.292,293]	(T)
	^{107}Cd (2679)	180(9) (at 438K)	615(60)	TDPAD	[5.293]	
	^{109}Cd	+ 89.0(57) (at mK)	+682(100)	NO	[5.294]	s.cry.
	^{109}Cd (463)	136.6(10) (at 483K)	614(60)	TDPAD	[5.293]	
	^{111}Cd (247)	+124.7(5)	+621(75)	TDPAC β-TDPAC TDPAD	[5.4] [5.295] [5.6,74,186, 277,296-299]	(T),(P)
	^{115}Cd (173)	- 50(15) (at mK)	+383(115)	NO	[5.294]	
	^{112}In (606)	(?)	(?)	TDPAD	[5.300]	(T)
	^{115}In (828)	-147(5)	+907(108)	TDPAC e-TDPAC	[5.186] [5.44]	s.cry.
	^{117}In (659)	-145.1(5)	952(45)	TDPAC	[5.6,186, 301,302]	s.cry. (T),(P)
	^{119}In (654)	147(3)	935(58)	TDPAC	[5.365]	
	^{112}Sn (2552)	66(3)	941(195)	TDPAD	[5.303]	(T)
	^{113}Sn (731)	96.2(10) (at 415K)	904(90)	TDPAD	[5.6,304-307]	(T)
	^{114}Sn (3088)	67.4(7) (at 480K)	871(30)	TDPAD	[5.304,307]	(T)
	^{115}Sn (619)	47.7(5) (at 550K)	856(112)	TDPAD	[5.307]	

(Appendix cont.)

Host	Probe nucleus	Coupling-constant eQV_{zz}/h [MHz]	EFG V_{zz} [10^{15} V/cm^2]	Method	Reference	Remarks
Cd (cont.)	^{116}Sn (2369)	51.7(9) (at 550K)	822(36)	TDPAD	[5.304-307]	(T)
	^{116}Sn (3548)	105.8(1.0) (at 480K)	875(80)	TDPAD	[5.304]	
	^{118}Sn (2575)	56.4(8) (at 550K)	833(80)	TDPAD	[5.307]	
	^{118}Sn (2321)	(?)	(?)		[5.304,309]	
	^{118}Sn (3112)	75.3(7) (at 550K)	842(80)	TDPAD	[5.307]	
	^{132}I (50)	430(140)	6350(2200)	IPAC	[5.310]	
	^{132}I (278)	310(30)	7865(780)	e-TDPAC	[5.310]	
	^{123}Xe	116(2) (at 278K)	(?)	TDPAD	[5.311]	
	^{152}Sm (122)	66(14)	192(41)	IMPAC	[5.282,283]	s.cry.
	^{154}Sm (82)	57(9)	177(28)	IMPAC	[5.282,283]	s.cry.
	^{182}W (100)	406(55)	819(120)	IMPAC	[5.282,283]	s.cry.
	^{184}W (111)	344(52)	890(150)	IMPAC	[5.282,283]	s.cry.
	^{186}W (123)	368(51)	928(160)	IMPAC	[5.282,283]	s.cry.
	^{188}Os (155)	218(19)	582(72)	IMPAC	[5.282,283]	s.cry.
	^{190}Os (187)	229(24)	707(80)	IMPAC	[5.282,283]	s.cry.
	^{192}Os (206)	153(89)	490(290)	IMPAC	[5.282,283]	s.cry.
	^{187}Ir (434)	650(8)	867(90)	TDPAC	[5.274]	
	^{184}Au (?)	158(8)	870	TDPAC	[5.308]	
	^{197}Hg (134)	23.0(6)	1174(87)	e-TDPAC	[5.209]	(T)
	^{197}Hg (299)	430(30) (at mK)	1210(120)	NO	[5.243]	
	^{204}Pb (1274)	118	720	TDPAC	[5.186]	

(Appendix cont.)

Host	Probe nucleus	Coupling-constant eQV_{zz}/h [MHz]	EFG V_{zz} [10^{15} V/cm^2]	Method	Reference	Remarks
Hg	^{107}Cd (845)	151.6(10)	667(60)	TDPAD	[5.64]	(T)
	^{111}Cd (247)	111(1) (at 77K)	553(67)	TDPAC	[5.186]	
	^{197}Hg (134)	126(3) (at 77K)	6430(475)	e-TDPAC	[5.313]	
	^{199}Hg (158)	205(15) (at 77K)	892(66)	TDPAC	[5.186,312]	
	^{196}Pb (2700)	249(5) (at 223K)	1584(100)	TDPAD	[5.315]	
	^{198}Pb (2800)	283(3) (at 223K)	1560(90)	TDPAD	[5.315]	
	^{200}Pb (3100)	322.8(16) (at 186K)	1690(65)	TDPAD	[5.314]	
	^{204}Pb (1274)	129 (at 77K)	785	TDPAC	[5.186]	
	^{206}Pb (4027)	201.6(8) (at 202K)	1635(100)	TDPAD	[5.314]	
α-Ga	^{66}Ga (1464)	104(5) (at 77K)	551(28)	TDPAD	[5.269]	
	^{68}Ga (1230)	96.2(24) (at 77K)	552(15)	TDPAD	[5.269]	
	^{69}Ga	21.648(1) (at 285K)	533(50)	NQR	[5.2,316-319]	(T),(P)
	^{71}Ga	13.641(1) (at 285K)	532(50)	NQR	[5.2,317,319]	(T)
	^{69}Ge (398)	70.8(8) (at 253K)	1170(500)	TDPAD	[5.271]	
	^{71}Ge (175)	15.5(4) (at 253K)	1280(600)	TDPAD	[5.273]	
	^{71}As (1001)	67(4) (at 288K)	(?)	TDPAD	[5.320]	
	^{73}As (428)	84.6(6) (at 288K)	(?)	TDPAD	[5.320]	
	^{79}Kr (147)	73.3(5)	740(55)	TDPAC	[5.323]	
	^{80}Rb (?)	50.8(1)	700(100)	TDPAD	[5.324]	
	^{111}Cd (247)	137.5(4)	685(82)	TDPAC	[5.186,321, 322]	(T),(T)
	^{119}In (654)	164(3)	1043(65)	TDPAC	[5.325]	
	^{119}Sn (24)	+ 28.8(5) (at 77K)	1267(55)	ME	[5.326]	

(Appendix cont.)

Host	Probe nucleus	Coupling-constant eQV_{zz}/h [MHz]	EFG V_{zz} [10^{15} V/cm^2]	Method	Reference	Remarks
β-Ga	^{69}Ga	9.9(-) (at 250K)	243(-)	NQR	[5.327,328]	(T),(P)
	^{71}Ga	7.82(-) (at 77K)	305(-)	NQR	[5.327]	(T)
	^{111}Cd	83(1) (at 20K)	414(50)	TDPAC	[5.322]	EFG at site 1 η = 0.92
		30(2) (at 20K)	149(18)	TDPAC	[5.322]	EFG at site 2 η = 0.8
In	^{19}F (198)	88(-)	3030(300?)	TDPAD	[5.165]	
	^{57}Fe (14)	6.6(4)	142(9)	ME	[5.193,330]	
	^{67}Ge (734)	22.6(6) (at 328K)	334(120)	TDPAD	[5.72,288]	(T)
	^{69}Ge (398)	21.4(1)	354(200)	TDPAD	[5.346]	
	^{111}Cd (247)	- 17.6(1)	- 88(10)	TDPAC	[5.6,78,186, 298,329, 331-335]	(T),(P)
	^{114}In (190)	- 50(15) (at mK)	-1290(400)	NO	[5.123]	
	^{115}In	- 30.54(1)	-147(18)	NQR	[5.37,122, 336-338]	(T),(P)
	^{117}In (659)	21.7(2)	142(7)	TDPAC	[5.6,186,301]	(T)
	^{116}Sn (2369)	14.8(14)	235(25)	TDPAC	[5.339]	
	^{112}Sb (796)	40(1)	330(130)	TDPAD	[5.340]	
	^{197}Hg (299)	104(5) (at mK)	292(28)	NO	[5.176]	
	^{204}Pb (1274)	41(1)	250(55)	TDPAC	[5.186]	
Tl	^{57}Fe (14)	8.1(8)	174(17)	ME	[5.193]	
	^{69}Ge (398)	6.30(3)	104(40)	TDPAD	[5.346]	
	^{111}Cd (247)	7.8(8)	39(5)	TDPAC	[5.78,186, 341]	(T),(P)
	^{204}Pb (1274)	13.4(3)	81(8)	TDPAC	[5.186,342, 343]	

(Appendix cont.)

Host	Probe nucleus	Coupling-constant eQV_{zz}/h [MHz]	EFG V_{zz} [10^{15} V/cm^2]	Method	Reference	Remarks
Tl (cont.)	^{206}Po (1582)	14.7(-)	(71)	TDPAD	[5.344]	(T)
	^{208}Po (1533)	7.7(-) (at 373K)	(50)	TDPAD	[5.344]	(T)
	^{210}Po (1557)	6.4(-) (at 373K)	46(10)	TDPAD	[5.344]	(T)
Sn	^{19}F (198)	4.95(-)	170(30)	TDPAD	[5.165]	
	^{57}Fe (14)			ME	[5.345]	QI at different sites in α-Sn
	^{68}Ga (1230)	42.3(20)	243(10)	TDPAD	[5.269]	
	^{67}Ge (734)	39.8(5) (at 448K)	590(200)	TDPAD	[5.288,290]	(T)
	^{69}Ge (398)	38.5(1)	637(400)	TDPAD SOPAD	[5.346] [5.347]	
	^{99}Ru (90)	24.0(15)	431(95)	TDPAC	[5.186]	
	^{111}Cd (247)	+ 36.7(1)	+183(22)	TDPAC e-TDPAC	[5.6,78,186, 279] [5.8]	(T),(P)
	^{117}In (659)	50.6(6)	333(16)	TDPAC	[5.186,348]	(T)
	^{113}Sn (731)	46.5(5)	437(44)	TDPAD	[5.349]	(T)
	^{116}Sn (2369)	25.0(10)	398(20)	TDPAC	[5.309]	
	^{118}Sn (2321)	15.6(11)	403(50)	TDPAC	[5.309]	
	^{119}Sn (24)	- 17.6(38)	+774(170)	e-TDPAC ME	[5.350] [5.117, 351-353]	(T)
	^{112}Sb (796)	71.4(4) (at 303K)	590(230)	TDPAD	[5.340]	
	^{120}Sb (88)	42.4(2) (at 305K)	(?)	TDPAD	[5.355,356]	(T)
	^{115}Te (279)	4.6(1)	(?)	TDPAD	[5.354]	(T)
	^{197}Hg (134)	24.7(6)	1260(95)	e-TDPAC	[5.209]	
	^{204}Pb (1274)	71.2(15)	433(95)	TDPAC	[5.186]	
	^{209}Po (1473)	36(4)	382(70)	TDPAD	[5.357]	

(Appendix cont.)

Host	Probe nucleus	Coupling-constant eQV_{zz}/h [MHz]	EFG V_{zz} [10^{15} V/cm^2]	Method	Reference	Remarks
Sn (cont.)	^{210}Po (1557)	50(8)	363(80)	TDPAD	[5.357]	
	^{210}Po (2849)	72(11)	363(80)	TDPAD	[5.357]	
	^{210}Po (4372)	60(9)	400(80)	TDPAD	[5.357]	
As	^{75}As	45.555(4)	648(-)	SH NQR	[5.358] [5.359-363]	(T)
	^{111}Cd (247)	118.7(5)	591(71)	TDPAC	[5.364,365]	(T)
	^{119}In (654)	8(1)	51(6)	TDPAC	[5.365]	
	^{204}Pb (1274)	65(15)	395(125)	TDPAC	[5.186]	
Sb	^{67}Ge (734)	20.3(3) (at 468K)	300(100)	TDPAD	[5.288,366]	
	^{69}Ge (398)	21.1(2)	350(100)	TDPAD	[5.346,366]	(T)
	^{99}Ru (90)	79.0(20)	1420(310)	TDPAC	[5.186]	
	^{111}Cd (247)	109(1)	543(65)	TDPAC	[5.149,186, 365,368,369]	(T) EFG depends strongly on (T) and (X) see [5.364]
	^{119}In (654)	17.9(5)	114(9)	TDPAC	[5.365]	
	^{118}Sn (2321)	12(1)	310(40)	TDPAC	[5.309]	
	^{112}Sb (796)	127.6(6) (at 540K)	1055(430)	TDPAD	[5.367]	
	^{121}Sb	71.613(1)	1139(440)	NQR	[5.115,145]	(T),(P)
	^{122}Sb	neg. QI (at mK)		NO	[5.176]	
	^{123}Sb	97.999(1) (at 4.2K)	1230(440)	NQR	[5.115]	
	^{124}Sb	neg. QI (at mK)		NO	[5.176]	
	^{197}Hg (134)	16.3(9)	832(62)	e-TDPAC	[5.370]	
	^{197}Hg (229)	+ 70(7) (at mK)	197(20)	NO	[5.370]	
	^{204}Pb (1274)	33.3(10)	202(45)	TDPAC	[5.186]	

(Appendix cont.)

Host	Probe nucleus	Coupling-constant eQV_{zz}/h [MHz]	EFG V_{zz} [10^{15} V/cm^2]	Method	Reference	Remarks
Bi	^{69}Ge (398)	10.35(5) (at 326K)	171(70)	TDPAD	[5.346]	(T)
	^{71}As (1001)	51(6)		TDPAD	[5.372]	
	^{111}Cd (247)	80.3(10)	400(48)	TDPAC	[5.321,365]	(T)
	^{119}In (654)	40(3)	254(25)	TDPAC	[5.365]	
	^{112}Sb (796)	88.8(6) (at 370K)	734(290)	TDPAD	[5.373]	
	^{204}Pb (1274)	17.6(5)	107(24)	TDPAC	[5.186]	
	^{209}Bi	48.8(5)	531(30)	NQR SM	[5.360,374] [5.375]	(T)
Se	^{125}Te (35)	596(3) (at 4.2K)	7950(500)	ME	[5.377]	
	^{129}I	-925(10) (at 4.2K)	+6920(100)	ME	[5.378]	EFG in trigonal Se
		-1183(10) (at 4.2K)	+8850(120)	ME	[5.378]	EFG in monoclinic Se
Te	^{57}Fe (14)	10.44(23)	224(10)	ME	[5.379]	(T)
	^{111}Cd (247)	+ 29.0(5)	+144(18)	TDPAC	[5.149,380, 381]	(T)
	^{115}In (828)	-565(62)	+3490(430)	e-TDPAC	[5.381]	s.cry.
	^{119}Sn (24)	6.36(18)	280(15)	e-TDPAC	[5.382]	
	^{125}Te (35)	+416(6) (at 4.2K)	-5550(350)	ME	[5.377,383]	(T)
	^{127}I (58)	693(200)	4036(1150)	IPAC	[5.384]	
	^{129}I	-385(5) (at 4.2K)	2880(100)	ME	[5.378,385]	
	^{129}I (28)	-373(6) (at 80K)	+2300(300)	ME	[5.385]	
	^{131}I (1797)	-4060(30)	22400	e-TDPAC	[5.186,310]	
	^{132}I (278)	-120(4)	3040(120)	e-TDPAC	[5.386]	
	^{132}I (50)	190(70)	2800(1030)	IPAC	[5.386]	

(Appendix cont.)

Host	Probe nucleus	Coupling-constant eQV_{zz}/h [MHz]	EFG V_{zz} [10^{15} V/cm^2]	Method	Reference	Remarks
Te (cont.)	^{133}Cs (81)	-430(26)		e-TDPAC	[5.381]	
	^{152}Sm (122)	82(19)	206(50)	IMPAC	[5.282,284 387]	s.cry.
	^{154}Sm (82)	65(6)	202(70)	IMPAC	[5.282,284, 387]	s.cry.
	^{181}Ta (482)	593.3(28)		TDPAC	[5.388]	EFG at lattice site I
		425(42)		TDPAC	[5.388]	EFG at lattice site II
	^{182}W (100)	284(50	625(110)	IMPAC	[5.282,284]	s.cry.
	^{184}W (111)	265(34)	616(80)	IMPAC	[5.282,284]	s.cry.
	^{186}W (123)	318(35)	787(90)	IMPAC	[5.282,284]	s.cry.
	^{188}Os (155)	234(19)	624(70)	IMPAC	[5.282,284]	s.cry.
	^{190}Os (187)	229(24)	707(70)	IMPAC	[5.282,284]	s.cry.
	^{192}Os (206)	153(90)	490(290)	IMPAC	[5.282,284]	s.cry.
	^{197}Au	263(37) (at 4.2K)		ME	[5.389]	
Sm	^{145}Eu (716)	12.5(4)		TDPAD	[5.390]	(T)
	^{147}Sm	- 4(6) (at mK)	+ 90(130)	SH	[5.391]	
	^{149}Sm	+ 1(1) (at mK)	+ 70(70)	SH	[5.391]	
Gd	^{44}Sc (68)	10.4(1)		TDPAC	[5.392]	
	^{68}Ga (1230)	83(4)	477(23)	TDPAD	[5.269]	
	^{99}Ru (90)	28.7(6) (at 308K)	516(110)	TDPAC	[5.393]	(T)
	^{111}Cd (247)	23.6(3)	118(14)	TDPAC	[5.394-398]	(T)
	^{125}Te (321)	42(11)	1480(900)	IPAC	[5.399]	
	^{152}Sm (122)	105(14)	306(45)	IMPAC	[5.282,284]	s.cry.

(Appendix cont.)

Host	Probe nucleus	Coupling-constant eQV_{zz}/h [MHz]	EFG V_{zz} [10^{15} V/cm^2]	Method	Reference	Remarks
Gd (cont.)	^{154}Sm (82)	90(10)	280(90)	IMPAC	[5.282,284]	s.cry.
	^{147}Gd (997)	61(6) (at 332K)	346(40)	TDPAD	[5.400]	s.cry.
	^{147}Gd (2582)	100(7) (at 416K)	328(25)	TDPAD	[5.400]	s.cry
	^{147}Gd (7550)	250(7) (at 413K)	329(20)	TDPAD	[5.400]	s.cry.
	^{155}Gd	+108(1) (at 4.2K)	+281(30)	ME	[5.197]	
	^{157}Gd	79.4(-) (at 1.6K)	164(20)	NMR	[5.401]	
	^{157}Gd (64)	51.6(10) (at 4.2K)	59(10)	ME	[5.402]	
	^{158}Gd (80)	34(17) (at 4.2K)	72(36)	ME	[5.403]	
	^{160}Gd (75)	41(10) (at 250K)	89(25)	IMPAC	[5.404]	
	^{170}Yb (84)	188(27) (at 4.2K)	367(70)	ME	[5.246]	
	^{181}Ta (482)	399(22)	657(40)	TDPAC	[5.355,405, 406]	(T)
	^{182}W (100)	-358(40)	+740(80)	IMPAC	[5.282,284, 407]	s.cry.
	^{184}W (111)	-267(22)	+690(70)	IMPAC	[5.282,284, 407]	s.cry.
	^{186}W (123)	-289(19)	+728(70)	IMPAC	[5.282,284, 407]	s.cry.
	^{188}Os (159)	461(42)	1230(130)	IMPAC	[5.282,284]	s.cry.
	^{190}Os (187)	458(53)	1410(180)	IMPAC	[5.282,284]	s.cry.
	^{192}Os (206)	401(85)	1275(290)	IMPAC	[5.282,284]	s.cry.
	^{193}Ir	+131(3) (at 4.2K)	+1850(470)	ME	[5.246,408]	
	^{197}Au	+207(10) (at 4.2K)	+1560(60)	ME	[5.409,410]	
Tb	^{99}Ru (90)	29.95(50)	540(120)	TDPAC	[5.393]	(T)
	^{111}Cd (247)	25.7(4)	128(15)	TDPAC	[5.394,397, 398,411]	(T),(P)

(Appendix cont.)

Host	Probe nucleus	Coupling-constant eQV_{zz}/h [MHz]	EFG V_{zz} [10^{15} V/cm^2]	Method	Reference	Remarks
Tb (cont.)	^{159}Tb	1344(-) (at 4.2K)	4150(-)	NMR	[5.412]	EFG at different lattice sites
		440(-) (at 1K)		SH	[5.51,413-416]	
	^{181}Ta (482)	360(6)	593(35)	TDPAC e-TDPAC	[5.405] [5.417]	
	^{193}Ir	259(5) (at 4.2K)	1530(390)	ME	[5.408]	
	^{197}Au	209(10) (at 4.2K)	2130(350)	ME	[5.410]	
Dy	^{99}Ru (90)	30.2(3)	543(120)	TDPAC	[5.393]	(T)
	^{111}Cd (247)	27.2(4)	136(16)	TDPAC	[5.397,398]	(T)
	^{161}Dy	+188(60) (at mK)	+330(100)	SH	[5.391]	
	^{161}Dy (26)	2060(160) (at 20K)	3460(280)	ME	[5.51,418]	(T)
	^{163}Dy	2720(-) (at 4.2K)	4380(-)	NMR	[5.412]	EFG at different sites
		108 (at mK)	335(-)	SH	[5.391]	
	^{181}Ta (482)	399(7)	657(40)	TDPAC	[5.355,405]	(T)
	^{193}Ir	313(5) (at 4.2K)	1850(480)	ME	[5.408]	
	^{197}Au	170(10) (at 4.2K)	1170(195)	ME	[5.410]	
Ho	^{99}Ru (90)	28.7(4)	516(110)	TDPAC	[5.393]	(T)
	^{111}Cd (247)	23.5(4)	117(14)	TDPAC	[5.355,394, 398,421]	(T),(P)
	^{165}Ho	150(30) (at mK)	229(50)	SH	[5.413,416, 419,420]	
	^{181}Ta (482)	408(7)	672(40)	TDPAC	[5.355,405]	(T)
	^{193}Ir	320(5) (at 4.2K)	1890(500)	ME	[5.408]	
	^{197}Au	205(15) (at 4.2K)	1410(240)	ME	[5.410]	

(Appendix cont.)

Host	Probe nucleus	Coupling-constant eQV_{zz}/h [MHz]	EFG V_{zz} [10^{15} V/cm^2]	Method	Reference	Remarks
Er	^{99}Ru (90)	28.3(4)	509(110)	TDPAC	[5.393]	(T)
	^{111}Cd (247)	19.6(4)	98(12)	TDPAC	[5.394,398]	(T)
	^{166}Er (81)	+905(90) (at 4.2K)	-1970(200)	ME	[5.422,423]	(T)
	^{167}Er	1484 (at 4.2K)	2170(-)	NMR	[5.412]	
	^{181}Ta (482)	387(7)	638(40)	TDPAC	[5.355,405]	
	^{193}Ir	302(5) (at 4.2K)	1780(460)	ME	[5.408]	
	^{197}Au	132(15) (at 4.2K)	910(150)	ME	[5.410]	
Tm	^{169}Tm (8)	217(-) (at 120K)	690(-)	ME	[5.424]	(T)
		+1975(-) (at 5K)	6280(-)	ME	[5.51,425]	(T)
	^{172}Yb (79)	267(6)	511(100)	TDPAC	[5.426-428]	(T)
	^{172}Yb (1172)	324(10)	462(95)	TDPAC	[5.426-428]	(T)
Lu	^{57}Fe (14)	7.0(9)	151(25)	ME NO	[5.193] [5.52]	
	^{99}Ru (90)	26.8(5)	482(105)	TDPAC	[5.393]	(T)
	^{111}Cd (247)	14(1)	70(8)	TDPAC	[5.237]	
	^{155}Gd	137(2) (at 4.2K)	354(35)	ME	[5.197]	
	^{175}Lu	429(-) (at mK)	311(-)	SH	[5.429]	
	^{176}Lu	613(-) (at mK)	317(-)	SH	[5.429]	
	^{176}Lu (127)	-128(16) (at mK)	+220(30)	NO	[5.52]	
	^{177}Lu	+144(6) (at mK)	+108(5)	NO	[5.123]	
		+294(37) (at mK)	-221(30)	NO	[5.52]	
	^{175}Hf	+364(24) (at mK)	+558(90)	NO	[5.52,237]	s.cry.

(Appendix cont.)

Host	Probe nucleus	Coupling-constant eQV_{zz}/h [MHz]	EFG V_{zz} [10^{15} V/cm^2]	Method	Reference	Remarks
Lu (cont.)	^{177}Hf (113)	310(310)	390(390)	IPAC	[5.430]	
	^{180}Os	+403(20) (at mK)	538(30)	NO	[5.237]	s.cry.
	^{193}Ir	281(3) (at 4.2K)	1660(430)	ME	[5.198]	
	^{197}Au	+300(3) (at 4.2K)	2070(340)	ME	[5.191]	
Pa	^{231}Pa	850(70) (at 4.2K	2070	ME	[5.431]	

References

5.1 W.D. Knight: Phys. Rev. *92*, 539 (1953)
5.2 W.D. Knight, R.R. Hewitt, M. Pomerantz: Phys. Rev. *104*, 271 (1956)
5.3 R.R. Hewitt, W.D. Knight: Phys. Rev. Lett. *3*, 18 (1959)
5.4 R.S. Raghavan, P. Raghavan: Phys. Lett. *36*A, 313 (1971)
5.5 R. Heubes, G. Hempel, H. Ingwersen, R. Keitel, W. Klinger, W. Loeffler, W. Witthuhn: In [5.19, p.208]
5.6 J. Christiansen, P. Heubes, R. Keitel, W. Klinger, W. Loeffler, W. Sandner, W. Witthuhn: Z. Phys. B*24*, 177 (1976)
5.7 R.S. Raghavan, E.N. Kaufmann, P. Raghavan: Phys. Rev. Lett. *34*, 1280 (1975)
5.8 P. Raghavan, E.N. Kaufmann, R.S. Raghavan, E.J. Ansaldo, R.A. Naumann: Phys. Rev. B*13*, 2835 (1976)
5.9 T.J. Rowland: Proc. Mater. Sci. *9*, 1 (1961)
5.10 L.E. Drain: Metall. Rev. *119*, 195 (1967)
5.11 R.G. Barnes: In *Magnetic Resonance*, ed. by C.K. Coogan, N.S. Ham, S.N. Stuart, J.R. Pillrow, G.V.H. Wilson (Plenum, New York 1970) pp.63-90
5.12 T.P. Das: Phys. Scri. *11*, 121 (1975)
5.13 R.S. Raghavan: Hyperfine Interact. *2*, 29-38 (1976)
5.14 K. Nishiyama, D. Riegel: Hyperfine Interact. *4*, 490-508 (1978)
5.15 E.N. Kaufmann: Hyperfine Interact. *9*, 219-234 (1981)
5.16 E.N. Kaufmann, R.J. Viaden: Rev. Mod. Phys. *51*, 161-214 (1979)
5.17 E. Matthias, D.A. Shirley (eds.): *Hyperfine Structure and Nuclear Radiations* (North-Holland, Amsterdam 1968)
5.18 G. Goldring, R. Kalish (eds.): *Hyperfine Interactions in Excited Nuclei* (Gordon and Breach, New York 1971)
5.19 E. Karlsson, R. Wäppling (eds.): *Hyperfine Interactions Studied in Nuclear Reactions and Decay* (Upplands Grafiska AB, Uppsala 1974) and Phys. Scri. *11*, 109-247 (1975)
5.20 R.S. Raghavan, D.E. Murnick (eds.): "Hyperfine Interactions IV", Hyperfine Interact. *4*, 1-986 (1978)
5.21 G. Kaindl, H. Haas (eds.): "Hyperfine Interactions V", Hyperfine Interact. *9/10*, 1-1298 (1981)
5.22 M.H. Cohen, F. Reif: In *Solid State Physics*, Vol.5, ed. by F. Seitz, D. Turnbull (Academic, New York 1957)

5.23 T.P. Das, E.L. Hahn: *Nuclear Quadrupole Resonance Spectroscopy* (Academic, New York 1958)
5.24 E.A.C. Lucken: *Nuclear Quadrupole Coupling Constants* (Academic, London 1969)
5.25 O.V. Lounasmaa: In *Hyperfine Interactions* (Academic, New York 1967) pp.467-496
5.26 P.E. Gregers-Hansen, M. Krusius, G.R. Pickett: Phys. Rev. Lett. *27*, 38 (1971)
5.27 D.R. Smith, D.H. Keesom: Phys. Rev. B*1*, 188 (1970)
5.28 S.R. de Groot, H.A. Tolhoek, W.J. Huiskamp: In *Alpha-, Beta- and Gamma-Ray Spectroscopy*, Vol.2, ed. by K. Siegbahn (North-Holland, Amsterdam 1968) pp.1199-1261
5.29 K.S. Krane: Nucl. Data Tables *11*, 407 (1973)
5.30 D.W. Murray, A.L. Allsop, N.J. Stone: Hyperfine Interact. *7*, 481 (1980)
5.31 D. Visser, L. Niesen, M. Postma, H. de Waard: Phys. Rev. Lett. *41*, 882 (1978)
5.32 R.L. Mössbauer: Z. Phys. *151*, 124 (1958)
5.33 G.K. Wertheim: *Mössbauer-Effekt: Principles and Applications* (Academic, New York 1964)
5.34 W. Potzel, A. Forster, G.M. Kalvins: Phys. Lett. A*67*, 421 (1978)
5.35 F. Bloch, W.W. Hansen, M. Packard: Phys. Rev. *70*, 474 (1946)
5.36 E.M. Purcell, H.C. Torrey, R.V. Pound: Phys. Rev. *79*, 37 (1946)
5.37 R.R. Hewitt, T.T. Taylor: Phys. Rev. *125*, 524 (1962)
5.38 W.H. Jones, F.J. Milford: Phys. Rev. *125*, 1259 (1962)
5.39 J. Buttet, P.K. Bailey: Phys. Rev. Lett. *24*, 1220 (1970)
5.40 H. Frauenfelder, R.M. Steffen: In *α-, β- and γ-Ray Spectroscopy*, Vol.2 (North-Holland, Amsterdam 1965) pp.997-1198
5.41 R. Böhm, J. Christiansen, W. Klinger, R. Keitel, W. Loeffler, W. Sandner, W. Witthuhn: Hyperfine Interact. *4*, 763-767 (1978)
5.42 R.S. Raghavan, P. Raghavan, E.N. Kaufmann: Phys. Rev. C*12*, 2022 (1975)
5.43 S.M. Harris: Nucl. Phys. *11*, 387 (1959)
5.44 R.S. Raghavan, P. Raghavan, E.N. Kaufmann: Phys. Rev. Lett. *31*, 111, 802 (1973)
5.45 E. Matthias, D.A. Shirley, M.P. Klein, N. Edelstein: Phys. Rev. Lett. *16*, 974 (1966)
5.46 J. Christiansen, H.-E. Mahnke, E. Recknagel, D. Riegel, G. Weyer, W. Witthuhn: Phys. Rev. Lett. *21*, 554 (1968)
5.47 J. Christiansen, H.-E. Mahnke, E. Recknagel, D. Riegel, G. Schatz, G. Weyer, W. Witthuhn: Phys. Rev. C*1*, 613 (1970)
5.48 G. Schatz, R. Brenn, D.B. Fossan: Phys. Lett. *57*B, 231 (1975)
5.49 W. Leitz, W. Semmler, R. Sielemann, T. Wichert: Phys. Rev. B*14*, 5228 (1976)
5.50 G.S. Collins: Hyperfine Interact. *4*, 523 (1978)
5.51 J. Pelz: Z. Phys. *251*, 13 (1972)
5.52 M. Ernst, E. Hagn, E. Zech, G. Eska: Phys. Rev. B*19*, 4460 (1979)
5.53 H. Ernst, E. Hagn, E. Zech: J. Phys. F*19*, 1701 (1979)
5.54 N.C. Mohapatra, P.C. Pattnaik, M.D. Thompson, T.P. Das: Phys. Rev. B*16*, 3001 (1977)
5.55 P.C. Pattnaik, M.D. Thompson, T.P. Das: Phys. Rev. B*19*, 4326 (1979)
5.56 M.D. Thompson, T.P. Das, G. Ciobanu: Phys. Rev. B*19*, 4328 (1979)
5.57 K.W. Lodge: Phys. Lett. A*64*, 315 (1977)
5.58 R.E. Watson, A.C. Gossard, Y. Yafet: Phys. Rev. *140*, A375 (1965)
5.59 R.C. Reno, R.L. Rasera, G. Schmidt: Phys. Lett. A*50*, 243 (1974)
5.60 E.N. Kaufmann, P. Raghavan, R.S. Raghavan, K. Krien, R.A. Naumann: Phys. Status Solidi *63*, 719 (1974)
5.61 P. Heubes: Dissertation, Erlangen (1975)
5.62 W. Witthuhn: Habilitationsschrift, Erlangen (1976)
5.63 F. Bloch: Z. Phys. *61*, 206 (1930)
5.64 H.E. Mahnke, E. Dafni, M.H. Rafailovich, G.D. Sprouse, E. Vapirev: Phys. Lett. A*71*, 112 (1979)
5.65 H. Bayer: Z. Phys. *130*, 227 (1951)
5.66 T. Kushida, G.B. Benedek, N. Bloembergen: Phys. Rev. *104*, 1364 (1956)
5.67 D. Quitmann, K. Nishiyama, D. Riegel: Proc. 18. Ampere Colloquium, Nottingham (1974) p.349
5.68 P. Jena: Phys. Rev. Lett. *36*, 418 (1976)

5.69 K. Nishiyama, F. Dimmling, T. Kornrumpf, D. Riegel: Phys. Rev. Lett. *37*, 357-360 (1976)
5.70 M.D. Thompson, P.C. Pattnaik, T.P. Das: Phys. Rev. B*18*, 5402 (1979)
5.71 W. Engel, W. Klinger, W. Witthuhn: Hyperfine Interact. *9*, 247-254 (1981)
5.72 W. Bartsch, B. Lamp, W. Leitz, H.E. Mahnke, W. Semmler, R. Sielemann, T. Wichert: Z. Phys. B*32*, 301 (1979)
5.73 A. Weidinger, O. Echt, E. Recknagel, G. Schatz, T. Wichert: Phys. Lett. A*65*, 247 (1978)
5.74 P. Raghavan, R.S. Raghavan, W.B. Holzapfel: Phys. Rev. Lett. *28*, 903 (1972)
5.75 T. Butz, G.M. Kalvius: Hyperfine Interact. *2*, 222 (1976)
5.76 H. Ernst, T. Butz, A. Vasquez: J. Phys. F*7*, 1329 (1977)
5.77 K. Nishiyama, D. Riegel: Phys. Lett. A*57*, 270 (1976)
5.78 J.A.H. Da Jornada, F.C. Zawislak: Phys. Rev. B*20*, 2617 (1979)
5.79 W.A. Harrison: *Pseudo-Potentials in the Theory of Metals* (Benjamin, New York 1966)
5.80 W. Harrison: *Solid State Theory* (McGraw-Hill, New York 1970)
5.81 K.W. Lodge, C.A. Sholl: J. Phys. F , 2073 (1974)
5.82 R.M. Sternheimer: Phys. Rev. *95*, 736 (1954)
5.83 R.M. Sternheimer: Phys. Rev. *96*, 951 (1954)
5.84 R.M. Sternheimer: Phys. Rev. *130*, 1423 (1963)
5.85 R.M. Sternheimer: Phys. Rev. *132*, 1637 (1963)
5.86 R.M. Sternheimer: Phys. Rev. *146*, 140 (1966)
5.87 R.M. Sternheimer: Phys. Rev. *159*, 266 (1967)
5.88 T.P. Das, R. Bersohn: Phys. Rev. *102*, 733 (1956)
5.89 H.M. Foley, R.M. Sternheimer, D. Tycko: Phys. Rev. *93*, 734 (1954)
5.90 R.M. Sternheimer, H.M. Foley: Phys. Rev. *92*, 140 (1953)
5.91 R.M. Sternheimer, H.M. Foley: Phys. Rev. *102*, 731 (1956)
5.92 A. Dalgarno: Adv. Phys. *11*, 281 (1962)
5.93 E.H. Hugh, T.P. Das: Phys. Rev. *143*, 452 (1966)
5.94 F.P. Feiock, W.R. Johnson: Phys. Rev. *187*, 39 (1969)
5.95 R.M. Sternheimer: Phys. Rev. *80*, 102 (1950)
5.96 R.M. Sternheimer: Phys. Rev. *84*, 244 (1951)
5.97 R.M. Sternheimer: Phys. Rev. *86*, 316 (1952)
5.98 R.M. Sternheimer: Phys. Rev. *105*, 158 (1957)
5.99 R.M. Sternheimer: Phys. Rev. *164*, 10 (1967)
5.100 R.M. Sternheimer: Phys. Rev. A*6*, 1702 (1972)
5.101 S.N. Ray, T. Lee, T.P. Das: Phys. Rev. A*8*, 1148 (1973)
5.102 J.E. Rodgers, R. Roy, T.P. Das: Phys. Rev. A*14*, 543 (1976)
5.103 R.M. Sternheimer, R.F. Peierls: Phys. Rev. A*3*, 837 (1971)
5.104 P.C. Pattnaik, M.D. Thompson, T.P. Das: Phys. Rev. B*16*, 5390 (1977)
5.105 M.D. Thompson, P.C. Pattnaik, T.P. Das: Hyperfine Interact. *4*, 515 (1978)
5.106 E.N. Kaufmann, R. Vianden: Hyperfine Interact. *4*, 532 (1978)
5.107 K.W. Lodge: J. Phys. F*6*, 1989 (1976)
5.108 B.R.A. Nijboer, F.W. DeWette: Physica *23*, 309 (1957)
5.109 B.R.A. Nijboer, F.W. DeWette: Physica *24*, 422 (1958)
5.110 F.W. DeWette, G.E. Schacher: Phys. Rev. *137*, A78 (1965)
5.111 F.W. DeWette, G.E. Schacher: Phys. Rev. *137*, A92 (1965)
5.112 T.P. Das, M. Pomerantz: Phys. Rev. *123*, 2070 (1961)
5.113 F.W. DeWette: Phys. Rev. *123*, 103 (1961)
5.114 E. Bodenstedt, B. Perscheid: Hyperfine Interact. *5*, 291 (1978)
5.115 R.R. Hewitt, B.F. Williams: Phys. Rev. *129*, 1188 (1963)
5.116 K.C. Das, D.K. Ray: Solid State Commun. *8*, 2025 (1970)
5.117 G.S. Collins, N. Benczer-Koller: Phys. Rev. B*17*, 2085 (1978)
5.118 N.C. Mohapatra, C.M. Singal, T.P. Das, P. Jena: Phys. Rev. Lett. *29*, 456 (1972)
5.119 P. Jena, S.D. Mahanti, T.P. Das: Phys. Rev. B*7*, 975 (1973)
5.120 N.C. Mohapatra, C.M. Singal, T.P. Das: Phys. Rev. Lett. *31*, 530 (1973)
5.121 B. Kolk: J. Phys. (Paris) Colloq. C*6*, 355 (1976)
5.122 F.C. Thatcher, R.R. Hewitt: Phys. Rev. B*1*, 454 (1970)
5.123 W.D. Brewer, G. Kaindl: Hyperfine Interact. *4*, 576 (1978)
5.124 G. Kaindl, D. Salomon: Phys. Lett. A*40*, 179 (1972)

5.125 G. Kaindl, F. Bacon, A.J. Soinski: Phys. Lett. *B46*, 62 (1973)
5.126 M. Piecuch, C. Janot: J. Phys. (Paris), Colloq. *C6*, 359 (1976)
5.127 M. Piecuch, C. Janot: Hyperfine Interact. *5*, 69 (1977)
5.128 L. D'Onofrio, R. Iraldi: Solid State Commun. *21*, 963 (1977)
5.129 C.A. Sholl: Proc. Phys. Soc. *91*, 130 (1967)
5.130 K. Nishiyama, F. Dimmling, D. Kornrumpf, D. Riegel: Phys. Lett. *62A*, 247 (1977)
5.131 K.W. Lodge: J. Phys. *F8*, 447 (1978)
5.132 K.W. Lodge: J. Phys. *F9*, 2035 (1979)
5.133 W. Engel: Dissertation, Universität Erlangen (1979)
5.134 E.G. Brovman, Yu. Kagan: Sov. Phys.-JETP *25*, 365 (1967)
5.135 T. Wang: Phys. Rev. *99*, 566 (1955)
5.136 E. Schempp, P.R.P. Silva: Phys. Rev. *B7*, 2983 (1973)
5.137 E. Schempp, P.R.P. Silva: J. Chem. Phys. *58*, 5116 (1973)
5.138 D.B. Utton: Metrologia *3*, 98 (1967)
5.139 D.B. Utton: J. Chem. Phys. *54*, 5441 (1971)
5.140 R.M. Lieder, N. Buttlar, K. Killig, K. Beck, E. Bodenstedt: In [5.18,p.449]
5.141 T. Butz, G.M. Kalvius: J. Phys. *F4*, 2331 (1974)
5.142 R.V. Kasowski: Phys. Rev. *187*, 891 (1969)
5.143 P. Jena: Phys. Rev. *B17*, 1046 (1978)
5.144 R.A. Cowley: Adv. Phys. *12*, 421 (1963)
5.145 W.J. O'Sullivan, J.E. Schirber: Phys. Rev. *135*, A1261 (1964)
5.146 T. Butz, H. Ernst: Phys. Lett. *53A*, 387 (1975)
5.147 S. Hoth, W. Engel, R. Keitel, W. Klinger, R. Seeböck, W. Witthuhn: Z. Phys. *B41*, 99-105 (1981)
5.148 R. Vianden, E.N. Kaufmann, R.A. Naumann, G. Schmidt: Hyperfine Interact. *7*, 247-252 (1979)
5.149 H. Barfuß, G. Böhnlein, P. Freunek, R. Hofmann, H. Hohenstein, W. Kreische, H. Niedrig, A. Reimer, W. Keppner, W. Körner: Hyperfine Interact. *9*, 235-240 (1981)
5.150 H. Barfuß, G. Böhnlein, P. Freunek, R. Hofmann, H. Hohenstein, W. Kreische, H. Niedrig, A. Reimer: Hyperfine Interact. *10*, 967-972 (1981)
5.151 W. Engel, D. Forkel, H. Föttinger, S. Hoth, M. Iwatschenko, R. Keitel, W. Klinger, N. Neusinger, R. Reichle, R. Seeböck, W. Witthuhn: Annual Report 1979/1980, Physikalisches Institut der Unversität Erlangen (1981) pp.62-64
5.152 R. Vianden: Quadrupole Interaction Frequencies in Metals. Hyperfine Interact. *2*, 169-185 (1976)
5.153 R. Vianden: Quadrupole Interaction Frequencies in Metals. Hyperfine Interact. *4*, 956-976 (1978)
5.154 E.N. Kaufmann, R. Vianden: Electric Field Gradient in Noncubic Metals, in 5.16, pp.203-208
5.155 R. Vianden: Quadrupole Interaction Frequencies in Metals. Hyperfine Interact. *10*, 1243-1272 (1981)
5.156 G. Kaindl, D. Salomon, G. Worthmann: Phys. Rev. Lett. *28*, 952 (1972)
5.157 W.T. Anderson, M. Ruhlig, R.R. Hewitt: Phys. Rev. *161*, 293 (1967)
5.158 D.E. Barnaal, R.G. Barnes, B.R. McCart, L.W. Mohn, D.R. Torgeson: Phys. Rev. *157*, 510 (1967)
5.159 H. Alloul, G. Froidevaux: J. Phys. Chem. Solids *29*, 1623 (1968)
5.160 B.R. McCart, R.G. Barnes: J. Chem. Phys. *48*, 127 (1968)
5.161 R.L. Williams, R.C. Haskell, L. Madansky: Phys. Rev. *C5*, 1435 (1972)
5.162 R.E. McDonald, T.K. McNab: Phys. Rev. Lett. *32*, 1133 (1974)
5.163 F.D. Correll: Hyperfine Interact. *4*, 544 (1978)
5.164 R.C. Haskell, F.D. Correll, L. Madansky: Phys. Rev. *B11*, 3268 (1975)
5.165 C. Budtz-Jorgensen, K. Bonde Nielsen, F. Abildskov, T.K. Laursen, W. Semmler, R. Sielemann, T. Wichert: Hyperfine Interact. *2*, 238 (1976)
5.166 R.M. Housley, J.G. Dash, R.H. Nussbaum: Phys. Rev. *136*, A464 (1964)
5.167 I. Campbell, B. Ferry, P. Imbert, F. Varret: Unpublished, quoted in [5.169]
5.168 C. Janot, P. Delcroix: Acta Metall. *20*, 637 (1972)
5.169 C. Janot, P. Delcroix, M. Piecuch: Phys. Rev. *B10*, 2661 (1974)
5.170 G. Hoy, J. Chappert, H.C. Benski: 4th Int. Conf. on Hyperfine Interactions, Madison, 1977

5.171 R.C. Reno, R.L. Rasera, G. Schmidt: Hyperfine Interact. *5*, 317 (1978)
5.172 K. Krien, J.C. Soares, A. Hanser, B. Feurer: Hyperfine Interact. *1*, 41 (1975)
5.173 K. Krien, J.C. Soares, K. Freitag, R. Tinschler, G.N. Rao, H.G. Müller, E.N. Kaufmann, A. Hanser, B. Feurer: Phys. Rev. B*14*, 4782 (1976)
5.174 E.N. Kaufmann, P. Raghavan, R.S. Raghavan, E.J. Ansaldo, R.A. Naumann: Phys. Rev. Lett. *34*, 1558 (1975)
5.175 P. Herzog, K. Freitag, Menschenbach, H. Walitzki: Z. Phys. A*294*, 13 (1980)
5.176 P. Herzog: Private communication, quoted in [5.155]
5.177 B. Perscheid, G. Kaindl: J. Phys. C*1*, 139 (1980)
5.178 E.N. Kaufmann, K. Krien, J.C. Soares, K. Freitag: Hyperfine Interact. *1*, 485 (1976)
5.179 K. Krien, F. Reuschenbach, J.C. Soares, P. Herzog, H.R. Folle, B. Perscheid, R. Trzcinski, K. Freitag: Hyperfine Interact. *7*, 401 (1980)
5.180 R.C. Haskell, L. Madansky: J. Phys. Soc. Jpn. Suppl. *34*, 167 (1973)
5.181 I. Tanihata, S. Kogo, K. Sugimoto: Phys. Lett. *67*B, 392 (1977)
5.182 P.D. Dougan, N.S. Sharma, D. Llewelyn Williams: Can. J. Phys. *47*, 1047 (1969)
5.183 E.M. Dickson, E.F.W. Seymour: J. Phys. C*3*, 666 (1970)
5.184 A. Andreff, H.J. Hunger, S. Unterricker: Phys. Status Solidi B*61*, 91 (1974)
5.185 U. De, S. Hoth, W. Klinger: Hyperfine Interact. *9*, 241-246 (1981)
5.186 H. Haas, D.A. Shirley: J. Chem. Phys. *58*, 3339 (1973)
5.187 R.G. Barnes, F. Borsa, S.L. Segel, D.R. Torgeson: Phys. Rev. *137*, A1828 (1965)
5.188 M. Forker, R. Trzcinski: Hyperfine Interact. *9*, 267-272 (1981)
5.189 K. Krusch, M. Forker: Z. Phys. B*37*, 225 (1980)
5.190 R. Vianden, J. Kotthaus, P. Winand: Z. Phys. B*37*, 221 (1980)
5.191 G. Wortmann, B. Perscheid, G. Kaindl, F.E. Wagner: Hyperfine Interact. *9*, 343-350 (1981)
5.192 T. Butz, W. Potzel, G. Wortmann: Hyperfine Interact. *1*, 377 (1976)
5.193 S.M. Qaim: J. Phys. C*2*,*2*, 1434 (1969)
5.194 J.S. Carpenter, N.N. Cathey: Phys. Lett. *64*A, 313 (1977)
5.195 R. Vianden, P.M.J. Winand: Hyperfine Interact. *9*, 339-342 (1981)
5.196 M. Forker, K. Kiessen: Hyperfine Interact. *7*, 135 (1979)
5.197 E.R. Bauminger, A. Diamant, I. Felner, I. Nowik, S. Ofer: Phys. Rev. Lett. *34*, 962 (1975)
5.198 M. Forker, K. Krusch: Phys. Rev. B*21*, 2090 (1980)
5.199 J. Kotthaus: Thesis, Bonn (1980)
5.200 B. Klemme, P. Herzog, G. Schäfer, R. Folle, G. Netz: Phys. Lett. *45*B, 38 (1973)
5.201 A. Narath: Phys. Rev. *179*, 359 (1969)
5.202 D.R. Torgeson, R.G. Barnes: Phys. Rev. *136*, A738 (1964)
5.203 W.M. Poteet, R.F. Tipsword, C.D. Williams: Phys. Rev. B*1*, 1265 (1970)
5.204 A. Narath: Phys. Rev. *162*, 320 (1967)
5.205 G. Wortmann, D.L. Williamson: Hyperfine Interact. *1*, 167 (1975)
5.206 H.G. Devare, H. de Waard: Hyperfine Interact. *3*, 63 (1977)
5.207 S. Unterricker: Annual Report, Zentr. Inst. für Kernforschung, Rossendorf, Dresden (1974) p.96
5.208 E.N. Kaufmann: Phys. Rev. B*8*, 1387 (1973)
5.209 K. Krien, J.C. Soares, K. Freitag, R. Vianden, A.G. Bibiloni: Hyperfine Interact. *1*, 217 (1975)
5.210 T. Hioki, M. Kontani, Y. Masuda: J. Phys. Soc. Jpn. *39*, 958 (1975)
5.211 C.V.K. Baba, D.B. Fossan, T. Faestermann, F. Feilitzsch, K.E.G. Löbner, C. Signorini: Phys. Lett. *48*B, 218 (1974)
5.212 E.N. Kaufmann: Phys. Rev. B*8*, 1328 (1973)
5.213 T. Butz, G. Wortmann, G.M. Kalvius, B. Holzapfel: Phys. Lett. *50*A,127 (1974)
5.214 R. Rasera, T. Butz, A. Vasques, H. Ernst, G.K. Shenoy, B.D. Dunlap, R.C. Reno, G. Schmidt: J. Phys. F*8*, 1579 (1978)
5.215 E.N. Kaufmann, D.B. McWhan: Phys. Rev. B*8*, 1390 (1973)
5.216 T. Butz, G.M. Kalvius, H. Göbel, W.B. Holzapfel: Hyperfine Interact. *1*, 1 (1975)
5.217 E.N. Kaufmann, R.S. Raghavan, P. Raghavan, E.J. Ansaldo, R.A. Naumann: Hyperfine Interact. *9*, 289-292
5.218 R.E. Snyder, J.W. Ross, D.S.P. Bunbury: J. Phys. C*2*,*1*, 1662 (1968)

5.219 E. Gerdau, P. Steiner, D. Steenken: In [5.17,p.261]
5.220 P. Boolchand, B.L. Robinson, S. Jha: Phys. Rev. *187*, 475 (1969)
5.221 P. Boolchand, D. Langhammer, Ching-Lu Lin, S. Jha: Phys. Rev. C*6*, 1093 (1972)
5.222 C.G. Jacobs, N. Hershkowitz: Phys. Rev. B*1*, 839 (1970)
5.223 O. Klepper, E.N. Kaufmann, D.E. Murnick: Phys. Rev. C*7*, 1691 (1973)
5.224 E. Gerdau, J. Wolf, H. Winkler, J. Braunsfurth: Proc. Roy. Soc. A*311*, 197 (1969)
5.225 J. Da Jornada, F.P. Liri, F.C. Zawislak: Phys. Status Solidi B*61*, K59 (1974)
5.226 R.W. Sommerfeld, T.W. Cannon, L.W. Coleman, L. Schecter: Phys. Rev. *138*, B763 (1965)
5.227 M. Salomon, L. Boström, T. Lindqvist, M. Perez, M. Zwanziger: Ark. Fys. *27*, 97 (1964)
5.228 J. Berthier, P. Boyer, J.I. Vargas: In [5.18,p.439]
5.229 N. Buttler: Dissertation, Bonn (1970)
5.230 L.E. Drain: Proc. Phys. Soc. *88*, 111 (1966)
5.231 V. Jaccarino, M. Peter, J.H. Wernick: Phys. Rev. Lett. *5*, 53 (1960)
5.232 Y. Masuda, K. Asayama, S. Kobayashi, J. Itoh: J. Phys. Soc. Jpn. *19*, 460 (1964)
5.233 T. Takabatake, M. Mazaki, T. Shinjo: Phys. Rev. Lett. *40*, 1051 (1978)
5.234 C. Trautmann, E. Hagn, H. Ernst, E. Zech: To be published, quoted in [5.155]
5.235 G. Netz, B. Bodenstedt: Nucl. Phys. A*208*, 503 (1973)
5.236 T. Butz, W. Potzel: Hyperfine Interact. *1*, 157 (1975)
5.237 H. Ernst, E. Hagn, E. Zech: Phys. Rev. B*22*, 2248-2256 (1980)
5.238 H. Ernst, E. Hagn, E. Zech, G. Eska: Hyperfine Interact. *4*, 581-584 (1978)
5.239 P. Keeson, C.A. Bryant: Phys. Rev. Lett. *2*, 260 (1959)
5.240 S.D. Rockwood, E.H. Gregory, D.L. Goodstein: Phys. Lett. *30*A, 225 (1969)
5.241 F.E. Wagner, H. Spieler, D. Kucheida, P. Kienle, R. Wäppling: Z. Phys. *254*, 112 (1972)
5.242 H. Ernst, W. Koch, F.E. Wagner, E. Bucher: Phys. Lett. *70*A, 246 (1979)
5.243 P. Herzog, K. Krien, J.C. Soares, H.-R. Folle, K. Freitag, F. Reuschenbach, M. Reuschenbach, R. Trzcinski: Phys. Lett. *66*A, 495 (1978)
5.244 D.L. Williamson, S. Bukshpan, R. Ingalls: Phys. Rev. B*6*, 4194 (1972)
5.245 D.I.C. Pearson, J.M. Williams: J. Phys. F*9*, 1797 (1979)
5.246 B. Perscheid, K. Krusch, M. Forker: J. Magn. Magn. Mater *9*, 14 (1978)
5.247 E. Recknagel, A. Weidinger, R. Wagner: Annual Report, Konstanz (1978) p.25
5.248 G. Netz: Dissertation, Bonn (1973)
5.249 L. Grodzins, Y.W. Chow: Phys. Rev. *142*, 86 (1966)
5.250 H. Haas: Annual Report HMI, Berlin (1977) p.51
5.251 F.E. Wagner, D. Kucheida, U. Zahn, G. Kaindl: Z. Phys. *266*, 223 (1974)
5.252 U. Atzmony, E.R. Bauminger, D. Lebenbaum, A. Mustachi, S. Ofer: Phys. Rev. *163*, 314 (1967)
5.253 F. Wagner, V. Zahn: Z. Phys. *233*, 1 (1970)
5.254 G.J. Perlow, C.E. Johnson, W. Marshall: Phys. Rev. *140*, A875 (1965)
5.255 M. Kawakami, T. Hihara, Y. Koi, T. Wakiyama: J. Phys. Soc. Jpn. *33*, 1591 (1972)
5.256 B. Lindgren, S. Bedi: 4th Int. Conf. on Hyperfine Interactions, Madison, 1977
5.257 R. Brenn, G.D. Sprouse, O. Klepper: J. Phys. Soc. Jpn. Suppl. *34*, 175 (1973)
5.258 R. Brenn, G.D. Sprouse, G. Yue, O. Klepper: Private communication
5.259 R. Brenn, G. Yue, G.D. Sprouse, O. Klepper: In [5.19,p.206]
5.260 W. Kündig, K. Ando, H. Bömmel: Phys. Rev. *139*, A889 (1965)
5.261 R.M. Housley, R.M. Nussbaum: Phys. Rev. *138*, A753 (1965)
5.262 H. Bertschat, E. Recknagel, B. Spellmeyer: Phys. Rev. Lett. *32*, 18 (1974)
5.263 W. Lien: Phys. Rev. *118*, 958 (1960)
5.264 W.T. Vetterling, R.V. Pound: Workshop on New Directions in Mössbauer Spectroscopy, Argonne; AIP Conf. Proc. *38*, 27 (1977)
5.265 E.N. Kaufmann, J.R. Brookeman, P.C. Canepa, T.A. Scott, Rasmussen, J.H. Perepezko: Solid State Commun. *29*, 375 (1979)
5.266 W. Potzel, A. Forster, G.M. Kalvius: J. Phys. C*2*, 29 (1979)
5.267 G. Seidel, P. Keesom: Phys. Rev. Lett. *2*, 261 (1959)
5.268 P. Herzog, H.R. Folle, E. Bodenstedt: Hyperfine Interact.*3*, 361 (1977)
5.269 P. Raghavan, R.S. Raghavan: Hyperfine Interact. *9*, 317-322 (1981)
5.270 D.A. Hutcheon, D.M. Sheppard, P. Kitching, J.M. Davidson, C.W. Luursema, L.E. Carlson: Nucl. Phys. A*245*, 306 (1975)

5.271 H. Haas, W. Leitz, H.E. Mahnke, W. Semmler, R. Sielemann, T. Wichert: Phys. Rev. Lett. *30*, 656 (1973)
5.272 W. Bartsch, W. Leitz, W. Semmler, R. Sielemann, T. Wichert: Phys. Lett. *54*A, 66 (1975)
5.273 R.E. Shroy, R. Brenn, D.B. Fossan, P.M.S. Lesser, G. Schatz: In [5.19] post deadline paper, unpublished
5.274 H. Haas: Annual Report HMI, Berlin (1976) pp.76-79
5.275 H. Haas, H.H. Bertschat: Hyperfine Interact. *9*, 273-276 (1981)
5.276 K. Krien, J.C. Soares, A.G. Bibiloni, R. Vianden, A. Hanser: Z. Phys. *266*, 195 (1974)
5.277 P. Raghavan, R.S. Raghavan, E.N. Kaufmann: Phys. Lett. *48*A, 131 (1974)
5.278 E. Hammer: Zulassungsarbeit, Erlangen (1976), unpublished
5.279 J.A.H. da Jornada, E.R. Froga, R.P. Livi, F.C. Zawislak: Hyperfine Interact. *4*, 589-593 (1978)
5.280 S.H. Devare, D.R.S. Somayajulu, H.G. Devare: Hyperfine Interact. *1*, 151 (1975)
5.281 H. Ooms, F. Namavar, J. Claes, H. Van de Voorde, R. Coussement, M. Rots: Hyperfine Interact. *4*, 559-563 (1978)
5.282 L.P. Rösch: Helv. Phys. Acta *48*, 287 (1975)
5.283 L.P. Rösch, R. Kulessa, F. Horber, H.P. Seiler, P. Marmier: Phys. Rev. B*9*, 4638 (1974)
5.284 L.P. Rösch, R. Kulessa, F. Horber: Phys. Status Solidi (b) *71*, 389 (1975)
5.285 S.C. Bedi, R.G. Pillay, S.H. Devare: Hyperfine Interact. *5*, 161 (1978)
5.286 K. Krien, F. Reuschenbach, J.C. Soares, R. Vianden, R. Trzcinski, K. Freitag, M. Hage-Ali, P. Siffert: Hyperfine Interact. *7*, 413 (1980)
5.287 M. Deicher, O. Echt, E. Recknagel, T. Wichert: Annual Report, Universität Konstanz (1978) p.15
5.288 W. Bartsch, W. Leitz, H.E. Mahnke, W. Semmler, R. Sielemann, T. Wichert: Z. Phys. B*21*, 131 (1975)
5.289 W. Bartsch, J. Golczewski, W. Leitz, H.E. Mahnke, W. Semmler, R. Sielemann, T. Wichert: Phys. Lett. *50*A, 413 (1975)
5.290 W. Bartsch, B. Lamp, W. Leitz, H.E. Mahnke, W. Semmler, R. Sielemann, T. Wichert: Int. Meeting on Hyperfine Interactions, University of Leuven, Haverlee, Belgium, 1975, p.9
5.291 K. Krien, J.C. Soares, R. Vianden, A.G. Bibiloni, A. Hanser: Hyperfine Interact. *1*, 295 (1975)
5.292 O. Echt, H. Haas, E. Ivanov, E. Recknagel, E. Schlodder, B. Spellmeyer: Hyperfine Interact. *2*, 230 (1976)
5.293 G.D. Sprouse, O. Häusser, H.R. Andrews, T. Faestermann, J.R. Beene, T.K. Alexander: Hyperfine Interact. *4*, 229 (1978)
5.294 S.S. Rosenblum, W.A. Steyert: Phys. Lett. *53*A, 34 (1975)
5.295 P. Raghavan, R.S. Raghavan: Phys. Rev. Lett. *27*, 724 (1971)
5.296 J.M. McDonald, P.M.S. Lesser, D.B. Fossan: Phys. Rev. Lett. *28*, 1057 (1972)
5.297 J. Bleck, R. Butt, H. Haas, W. Ribbe, W. Zeitz: Phys. Rev. Lett. *29*, 1371 (1972)
5.298 E. Bodenstedt, U. Ortabasi, W.H. Ellis: Phys. Rev. B*6*, 2909 (1972)
5.299 T. Butz, B. Lindgren, H. Saitovich: Hyperfine Interact. *7*, 81 (1979)
5.300 E.A. Ivanow, E. Recknagel: Private communication
5.301 R.S. Raghavan, P. Raghavan: Phys. Rev. Lett. *28*, 54 (1972)
5.302 R. Butt, H. Haas, S.S. Rosenblum: Phys. Rev. B*9*, 3705 (1974)
5.303 R. Vianden, K. Krien, H.U. Schmidt: Nucl. Phys. A*243*, 29 (1975)
5.304 F. Dimmling, D. Riegel, K.G. Rensfelt, C.J. Jerrlande: Phys. Lett. *55*B, 293 (1975)
5.305 R. Keitel, W. Engel, S. Hoth, W. Klinger, W. Witthuhn: Hyperfine Interact. *10*, 681-686 (1981)
5.306 O. Echt, H. Haas, E. Ivanov, E. Recknagel, E. Schlodder, B. Spellmeyer: Hyperfine Interact. *2*, 232 (1976)
5.307 H. Bertschat, O. Echt, H. Haas, E. Ivanov, E. Recknagel, E. Schlodeer, B. Spellmeyer: Hyperfine Interact. *2*, 326 (1976)
5.308 H. Haas: Annual Report HMI, Berlin (1977) p.52
5.309 K. Krien, B. Klemme, R. Folle, E. Bodenstedt: Nucl. Phys. A*228*, 15 (1974)

5.310 H. Ooms, F. Namavar, J. Claes, H. Van de Voorde, R. Coussement, M. Rots: Hyperfine Interact. *4*, 559 (1978)
5.311 W.D. Zeitz, H. Bertschat, H. Grawe, H. Haas, R. Sielemann: Annual Report HMI, Berlin (1979) p.76
5.312 W.A. Edelstein, R.V. Pound: Phys. Rev. B*11*, 985 (1975)
5.313 R. Vianden, K. Krien: Nucl. Phys. A*277*, 492 (1977)
5.314 H.E. Mahnke, T.K. Alexander, H.R. Andrews, O. Häusser, P. Taras, D. Ward, E. Dafni, G.D. Sprouse: Phys. Lett. *88*B, 48-50 (1979)
5.315 S. Zywietz, H. Grawe, H. Haas, M. Menningen: Hyperfine Interact. *9*, 109-112 (1981)
5.316 T. Kushida, G.B. Benedek: Bull. Am. Phys. Soc. *II*,*3*, 167 (1958)
5.317 R.H. Hammond, W.D. Knight: Phys. Rev. *120*, 762 (1960)
5.318 J.J. Hwang, P.C. Canepa, T.A. Scott: J. Phys. Chem. Solids *38*, 1403 (1977)
5.319 M.I. Valic, D.L. Williams: J. Phys. Chem. Solids *30*, 2337 (1969)
5.320 O. Echt, H. Haas, W. Leitz, E. Recknagel, W. Semmler, E. Schlodder: Hyperfine Interact. *2*, 228 (1976)
5.321 W. Keppner, W. Körner, P. Heubes, G. Schatz: Hyperfine Interact. *9*, 293-296 (1981)
5.322 P. Heubes, D. Korn, G. Schatz, G. Zibold: Phys. Lett. A*74*, 267 (1979)
5.323 M. Menningen, H. Haas: Annual Report HMI, Berlin (1978) p.61
5.324 M. Menningen, H. Grove, H. Haas, W.-D. Zeitz, R. Keitel: Annual Report HMI, Berlin (1979) p.80
5.325 H. Haas, M. Menningen: Annual Report HMI, Berlin (1979) p.81
5.326 H. Haas, M. Menningen, G. Weyer: Annual Report HMI, Berlin (1979) p.81
5.327 S.L. Segel, R.D. Heyding, E.F.W. Seymour: Phys. Rev. Lett. *28*, 970 (1972)
5.328 R.J. Brown, S. Segel: J. Phys. F*5*, 1073 (1975)
5.329 C. Budtz-Jorgensen, F. Abildskov, K. Bonde Nielsen, H. Ravn, U. Willumsen: Phys. Rev. B*8*, 5411 (1973)
5.330 P.A. Flinn, U. Gonser, R.W. Grant, R.M. Housley: Phys. Rev. *157*, 538 (1967)
5.331 P. Lehmann, J. Miller: J. Phys. Radium *17*, 526 (1956)
5.332 D. Brandt, S.S. Rosenblum: J. Phys. Soc. Jpn. Suppl. *34*, 178 (1973)
5.333 H.J. Behrend, D. Budnick: Z. Phys. *168*, 155 (1962)
5.334 C. Budtz-Jorgensen, K. Bonde Nielsen: Hyperfine Interact. *1*, 81 (1975)
5.335 B. Lindgren: Phys. Lett. *66*A, 241 (1978)
5.336 W.W. Simmons, C.P. Slichter: Phys. Rev. *121*, 1580 (1961)
5.337 D.R. Torgeson, R.G. Barnes: Phys. Rev. Lett. *9*, 255 (1962)
5.338 W.J. O'Sullivan, J.E. Schirber: Phys. Rev. *135*, A1261 (1964)
5.339 H.R. Folle: Diplomarbeit, University Bonn (1974)
5.340 H.E. Mahnke, E. Dafni, M.H. Rafailovich, G.D. Sprouse, E. Vapirev: Annual Report HMI, Berlin (1979), p.74
5.341 P. Heubes, W. Keppner, G. Schatz: Hyperfine Interact. *7*, 93 (1979)
5.342 G.K. Wertheim, R.V. Pound: Phys. Rev. *102*, 185 (1956)
5.343 R.M. Lieder, W. Delang, M. Fleck: Z. Phys. *206*, 29 (1967)
5.344 F.v.Feilitzsch, T. Faestermann, O. Häusser, K.E.G. Löbner, R. Lutter, H. Bohn, D.J. Donahue, R.L. Hershberger, F. Riess: In [5.19,p.142]
5.345 M. Shimatomai, M. Doyama: Hyperfine Interact. *9*, 329-332 (1981)
5.346 R. Raghavan, R.S. Raghavan: Hyperfine Interact. *4*, 569 (1978)
5.347 P. Raghavan, R.S. Raghavan: Hyperfine Interact. *3*, 371 (1977)
5.348 S.H. Devare, H.G. Devare: Hyperfine Interact. *2*, 235 (1976)
5.349 W. Semmler, P. Raghavan, M. Senba, R.S. Raghavan: Hyperfine Interact. *9*, 323-328 (1981)
5.350 J.C. Soares, K. Krien, A.G. Bibiloni, K. Freitag, R. Vianden: Phys. Lett. *45*A, 465 (1973)
5.351 K.P. Mitrofanov, M.V. Plotnikova, V.S. Shpinell: JETP *21*, 524 (1965)
5.352 B.A. Komissarova, A.A. Sorokin, V.S. Shpinell: JETP *23*, 800 (1966)
5.353 V.N. Panyushkin, L. Bogner, G. Wortmann: In *Proceedings of the Intern. Conf. on Mössbauer Spectroscoy*, ed. by A.Z. Hrynkiewicz, J.A. Zawicki (Krakow, Poland 1975) p.99
5.354 E. Ivanov, W. Leitz, E. Recknagel, E. Schlodder: Annual Report, HMI Berlin (1974) p.81
5.355 M. Forker, S. Scholz: Hyperfine Interact. *7*, 353 (1979)

5.356 R. Seeböck: Diplomarbeit, Erlangen (1978)
5.357 E. Dafni, M.H. Rafailovich, T. Marshall, G. Schatz, G.D. Sprouse: 5th Intern. Conf. on Hyperfine Interactions, Berlin (1980), Contr.A23
5.358 W.A. Taylor: Phys. Rev. *161*, 652 (1967)
5.359 S.N. Sharma: Phys. Lett. A*57*, 379 (1976)
5.360 T.J. Bastow, K. Whitefield: Solid State Commun. *18*, 955 (1976)
5.361 T.J. Bastow, R.J.C. Brown, H.J. Whitefield: Phys. Lett. A*78*, 198 (1980)
5.362 T.T. Taylor, E.H. Hygh: Phys. Rev. *129*, 1193 (1963)
5.363 M. Krusius, G. Pickett: Solid State Commun. *9*, 1917 (1971)
5.364 H. Barfuß, G. Böhnlein, P. Freunek, R. Hofmann, H. Hohenstein, W. Kreische, H. Niedrig, A. Reimer: Hyperfine Interact. *10*, 967-972 (1981)
5.365 H. Haas, M. Menningen: Hyperfine Interact. *9*, 277-282 (1981)
5.366 W. Bartsch, W. Leitz, H.E. Mahnke, W. Semmler, R. Sielemann, T. Wichert: Annual Report, HMI Berlin (1974) p.72
5.367 H.E. Mahnke, E. Dafni, M.H. Rafailovich, G.D. Sprouse, E. Vapirev: Bull. Am. Phys. Soc. *23*, 962 (1978)
5.368 S.A. Lis, R.A. Naumann: Hyperfine Interact. *3*, 283 (1977)
5.369 U. Frey, H. Hohenstein, W. Kreische, M. Meinhold, H. Niedrig, U. Pechtl, K. Reuter: Z. Phys. B*33*, 141 (1979)
5.370 J.C. Soares, K. Krien, P. Herzog, H.R. Folle, K. Freitag, F. Reuschenbach, M. Reuschenbach, R. Trzcinski: Z. Phys. B*31*, 395 (1978)
5.371 B. Perscheid, H.W. Geyer, K. Krien, K. Freitag, J.C. Soares, E.N. Kaufmann, R. Vianden: Hyperfine Interact. *4*, 554 (1978)
5.372 W. Semmler, M. Haas, H.E. Mahnke, R. Sielemann, W.D. Zeitz: Annual Report HMI (1979) p.82
5.373 H.E. Mahnke, M. Haas, W. Semmler, R. Sielemann, W.D. Zeitz: Hyperfine Interact. *9*, 311-314 (1981)
5.374 B.F. Williams, R.R. Hewitt: Phys. Rev. *146*, 286 (1966)
5.375 N.E. Phillips: Phys. Rev. *118*, 644 (1960)
5.376 K. Krien, M. Saitovitch, K. Freitag, F. Reuschenbach, J.C. Soares, E.N. Kaufmann: Hyperfine Interact. *4*, 549 (1978)
5.377 P. Boolchand, T. Henneberger, J. Oberschmidt: Phys. Rev. Lett. *30*, 1292 (1973)
5.378 C.S. Kim, P. Boolchand: Phys. Rev. B*19*, 3187 (1979)
5.379 M.C. Verma, G.N. Rao: Phys. Rev. Abstract *11*, No.12,6 (1980)
5.380 S.A. Lis, R.A. Naumann: Hyperfine Interact. *3*, 279 (1977)
5.381 R. Kulessa, N. Tung: Hyperfine Interact. *4*, 564 (1978)
5.382 M. Roots, F. Namavar, G. Langouche, R. Coussement, M. van Rossum, P. Boolchand: J. Phys. F*5*, L117 (1978)
5.383 P. Boolchand, B.L. Robinson, S. Jha: Phys. Rev. B*2*, 3463 (1970)
5.384 K.V. Makryunas, E.K. Makryuene: Sov. Phys. Solid State *10*, 514 (1968)
5.385 M. Pasternak, S. Bukshpan: Phys. Rev. *163*, 297 (1967)
5.386 H. Oøms, J. Claes, F. Namavar, H. Van de Voorde, M. Rots: Hyperfine Interact. *4*, 226 (1978)
5.387 R. Kulessa, L.P. Roesch, F. Horber: In [5.19,p.224]
5.388 S.H. Devare, H.G. Devare: Hyperfine Interact. *7*, 313 (1979)
5.389 T.K. Sham, R.E. Watson, M.C. Permann: Phys. Rev. B*20*, 3552 (1979)
5.390 R. Böhm, W. Engel, S. Hoth, W. Klinger, R. Seeböck, W. Witthuhn: Z. Phys. *39*, 115 (1980)
5.391 A.C. Anderson, B. Holström, M. Krusius, G.R. Pickett: Phys. Rev. *183*, 546 (1969)
5.392 M.G. Colley, D.M. Chaplin, D.E. Swan, G.V.M. Wilson: J. Phys. F*6*, 131 (1976)
5.393 M. Forken, S. Scholz: Hyperfine Interact. *9*, 261-266 (1981)
5.394 J.B. Fechner, M. Forker, G. Schäfer: Z. Phys. *265*, 197 (1973)
5.395 L. Boström, B. Jouson, E. Karson, G. Liljegren, S. Zetterlund: Phys. Lett. *31*A, 436 (1970)
5.396 L. Boström, G. Liljegren, B. Johnson, E. Karlson: Phys. Scr. *3*, 75 (1971)
5.397 M. Forker, A. Hammesfahr: Z. Phys. *260*, 131 (1973)
5.398 S.A. Lis, R.A. Naumann, G. Schmidt: Hyperfine Interact. *5*, 431 (1978)
5.399 D.W. Cruse, K. Johanson, E. Karlson: Nucl. Phys. A*154*, 369 (1970)
5.400 O. Häusser, M.-E. Mahnke, J.F. Sharey-Schäfer, M.L. Swanson, P. Taras, D. Ward, M.R. Andrews, T.K. Alexander: Phys. Rev. Lett. *44*, 132 (1980)

5.401 Y. Seiwa, T. Tsuda, A. Mirai, C.W. Searle: Phys. Lett. *43*A, 23 (1973)
5.402 J. Göring: Z. Phys. *251*, 185 (1972)
5.403 J. Fink: Z. Phys. *207*, 225 (1967)
5.404 M. Spehl, N. Wirtz: Z. Phys. *243*, 431 (1971)
5.405 M. Forker, J.B. Fechner, M. Haverkamp: Z. Phys. *269*, 279 (1974)
5.406 K. Krin, M. Forker, F. Reuschenbach, R. Trczinski: Hyperfine Interact. *7*, 19 (1979)
5.407 E.N. Kaufmann, D.E. Murnick, C.T. Alonso, L. Grodzins: In [5.18,p.1033]
5.408 M. Forker, K. Krusch: Hyperfine Interact. *9*, 399 (1981)
5.409 B. Perscheid, M. Büchsler, M. Forker: Phys. Rev. B*14*, 4803 (1976)
5.410 B. Perscheid, M. Forker: Z. Phys. B*31*, 49 (1978)
5.411 T. Butz, B. Lindgren: Phys. Scr. *21*, 836 (1980)
5.412 K. Shimizo, H. Mizotani, J. Itoh: J. Phys. Soc. Jpn. *43*, 57 (1977)
5.413 M. Krusius, A.C. Anderson, B. Holström: Phys. Rev. *177*, 910 (1969)
5.414 O.V. Lounasmaa, P.R. Roach: Phys. Rev. *128*, 622 (1962)
5.415 B. Bleaney, R.W. Hill: Proc. Phys. Soc. London *78*, 313 (1961)
5.416 H. Van Kempen, A.R. Miedema, W. Kruiskamp: Physica *30*, 229 (1964)
5.417 M. Forker, K. Krin, F. Reuschenbach: Hyperfine Interact. *9*, 255-260 (1981)
5.418 S.C. Sylvester, D. Schroeer: Phys. Rev. C*7*, 2056 (1973)
5.419 G. Brunhardt, H. Postuma, V.L. Soritor: Phys. Rev. *137*, 1484 (1965)
5.420 O. Lunasmaa: Phys. Rev. *128*, 1136 (1962)
5.421 B. Lindgren, A.K. Bhati: Hyperfine Interact. *9*, 303-310 (1981)
5.422 P. Kienle: Rev. Mod. Phys. *36*, 372 (1964)
5.423 R.A. Reese, R.G. Barnes: Phys. Rev. *163*, 465 (1967)
5.424 D.L. Uhrich, R.G. Barnes: Phys. Rev. *164*, 428 (1967)
5.425 M. Kalvius, P. Kienle, M. Eichner, W. Wiedemann, C. Schüler: Z. Phys. *172*, 231 (1963)
5.426 A.R. Chuhran-Long, A. Li-Scholz, R.L. Rasera: Phys. Rev. B*8*, 1791 (1973)
5.427 R.L. Rasera, A. Li-Scholz: Phys. Rev. B*1*, 1995 (1970)
5.428 A. Li-Scholz, R.L. Rasera: Phys. Rev. Lett. *23*, 181 (1969)
5.429 O.V. Lounasmaa: Phys. Rev. *133*, A219 (1964)
5.430 K.P. Gopinathan: Hyperfine Interact. *9*, 351 (1981)
5.431 J.M. Friedt, G.M. Kalvius, R. Poinsot, J. Rebizont, J.C. Spirlet: Phys. Lett. *69*A, 225 (1978)
5.432 C.M. Lederer, V.S. Shirley: *Table of Isotopes* (Wiley, New York 1978) Appendix VII: Table of Nuclear Moments

6. β Emitters and Isomeric Nuclei as Probes in Condensed Matter

H. Ackermann, P. Heitjans, and H.-J. Stöckmann

With 30 Figures

This chapter surveys the behaviour of radioactive probe nuclei with lifetimes in the range of $\tau_N \approx 10^{-5} - 10^3$ s in condensed matter. Not only hyperfine interactions (hfi) play a role in the experiments to be reported here, but all interactions of the probe's nuclear moments with electromagnetic fields. An important example is the nuclear dipole-dipole interaction. In order to restrict the length of this review we consider only experiments where the probe nuclei have been created *and* oriented "in-beam", i.e., by reactions with a particle beam during the course of the measurement. Thus, the important area of low temperature nuclear orientation ($\tau_N \gtrsim 1$ h) is omitted here, but this has been excellently treated in several recent reviews [6.1,2]. We discuss here all in-beam experiments using β emitting nuclides ($\tau_N \approx 10^{-2} - 10^3$ s) in condensed matter. Further, those in-beam experiments using isomeric γ emitters are reviewed, where nuclear magnetic resonance (NMR) signals were observed or where spin-lattice relaxation phenomena - mostly in liquid metals - were investigated ($\tau_N \approx 10^{-5} - 10^{-2}$ s). The many other experiments applying short-lived γ active nuclei to solid state studies have been discussed in Chaps.4 and 5.

The comparatively long lifetimes of the applied nuclides have three major consequences:

i) NMR experiments can be performed in a rather straightforward manner, as the radio frequency (RF) fields necessary to produce transitions during τ_N can still be handled conventionally: the RF-magnetic induction B_1 usually lies below 2 mT.

ii) The natural line broadening does not generally restrict the obtainable accuracy of NMR measurements. To give an example: $\tau_N = 1$ ms causes $\Delta\nu_{nat} = (\pi\tau_N)^{-1} \approx 300$ Hz which amounts only to about 10% of the typical dipolar broadening of a NMR line in nonmagnetic solid samples. Thus, one observes similar line widths, as in the case of conventional NMR on stable nuclides in solids. Clearly, this statement does not hold for liquid samples, where dipolar widths are motionally narrowed.

iii) Transients of the nuclear polarization are only observable in time scales not too different from τ_N (rule of thumb: $0.01\ \tau_N <$ observable $\tau < 100\ \tau_N$). Therefore, relaxation transients with time constants τ between μs and hours can be measured with in-beam methods using isomeric and β active nuclides.

The first in-beam NMR experiment was performed on the β emitter ^{8}Li by CONNOR (1959) [6.3] and ten years later the technique was extended to isomeric γ emitters by QUITMANN et al. [6.4]. Again, β emitters were applied in the first observation of quadrupolar split NMR spectra by SUGIMOTO et al. [6.5] in 1970, and further, in making systematic studies of relaxation phenomena by WINNACKER et al. one year later [6.6]. In 1972, RIEGEL et al. [6.7] applied isomeric γ emitters to relaxation studies in liquid metals. Finally, HAMAGAKI et al. demonstrated in 1976 the usefulness of β active nuclides in probing hyperfine fields in ferromagnets [6.8].

All applications devoted to condensed matter studies have been performed within the last fifteen years. Much of the material has been presented in three conference proceedings [6.9,10,175]. The application of β emitters to the study of quadrupole interactions, in particular, has been discussed in a recent review article [6.11]. In addition, two recent review articles should be mentioned: one on electric field gradients in noncubic metals by KAUFMANN and VIANDEN [6.173], the other one by the present authors, covering the application of nuclear in-beam methods to internal fields, defects and motions [6.174].

Section 6.1 summarizes the theory necessary to interprete the experiments reported in this chapter. The experimental methods are described in Sect.6.2. Measurements and results are discussed in Sects.6.3,4. Finally, the experiments are listed in the form of a tabular summary (Sect.6.5).

6.1 Theory

The behaviour of a nuclear ensemble of a spin species I is described most effectively by the density matrix formalism. In this section we study the development of the density matrix due to static and fluctuating magnetic and electric interactions. We start with the determination of the eigenvalues of the relevant static interaction Hamiltonian. Next, the influence of radiofrequency fields is studied. Finally, we discuss relaxation phenomena due to fluctuating interactions. As all these topics are already treated extensively in well-known review articles and monographs (see, e.g., [6.12-14]), we restrict ourselves to a survey of the most essential ideas and formulas and do not try to give a comprehensive description.

6.1.1 Hamiltonian and Energy Levels

The total Hamiltonian of a spin system I can be written as

$$\mathcal{H} = \mathcal{H}_D + \mathcal{H}_Q \quad . \tag{6.1}$$

Here $\mathcal{H}_D$ is the operator of the magnetic dipole interaction and $\mathcal{H}_Q$ is the electric quadrupole (Q) operator. The interaction with higher multipoles (which has been demonstrated experimentally only in few cases up to now) is disregarded here.

The magnetic dipole operator $\mathcal{H}_D$ is given by

$$\mathcal{H}_D = -\hbar\omega_I I_Z \quad , \tag{6.2}$$

where

$$\omega_I = g_I \mu_N B/\hbar \quad .$$

In (6.2) the z-axis is chosen parallel to the applied magnetic induction **B**. The Hamiltonian of the electric Q-interaction depends on the orientation of the electric field gradient (EFG) tensor relative to **B**. If the principal axis of the EFG-tensor is parallel to **B**, we get for the Q-Hamiltonian,

$$\mathcal{H}_Q = \frac{e^2qQ}{4I(2I-1)}\left[3I_Z^2 - \mathbf{I}^2 + \frac{\eta}{2}\left(I_+^2 + I_-^2\right)\right] \quad . \tag{6.3}$$

Here eQ is the nuclear Q-moment, $eq = V_{zz}$ is the largest principal axis of the EFG-tensor and $\eta = (V_{xx} - V_{yy})/V_{zz}$ is the asymmetry parameter. e^2qQ is the quadrupole coupling constant (QCC). By definition we have

$$|V_{xx}| \le |V_{yy}| \le |V_{zz}| \quad . \tag{6.4}$$

From (6.4) we obtain the relation $0 \le \eta \le 1$. Because of the Poisson relation $\Delta V = 0$, the EFG tensor is, apart from orientation, completely determined by the two parameters eq and η. (A possible nonvanishing charge density at the site of the nucleus would shift all m-sublevels by the same amount and is not considered here.) In the case of arbitrary orientation, the Q-Hamiltonian can be obtained by use of the rotation matrixes D^{ℓ}_{pq} (see, e.g., [6.15]). We do not perform these rotations explicitly here. Only the secular part of $\mathcal{H}_Q$ shall be specified, since it is interesting in the case where the Q-interactions can be treated as a perturbation. One obtains

$$\mathcal{H}_Q(\vartheta,\varphi)_s = \frac{e^2qQ}{4I(2I-1)}\left(\frac{3}{2}\cos^2\vartheta - \frac{1}{2} + \frac{\eta}{2}\sin^2\vartheta\cos 2\varphi\right)\left(3I_z^2 - \mathbf{I}^2\right) \quad . \tag{6.5}$$

ϑ,φ are the polar angles describing the orientation of the EFG-tensor with respect to **B**. In the case $\eta = 0$, one gets the well-known formula

$$\mathcal{H}_Q(\vartheta)_s = \frac{e^2qQ}{4I(2I-1)}\left(\frac{3}{2}\cos^2\vartheta - \frac{1}{2}\right)\left(3I_z^2 - \mathbf{I}^2\right) \quad . \tag{6.5a}$$

If $\vartheta = 54.7^\circ$, the Q-interaction vanishes in first order perturbation theory and only nondiagonal terms are left.

In order to determine the energy splitting of a certain nuclear state due to magnetic and Q-interactions, the total Hamiltonian in (6.1) has to be diagonalized. Standard computer software is available for this purpose (e.g., the FORTRAN-subroutine EIGEN [6.16]). Figure 6.1 shows for the special case $I = 2$, $\eta = 0.36$ energy level schemes for the three main orientations, where one EFG-principal axis is parallel to **B**.

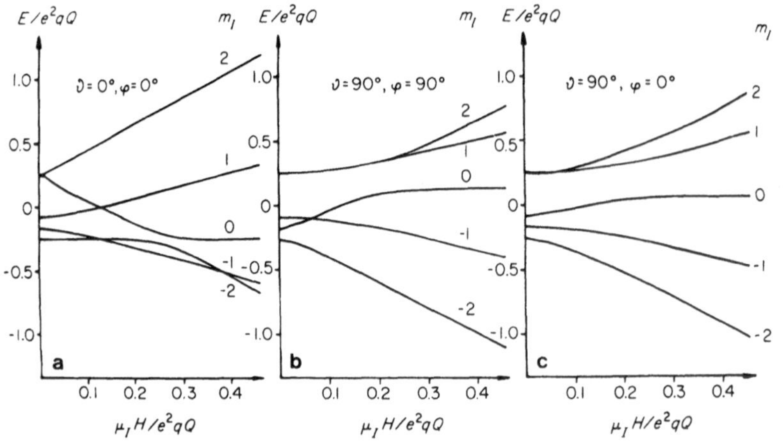

Fig.6.1. Quadrupole splitting of a state with I = 2 for the three orientations $(\vartheta,\varphi) = (0°,0°)$; $(90°,90°)$; $(90°,0°)$. These are the three cases where one EFG-principal axis is parallel to **B**. The figure shows the energies of the five m-levels as a function of $g_I\mu_N B$. Both coordinates are normalized to e^2qQ. The asymmetry parameter is $\eta = 0.36$ [6.17]

6.1.2 Reorientation in Electromagnetic Fields

The time dependence of the different nuclear orientations such as polarization, alignment, etc., can be calculated by means of the Schrödinger equation for the density matrix $\rho(t)$:

$$\dot{\rho} = \frac{i}{\hbar}[\rho, \mathcal{H}] \quad . \tag{6.6}$$

Equation (6.6) is often called the Liouville equation and is discussed in detail in [6.18]. If $\mathcal{H}$ does not depend explicitly on time, (6.6) can be integrated directly:

$$\rho(t) = \exp(-\frac{i}{\hbar}\mathcal{H}t)\rho(0)\exp(\frac{i}{\hbar}\mathcal{H}t) \quad . \tag{6.7}$$

Using the diagonal representation of $\mathcal{H}$

$$\mathcal{H} = \hbar A\Omega A^+ \tag{6.8}$$

where Ω is a diagonal matrix, we obtain from (6.7)

$$\rho(t) = A\,e^{-i\Omega t}\,A^+\rho(0)A\,e^{i\Omega t}\,A^+ \quad . \tag{6.9}$$

Equation (6.9) describes how the initial density matrix $\rho(0)$ changes under the influence of the Hamiltonian $\mathcal{H}$ in the course of time. This process is called reorientation. From (6.9) the expectation value $\langle K\rangle$ of any quantum mechanical operator K can be determined, using the relation

$$\langle K\rangle = \mathrm{Tr}_I\{K\cdot\rho\} \quad .$$

Tr_I means the trace with respect to all substates $|m_I\rangle$ of the nuclear state $|I\rangle$ in question.

The above-sketched method of solving the Liouville equation (6.6) is straightforward and can be applied in any case where the Hamiltonian $\mathcal{H}$ does not depend explictly on time. A deeper insight into the physical process can be gained, however, if the irreducible tensor formalism is used. To this end we define the irreducible tensor set $T_q^{(k)}(I)$ by its matrix elements

$$\langle Im'|T_q^{(k)}(I)|Im\rangle = (-1)^{I-m'}(ImI - m'|IIkq) \quad . \tag{6.10}$$

Here, $(ImI - m'|IIkq)$ is a Clebsch-Gordan coefficient (in the notation of EDMONDS [6.19]). The $T_q^{(k)}$ obey the following orthogonality relation:

$$Tr_I\{T_q^{(k)}T_{q'}^{(k')+}\} = \delta_{kk'}\delta_{qq'} \quad . \tag{6.11}$$

$T_q^{(1)}$ and $T_q^{(2)}$ are proportional to the irreducible components of the nuclear spin vector $\mathbf{I}$ and the Q-moment tensor, respectively:

$$I_q = \sqrt{\frac{I(I+1)(2I+1)}{3}}\; T_q^{(1)}(I) \quad , \tag{6.12}$$

where

$$I_0 = I_z \quad , \qquad I_{\pm 1} = \mp\frac{1}{\sqrt{2}} I_\pm \quad ,$$

and

$$Q_q = \sqrt{\frac{(2I+3)!}{5(2I-2)!}}\;\frac{eQ}{4I(2I-1)}\; T_q^{(2)}(I) \quad , \tag{6.13}$$

where

$$Q_0 = \frac{eQ}{2I(2I-1)}(3I_z^2 - \mathbf{I}^2) \quad , \quad Q_{\pm 1} = \mp\frac{eQ}{2I(2I-1)}\frac{\sqrt{6}}{2}(I_\pm I_z + I_z I_\pm) \quad ,$$

$$Q_{\pm 2} = \frac{eQ}{2I(2I-1)}\frac{1}{2} I_\pm^2 \quad .$$

Because of the orthogonality relation (6.11), ρ can be expanded into a multipole series

$$\rho = \sum_{q,k} \rho_q^{(k)} T_q^{(k)} \tag{6.14}$$

where

$$\rho_q^{(k)} = Tr_I\{\rho\cdot T_q^{(k)+}\} \quad . \tag{6.15}$$

The $\rho_q^{(k)}(t)$ are the so-called statistical tensors[1]. $\rho_0^{(1)}$ describes the longitudinal

1 Other authors define $\rho_q^{(k)}$ by $\rho_q^{(k)} = Tr_I\{\rho\cdot T_q^{(k)}\}$; see, e.g., [6.20]. The definition adopted here is in accordance with [6.15,18,22].

dipolar polarization, $\rho_0^{(2)}$ the longitudinal quadrupolar alignment, etc. The $\rho_q^{(k)}$ with $q \neq 0$ stand for the transversal orientations. Besides the $\rho_0^{(k)}$, a different set f_k is also used. The connection of the f_k with $\rho_0^{(k)}$ is as follows [6.21]:

$$f_k = \binom{2k}{k}^{-1} I^{-k} \sqrt{\frac{(2I + k + 1)!}{(2k + 1)(2I - k)!}} \rho_0^{(k)} \quad . \tag{6.16}$$

The Hamiltonian $\mathcal{H}$ can also be expanded into a multipole series

$$\mathcal{H} = \hbar \sum_{q,k} \Omega_q^{(k)} T_q^{(k)} \quad . \tag{6.17}$$

Normally, the series (6.17) contains only terms with k = 1 and k = 2, i.e., magnetic dipole and electric quadrupole interactions, respectively. Using (6.6,15,17) one obtains the following differential equation system for the $\rho_q^{(k)}$ [6.22]:

$$\dot{\rho}_q^{(k)} = -i \sum_{q'q''k'k''} (-1)^{2I+k} [1-(-1)^{k'+k''-k}] \sqrt{(2k'+1)(2k''+1)} \begin{Bmatrix} k'' & k' & k \\ I & I & I \end{Bmatrix}$$

$$\cdot (k''q''k'q' | k''k'kq) \Omega_{q''}^{(k'')} \rho_{q'}^{(k')} \quad . \tag{6.18}$$

The curled bracket is a Wigner 6j-symbol. Using properties of the Clebsch-Gordon coefficients and the 6j-symbols, the following simple selection rules can be derived from (6.18): The derivative $\dot{\rho}_q^{(k)}$ is coupled to the component $\rho_{q'}^{(k')}$ by the field component $\Omega_{q''}^{(k'')}$ only if

$$k' + k'' - k = \text{odd} \quad ,$$

$$|k' - k''| \leq k \leq k' + k'' \quad ,$$

$$q = q' + q'' \quad ,$$

$$k,k',k'' \leq 2I \quad . \tag{6.19}$$

These selection rules are the very success of the irreducible tensor formalism. They are not obvious in the Cartesian representation of the Liouville equation (6.6).

If one is only interested in the time average of the density matrix, e.g., over the nuclear lifetime τ_N, the differential equation system (6.6) can be transformed into an ordinary linear equation system. We take the time average of (6.6),

$$\frac{1}{\tau_N} \int_0^\infty \exp(-t/\tau_N) \dot{\rho} \, dt = \frac{1}{\tau_N} \int_0^\infty \frac{i}{\hbar} [\rho, \mathcal{H}] \exp(-t/\tau_N) \, dt \tag{6.20}$$

and obtain after a straightforward partial integration

$$\frac{1}{\tau_N} [\bar{\rho} - \rho(0)] = \frac{i}{\hbar} [\bar{\rho}, \mathcal{H}] \quad , \tag{6.21}$$

where

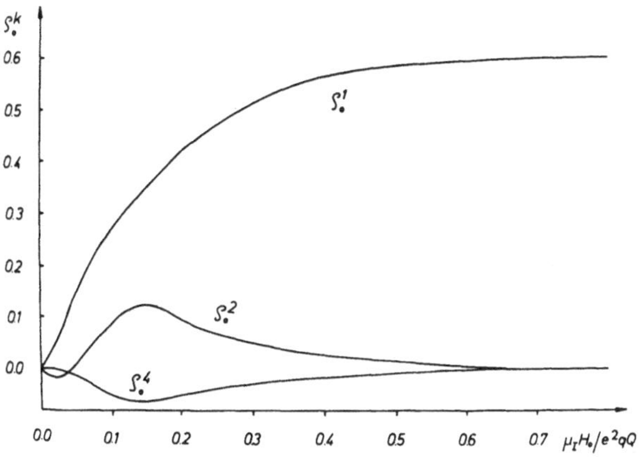

Fig.6.2. B dependence of the nuclear orientations $\rho_0^{(k)}$ (k = 1,2,4) of a I = 2 spin ensemble. The orientation of the EFG tensor is $(\vartheta,\varphi) = (90°,0°)$. $\rho_0^{(1)}$ is proportional to the nuclear polarization, $\rho_0^{(2)}$, $\rho_0^{(4)}$, ... describe the nuclear alignment. An initially purely dipolar orientation is assumed [6.17]

$$\bar{\rho} = \frac{1}{\tau_N} \int_0^\infty \exp(-t/\tau_N)\rho \, dt \quad . \tag{6.22}$$

The same procedure may be performed with (6.18). Whereas the solving of the differential equation system (6.6) or (6.18) requires the diagonalization of a matrix, only matrix inversion is necessary to solve the linear equation system (6.21). Figure 6.2 shows the B-dependence of $\bar{\rho}_0^{(1)}$, $\bar{\rho}_0^{(2)}$, $\bar{\rho}_0^{(4)}$, if magnetic and electric interactions are present simultaneously. An initially purely dipolar orientation $\rho_0^{(1)}(0)$ is assumed. At low inductions the different degrees of orientation are mixed by the Q-interaction, whereas at high inductions the Q-interaction is decoupled, and only the initial orientation $\rho_0^{(1)}(0)$ is observed.

Now the influence of a magnetic RF field on the density matrix shall be studied. The Hamiltonian describing an oscillating RF field along the x-axis of the coordinate system is given by

$$\mathcal{H}_1(t) = -2\hbar\omega_1 I_x \cos\omega t \quad , \tag{6.23}$$

where

$$\omega_1 = g_I \cdot \mu_N \cdot B_1/\hbar \quad .$$

$2B_1$ is the amplitude of the oscillating RF induction. The Liouville equation of the system now reads

$$\rho = \frac{i}{\hbar} [\rho, \mathcal{H} + \mathcal{H}_1(t)] \quad , \tag{6.24}$$

where $\mathcal{H}$ is the static Hamiltonian of the system. Since now the Hamiltonian depends explicitly on time, a direct solution of (6.24) is not possible. If $\mathcal{H}$ commutes with I_z, the standard procedure is the transformation to a rotating coordinate system. To this end we define

$$\rho_R = \exp(-i\omega I_z t)\rho \exp(i\omega I_z t) \quad . \tag{6.25}$$

ρ_R obeys the Liouville equation

$$\dot{\rho}_R = i\left[\rho_R, \frac{1}{\hbar}\mathcal{H} + \omega I_z - \omega_1 I_x\right] \quad . \tag{6.26}$$

In (6.26) the counter rotating part of the RF induction has been neglected. The Hamiltonian involved in (6.26) is independent of time, and the solving procedures described above can be applied.

If $\mathcal{H}$ does not commute with I_z, this method breaks down and the Liouville equation (6.24) can be solved only by perturbation methods.

6.1.3 Relaxation

In the last section the reorientation of the density matrix under the influence of static and RF interactions was studied. Now we want to discuss how the original nuclear orientation is lost by spin-lattice relaxation. We restrict ourselves to a listing of those facts which are needed in the discussion of the experimental results. We again use the irreducible tensor formalism and follow the presentation of HARTMANN-BOUTRON [6.20].

The evolution of a nuclear spin ensemble I due to spin-lattice interactions is described by the Liouville equation

$$\dot{\rho}(I,L) = \frac{i}{\hbar}\left[\rho(I,L), \mathcal{H}_I + \mathcal{H}_{IL} + \mathcal{H}_L\right] \quad . \tag{6.27}$$

$\rho(I,L)$ depends both on spin and lattice variables. $\mathcal{H}_I$ describes the interaction of the spin I with external fields. We assume for simplicity that $\mathcal{H}_I$ is given by the Zeeman interaction alone,

$$\mathcal{H}_I = \mathcal{H}_D = -\hbar\omega_I I_z \quad . \tag{6.28}$$

$\mathcal{H}_L$ is the lattice Hamiltonian and contains all interactions which do not involve I. $\mathcal{H}_{IL}$ describes the spin-lattice interactions. $\mathcal{H}_{IL}$ shall be given only by magnetic dipole or electric quadrupole interactions. It can be written analogously to (6.17) as

$$\mathcal{H}_{IL} = \hbar\alpha \sum_p F_p^{(\ell)}(L)T_p^{(\ell)}(I) \quad , \tag{6.29}$$

where $\ell = 1$ stands for magnetic dipole and $\ell = 2$ for electric quadrupole interactions, α is a prefactor containing nuclear moments, spin factors, etc. The $F_p^{(\ell)}(L)$ depend on the lattice variables alone. If $\mathcal{H}_{IL}$ is treated in second-order perturbation theory and if the lattice variables being unobservable are averaged by forming the trace, we obtain from (6.27) the following master equation [6.12,20]

$$\dot{\rho}(I) = \frac{i}{\hbar}\left[\rho(I), \mathcal{H}_I\right] - \frac{\alpha^2}{2}\sum_p j_p(p\omega_I)\left[\left[\rho(I), T_p^{(\ell)}(I)\right], T_p^{(\ell)}(I)^+\right] \quad , \tag{6.30}$$

where $\rho(I)$ now depends on the spin variables alone. $j_p(\omega)$ is called spectral density and is the Fourier transform of the correlation function $g_p(\tau)$:

$$j_p(\omega) = \int_0^\infty g_p(\tau)\, e^{-i\omega\tau} d\tau \quad , \tag{6.31}$$

where $g_p(\tau)$ is defined as

$$g_p(\tau) = \frac{\frac{1}{2}\,\mathrm{Tr}_L\{\exp(-\mathcal{H}_L/k_BT)[F_p^{(\ell)}(\tau)F_p^{(\ell)}(0)^+ + F_p^{(\ell)}(0)^+F_p^{(\ell)}(\tau)]\}}{\mathrm{Tr}_L\ \exp(-\mathcal{H}_L/k_BT)} \tag{6.32}$$

or, more concisely,

$$g_p(\tau) = \overline{F_p^{(\ell)}(\tau)F_p^{(\ell)}(0)^+} \quad , \tag{6.33}$$

where the bar denotes averaging over the lattice variables. $F_p^{(\ell)}(\tau)$ is defined by

$$F_p^{(\ell)}(\tau) = \exp\left(\frac{i}{\hbar}\mathcal{H}_L\tau\right)F_p^{(\ell)}\exp\left(-\frac{i}{\hbar}\mathcal{H}_L\tau\right) \quad . \tag{6.34}$$

in many cases $g_p(\tau)$ is independent of p and depends exponentially on τ [6.23]:

$$g_p(\tau) = g(0)\exp(-|\tau|/\tau_c) \quad . \tag{6.35}$$

τ_c is called correlation time. The corresponding spectral density is

$$j_p(\omega) = g(0)\frac{2\tau_c}{1 + \omega^2\tau_c^2} \quad . \tag{6.36}$$

In the two limiting cases $\omega\tau_c \ll 1$ and $\omega\tau_c \gg 1$, we get from (6.36)

$$j_p(\omega) \propto \begin{cases} \tau_c & \text{if} \quad \omega\tau_c \ll 1 \qquad (6.37a) \\ 1/(\omega^2\tau_c) & \text{if} \quad \omega\tau_c \gg 1 \quad . \qquad (6.37b) \end{cases}$$

In the first case, $j_p(\omega)$ depends neither on p nor on ω. This situation is called "the extreme narrowing case".

If $\rho(I)$ is expanded into a multipole series, we obtain from (6.30) the following equation system for the $\rho_q^{(k)}$:

$$\dot{\rho}_q^{(k)} = i\omega_I q\rho_q^{(k)} - \frac{\alpha^2}{2}\sum_{p,k'} j_p(p\omega_I)\,\mathrm{Tr}_L\left\{\left[T_q^{(k)+},T_p^{(\ell)}\right]\left[T_p^{(\ell)+},T_q^{(k)}\right]\right\} \quad . \tag{6.38}$$

In the most general case (6.38) is a coupled differential equation system. In some important special cases, however, the equations for the different $\rho_q^{(k)}$ are decoupled:

i) in the case of relaxation due to magnetic dipole-dipole interactions. If

$$\mathcal{H}_{IL} = \hbar\gamma_I\sum_p B_q^* I_q = \hbar\alpha\sum_q B_q^* T_q^{(1)} \quad , \tag{6.39}$$

$$\alpha = \gamma_I \sqrt{\frac{I(I + 1)(2I + 1)}{3}}$$

where the B_q are the irreducible components of the magnetic inductions produced by the neighbouring nuclei at the site of the I spin, we obtain from (6.38)

$$\dot{\rho}_q^{(k)} = (i\omega_I q - \lambda_q^{(k)})\rho_q^{(k)} \quad , \tag{6.40}$$

where

$$\lambda_q^{(k)} = \frac{\gamma_I^2}{2} [j_0(0)q^2 + j_1(\omega_I)(k(k + 1) - q^2)] \quad .$$

ii) In the extreme narrowing case [6.24]. Then we have $j_p(\omega) = j(0)$ for all p,ω and one obtains from (6.38) and (6.10), using relations of the Clebsch-Gordan coefficients.

$$\dot{\rho}_q^{(k)} = (i\omega_I q - \lambda^{(k)})\rho_q^{(k)} \quad , \tag{6.41}$$

where

$$\lambda^{(k)} = \alpha^2 j(0)(2\ell + 1) \left[\frac{1}{2I + 1} - (-1)^{2I+\ell+k} \left\{ \begin{matrix} I\ell I \\ IkI \end{matrix} \right\} \right] \quad .$$

In the case of magnetic dipole interactions (6.39), i.e., $\ell = 1$, we obtain

$$\lambda^{(k)} = j(0)\gamma_I^2 \frac{k(k + 1)}{2} \quad , \tag{6.42}$$

which also follows directly from (6.40). In the case of electric interactions, i.e., $\ell = 2$,

$$\begin{aligned} \mathcal{H}_{IL} &= \sum_q V_q^* Q_q = \hbar\alpha \sum_q V_q^* T_q^{(2)} \quad , \\ \alpha &= \frac{eQ}{\hbar} \frac{1}{4I(2I - 1)} \sqrt{\frac{(2I + 3)!}{(2I - 2)!5}} \quad , \end{aligned} \tag{6.43}$$

where the V_q are the components of the EFG-tensor, we get

$$\lambda^{(k)} = \frac{3}{8} j(0)\left(\frac{eQ}{\hbar}\right)^2 \frac{k(k + 1)[4I(I + 1) - k(k + 1) - 1]}{I^2(2I - 1)^2} \quad . \tag{6.44}$$

In the cases (i) and (ii), each component $\rho_q^{(k)}$ decays exponentially with its own decay constant $\lambda_q^{(k)}$. The reciprocals of $\lambda_0^{(k)}$ and $\lambda_1^{(k)}$ are called $T_1^{(k)}$ and $T_2^{(k)}$, the longitudinal and transversal relaxation times of kth multipole order, respectively.

In the extreme narrowing case, only one decay constant $\lambda^{(k)}$ for each tensor degree exists, i.e., $T_1^{(k)} = T_2^{(k)}$. As will be pointed out in Sect.6.2, in the experiments using β-emitters only $\rho_0^{(1)}$ is observed, and we have $T_1^{(1)} = T_1$ which is the spin-lattice relaxation time of classical NMR. In the case of γ-isomers, $k = 2$ usually holds and one observes the second-rank relaxation time $T_1^{(2)}$. We shall denote the relaxation times with the index "D" or "Q" in order to distinguish between relaxation processes induced by dipolar and quadrupolar interactions.

Equation (6.38) and its special cases (6.40,41) have been derived mainly by means of symmetry consideration. The relevant physical information is contained in the spectral densities $j_q(\omega)$. We are now going to discuss some special experimental situations, where the spectral densities can be specified in a more detailed way.

a) *Nuclear Relaxation by Magnetic Coupling to Conduction Electrons*

This type of relaxation usually dominates in metals and is caused by the contact interaction of the nuclear spins with the spins of the conduction electrons. Because of the short correlation times of the fluctuating electronic spins, the extreme narrowing case holds and we obtain as a special case of (6.42) for the relaxation rate $1/T_{1e}^{(k)}$ due to the conduction electrons [Ref.6.12,Chap.IX A]:

$$1/T_{1e}^{(k)} = \frac{k(k+1)}{2}\,\frac{64\pi^3}{9}\,\gamma_I^2\gamma_e^2\hbar^3\,|\langle|\psi(0)|^2\rangle_F|^2\rho^2(E_F)k_BT\ . \tag{6.45}$$

γ_e is the gyromagnetic ratio of the electron, $\langle|\psi(0)|^2\rangle_F$ the probability density of the conduction electrons at a nuclear site averaged over the Fermi surface, and $\rho(E_F)$ the density of states at the Fermi level.

$T_{1e}^{(k)}$ is related to the Knight shift K_s by the Korringa relation

$$T_{1e}^{(k)}TK_s^2 = \frac{2}{k(k+1)}\,\frac{\hbar}{4\pi k_B}\left(\frac{\gamma_e}{\gamma_I}\right)^2\frac{1}{K(\alpha)} \tag{6.45a}$$

which is modified here by a correction factor $K(\alpha)$ [6.25]. $K(\alpha)$ accounts for the electron-electron interaction (measured by α) and normally takes values in the range 0.6...0.9 in pure metals [6.26].

b) *Relaxation by Fluctuating Nuclear Dipole-Dipole Interactions*

This case is important if the relaxation is caused by motion of the lattice atoms. The dipole-dipole interaction between the relaxing I-spin and the S^i-spins of the surrounding nuclei can be written in the form (6.39) [Ref.6.12,Chap.VIII E] where the components B_q are

$$\begin{aligned} B_0 &= \hbar\gamma_S\sum_i\left(S_0^iF_i^{(0)} + \frac{3}{2}\sqrt{2}\,S_1^iF_i^{(1)} - \frac{3}{2}\sqrt{2}\,S_{-1}^iF_i^{(-1)}\right)\ ,\\ B_{\pm1} &= \hbar\gamma_S\sum_i\left(S_{\pm1}^iF_i^{(0)} \pm \frac{3}{2}\sqrt{2}\,S_0^iF_i^{(\mp1)} + \frac{3}{2}\,S_{\mp1}^iF_i^{(\mp2)}\right)\ . \end{aligned} \tag{6.46}$$

With

$$F_i^{(0)} = \frac{1-3\cos^2\vartheta_i}{|r_i|^3}\ ,\qquad F_i^{(\pm1)} = \frac{\sin\vartheta_i\,\cos\vartheta_i\,e^{\mp i\varphi_i}}{|r_i|^3}\ ,$$

$$F_i^{(\pm 2)} = \frac{\sin^2\vartheta_i \; e^{\mp 2i\varphi_i}}{|r_i|^3}$$

$\mathbf{r}_i = (r_i, \vartheta_i, \varphi_i)$ is the vector pointing from the I spin site to the S^i spin site. As we are concerned here with the relaxation of radioactive probes, we assume that all S^i spins are different from the I spin. For simplicity, all S^i spins are assumed to be equal. Defining the new correlation functions

$$G_i(t) = \overline{F^{(i)}(t)F^{(i)}(0)^+} \tag{6.47}$$

and the corresponding spectral densities $\mathcal{J}_i(\omega)$, one obtains from (6.40,46), if we specialize to the case q = 0,

$$1/T_{1D}^{(k)} = \frac{k(k+1)}{2} \frac{\gamma_I^2\gamma_S^2\hbar^2 S(S+1)}{3} \frac{1}{4}\left\{\mathcal{J}_0(\omega_I - \omega_S) + \frac{9}{2}\mathcal{J}_1(\omega_I) + \frac{9}{4}\mathcal{J}_2(\omega_I + \omega_S)\right\} \quad . \tag{6.48}$$

It should be noted that the $F_i^{(q)}$ defined in (6.46) are not properly normalized and therefore do *not* form the components of an irreducible tensor. We have adopted here this somewhat unsuitable notation as it is widely used in the literature. As a consequence, the $\mathcal{J}_q$ are not equal in the extreme narrowing case but obey the relation $\mathcal{J}_0 : \mathcal{J}_1 : \mathcal{J}_2 = 6:1:4$. The detailed form of the $\mathcal{J}_q$ depends on the diffusion process being involved (Sect.6.3.2a).

c) *Quadrupolar Relaxation Induced by Atomic Motion*

The motion of impurities in a crystal lattice causes fluctuating EFG's which give rise to relaxation. In this case the differential equation system (6.38) is not decoupled in general, and the relaxation is described by a superposition of several exponentials. If this process is studied in liquids, however, we again have the extreme narrowing case, and (6.44) is valid. The spectral density j(0) can be expressed in terms of the dynamic structure factor (scattering function) $S(q,\omega)$ which can be measured in neutron scattering experiments. Under certain approximations not discussed here (see e.g., [6.27-29]), (6.44) reads

$$\frac{1}{T_{1Q}^{(k)}} = \frac{3}{8}\left(\frac{eQ}{\hbar}\right)^2 \frac{k(k+1)[4I(I+1) - k(k+1) - 1]}{I^2(2I-1)^2} \, n \int_0^\infty S(q,0)I(q)dq \quad . \tag{6.49}$$

Here n is the number density, S(q,0) is the dynamic structure factor in the limit $\omega = 0$, and I(q) is a function depending on the EFG being produced by the impurity.

d) *Relaxation by Quadrupolar Spin-Phonon Coupling*

In this case, relaxation is accomplished by the interaction of the relaxing nuclei with fluctuating EFGs caused by lattice vibrations. This process is often the dominant one in insulators and semiconductors. The relaxation mechanism is a 2-phonon Raman process, i.e., one phonon with energy $\hbar\omega_i$ is absorbed, another one with energy $\hbar\omega_j$ is emitted and the energy difference $\hbar\omega = \hbar(\omega_i - \omega_j)$ is used to flip the nuclear spin. As the angular frequencies of the lattice vibrations ω_i, ω_j are of the order of $10^{13}\ s^{-1}$ and the nuclear Larmor frequency ω is of the order of $10^6\ s^{-1}$, the relaxation rate is practically independent of ω. Thus we have the extreme narrowing case and obtain from (6.44):

$$\frac{1}{T_{1Q}^{(k)}} = \frac{k(k+1)[4I(I+1) - k(k+1) - 1]}{2I^2(2I-1)^2}\left(\frac{eQ}{\hbar}\right)^2 F(T/\theta) \quad . \tag{6.50}$$

θ is the Debye temperature. $F(T/\theta)$ is a complicated function of temperature. In the high and low temperature limit, $F(T/\theta)$ shows the following temperature dependence [6.30]:

$$F(T/\theta) \propto \begin{cases} (T/\theta)^7 & \text{if} \quad T \lesssim 0.02\ \theta \\ (T/\theta)^2 & \phantom{\text{if}} \quad T \gtrsim 0.5\ \theta \quad . \end{cases} \tag{6.51}$$

The relaxation processes normally relevant in metals and insulators have been discussed above. Though in many cases one process is dominant, often different types of relaxation are observed simultaneously. Then the relaxation rates belonging to the different mechanisms can simply be added, provided that the processes are not correlated, as is normally the case.

6.2 Experimental Methods

All in-beam experiments on β emitters or γ isomers proceed along the same scheme: in the first step a nuclear reaction with the beam particles creates oriented, radioactive probes in the sample. In the second step the anisotropic nuclear radiation monitors any change of the probe polarization due to spin-lattice relaxation, NMR transitions, spin rotation, etc.

In this section we classify the experimental techniques according to their different production methods of the oriented probe nuclei.

6.2.1 Experiments Using β Emitters

The following methods have many common features and use partly the same nuclides. Thus we explain the principles in some detail for the neutron-activation technique and discuss later on only the modifications in the accelerator experiments.

a) *Probe Creation by Capture of Polarized Thermal Neutrons*

Thermal neutron capture ${}^{A-1}_{Z}X + n_{pol} \rightarrow {}^{A}_{Z}X^{*}_{pol}$ leads in the first instance to an excited compound state with an energy of 2 MeV for ${}^{8}Li^{*}$, 3.4 MeV for ${}^{12}B^{*}$ and 6...7 MeV for the heavier nuclides. As can be understood by coupling of angular momenta [6.31], the compound state polarization f_1^{*} is given by $(I_1^{*} + 1)/3\, I^{*}$ for $I^{*} = I_t + 1/2$, and $-1/3$ for $I^{*} = I_t - 1/2$, if unpolarized target nuclei with spin I_t capture completely polarized neutrons. Thus, the polarization f_1^{*} is high unless the contributions of both capture channels cancel accidentally. The states ${}^{A}_{Z}X^{*}_{pol}$ are purely dipolarly polarized (i.e., occupation probability depends linearly on the magnetic quantum number m) [6.32] due to the capture of thermal spin 1/2 particles. ${}^{A}_{Z}X^{*}_{pol}$ de-excites promptly by emission of several γ quanta to its ground state ${}^{A}_{Z}X_{pol}$, representing the probe (an exception is ${}^{24}Na$ where the first excited 20 ms state is used instead of the 15 h ground state). The γ cascade diminishes the degree of polarization to a value f_1^{nr}, but does not change its purely dipolar order. Usually, the *n*uclear *r*eaction produced polarization f_1^{nr} exceeds 10%. Up to now the following β-emitters have been produced by this method (half-lifes in parentheses): ${}^{8}Li$ (0.84 s), ${}^{12}B$ (20 ms), ${}^{20}F$ (11 s), ${}^{24}Na^{m}$(20 ms), ${}^{28}Al$ (2.2 m), ${}^{38}Cl$ (37 m), ${}^{66}Cu$ (5.1 m), ${}^{108}Ag$(2.4 m), ${}^{110}Ag$ (24.6 s), and ${}^{116}In$ (14 s).

Irradiation of a crystal with a polarized, thermal n-beam does not produce any detectable bulk radiation damage during the course of an experiment. However, the mechanical recoil, imparted to the activated nuclei as a consequence of the capture-γ radiation, may produce some local point defects near the probes. From the assumption that the γ cascade consists only of one transition (instead of actually 3-4 steps), one gets the upper limits for the recoil energies $E_{R\,max}$. Some examples for $E_{R\,max}$/eV are: 275 (${}^{8}Li$), 1200 (${}^{20}F$), 230 (${}^{110}Ag$), and 200 (${}^{116}In$). Partial cancellation of γ-recoil momenta causes a distribution of E_R between 0 and $E_{R\,max}$ [6.33]. But, nevertheless, nearly all probes obtain resultant recoil energies above 0.1 $E_{R\,max}$, being sufficient for a displacement in the lattice. In some cases further defect creation may be induced by internal conversion of the γ radiation leading to strongly ionized electronic shells.

The initial polarization f_1^{nr} is not preserved. It may diminish to a value $f_1(t)$ or be transferred to higher orders $f_2(t)$, $f_3(t)$,... by interactions with static or RF fields. If the probe decays by allowed β radiation, f_1 is reflected in the asymmetric angular distribution [6.21]

$$W_\beta(\vartheta) = 1 + f_1 \frac{v}{c} A \cos\vartheta \qquad (6.52)$$

(v: electron velocity, A: asymmetry coefficient, ϑ: angle between **v** and $\mathbf{f}_1$). The even orders f_2, f_4, ... can be observed in the anisotropic directional distribution of γ radiation [6.21]:

$$W_\gamma(\vartheta) = 1 + f_2 A_2 P_2(\cos\vartheta) + f_4 A_4 P_4(\cos\vartheta) + \ldots \qquad (6.53)$$

(P_2,P_4, ... Legendre polynomials; A_2,A_4, ... anisotropy coefficients). In some cases the β transition is followed in the daughter nuclide by a γ decay which is suited to measure $W_\gamma(\vartheta)$. Examples are $^{20}F \xrightarrow{\beta^-} {}^{20}Ne^* \xrightarrow{\gamma} {}^{20}Ne$ and $^{28}Al \xrightarrow{\beta^-} {}^{28}Si^* \xrightarrow{\gamma} {}^{28}Si$

The reorientation mechanisms are discussed in detail in Sect.6.1.2. We mention here only three facts following from the selection rules (6.19): (i) magnetic-dipole coupling does not change the order f_1 of the probe polarization; (ii) electric-quadrupole coupling can convert f_1 to an alignment f_2 (and to any higher order $\leq 2I$); (iii) the sign of f_2 depends on the signs of f_1 and e^2qQ and can thus be used to determine the sign of the QCC of a quadrupolar reorientation.

Figure 6.3 shows an experimental set-up. Thermal neutrons from a neutron guide (10^9 neutrons/cm^{2}s) are polarized by mirror reflection (5.10^7 polarized neutrons/cm^{2}s at the target) with a degree above 80%. The β particles from the activated

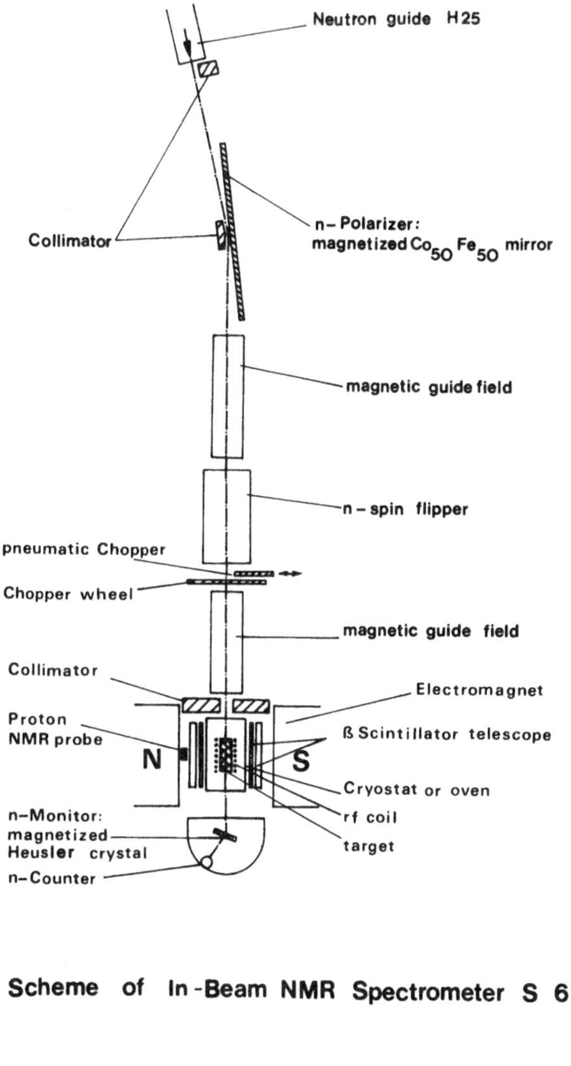

Fig.6.3. Experimental set-up of the in-beam neutron-activation NMR spectrometer S6 at the High Flux Reactor, Grenoble

sample are registered in two double coincidence telescopes mounted between sample holder and pole face. In the case of additional counting of γ particles, one NaJ(Tl) scintillator is mounted below the sample and another one between one of the β telescopes and the adjacent pole face. Some experimental details are described in [6.6,11,17].

The nuclear polarization f_1 is measured by observation of the asymmetry a(t) of the β counting rates Z in the two telescopes:

$$f_1 \propto a(t) = [Z(0^\circ) - Z(180^\circ)]/[Z(0^\circ) + Z(180^\circ)] \quad . \tag{6.54}$$

The experimentally observed value for a is by a factor of 5-10 smaller than the value $f_1 vA/c$ expected from (6.52). The reasons are backscattering of the electrons in the sample, background counting rates and large solid angles for β detection due to focusing in the magnetic field. On the other hand, usually the n polarization is periodically reversed in order to avoid instrumental asymmetries. This procedure doubles a(t).

The β-decay asymmetry can be observed in a time-differential or in a stationary way. In the case of spin-lattice relaxation with a single relaxation time $T_1(\equiv T_1^{(1)})$, as defined in Sect.6.1.3), a(t) decreases after pulsed n activation according to

$$a(t) = a(0)\ e^{-t/T_1} \tag{6.55}$$

[a(0) is the asymmetry at the end of the activation pulse]. For continuous activation the average asymmetry is given by

$$a = \bar{a} = \frac{1}{\tau_\beta}\int_0^\infty a^0\ e^{-t/T_1}\ e^{-t/\tau_\beta}\ dt = \frac{a^0}{1 + \tau_\beta/T_1} \quad , \tag{6.56}$$

where a^0 is the β-decay asymmetry prior to the onset of spin-lattice relaxation.

From (6.55) and/or (6.56), T_1 can be determined without the application of any RF field. In order to detect NMR signals, one measures $\bar{a}$ as a function of the frequency of the RF field.

In (6.53) we discussed the anisotropy of γ radiation following the β decay in some cases. If alignments $f_2, f_4, \ldots$ have been produced by reorientation, they can be monitored by the 0°-90° γ-decay anisotropy

$$a_\gamma(t) = [Z_\gamma(0^\circ) - Z_\gamma(90^\circ)]/[Z_\gamma(0^\circ) + Z_\gamma(90^\circ)] \quad . \tag{6.57}$$

If only f_2 and f_4 are considered, $a_\gamma(t)$ is proportional to $(-12A_2f_2-5A_4f_4)/(16 + 4A_2f_2 + 11A_4f_4) \approx -(3/4)A_2f_2-(5/16)A_4f_4$.

b) *Probe Creation by Fast Particle Reactions with Selected Recoil Angle*

The experiments we are going to discuss in the following two sections differ from the neutron-activation technique described above mainly regarding the creation of the probes. β-ray detection and measurement procedures are quite similar. The method is best explained by means of Fig.6.4 published by SUGIMOTO et al. [6.34]. In this example a fast ^{2}H beam hits a thin SiO_2 target and produces ^{17}F by the reaction $^{16}O(d,n)\,^{17}F$. After having passed the target, only those ^{17}F ions ejected in a preselected solid angle are stopped in the sample. The polarization of the recoiled nuclei is normal to the reaction plane and depends on beam energy and recoil angle. This behaviour was studied in detail for the case $^{11}B(d,p)^{12}B$ by PFEIFFER and MADANSKI [6.35].

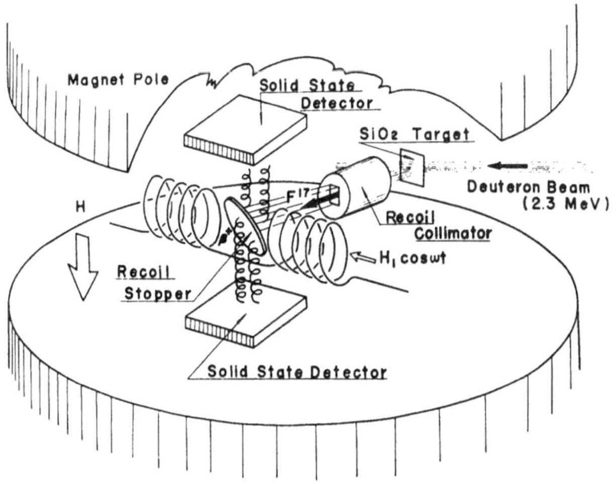

Fig.6.4. Experimental set-up for NMR measurements on ^{17}F produced via the reaction $^{16}O\,(d,n)^{17}F$ [6.34]

The following reactions have been used to create polarized β emitters (half-lives in parentheses after the probe nuclide):

$^{7}Li(d,p)^{8}Li$ (0.84 s), $^{6}Li(^{3}He,n)^{8}B$(0.76 s),
$^{11}B(d,p)^{12}B$(20 ms), $^{11}B(t,p)^{13}B$(17 ms),
$^{10}B(^{3}He,n)^{12}N$(11 ms), $^{16}O(d,n)^{17}F$(64 s),
$^{28}Si(d,n)^{29}P$(4.2 s), $^{40}Ca(d,n)^{41}Sc$(0.6 s)
$^{100}Mo(^{14}N,^{12}B$(20 ms)), $^{232}Th(^{14}N,^{12}B$(20 ms)),
$^{27}Al(^{14}N,^{12}B$(20 ms)), $^{nat}Cu(^{14}N,^{12}B$(20 ms)).

In all these cases the experimental principles are quite similar to the example explained above. The four last mentioned reactions are, however, an exception insofar as the ^{12}B probes are created as recoil nuclei in heavy-ion reactions.

Compared with the neutron-activation method, a greater number of β active nuclides is accessible. Furthermore, the stopper (sample) and target can be chosen independently. On the other hand, one is restricted to the light nuclides which can leave the target and one runs a much higher risk of producing radiation damage in the sample due to the about one thousand times higher recoil energy. Thus, for example, in the ^{17}F experiment shown in Fig.6.4, the CaF_2 stopper had to be exchanged after 1-4 hours of normal use at room temperature (RT) because of radiation damage.

c) *Probe Creation by Reactions with Polarized Fast Particles*

The experimental technique to be described now uses *polarized* fast beams of protons or deuterons to create β active nuclides which then are also polarized irrespective of the recoil angle. Thus, the target and stopper can be combined into one thick sample. The polarization transfer can be positive or negative and its degree depends strongly on the beam energy, as was demonstrated by MINAMISONO et al. [6.36] for the reaction ^{31}P $(\vec{p},n)^{31}$S.

Compared with the recoil-selection method of Sect.6.2.1b, the counting rates are much higher. Two disadvantages are the higher risk of radiation damage, as the primary beam now bombards the target, and the loss of the possibility of choosing the sample independently of the target. The experimental arrangements are similar to those shown in Fig.6.4 with the difference that the beam hits the stopper directly.

This technique has been applied up to now in most cases at the Stanford University (USA) using the following reactions (half-lives in parentheses)[2]:

^{7}Li$(\vec{d},p)^{8}$Li(0.84 s) ^{11}B$(\vec{d},p)^{12}$B(20 ms),
^{16}O$(\vec{d},n)^{17}$F(64 s), ^{19}F$(\vec{d},p)^{20}$F(11 s),
^{28}Si$(\vec{p},\alpha)^{25}$Al(7.2 s), ^{28}Si$(\vec{d},n)^{29}$P(4.2 s),
^{31}P$(\vec{p},n)^{31}$S(2.6 s), ^{39}K$(\vec{p},n)^{39}$Ca(0.86 s),
^{40}Ca$(\vec{d},n)^{41}$Sc(0.6 s).

One further reaction, ^{28}Si$(\vec{p},n)^{28}$P(0.27 s), has been reported by the Osaka group [6.176]. At the Heidelberg EN-Tandem polarized ^{7}Li ions have been applied in the reactions ^{7}Li$(\overrightarrow{^{7}Li}$, ^{8}Li(0.84 s)), ^{7}Li$(\overrightarrow{^{7}Li}$, ^{12}B(20 ms)) and ^{9}Be$(\overrightarrow{^{7}Li}$, ^{8}Li(0.84 s)). The polarized ^{8}Li and ^{12}B nuclei were created in thin targets and stopped in gold backings [6.177].

d) *Probe Creation by Fast Particle Reactions and Subsequent Polarization in the Stopper*

The one experiment to be reported here deviates from the common scheme of the experiments of Sect.6.2 insofar as creation and polarization of the probes proceeds in two separate steps. ^{28}Al nuclei, produced in the reaction ^{27}Al(d,p), were ejected

2 References not listed in Sect.6.5 were taken from [6.178,179].

from a thin Al target and implanted into a metallic Li stopper at 1.5-4.2 K. By means of the Overhauser effect the ^{28}Al nuclei could be polarized through contact interaction with the Li conduction electrons. The method requires the application of a RF field saturating the spin resonance of the conduction electrons. Besides ^{28}Al, the β emitters ^{20}F, ^{25}Al, ^{30}P were polarized by this method [6.149,180].

6.2.2 Experiments Using Isomeric γ Emitters

The two preceding articles of this volume dealt with the static hfi of γ emitters in solids. There the γ-PAD methods are explained in detail. In this review the spin-lattice relaxation experiments using isomeric nuclei ($\tau_N \approx 10^{-5}...10^{-2}$ s), performed almost exclusively in liquid metals, remain to be discussed. The standard technique for this is the pulsed-beam time-differential spin rotation (TDPAD). Additionally, in some cases γ-ray detected nuclear magnetic resonance (γ-NMR-PAD) has been applied.

In the spin-rotation technique the relaxation rate is obtained from the damping of the modulated γ-ray anisotropy in an external B field perpendicular to the reaction plane[3]. All experiments to be reported here can be described using only the multipole order k = 2. Under further restrictions to the extreme narrowing case (Sect.6.1.3) valid in liquid metals, the γ-ray intensity at an angle ϑ with respect to the beam axis is given by [6.37]

$$I(\vartheta,t) = I_0\, e^{-t/\tau_N} \left\{1 + A_2\, e^{-\lambda^{(2)}t} \left[\frac{1}{4} + \frac{3}{4}\cos 2\,(\vartheta - \omega_L t)\right]\right\} \quad . \tag{6.58}$$

From the fit of (6.58) to the experimental signal, one obtains the Larmor frequency ω_L and the relaxation rate $\lambda^{(2)} = 1/T_1^{(2)}$.

The following reactions have been used to create aligned isomeric states in the TDPAD experiments (half-lives in parentheses):

^{71}Ga(p,n)^{71}Ge(20 ms), ^{85}Rb(α,n)^{88}Y(14 ms),
^{113}In(α,3n)^{114}Sb(0.21 ms), ^{114}Sn(d,2n)^{114}Sb(0.21 ms),
^{115}In(p,n)^{115}Sn(0.16 ms), ^{115}In(α,2n)^{117}Sb(0.34 ms),
^{118}Sn(d,3n)^{117}Sb(0.34 ms), ^{130}Te(α,2n)^{132}Xe(8.4 ms),
^{124}Sn(^{16}O,4n)^{136}Ce(2 μs), ^{138}Ba(α,4n)^{138}Ce(0.08 μs),
^{138}Ba(α,3n)^{139}Ce(0.07 μs), ^{139}La(^{12}C,4n)^{147}Eu(0.77 μs),
^{139}La(^{12}C,3n)^{148}Eu(0.17 μs), ^{122}Sn(^{20}Ne,4n)^{138}Nd(0.35 μs),
^{139}La(α,4n)^{139}Pr(0.04 μs), ^{202}Hg(d,2n)^{202}Tl(0.57 ms),
^{204}Hg(α,3n)^{205}Pb(5.5 ms), ^{204}Hg(α,2n)^{206}Pb(0.12 ms),
^{205}Tl(α,2n)^{207}Bi(0.18 ms), ^{207}Pb(d,2n)^{207}Bi(0.18 ms),
^{206}Pb(α,3n)^{207}Po(47 μs), ^{209}Bi(d,4n)^{207}Po(47 μs),

3 The spin rotation technique is explained in Chap.1.

$^{203}Tl(^{13}C,4n)^{212}Fr(27\ \mu s)$, $^{205}Tl(^{12}C,4n)^{213}Fr(0.24\ \mu s)$,
$^{206}Pb(^{12}C,4n)^{214}Ra(67\ \mu s)$.

In the γ-NMR-PAD method, a RF field produces NMR transitions in the isomeric state which are detected through their effect on the angular distribution of the γ rays. The combination of pulsed-beam excitation with RF irradiation gives rise to a number of interesting resonance phenomena, those being functions of the timing conditions, of the phase between RF field and isomer state creation, etc. [6.38]. The study of these phenomena as such and the determination of magnetic moments have been the main aim of the γ-NMR-PAD experiments up to now. Further, some spin-lattice relaxation rates were determined by this method.

The following reactions have been used to create aligned isomeric states in the γ-NMR experiments (half-lives in parentheses)

$^{71}Ga(p,n)^{71}Ge(20\ ms)$, $^{71}Ga(\alpha,2n)^{73}As(5.8\ \mu s)$,
$^{78}Se(p,n)^{78}Br(0.12\ ms)$, $^{80}Se(d,n)^{81}Br(37\ \mu s)$,
$^{115}In(p,n)^{115}Sn(0.16\ ms)$, $^{205}Tl(\alpha,3n)^{206}Bi(0.88\ ms)$,
$^{208}Pb(d,2n)^{208}Bi(2.5\ ms)$, $^{206}Pb(^{12}C,4n)^{214}Ra(67\ \mu s)$.

6.3 Metals

The number of β emitting nuclides which can be used as probes in metals is restricted, because in most cases they lose their polarization quickly compared with the nuclear lifetime τ_N for $T \gtrsim 1$ K due to the interaction of their magnetic moments with the conduction electrons. Exceptions are some light nuclides such as ^{8}Li (0.84 s), ^{8}B (0.76 s), ^{12}B (20 ms), ^{13}B (17 ms), ^{12}N(11 ms), $^{24}Na^{m}$ (20 ms) and ^{41}Sc (0.6 s). Implantation or creation of these light nuclides in the target does not, in general, damage the lattice too strongly and the measurements are not seriously disturbed. In particular, ^{8}Li, ^{12}B, and ^{12}N represent excellent probes which have been applied in many experiments.

The isomeric γ emitters used in the in-beam experiments with $\tau_N \approx 10^{-5}...10^{-2}$ s fulfill the condition $T_1^{(2)} \gtrsim \tau_N$ in a broad T range if only relaxation caused by the conduction electrons is considered. However, these heavier nuclides produce many recoil defects which depolarize the probes in solids by quadrupolar interaction within about 1 μs. Therefore most in-beam experiments on γ isomers were performed in liquid metals where no static quadrupole interaction was present. Here, the γ isomers are ideal probes for relaxation studies as $T_1^{(2)}$ is often comparable in magnitude to τ_N in the liquid host.

Very recently RIEGEL et al. [6.181-184] started TDPAD measurements on the rare earth ions Ce, Pr, Nd, and Eu implanted in solid and liquid metals. The observed nuclear relaxation times give insight into the 4f spin dynamics of the rare earth impurities. The nuclear lifetimes of the applied rare earth nuclides lie outside the range to be considered in this chapter. The experiments are listed, however, in lines 99a-99ℓ of Sect.6.5.

6.3.1 Static Interactions, NMR Spectra

a) *Magnetic Energy Splitting*

First we look at the γ-NMR experiments and discuss subsequently the β-NMR measurements.

The γ-NMR experiments on isomeric nuclides in *liquid* metals were mainly devoted to the determination of magnetic dipole moments and to the investigation of sophisticated resonance phenomena per se. These experiments are listed in lines 77,79-81, 89,103,105 of Sect.6.5. The only static interaction which played a role was the magnetic interaction with the external B field. The same was true for the only γ-NMR application in a *solid* host. In this exceptional case, ^{214}Ra(67 μs,8^+) retains its polarization in solid Pb during its lifetime [Sect.6.3.2a(ii)]. The reason may be that the 8^+ state of ^{214}Ra has a very small quadrupole moment and the electric interaction with radiation defects is not effective [6.39].

We are now going to discuss the experiments which used the β-NMR method. In all cases the probes were created by fast particle reactions.

First, there is a number of measurements, where again the static interaction of the nuclear dipoles with the external B field was observed. The aim was mainly to detect a nuclear polarization or to determine dipole moments (lines 14,16-21,23, 43,44,46,66,76 of Sect.6.5). The NMR signals of ^{8}Li, ^{12,13}B and ^{12}N have been observed in several host materials for each probe nuclide which resulted in Knight shift determinations.

The observation of undisturbed NMR signals in various solid metals and semiconductors proved that all, or a considerable part of, the implanted probes reach undisturbed sites with cubic crystal symmetry.

More information on the probe sites could be obtained by comparing the experimental line widths with calculated dipolar line broadenings which depend on the probe site and on the orientation of the crystal relative to **B**. Using this method, SUGIMOTO et al. concluded that implanted ^{12}B and ^{12}N nuclei reach undisturbed interstitial sites in the fcc metals Al,Cu,Pt,Au and in Al,Cu,Pt, respectively [6.51,52].

There are two types of cubic interstitial sites in fcc metals having octahedral or tetrahedral symmetry. McNAB and McDONALD [6.56] found from detailed investigations on linewidths and NMR signal heights that at 77 K, 90-100% and 76-100% of the implanted ^{12}B probes occupied the octahedral interstitial sites in single crystals

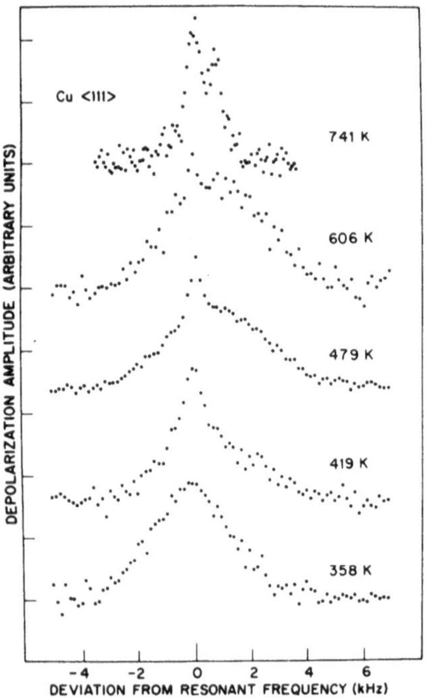

Fig.6.5. NMR curves at various temperatures of ^{12}B in a Cu single crystal oriented <111>$\|$**B**. B = 0.5 T, resonance frequency $\nu_L \approx 3.8$ MHz [6.57]

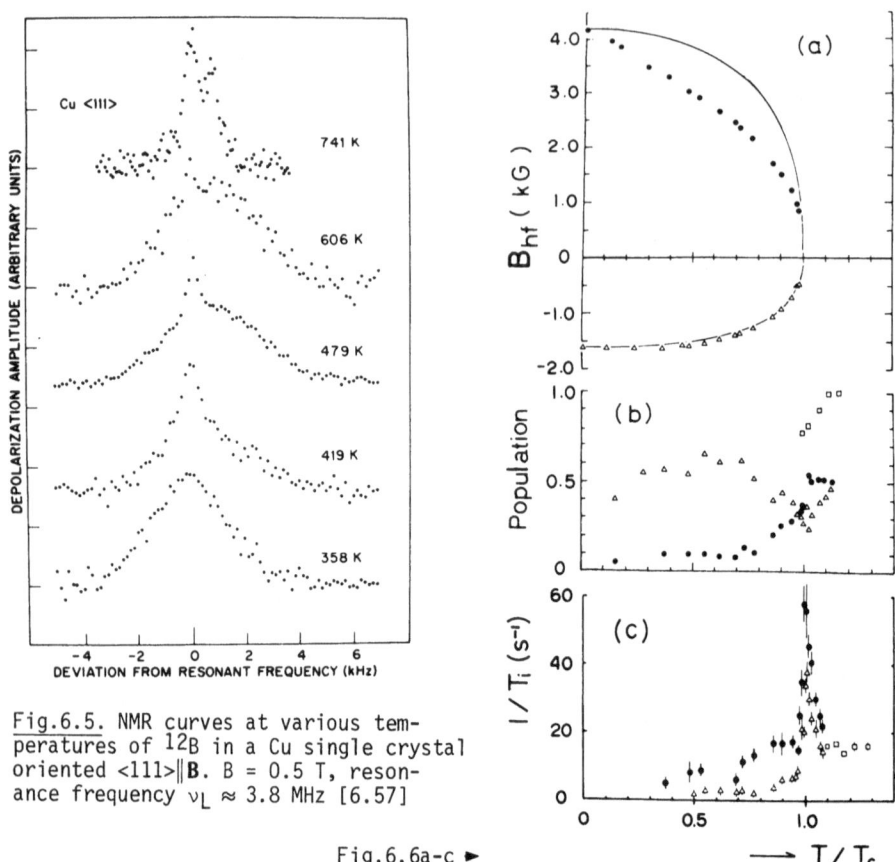

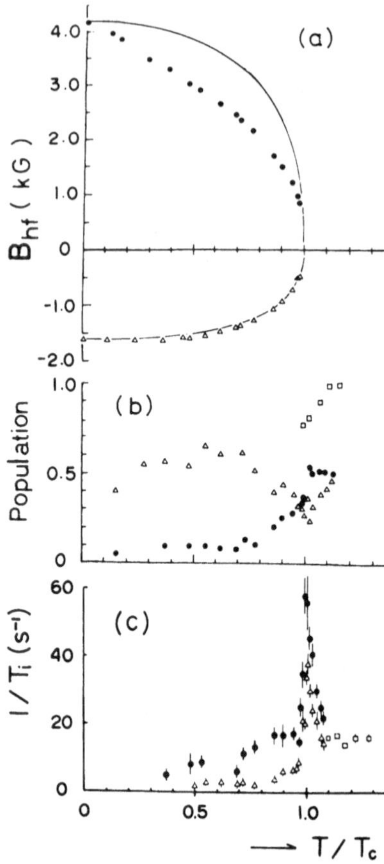

Fig.6.6a-c ►

Fig.6.6a-c. Polarized ^{12}B implanted in Ni at B = 0.3 T. (a) The two hf fields $B_{hf}(\pm)$ vs T/T_c (T_c = 627 K). The solid curves represent the bulk magnetization M(T) of Ni normalized at 6 K; (b) site populations; (c) relaxation rates. (●) $B_{hf}(+)$ site, (Δ) $B_{hf}(-)$ site, (□) both sites [6.61]

of Al and Cu, respectively. In a subsequent study, the Cu crystal was heated [6.57]. Up to 358 K nothing changed. As is shown in Fig.6.5, at higher temperatures the NMR curve narrowed and a second broader one appeared at the high frequency side. Its width, measured at 480 K and at three orientations, agreed best with calculated dipolar widths for substitutional sites. Thus, the authors concluded that for $T \gtrsim 350$ K, ^{12}B takes up either an octahedral interstitial site or a substitutional one.

A similar experiment using single crystalline Si and Ge stoppers (diamond structure) showed that at RT, 67(6)% and 42(4)% of the ^{12}B ions occupied substitutional sites, respectively. The sites of the residual ^{12}B ions remained unspecified [6.58]. An extension of this study investigating interactions with radiation defects of ^{12}B in Ge in a broad T range [6.59] is discussed in Sect.6.3.1b(iii).

Finally, we discuss in this section an application of the in-beam β-NMR technique to investigate magnetic hyperfine fields in a ferromagnetic host. The Osaka group implanted polarized ^{12}B and ^{12}N nuclei in a polycrystalline Ni foil. From the position of the NMR signals, the local fields were determined for ^{12}N between 98 and 773 K [6.8,185] and for ^{12}B between 6 and 730 K [6.8,60,61]. For ^{12}B two hf fields $B_{hf}^{(\pm)}$ with opposite signs were found.

Octahedral and tetrahedral interstitial sites are suggested as the positions of the probe nuclei corresponding to $B_{hf}^{(-)}$ and $B_{hf}^{(+)}$, respectively. In Fig.6.6, $B_{hf}^{(\pm)}$ and the population of the two ^{12}B sites are shown as functions of T. As can be seen, $B_{hf}^{(+)}$ (T) deviates strongly from the host magnetization M(T), suggesting the presence of a localized moment. Further, the critical behaviour of $B_{hf}^{(-)}$ around the Curie temperature T_C was studied in detail. For ^{12}N in Ni only one stopping site was found to be populated. Its nature remained uncertain, but arguments are given in favour of the octahedral interstitial site.

The probe nucleus ^{12}B has been implanted in single- and polycrystalline iron. Again two NMR signals could be distinguished. They were attributed to two different final sites, probably the octahedral and tetrahedral interstitial ones [6.188]. The observed relaxation phenomena are reported in Sect.6.3.2a(i).

b) *Energy Splittings in the Presence of Quadrupole Interactions*

No in-beam γ-NMR experiments have been performed up to now in order to study static quadrupolar interactions in solid metals, reasons for which we have already explained in the introduction to this Section.

The in-beam β-NMR studies to be reported here used almost exclusively ^{12}B and ^{12}N as probes. Only two minor studies applied ^{13}B and ^{8}Li. In all cases the polarized probes were implanted in the sample under selected recoil angles (Sect. 6.2.1b). Further, in most measurements in this section, the Zeeman energy is large compared with the quadrupole coupling, i.e., $g_I\mu_N B \gg e^2qQ$, and we obtain from (6.2,5) for the transition frequencies

$$\nu_{m\to m+1} = \nu_L + \frac{e^2qQ}{4I(2I-1)h}\left[\left(\frac{3}{2}\cos^2\vartheta - \frac{1}{2}\right) + \frac{\eta}{2}\sin^2\vartheta\cos 2\varphi\right]3(2m+1) \qquad (6.59)$$

with $\nu_L = g_I\mu_N B/h$. For ^{12}B and ^{12}N with I = 1, (6.59) gives

$$\nu_{1,2} = \nu_L \mp \frac{3e^2qQ}{8h}\left[(3\cos^2\vartheta - 1) + \eta\sin^2\vartheta\cos 2\varphi\right] \quad . \qquad (6.60)$$

For axially symmetric field gradients, η is zero and (6.60) reduces to

$$\nu_{1,2} = \nu_L \mp \frac{3e^2qQ}{8h}(3\cos^2\vartheta - 1) \quad . \qquad (6.61)$$

In polycrystalline samples, the quadrupolar term has to be averaged over all orientations which gives rise to a spread of the transitions frequencies. Clearly, this strongly complicates the interpretation of NMR patterns, especially in the present cases with the additional risks of defect creation near the implanted probes.

i) Host Metals with Body-Centred-Cubic Structure. We discussed in the preceding section that implanted ^{12}B and ^{12}N ions occupy predominantly interstitial sites in the fcc metals Al,Cu,Pt,Au. This observation suggests that also in quite heavy bcc metals, ^{12}B and ^{12}N come to rest on interstitial sites. In bcc metals, however, interstitial sites have no cubic symmetry and possess, therefore, a nonzero EFG even in the case of an undamaged neighbourhood. Indeed, experiments by the Osaka group [6.52] showed for ^{12}B in polycrystalline Nb,Mo,Ta,W and for ^{12}N in Nb and Ta that the major sites were interstitial. It is true, the interpretation of the observed powder spectra required several ad hoc assumptions (e.g., $\eta \neq 0$, broadly distributed q-values) which might be explained by lattice defects. Nevertheless, the quadrupolar line splittings are doubtless and allowed the determination of QCC's (with relative signs) for interstitial ^{12}B and ^{12}N. In a subsequent paper [6.62], the sign of the initial nuclear alignment f_2^{nr} of ^{12}B in Ta was determined from the observation of the angular distribution of the γ decay populating the ^{12}B ground state. Using this information it was possible to conclude from the NMR-line asymmetry that $e^2qQ < 0$ in that case.

^{28}Al nuclei were implanted in ^{7}Li metal at 1.5-4.2 K and subsequently polarized by means of the Overhauser effect. A quadrupolar broadened NMR line was observed at ν_L and the magnetic dipole moment of ^{28}Al was determined. It was concluded from the line broadening that the ^{28}Al nuclei were stopped on equivalent sites whose structure, however, remained uncertain [6.149,180].

ii) Metallic Hosts with Hexagonal Structure. The experiments to be discussed now used Be,Mg,Zn,ZrB_2 and TiB_2 as stoppers. They possess a nonvanishing EFG at both lattice and interstitial positions. Some early measurements on polycrystals of Be, ZrB_2 and TiB_2 (lines 31,27 of Sect.6.5) have been superseded by single crystal experiments and will not be considered here further.

A group at the Johns Hopkins University, USA, implanted ^{12}N in single crystals of Be and Mg [6.63] and ^{12}B in these crystals [6.64,65] and also in Zn [6.63] with the aim of deducing from quadrupole split β-NMR lines, reliable values of the ^{12}B and ^{12}N quadrupole moments. This could not be accomplished satisfactorily. It would have required answers to the difficult questions concerning charge states and locations of the probes in the lattice and calculation of the EFG at the probe nuclei. Nevertheless, in all four cases of ^{12}B or ^{12}N in Be or Mg, quadrupole split patterns were observed showing the correct ϑ dependence of (6.61) which could be explained by a definite QCC in each case. This gave evidence of a single—although

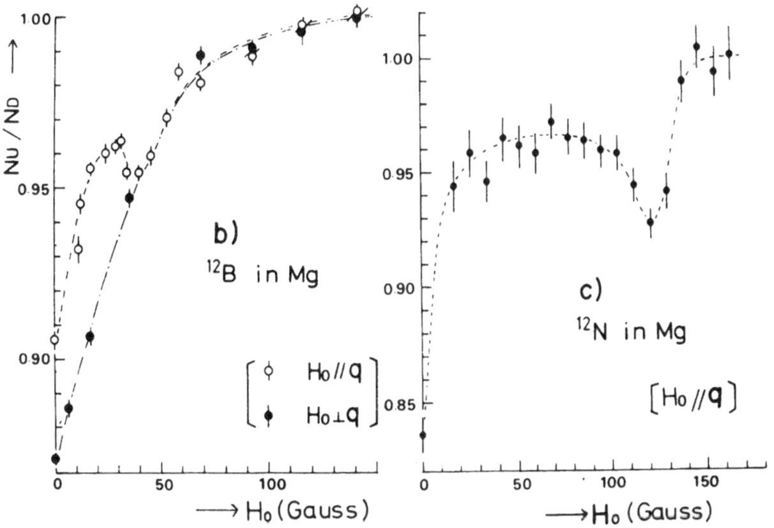

Fig.6.7. Polarization of implanted ^{12}B and ^{12}N in a Mg single crystal versus the external magnetic field. The counting-rate ratio N_U/N_D at the ordinate is proportional to the nuclear polarization [6.67]

undetermined—probe location. In Zn, however, the ^{12}B NMR patterns proved the existence of two different QCCs which were ascribed to two inequivalent probe positions. CORRELL [6.66] recently reconsidered these experiments and gave arguments in favor of the tetrahedral interstitial site as a probe position in Be,Mg,Zn and suggested for ^{12}B in Zn the substitutional lattice site as an additional possibility.

The same group observed in a Mg single crystal the quadrupole split NMR spectrum of ^{13}B yielding the ratio $|Q\ (^{13}B)/Q\ (^{12}B)|$ [6.65].

A method of measuring QCCs without RF irradiation was demonstrated by TANIHATA et al. [6.67] for ^{12}B and ^{12}N implanted in a Mg single crystal. Its c-axis was oriented parallel to **B** ($\vartheta = 0$). In this case, (6.61) is valid for any value of B and immediately yields a crossing of the sublevels m = 0 and 1 for $g_I \mu_N B = \frac{3}{4} e^2 qQ$. At the crossing point the two levels equalize their populations and a dip in the nuclear polarization occurs. In Fig.6.7 this dip is shown for ^{12}B and ^{12}N. From the corresponding B values followed QCCs in accordance with the values from NMR patterns. The decrease of polarization for B towards zero was suspected to be due to lattice defects.

For $\vartheta \neq 0$ the Zeeman levels mix and transitions ($m = +1 \leftrightarrow -1$) in a one-photon process become allowed. MAIO [6.186] observed such $\Delta m = 2$ transitions for ^{12}B in a Mg single crystal. Their linewidth was markedly smaller than that of ordinary $\Delta m = 1$ transitions. This was to be expected, because quadrupolar level shifts are proportional to m^2 and hence quadrupolar broadening cancelled in first order ($m = +1 \leftrightarrow -1$) transitions. MAIO registered $\Delta m = 2$ transitions for ^{12}B in single crystalline Au, too. The explanation for the occurrence of $\Delta m = 2$ one-photon transitions in a fcc crystal remained open.

Mg was also used as host material for ^{12}B [6.68-71] and ^{12}N [6.68,69] in studies looking for a correlation between the β-decay asymmetry (or/and the β-ray energy in [6.68,69]) and the nuclear alignment as a possible indication of second class axial-vector currents in β decay. These experiments applied RF transitions or sublevel crossings to induce the required alignment. The objectives of these investigations lie outside the scope of this report.

Finally, Mg and Be were used as stoppers for ^{8}Li, but no quadrupole splitting could be resolved in the NMR signal [6.44].

MINAMISONO et al. [6.72] used a single crystal of the metallic compound ZrB_2 as a stopper for ^{12}B. Those ^{12}B nuclei contributing to the NMR signal occupied substitutional boron sites. This could be deduced from the orientational dependences of quadrupolar splitting and the linewidth of the central double-quantum transition. The temperature dependence of the QCC was found to be $\lesssim 2\%$ between 200-400 K.

Double-quantum transitions of ^{12}B implanted in a single-crystal Be host were observed by McDONALD and McNAB [6.73].

iii) Semiconducting Host with Diamond Structure. We describe here one experiment in which the quadrupole interaction arose from radiation defects created by the 100 keV-implantation process of ^{12}B in a Ge single crystal [6.59]. It was reported in Sect.6.3.1a that at RT, about 40% of the ^{12}B probes occupied substitutional sites. As is shown in Fig.6.8, the polarization is about 4% at RT and varies strongly with T. The authors gave the following interpretation. Each ^{12}B produces up to about two thousand Frenkel pairs of which a few hundred interstitials (I) and vacancies (V) are assumed not to recombine. Below 200 K quadrupole interacting ^{12}B-I defects and between 200 K and 325 K magnetic-hf interacting ^{12}B-V defects are suggested as causing the depolarization. Above 325 K the polarization increases with temperature. This was explained by thermal dissociation of these defects and the return of the ^{12}B probes to substitutional sites.

6.3.2 Relaxation of Nuclear Orientation

In the absence of radiation defects, nuclear spin-lattice relaxation (SLR) in most metals, whether solid or liquid, is governed by the magnetic hyperfine coupling between the nuclear and the conduction s-electron spins. Another important relaxation mechanism in solids is due to the fluctuating dipole-dipole interaction between the nuclear spins of the diffusing atoms. Going to the liquid state, the correlation time of thermal motion is shortened by a factor of 10^{-2} to 10^{-5} depending on the metal [6.74,75] and the dipolar relaxation rate is reduced proportionally according to the extreme narrowing approximation, (6.37a). In principle, the same is true for the relaxation by quadrupolar coupling of probe nuclei having $I > 1/2$ with fluctuating EFG's from ionic diffusion. Since, however, in most cases quadrupole

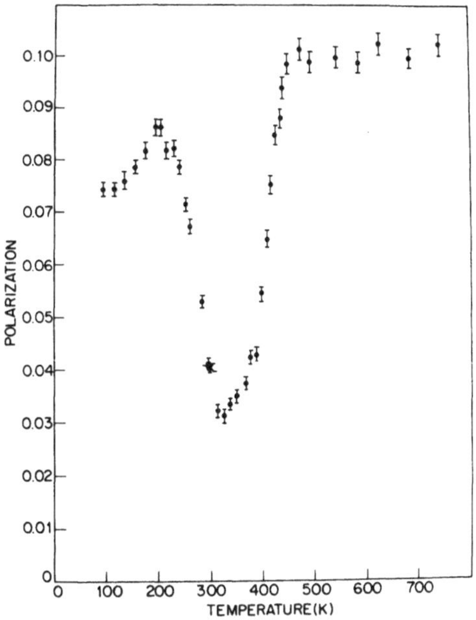

Fig.6.8. Polarization versus temperature for ^{12}B implanted in a Ge single crystal [6.59]

coupling constants (1...100 MHz [6.76]) are much larger than dipolar ones (≈ 10 kHz), quadrupolar relaxation rates may also be quite sizeable in the liquid state. This holds especially for alloys and heavy polyvalent metals. In transition metals, relaxation by conduction electrons is characterized by orbital and core polarization contributions from the unpaired d-electrons which may become even larger than that due to the direct contact interaction [6.77]. Another mechanism is caused by the fluctuations of the d-spins which in ferromagnets critically slow down at the Curie temperature and so give rise to a maximum in the relaxation rate [6.78].

Relaxation phenomena in metals associated with the above-mentioned interactions have been extensively studied by conventional NMR methods [6.79-81]. Often, however, in-beam techniques give complementary information or have technical advantages. For example, measurements on highly diluted impurities, inherent in these techniques, are not feasible with conventional NMR due to the lack of sensitivity. Also, in some cases the radioactive probes contrary to the stable isotopes have the appropriate nuclear properties (e.g., $Q \neq 0$ or large μ_I value). A basic distinction, as compared to classical NMR, is that these techniques are a priori not less sensitive at high temperatures and low magnetic fields because the orientation created by the nuclear reaction is not governed by a Boltzmann factor. Another advantage, particularly in relaxation studies in metals, is that T_1 is usually measured without applying a RF field. Therefore, skin effect problems are avoided and the measurements can be done on bulk metallic samples.

Systematic investigations of SLR in metals by in-beam methods over wide ranges of temperature and/or magnetic induction were first done [6.7] using the γ-PAD method which in practice is confined to liquids where radiation defects are negligible. Measurements covering essentially the liquid *and* the solid phase of a metal have only been done by the neutron-activation, β-asymmetry method up to now [6.82, 83]. The methods using β emitters produced by fast particle reactions have only recently been employed in comprehensive relaxation studies in solids [6.61,84,185,188].

a) *Solid Metals*

i) Spin-Lattice Relaxation of β Emitters. We discuss first the results obtained by the neutron-activation method. In this class of experiments, ^{8}Li turned out to be an excellent probe for several reasons. It has a convenient half-life of $T_{1/2}=0.84$ s being of the order of T_1 at RT in Li metal and its alloys. The high end-point energy of 13 MeV enables the β particles to penetrate easily both cryostat or oven windows and the stainless steel boxes containing the samples.

T_1 of ^{8}Li was measured as a function of T and of B by HEITJANS et al. in ^{7}Li metal [6.82,86] and in series of ^{7}Li-Mg [6.85,89] and ^{7}Li-Ag alloys [6.89], both with up to 11 at% solute concentration. Figure 6.9 gives a survey of the T dependence of $1/T_1$ for two samples. The general trend is determined by a contribution $1/T_{1e}$ due to the conduction electrons. The peak below the melting temperature, manifesting an additional rate $1/T_{1\,diff}$, is caused by thermal diffusion which contributes most to $1/T_1$ when the effective atomic jump rate is of the order of the ^{8}Li Larmor frequency. The discussion of the data in the liquid state is postponed to Sect.6.3.2b.

In accordance with the theory of relaxation by conduction electrons (Sect. 6.1.3a) $1/T_{1e}$, as inferred from the data below 250 K, increases fairly linear with T and is independent of B down to values as low as 5 mT. In the case of pure Li, a slight T dependence of $1/(T_{1e}T)$ was found which was also adopted for the alloys in evaluating $1/T_{1\,diff}$ (see below). In the two alloy systems, $1/(T_{1e}T)$ decreases with increasing solute concentration, the effect being larger in Li-Mg than in Li-Ag. These $1/(T_{1e}T)$ values combined with the ^{7}Li Knight shift K_s known in these alloys [6.87,88] fulfil the Korringa relation $T_{1e}TK_s^2$ = const. in the concentration range covered.

The diffusional contribution $1/T_{1\,diff}$ is obtained by subtracting $1/T_{1e}$ from the measured rate $1/T_1$. In pure Li, $1/T_{1\,diff}$, as shown in Fig.6.10 for different B, is explained solely by the fluctuating dipole-dipole interaction, (6.48), of ^{8}Li with the surrounding ^{7}Li nuclei. The solid lines represent a theoretical fit based on a model of correlated self-diffusion via monovacancies by WOLF [6.90] which yielded the activation energy E_A and the pre-exponential factor τ_0^{-1} in the Arrhenius relation $\tau^{-1} = \tau_0^{-1}\exp(-E_A/k_BT)$. The atomic jump rate τ^{-1} (and thus the self-diffusion coefficient $D = d^2/6\tau$, d = nearest-neighbour distance) was determined in this experiment over almost 7 decades ranging from $\tau^{-1} \approx 10^9\ s^{-1}$ near the melting point (T_m = 454 K) down to some $10^2\ s^{-1}$ at about 220 K ($D \approx 10^{-7} \dots 10^{-14}\ cm^2s^{-1}$). A slight curvature in the Arrhenius plot log τ^{-1} versus T^{-1} at elevated T was tenta-

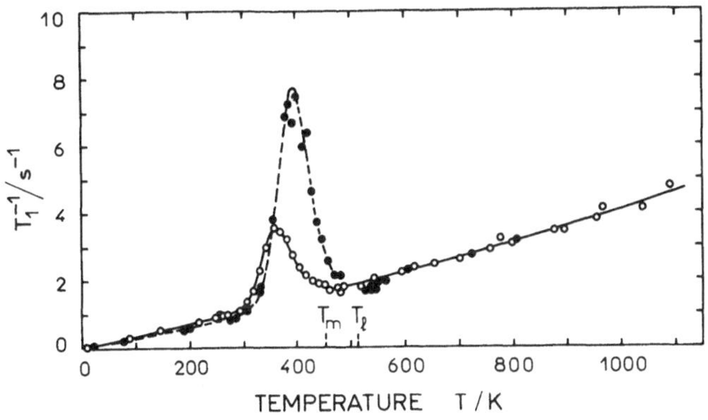

Fig.6.9. Spin-lattice relaxation rate $1/T_1$ of ^8Li in ^7Li (o) and ^7Li-11 at% Mg (•) versus temperature at B = 0.3385 T. T_m is the melting temperature of pure Li, T_ℓ the liquidus temperature of the alloy. The lines through the data represent best fits of the electronic and diffusional contributions, respectively [6.85,86]

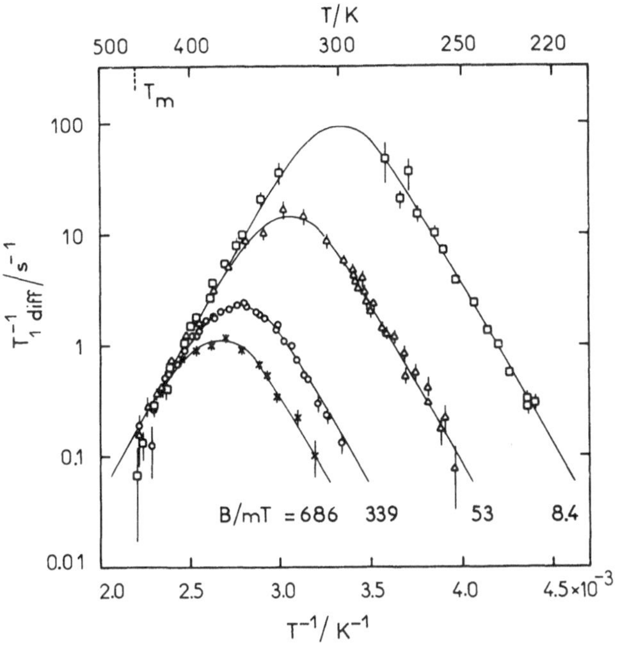

Fig.6.10. Diffusion-induced relaxation rate $1/T_{1\,diff}$ of ^8Li in ^7Li versus reciprocal temperature for different magnetic inductions B (corresponding to ^8Li Larmor frequencies ω_L in the range 0.33..27.1 × 10^6 s^{-1}). The solid lines represent a fit of WOLF's model of correlated self-diffusion via monovacancies assuming purely dipolar relaxation [6.86]

tively ascribed to partial diffusion via divacancies. The corresponding theory [6.91] allowed a somewhat better fit. Similar results were also obtained for stable ^7Li by a combined analysis of data from different conventional NMR techniques [6.92,93].

In alloying Li with Mg and Ag, the $1/T_{1\,diff}(T^{-1})$ curves, shown in Fig.6.11 for Li-Mg at one B value, increase in amplitude and shift in temperature [6.85,89]. This is due to the quadrupolar interaction of the ^8Li probes with the fluctuating EFG of the diffusing solute ions. The measured curves could be quantitatively explained by the superposition of a quadrupolar contribution $1/T_{1Q}(T^{-1})$, and the dipolar contribution $1/T_{1D}(T^{-1})$ already present in pure Li. Strictly speaking, quadrupolar relaxation should not be describable by a single time constant T_{1Q} in the pre-

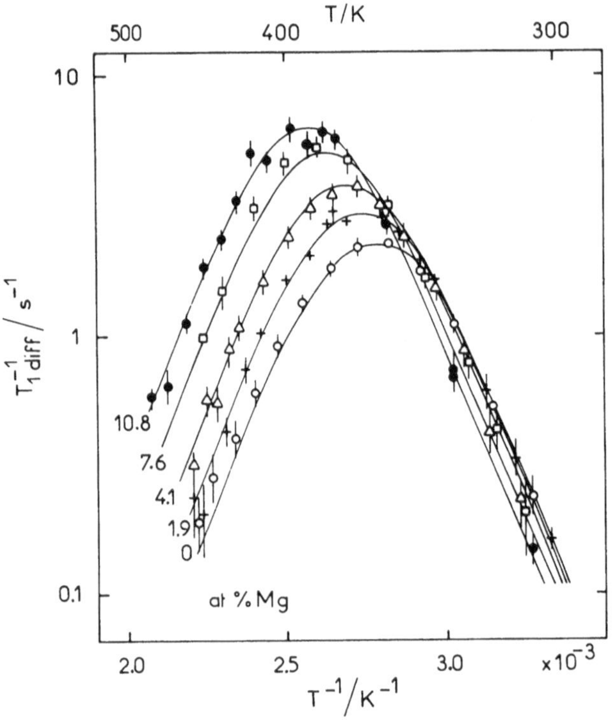

Fig.6.11. Diffusion-induced relaxation rate $1/T_{1\,diff}$ of ^{8}Li in various ^{7}Li-Mg alloys versus reciprocal temperature at B = 0.3385 T. The solid lines represent a fit of SHOLL's model of uncorrelated diffusion assuming dipolar as well as quadrupolar relaxation. The purely dipolar relaxation rate in ^{7}Li (Fig.6.10) is also shown [6.89]

sent case where the extreme narrowing approximation is a priori not valid (Sect. 6.1.3c). An analysis taking account of this fact, however, yielded the same diffusion results as an analysis based on the concept of monoexponential relaxation.

Within a model of uncorrelated diffusion in a lattice given by SHOLL et al. [6.94] and under some assumptions reducing the number of fit parameters, it was found that in Li-Mg the jump rates of Li and Mg decrease exponentially with Mg concentration and that Mg jumps less frequently than Li. The ratio $\tau_{Mg}^{-1}/\tau_{Li}^{-1}$ turned out to be compatible with $(m_{Li}/m_{Mg})^{\frac{1}{2}}$ suggested by a simple oscillator model (m: atomic mass). In Li-Ag an analysis based on solute diffusion by substitutional jumps via vacancies, which is the most likely mechanism in Li-Mg, gave results which differed from tracer measurements on Ag in Li [6.95]. Consistent results were obtained when a combined interstitial-vacancy mechanism [6.96] was adopted. In that case it was found that Ag was jumping more frequently than Li. The measurement on ^{8}Li can be compared with an analogous NMR investigation on ^{7}Li [6.97] where, however, a detailed analysis was not achieved, apparently on account of the poorer data which were taken at one B value only.

The relaxation studies on ^{8}Li probes in solids, reported up to now, were done on samples with isotopically pure ^{7}Li. In another experiment with natural Li metal, which contains 7.4% ^{6}Li, in addition, cross-relaxation between the ^{8}Li and ^{6}Li nuclei showed up [6.98]. As the gyromagnetic ratios of ^{8}Li and ^{6}Li differ by only 0.6%, the dipolar NMR lines in the rigid lattice ($T < 250$ K [6.99]) overlap for $B \lesssim 100$ mT and the ^{8}Li polarization leaks off to the ^{6}Li reservoir. Below 2 mT cross-relaxation between ^{8}Li and ^{7}Li could be observed as well (line 9, Sect.6.5).

Though ^{8}Li is unique for the application of the neutron-activation method to metals, there seem to be other probes such as ^{12}B(20 ms), ^{24}Na^m(20 ms) and ^{116}In (14 s). For ^{24}Na^m this has been demonstrated in a recent experiment on Na metal [6.147] (line 65, Sect.6.5). ^{116}In has been used in preliminary experiments on In metal before [6.100,101] (lines 95,94, Sect.6.5).

We now turn to the experiments using β emitters created by fast particle reactions. Again ^{8}Li was used as a probe which in this case was produced by bombarding ^{7}Li with polarized [6.102] or unpolarized [6.44] deuterons. No systematic measurements have, however, been made. One T_1 value at RT in each case was determined for ^{8}Li in Li [6.102], Au, Pd and Pt [6.44] and was used in the latter three cases to estimate the Knight shifts via the Korringa relation (6.45a) in order to correct the magnetic dipole moments. To this end, T_1 also for ^{8}B in Pt [6.45] and ^{12}B in Au, Pd and Pt [6.49] were measured (lines 16,21, Sect.6.5). Both for ^{8}Li and ^{12}B, positive Knight shifts were deduced in Au and negative ones in Pd and Pt and ascribed to contact interaction and core polarization, respectively, as the dominant mechanisms. The deduced Knight shifts for ^{12}B can be compared with the shifts of the ^{12}B resonance frequencies in the three metals with respect to Si [6.50]. Agreement is found for Au and Pt. An additional important mechanism besides relaxation by conduction electrons is suggested by the strong T dependence of T_1T found for ^{12}B between 200 and 600 K [6.49]. In the light of the experiment on ^{12}B in Al to be discussed next, the variation with T might be partly due to atomic motion.

In a recent experiment by the Palo Alto group on ^{12}B in an Al single crystal [6.84], it was shown what a variety of phenomena may, in fact, become apparent from T_1 measurements over a large T range (line 26, Sect.6.5). The data shown in Fig.6.12 were quantitatively explained by decomposing $1/T_1$ into four different contributions. According to a fit, the relaxation rate $1/T_{1e}$ due to the conduction electrons is dominant in nearly the whole temperature range. $1/T_{1e}$ agrees with the rate estimated from the experimental Knight shift [6.50] when the influence of electron-electron interactions is taken into account (6.45a). The minimum in T_1 below RT is attributed to an additional quadrupolar interaction of ^{12}B with fluctuating dislocation loops formed by the agglomeration of migrating Al interstitials, which in turn were created in the implantation process. The T_1 minimum between 500 and 600 K is explained by the fluctuating dipole-dipole interaction experienced by ^{12}B diffusing from one octahedral interstitial site to the other. This process was analysed in terms of the random-walk model of TORREY [6.103]. The fourth mechanism, discussed by the authors, results from the trapping of ^{12}B at high T. For $T > 600$ K, ^{12}B diffuses so rapidly that it is trapped with high probability during its lifetime and depolarized at low symmetry sites such as surfaces, impurities and dislo-

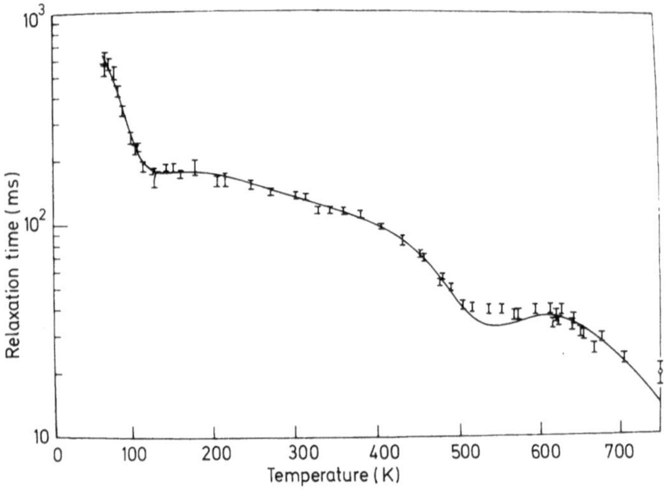

Fig.6.12. Relaxation time T_1 of ^{12}B implanted in an Al single crystal versus temperature at $B = 0.5$ T. The solid line is the best fit of four relaxation mechanisms operating in different temperature ranges (see text) [6.84]

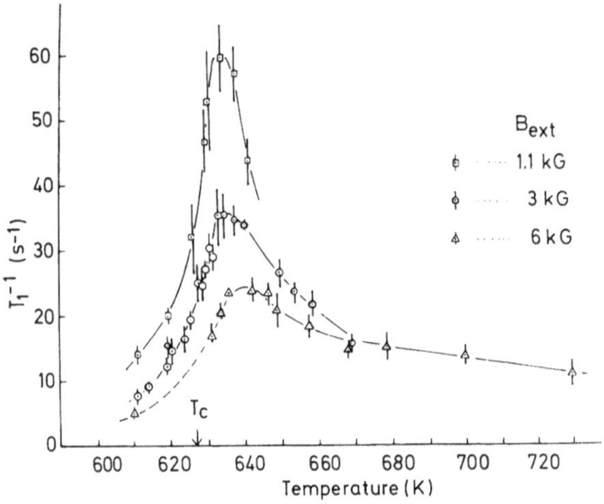

Fig.6.13. Relaxation rate $1/T_1$ of those ^{12}B nuclei in Ni which are subject to the hf field $B_{hf}^{(-)}$ (cf. Fig.6.6) plotted versus temperature around the Curie temperature T_C for different external fields. The solid lines are not based on a theoretical analysis and are drawn to guide the eye [6.61]

cations. In a subsequent paper by the same group [6.187] the orientation dependence of T_1 of ^{12}B in single-crystal Al was studied for $T > 300$ K. In the range 400 K $< T <$ 500 K, i.e., on the low-temperature side of the diffusion-induced T_1 minimum (Fig.6.12), an orientation dependence was found which turned out to be considerably stronger than expected from calculations based on the random-walk model. This was regarded as indicative of a departure from uncorrelated diffusion of the boron impurities.

We now resume the discussion of Sect.6.3.1a, on the experiments of the Osaka group using ^{12}B and ^{12}N in Ni [6.60,61,185] and ^{12}B in Fe [6.188] concentrating on the relaxation measurements. In Fig.6.6c the $1/T_1(T)$ curves for ^{12}B on the $B_{hf}^{(+)}$

and the $B_{hf}^{(-)}$ sites in Ni both show the critical behaviour near the Curie temperature T_c due to the slowing down of the d-spin fluctuations. The relaxation of the ^{12}B nuclei which experience the hf field $B_{hf}^{(-)}$ was measured in more detail in the transition region from the ferro to the paramagnetic state for different external fields B. As shown in Fig.6.13, the maxima of $1/T_1(T)$ increase with decreasing B and occur at temperatures slightly above T_c. The field dependence of $1/T_1$ at $T = T_c$ was found to be compatible with the theoretical prediction for weakly ferromagnetic metals by MORIYA and KAWABATA [6.166]. For ^{12}N in Ni and ^{12}B in Fe $1/T_1$ was measured as a function of T only. Near the respective Curie temperatures, a maximum in $1/T_1(T)$ due to critical slowing down was again observed. In both cases T_1T at $T \gtrsim 300$ K decreases with increasing T. For ^{12}B in Fe two sets of T_1 data corresponding to the different stopping sites were observed at $T \lesssim 500$ K. The data in the T range up to 900 K indicate a deep minimum in T_1T which, however, was only partly accessible. The question of the underlying relaxation mechanism had to be left open.

The experiment on ^{28}Al in Li using the Overhauser effect [6.149,180] whose NMR results have been discussed in Sect.6.3.1b also yielded a T_1 value at T = 4.2 K. It was found to be about twice the value for ^{28}Al in Al calculated from T_1T for ^{27}Al in Al [6.189] assuming a pure magnetic hyperfine interaction. A comparison of the spin susceptibilities of Li and Al [6.190], though not made in [6.149,180], shows that the reduction of $1/T_1$ of ^{28}Al in going from the pure Al to the pure Li host can be explained in terms of (6.45) by an increase of the density of states $\rho(E_F)$ and a decrease of the probability density $\langle|\Psi(0)|^2\rangle_F$ by factors of about 1.4 and 0.5, respectively.

To conclude this subsection we mention the few experiments where fast ions other than deuterons have been used to produce β-emitting probes for relaxation studies in solids. There are two experiments with polarized protons where one T_1 value at RT was taken in each case for ^{25}Al and ^{29}P in a Si single crystal [6.104] and for ^{31}S in red phosphorous [6.36]. T_1 of ^{25}Al turned out to be affected by radiation damage.

Using the $(^7\vec{Li}, {}^8Li)$ reaction, T_1 at RT of ^8Li implanted in Au, Pd, Pt and Pb as well as of ^{12}B in Au was measured [6.177]. T_1 was found to disagree with the results of [6.44,49], except for ^8Li in Pd and ^{12}B in Au.

ii) Spin-Lattice Relaxation of Isomeric γ Emitters. As already mentioned at the beginning of Sect.6.3, due to the rapid depolarization by recoil defects in in-beam experiments on isomeric γ emitters in solids, there is little material to be reported here. In a γ-NMR investigation on $^{214}Ra(67\ \mu s, 8^+)$ in solid Pb [6.39], a rough value for $T_1^{(2)}$ at RT was obtained which amounts to about four times the half-life of the state. $T_1^{(2)}$ was ascribed to the interaction with the conduction electrons alone, since the comparison with NMR patterns at 543 K revealed that quadrupole interaction is negligible.

We further mention experiments by the spin-rotation method on ^{207}Po(47 μs, $13/2^+$) in Pb [6.37] and on ^{207}Bi(0.18 ms, $21/2^+$) in Tl and Pb [6.105], intentionally conducted in the liquid hosts [Sect.6.3.2b(ii)]. A few $T_1^{(2)}$ values were also measured in the solid state, but not further analysed by the authors. In the case of ^{207}Bi the relaxation should primarily be magnetic because of the small quadrupole moment.

b) *Liquid Metals*

i) Spin-Lattice Relaxation of β Emitters. In liquid metals SLR of β emitters has been studied up to now only by the neutron-activation method.

$1/T_1$ of ^8Li was measured by HEITJANS et al. [6.83,86] in liquid Li from the melting point (454 K) up to 1100 K, thus exceeding the T range covered in this metal by conventional NMR. $1/(T_1T)$ was found to be independent of B(5...700 mT) in accordance with the theory of relaxation by conduction electrons, (6.45), but increases with T by nearly 20%, as shown in Fig.6.14. Assuming purely magnetic hyperfine interaction, the observed T dependence cannot be explained by volume expansion. The positive slope of $1/(T_1T)$ versus T can, however, be qualitatively understood if the change of liquid structure with T is taken into account, as has been demonstrated by a recent pseudopotential calculation which involves the static structure factor S(q) of liquid Li measured at different T by neutron diffraction [6.167]. In going from liquid to solid Li, an increase in $1/(T_1T)$ of 3% is observed which is consistent with the measured step of the ^7Li Knight shift K_s at the melting point [6.106]. The sign of the step is opposite to that in the other alkali metals except Cs [6.81]. On the other hand, the sign of the temperature coefficient of K_s in liquid Li, as inferred from the $1/(T_1T)$ data through the Korringa relation (6.45a), is the same as in Na, K and Rb but opposite to that in Cs. Thus, Li and Cs appear to be anomalous among the alkali metals, as far as these hyperfine properties are concerned.

These measurements have recently been extended to the liquid alloys of Li with Na, Mg, Pb [6.168] and Bi [6.191]. $1/T_1$ of ^8Li is again dominated by the magnetic hyperfine interaction with conduction electrons. $(T_1T)^{-1/2}$ shows an approximately linear dependence on composition in Li-Na and Li-Mg which in the former case can be attributed almost completely to the change of the s-electron density $\langle|\Psi(0)|^2\rangle_F$. The experimental Korringa product $T_1TK_s^2$ is independent of Na concentration and, using very recent ^7Li Knight shift data [6.192], slightly dependent on Mg concentration which in the latter case might be related to recent neutron diffraction results on liquid Li-Mg [6.193]. Whereas Li-Na and Li-Mg may reasonably be regarded as nearly-free-electron alloys, this is not true for Li-Pb and Li-Bi. Near the compositions $Li_{0.8}Pb_{0.2}$ and $Li_{0.75}Bi_{0.25}$, respectively, $(T_1T)^{-1/2}$ has a marked minimum, as shown in Fig.6.15 for Li-Bi. Though ^7Li Knight shift data, available for Li-Pb, display a similar overall behaviour, the Korringa product, plotted on reciprocal scale in Fig.6.16, is not constant. The measured rate near 20at%Pb is enhanced by a factor $\eta_W > 1$ (Fig.6.16, right-hand scale) over the Korringa rate calculated from the ex-

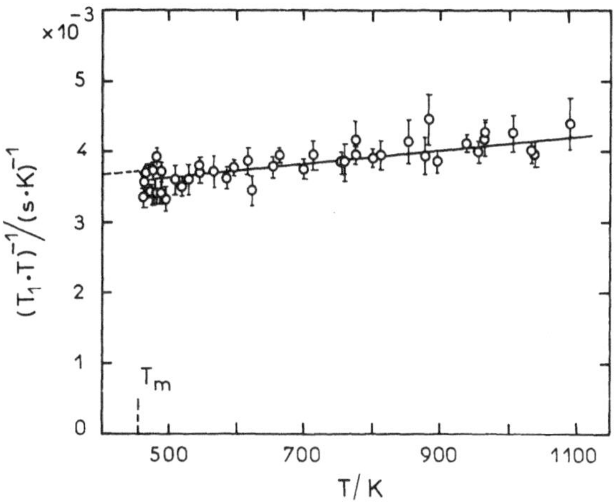

Fig.6.14. $1/(T_1T)$ of ^{8}Li in liquid ^{7}Li versus temperature for $B > 5$ mT. The full line represents a linear fit to the data above the melting point T_m. The broken line below T_m is the extrapolation of a linear fit to the data for $T \lesssim 200$ K where T_1 is due to conduction electrons (cf. Fig.6.9) [6.83,86]

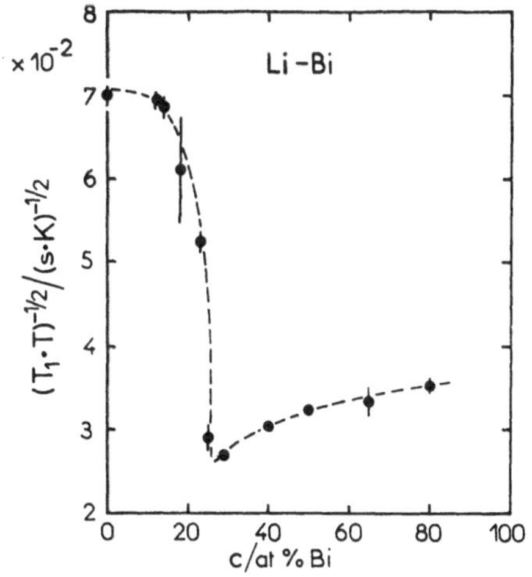

Fig.6.15. $(T_1T)^{-\frac{1}{2}}$ of ^{8}Li in liquid $Li_{1-c}Bi_c$ versus c at T = 150 K. T_1 is due to magnetic hyperfine interaction with conduction electrons. The broken line is drawn to guide the eye [6.191]

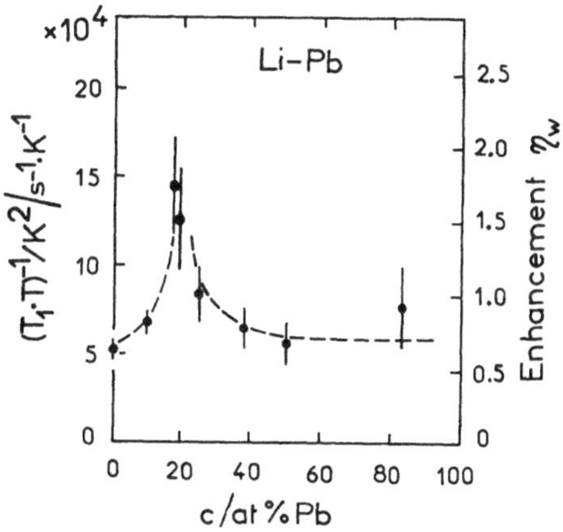

Fig.6.16. $1/(T_1T)$ of ^{8}Li normalized to the square of the Knight shift (left-hand scale) and enhancement η_W over the Korringa rate (right-hand scale) in liquid $Li_{1-c}Pb_c$ versus c at T = 1000 K. The broken line is drawn to guide the eye [6.168]

perimental Knight shift. This can be ascribed to the tendency towards electron localization and compound formation in the melt. As pointed out by WARREN [6.125,194], an enhancement $\eta_W > 1$ is directly related to the mean "residence time" of the conduction electrons on a nuclear site and to the electrical resistivity in the strong-scattering regime. An analysis along these lines showed that the compound-forming tendency is significantly stronger in liquid Li-Bi than Li-Pb.

ii) Spin-Lattice Relaxation of Isomeric γ Emitters. Besides the few exceptions presented in Sect.6.3.2a(ii) (lines 104,106,110, Sect.6.5), relaxation of γ isomers has been studied exclusively in liquids. The experimental technique used in nearly all cases has been the spin-rotation method. γ NMR was applied only to ^{71}Ge(20 ms, $9/2^+$) in Ga [6.38], ^{78}Br(0.12 ms, $4^{(+)}$) in Se-Tl [6.38] and ^{115}Sn(0.16 ms, $11/2^-$) in In [6.41] for the measurement of one T_1 value in each case (lines 77,80,89, Sect. 6.5).

The activities in the field under discussion up to the Uppsala Conference in 1974 were reviewed by RIEGEL [6.37]. The systems considered there are listed in lines 76,78,90,98,101,102,104 and 106 of Sect.6.5 and only some general remarks on these early experiments will be added. Apart from the In-Ga alloys (line 78, Sect.6.5), only pure metals had been used as hosts at that time. Generally, $1/T_1^{(2)}$ consisted of a contribution $1/T_{1e}^{(2)}$ due to the conduction electrons and a quadrupolar rate $1/T_{1Q}^{(2)}$. The T dependence of $1/T_1^{(2)}$ could mostly by described by a superposition of the functions $1/T_{1Q}^{(2)} = a \cdot D^{-1}(T)$ and $1/T_{1e}^{(2)} = b \cdot T$, where a and b are fit parameters and D(T) the self-diffusion coefficient of the matrix (lines 76,90,101, 102, Sect.6.5).

The relation $1/T_{1Q}^{(2)} \sim D^{-1}$, valid in the extreme narrowing case (6.37a), follows from (6.49) if the dynamic structure factor $S(q,\omega)$ in the limit $\omega = 0$ is approximated by the self-diffusion coefficient and the static structure factor according to $S(q,0) \sim D^{-1}S(q)q^{-2}$. Thus, only uncorrelated single-particle motion is taken into account [6.27]. Comparing the diffusion induced rates $1/T_{1Q}^{(2)}$ with those of stable nuclei in the same hosts, some quadrupole moments could be deduced (lines 76,101, 102, Sect.6.5). $1/T_{1e}^{(2)}$ was utilized to derive Knight shifts K_s of the impurities with the help of (6.45a) adopting $K(\alpha) = 0.76$ for the electron-electron interaction factor (lines 76,90,101,104, Sect.6.5). In this analysis, $1/(T_{1e}^{(2)}T)$ was assumed to be independent of T. This could be verified experimentally for ^{207}Bi(0.18 ms, $21/2^+$) in Tl in the T range 600 - 1100 K, where relaxation is supposed to be due only to the conduction electrons because of the small quadrupole moment of the probe [6.37,105]. In general, however, $1/(T_{1e}^{(2)}T)$ is not necessarily constant over the T range covered in these experiments as is demonstrated by the measurement on ^{8}Li in liquid Li [6.83,167] reported in the preceding subsection.

At the time of the Uppsala Conference [6.9], only one monatomic liquid metal was known where the quadrupolar rate displayed a temperature dependence being definitely inconsistent with purely diffusion induced relaxation: $1/T_{1Q}^{(2)}$ of ^{207}Po(47 μs, $13/2^+$) in Pb was not found to decrease with increasing T as quickly as the inverse of the diffusion coefficient $D^{-1}(T)$. The more recent development in this field, reviewed by GABRIEL up to 1975 [6.107], is characterized by the study of partly "nondiffusive systems". Examples are ^{114}Sb(0.21 ms, 8^-) in Sn [6.108], ^{202}Tl(0.57 ms,7^+) in Hg [6.108], ^{207}Po(47 μs, $13/2^+$) in Bi [6.109] and ^{132}Xe(8 ms, 10^+) in Te [6.110]. In these cases, except the last one where the

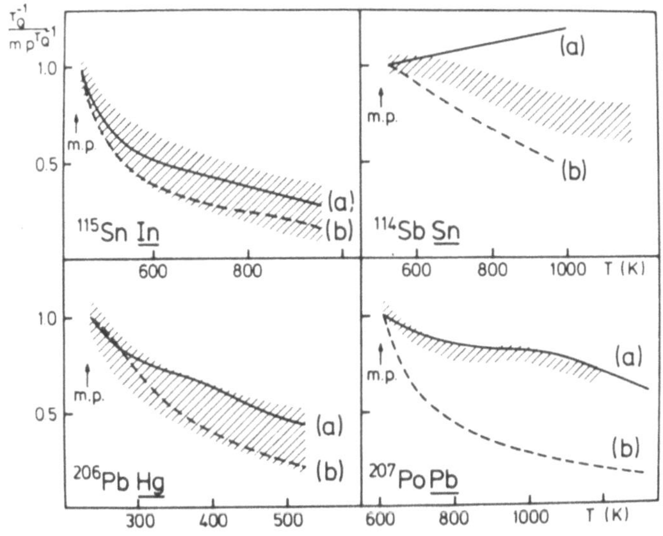

Fig.6.17. Quadrupolar relaxation rates $1/T_{1Q}^{(2)}$ normalized at the melting point versus temperature for two "diffusive" (*left*) and two partly "nondiffusive" systems (*right*). (IIIIII) experimental data. (Curves a, b) predictions of theory based on the Cocking-Egelstaff [6.112] and the diffusion approximation, respectively (see text) [6.111]

quadrupolar contribution was estimated directly, $1/T_{1Q}^{(2)}$ was separated from $1/T_1^{(2)}$ by evaluating $1/T_{1e}^{(2)}$ through the Korringa relation with $K(\alpha) = 0.75$ and a guessed value for the impurity Knight shift K_s.

NISHIYAMA and RIEGEL [6.111] compared the experimental T dependence of $1/T_{1Q}^{(2)}$ for several examples with theory using (6.49). They used for the dynamic structure factor $S(q,\omega)$ in the $\omega = 0$ limit the parametrization $S(q,0) \sim T^{-1/2}[S(q)]^{3/2}q^{-1}$ which had been reported as describing inelastic neutron scattering data in liquid Pb [6.112]. Considering also the influence of volume expansion on the static structure factor S(q) and the EFG form factor I(q), the curves (a) in Fig.6.17 were obtained. For comparison, the results based on the diffusion approximation $S(q,0) \sim D^{-1}S(q)q^{-2}$ are also shown (curves b). According to these authors the influence of structure is, at least for ^{207}Po<u>Pb</u>, of similar importance for the T variation of $1/T_{1Q}^{(2)}$ as liquid dynamics.

SCHIRMACHER [6.113] also evaluated the T dependence of $1/T_{1Q}^{(2)}$ starting not from (6.49) but from a more general form given by SHOLL [6.27] and WARREN [6.28] which contains $S(q,\omega)$ and the self-part $S_s(q,\omega)$. Using various models for $S(q,\omega)$ and $S_s(q,\omega)$ the experimental T dependence could approximately be reproduced by considering solely the dynamics. According to this approach the T dependence of $1/T_{1Q}^{(2)}$ results from the relative importance of contributions due to diffusive and collective motions being described by $D^{-1}(T)$ and $D(T)/T$, respectively.

Recently, BOSSE et al. [6.195,196] made a theoretical investigation of the T dependence of $1/T_{1Q}^{(2)}$ in terms of a mode-coupling approximation. The theory was applied

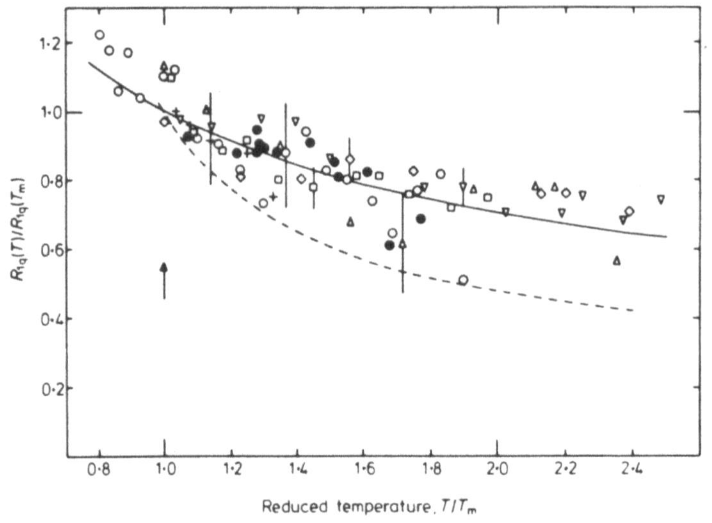

Fig.6.18. Quadrupolar relaxation rates $1/T_{1Q}(\equiv R_{1q})$ normalized at the melting point versus reduced temperature T/T_m in monatomic liquids. Conventional NMR: (●) Hg [6.114], (o) Ga [6.126], (+) Sb [6.127]. γ-PAD: (△) ^{117}SbIn [6.117], (□) ^{207}PoPb [6.37], (▽) ^{114}SbSn [6.108], (◊) ^{207}PoBi [6.109]. The full line represents $1/T_{1Q} \sim T^{-1/2}$ (the broken line represents a theoretical prediction not discussed here) [6.114]

to liquid Ga. According to these calculations the quadrupolar relaxation rate of Ga in liquid Ga decreases more slowly than predicted by the simple diffusion model at low T and even increases slightly at high T. A similar result was obtained for Ge in Ga, provided that the activation energy for impurity diffusion was put equal to or greater than that for self-diffusion in liquid Ga. Contrary to the early $1/T_{1Q}^{(2)}$ data for ^{71}Ge (20 ms, $9/2^+$) in Ga [6.7], recent unpublished data by NISHIYAMA, given in [6.196], indeed show such a behaviour. A minimum in $1/T_{1Q}^{(2)}(T)$ is displayed also by the data measured on ^{132}Xe(8 ms, 10^+) in liquid Te [6.110].

In a paper by MARSDEN et al. [6.114], it was pointed out that a good deal of the $1/T_{1Q}$ data in pure liquid metals obtained both by conventional NMR and γ-PAD can be described approximately by a $T^{-1/2}$ law (see Fig.6.18). In any case, systems like ^{115}SnIn, ^{205,206}PbHg and perhaps ^{71}GeGa, classified above as purely "diffusive systems", rather appear to be exceptions.

An interesting phenomenon which needs further consideration was found for ^{88}Y (14 ms, 8^+) in Rb [6.115]: $1/(T_1^{(2)}T)$, determined for T up to 100 K above the melting point, increases by about 100% and is two orders of magnitude smaller than expected on account of the Knight shift of the host.

Finally, among the studies concerned with pure metals, we mention $1/T_1^{(2)}$ measurements on ^{212}Fr(0.18 ms, 11^+) in Tl and ^{214}Ra(67 μs, 8^+) in Pb at one temperature each [6.116]. As the quadrupole moments are expected to be small, only relaxation by conduction electrons was taken into account. In each case the Knight shift was deduced from the Korringa relation using $K(\alpha) = 0.76$, and good agreement was found with values obtained directly by measuring spin precession frequencies.

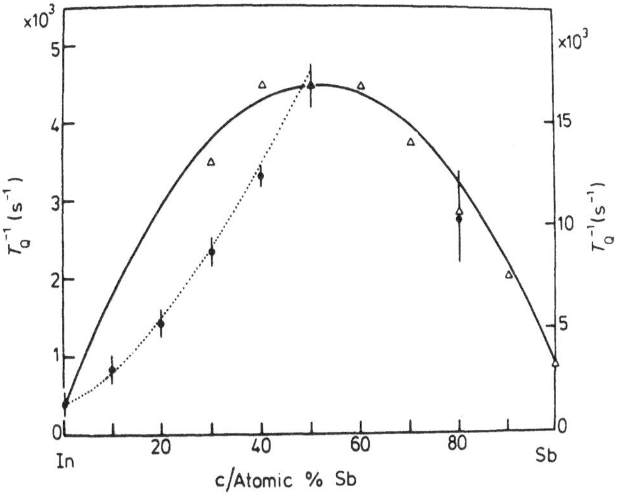

Fig.6.19. Quadrupolar relaxation rate $1/T_{1Q}$ in liquid $In_{1-c}Sb_c$ versus c at T = 900 K. (●) left-hand scale: ^{117}Sb(0.34 ms) [6.117]. (Δ) right-hand scale: ^{121}Sb(stable) [6.128]. The full curve essentially reflects a c(1-c) dependence fixed by the data at c = 0, 0.5, 1.0. The dotted line represents an empirical description by a power law [6.117]

We now turn to the experiments on γ isomers in liquid alloys. Here again the main point of interest has been the quadrupolar relaxation rate. Besides temperature, the alloy partners and their concentration were varied in these experiments which were done essentially in In-, Pb- and Te-based binary systems by v. HARTROTT, QUITMANN and coworkers ([6.197,198] and references therein).

In In alloys with 15 different partner elements (line 97, Sect.6.5) ^{117}Sb (0.34 ms, $25/2^+$) was used as probe. For the magnetic contribution to the relaxation rate of ^{117}Sb, generally considered to be small as compared to $1/T_{1Q}^{(2)}$, a common value was adopted for all the In alloys except In-Te [6.119]. The latter will be discussed together with the Te alloys. In studying the alloys $In_{1-c}X_c$ with c = 0.5, it was found that $1/T_{1Q}^{(2)}$ is enhanced over the value in pure In by factors of up to about 10. On the average the enhancement is correlated with the valence difference between In and the partner element, though this is evidently not the only ingredient [6.117,118]. The temperature dependence of $1/T_{1Q}^{(2)}$, studied explicitly for the equicomposition alloys In-Sb [6.117], In-Sn [6.199], In-Bi and In-Hg [6.118, 122], is stronger than in pure In [6.117] and can be described by an Arrhenius-type relation $1/T_{1Q}^{(2)} \sim \exp(E_A/k_BT)$. The parameter E_A, though correlated with the melting temperature T_m of the alloy, turned out to be definitely smaller than the empirical value 3.7 k_BT_m for the activation energy of self-diffusion in liquids. $1/T_{1Q}^{(2)}$ as a function of concentration roughly follows the c(1-c) rule for substitutional alloys, see, e.g., [6.81]. However, in the In-rich region, $1/T_{1Q}^{(2)}$ for In-Sb substantially deviates from the ideal behaviour, as shown in Fig.6.19 [6.117]. The same was found for In-As, In-Bi, In-Au and In-Hg [6.122]. The results have been discussed in terms of a formalism presented by the authors [6.118] which extends the theory by SHOLL [6.27] and WARREN [6.28] to alloys using partial dynamic structure factors. Disregarding the dynamics, an empirical correlation was found between the partial static structure factor $S_{CC}(q)$ [6.200] in the limit $q = 0$, which describes the mean square fluctuation in concentration, and the measured $1/T_{1Q}^{(2)}$ in the In alloys [6.123,199]. $S_{CC}(0)$ was evaluated from measured values of the enthalpy of mixing.

In the experiments on ^{117}Sb in In alloys reported up to now, the probe was produced from the In component. In the case of In-Sn it was possible to produce the same probe also from the Sn component thus allowing the measurement of a complete composition dependence of $1/T_{1Q}^{(2)}$ [6.199]. In the same paper measurements for ^{117}Sb in Sn rich Sn-Sb alloys are shown, too.

Probes which were used besides ^{117}Sb in In alloys are ^{115}Sn(0.16 ms, $11/2^-$) [6.121], ^{132}Xe(8 ms, 10^+) [6.119,120], ^{114}Sb(0.21 ms, 8^-) [6.201a] and ^{207}Po (47 μs, $13/2^+$) [6.172,201a]. The latter two isomers were employed in liquid InPb to study the probe-specific effect by measuring the T dependence of their respective $1/T_{1Q}^{(2)}$. The difference is small compared with the differences observed for various hosts and can be related to the difference in atomic radii [6.201a].

We now turn to another systematic study of quadrupolar relaxation by the same group [6.172] which was done on liquid Pb alloys with numerous partner elements using ^{207}Po as probe (line 106a, Sect.6.5). Contrary to the In alloys the increase of $1/T_{1Q}^{(2)}$ at $c = 0.5$ over the value at $c = 0$ remains within a factor of two. The temperature dependence was found to be weak, similar to that in pure metals and different from the strong T dependence in the In alloys with pronounced alloy enhancement. This correlation between $1/T_{1Q}^{(2)}(c)$ and $1/T_{1Q}^{(2)}(T)$ seems to be a general feature. A qualitative explanation was given which starts from the decomposition [6.118,172]

$$\frac{1}{T_{1Q}^{(2)}} = \left(\frac{1}{T_{1Q}^{(2)}}\right)_{NN} + \left(\frac{1}{T_{1Q}^{(2)}}\right)_{CC} \quad . \tag{6.62}$$

The two contributions are due to (number) density and concentration fluctuations and involve the partial structure factors $S_{NN}(q)$ and $S_{CC}(q)$, respectively [6.200].

Model calculations showed that contrary to the In alloys in the Pb alloys the contribution $(1/T_{1Q}^{(2)})_{NN}$ dominates. Since $(1/T_{1Q}^{(2)})_{NN}$, which refers to the overall structure of the alloy, is a monotonic function of c whereas $(1/T_{1Q}^{(2)})_{CC}$ has a maximum at intermediate c, it is clear that the enhancement of $1/T_{1Q}$ at $c \approx 0.5$ decreases when $(1/T_{1Q}^{(2)})_{NN}$ increases. The T dependences of the two contributions were considered to be related by $(1/T_{1Q}^{(2)})_{NN} \sim T(1/T_{1Q}^{(2)})_{CC}$. In fitting the data for the Pb alloys by $1/T_{1Q}^{(2)} \sim T\exp(E_A/k_BT)$, the parameter E_A was found to have approximately the same value in units of k_BT_m as in the In alloys. This is regarded as a support for the internal consistency of the approach. However, the deviation of E_A from the activation energy for self-diffusion, mentioned above, is still an open question.

The alloy system to be discussed next is Te-X where X stands for Sb, Bi, In, Ga, As, and Au. The concentration dependence of $T_1^{(2)}$ was measured using ^{73}As(6 μs, $9/2^+$) [6.197], ^{117}Sb [6.119] and especially ^{132}Xe(8 ms, 10^+) [6.120] as probes. It is argued that the magnetic relaxation rate, though strongly dependent on concentration (e.g., in In-Te [6.119]), is only a small fraction of the quadrupolar rate. In Sb_cTe_{1-c} and Bi_cTe_{1-c}, the first two of the above series, $1/T_{1Q}^{(2)}$ shows an increase with c, similar to that in other alloys with high electrical conductivity, indicating an essentially metallic character. In In_cTe_{1-c} and Ga_cTe_{1-c}, on the other hand, $1/T_{1Q}^{(2)}$, over a limited c range, is much more enhanced than expected for a metallic alloy and shows a pronounced maximum at c ≈ 40 at% which can be assigned to the known liquid semiconductors In_2Te_3 and Ga_2Te_3, respectively [6.124]. The results were interpreted in terms of a reduced density of states and electron localization. In the case of Ga-Te also the influence of ion dynamics was estimated using viscosity data. Whereas in the Te alloys discussed up to now there is some correlation between the enhancement in $1/T_{1Q}^{(2)}$ and the reduction in conductivity relative to pure Te, this seems not to be true for the remaining alloys As-Te and Au-Te. Here $1/T_{1Q}^{(2)}$, measured in the Te-rich region, was found to be essentially unaltered in contrast to a de- and an increase, respectively, in conductivity. It seems not yet clear how these findings can be related to the notion of covalent and ionic bonding in As-Te and Au-Te, respectively.

Of the Te alloys mentioned here, In-Te, Ga-Te and Sb-Te have also been investigated by classical NMR [6.125,194]. There, however, the magnetic hyperfine interaction was studied in detail. As mentioned in Sect.6.3.2b for liquid Li-Pb and Li-Bi, the enhancement of the magnetic relaxation rate over the Korringa rate, which can very well accompany a reduction of $1/T_1$ relative to the pure host, allows a straightforward analysis in terms of the mean "residence time" of localizing electrons.

Finally in this subsection we mention liquid alloy systems where the magnetic hyperfine interaction is predominant. This is the case for the relaxation of ^{207}Bi (0.18 ms, $21/2^+$) in Tl-Pb, Tl-Bi and Tl-Au which was measured as a function of temperature and of composition [6.201b]. In agreement with conductivity data, which indicate negligible compound-forming tendencies in these melts, the $1/T_1^{(2)}$ results could roughly be explained in the frame of the free-electron model assuming that only the density of states $\rho(E_F)$ changes in alloying.

6.4 Insulators

There is one important difference between the experiments to be reported now and those using metallic hosts, discussed in the preceding section. In the latter case, the probe nuclide represents a foreign element, i.e., an impurity (exceptions: ^{12}B in TiB_2 and ZrB_2, ^8Li in Li and Li alloys, ^{117}Sb in In-Sb and ^{207}Bi in Tl-Bi).

Therefore, even at high temperatures where possible radiation defects anneal quickly compared with τ_N, it is generally an open question at which position the probe comes to rest in the lattice. In noncubic hosts it was not possible in all cases to determine the probe sites definitely. Further, the probes are (with the exception of ^8Li in Li and Li alloys) created by accelerated beams, resulting in an additional high risk of radiation damage.

In the experiments performed in insulators the situation is different. With few exceptions the probes do not represent an impurity, but are an isotope of an element being present in the host material. Therefore the probe nuclei, in most cases, occupy substitutional sites on undisturbed lattice positions. Furthermore, in most experiments nuclides are produced by the relatively soft $(\vec{n},\gamma)$ process, and the risk of radiation damage is low. Therefore, a general discussion of probe positions is not necessary within this section. Exceptions are investigations which have been performed especially to study structure and annealing behaviour of $(\vec{n},\gamma)$-induced point defects at temperatures low enough to freeze these defects for times $\geq \tau_N$.

6.4.1 Static Interactions, NMR Spectra

a) *Lithium Compounds*

The first β-NMR experiment, already mentioned in the introduction, was performed by CONNOR in 1959 [6.3]. He produced polarized ^8Li (0.8 s, 2^+) nuclei in a ^7LiF single crystal by capturing polarized neutrons. Figure 6.20 shows the measured NMR curve from which the magnetic moment of ^8Li was determined. In a second experiment using a rotating RF field, the sign of the magnetic moment was found to be positive [6.129].

In the same crystal the dependence of the nuclear polarization of ^8Li on the external magnetic induction B was measured [6.130]. The authors found a constant polarization above 0.1 T and a strong decrease at lower inductions. This behaviour was assumed to be caused by paramagnetic relaxation. More recent measurements in LiF single crystals showed, however, that in ^7LiF no B dependent relaxation exists down to fields of some mT. Below this field, cross-relaxation takes place between the ^8Li spin and the ^7Li and ^{19}F spins in the surroundings. In LiF containing Li in its natural isotopical abundance, ^8Li-^6Li cross-relaxation is possible up to comparably high fields of 0.1-0.4 T, depending on the orientation of the crystal [6.98]. This is due to the nearly identical gyromagnetic ratios of ^6Li and ^8Li. A theoretical description of the ^8Li-^6Li cross-relaxation in LiF has been given by DZHEPAROV et al. [6.202,203].

The measurements were performed in single crystals of different orientations. The experiments on ^8Li in ^7LiF were later repeated by GUL'KO et al. [6.131,132]. They found the same value for μ as CONNOR. The same group studied in detail in LiF single crystals the dependence of the ^8Li resonance linewidths on the crystal orientation and found good agreement with calculated dipolar line widths [6.133]. New NMR measurements on a <111>-oriented LiF single crystal showed that at 13 K, no complete reson-

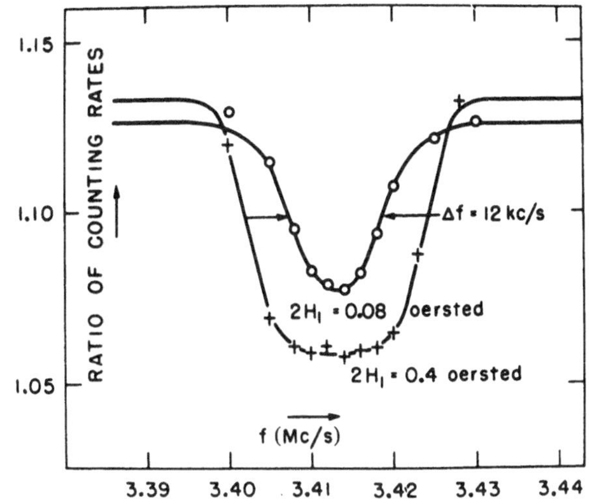

Fig.6.20. ^{8}Li NMR signals versus RF frequency for two RF field strengths. B = 541.8 mT [6.3]

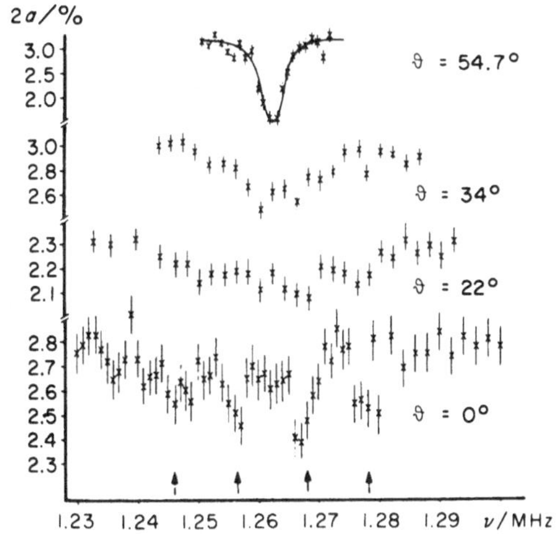

Fig.6.21. ^{8}Li NMR signal in a $LiNbO_3$ single crystal for different angles ϑ. At ϑ = 54.7° the four quadrupolar split (Δm = 1) transitions collapse to a single line. 2 B_1 = 8 μT, B = 0.2 T [6.136]

ant depolarization of the ^{8}Li nuclei was possible [6.134]. The full depth of the resonance curve was obtained only if B was modulated. This behaviour was ascribed to a quadrupolar line splitting induced by radiation defects. The modulation amplitude necessary to get the full resonance effect yielded the defect EFG. From the annealing behaviour of the defects, the corresponding migration energy was determined. Measurements on a <100> oriented crystal showed no quadrupolar effects. From this it could be concluded that the principal axis of the defect EFG coincides with the crystallographic <111> axis. Similar experiments had been performed earlier on CaF_2. They are discussed below.

In a recent experiment a precision value of the ^{8}Li magnetic moment was obtained by NMR in $^7Li_2S$ and $^7Li_2O$ powder samples [6.135]. The contributions of the ^{7}Li nuclei to the ^{8}Li dipolar linewidths (^{7}Li is the *only* nuclide with $I \neq 0$ in the used samples, apart from the 0.04% isotope ^{17}O) were suppressed by additional irradiation of the ^{7}Li NMR frequency. By this method a linewidth as small as 32 Hz was obtained in the case of $^{7}Li_2S$.

In a $LiNbO_3$ single crystal, e^2qQ of ^{8}Li was determined by ACKERMANN et al. using neutron activation [6.136]. Figure 6.21 shows the ^{8}Li NMR spectrum for different crystal orientations. For the orientation $\vartheta = 0^\circ$, four NMR signals corresponding to the ($\Delta m = 1$)-transitions in the $I = 2$ groundstate of ^{8}Li were observed. At the angle $\vartheta = 54.7^\circ$, all lines collapsed to a single line, since here the Q interaction disappears in first-order perturbation theory, (6.5a). Using the known QCC of ^{7}Li in $LiNbO_3$, the Q moment ratio $|Q(^{8}Li)/Q(^{7}Li)|$ could be determined and by this the Q moment of ^{8}Li. In a similar experiment the QCC of ^{8}Li in $LiTaO_3$ was measured [6.137]. In this case the method of multiquantum transitions (first studied with β NMR by McDONALD and McNAB [6.73], see Sect.6.3.1) was used to improve the signal-to-noise ratio of the resonance lines by one order of magnitude. The resulting Q moment ratio was in agreement with that obtained from the $LiNbO_3$ measurement.

In a further experiment in $LiNbO_3$ cross-relaxation processes (Sect.6.4.1b) between ^{8}Li and the stable isotope ^{93}Nb were studied [6.98]. There the ^{8}Li polarization could be used to measure the QCC of ^{93}Nb as a function of temperature. Their temperature dependence known from conventional NQR in the range 4-500 K was extended up to 820 K.

The QCC of ^{8}Li in $LiIO_3$ was determined by MINAMISONO et al. [6.102]. In this case the polarized ^{8}Li nuclei were produced by the reaction $^{7}Li\,(d,p)^{8}Li$.

b) *Fluorine Compounds*

Another favorite β-NMR nuclide is ^{20}F (11.4 s, 2^+). The magnetic moment of ^{20}F and its sign were determined in 1963 [6.138]. ^{20}F was produced by the $(\vec{n},\gamma)$-reaction in a CaF_2 single crystal. As in the case of LiF, the B dependence of the ^{20}F polarization was measured at RT [6.130]. The dependence was again explained by paramagnetic relaxation, but as in the case of LiF, later measurements showed that no B dependent relaxation is present in CaF_2 at RT [6.17,139,159].

The selected recoil angle method (Sect.6.2.1b) was used by SUGIMOTO et al. to measure the magnetic moment of ^{17}F (66 s, $5/2^+$) [6.34,140]. ^{17}F was produced in a SiO_2 target by the reaction $^{16}O(d,n)^{17}F$; the nuclei were stopped in a thin layer of CaF_2. GUL'KO et al. studied the ^{20}F NMR in CaF_2 [6.131,132]. They got the interesting result that the resonance signal, being very pronounced at RT, nearly vanished at 4.2 K and showed diffuse structures. This behaviour was examined further by STÖCKMANN et al. [6.141,142]. They found that the single NMR line, which was observed at RT, was reduced to about 20% intensity below 75 K. Furthermore, four satellite lines appeared. This fact could be explained by an additional quadrupole interaction due to EFGs of stable diamagnetic point defects. Magnitude, sign, and symmetry of the EFG tensor could be determined. By this it was shown that the defects

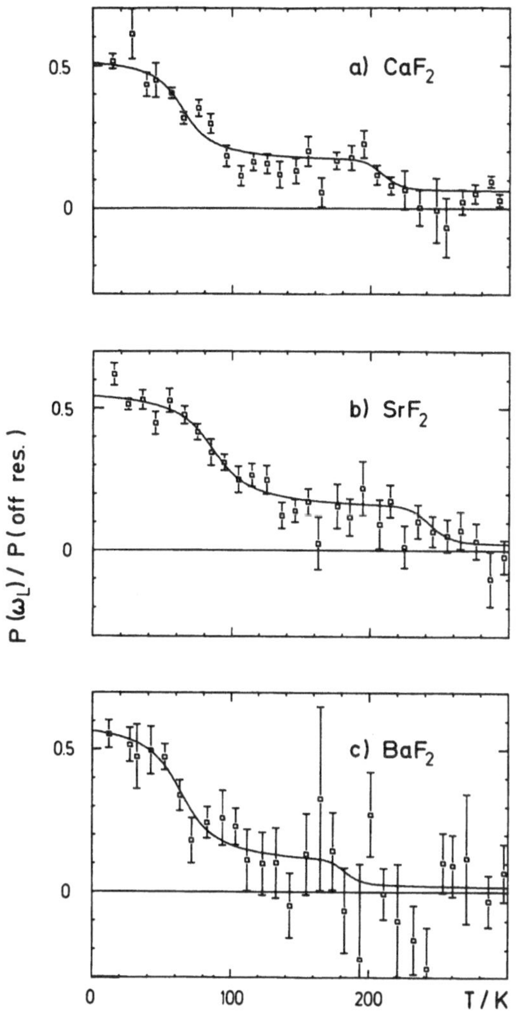

Fig.6.22. Part of defect interacting ^{20}F nuclei in CaF_2, SrF_2, BaF_2. Annealing stages occur at temperatures where the annealing time equals the ^{20}F β lifetime [6.170]

consist of F^- ions on interstitial sites. The annealing time τ of the defects was governed by an Arrhenius relation $\tau = \tau_0 \exp(E_M/k_BT)$. In the range 70-80 K, τ and the nuclear lifetime τ_N were of the same order of magnitude and the annealing process could be studied directly via the β-decay asymmetry.

This work in CaF_2 was extended to SrF_2 and BaF_2 [6.170]. In all three fluorites the annealing stage at about 80 K was observed, too (Fig.6.22). It was ascribed to correlated recombination of close Frenkel pairs with a migration enthalpy E_M of about 0.2 eV. An additional less marked stage was seen at about 220 K again in all three components, corresponding to $E_M \simeq 0.5$ eV. In this case the underlying annealing process could not be identified.

Another possibility to study quadrupolar interactions in cubic compounds is the application of pressure to the crystal. By this method a pressure broadening of the ^{20}F NMR signal in CaF_2 could be observed, yielding a value of $\delta(\Delta\nu)/\delta p = 0.34(8)$

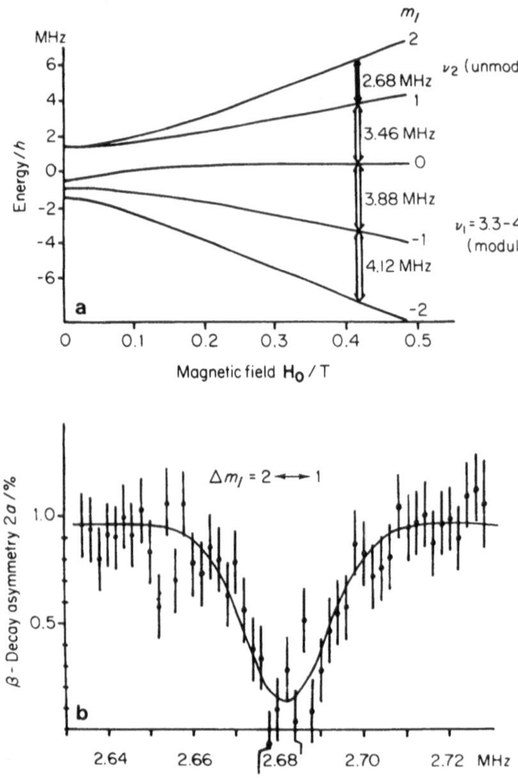

Fig.6.23. a) Energy levels of ^{20}F in tetragonal MgF_2 with <001> axis $\|$**B**. b) β-decay asymmetry versus frequency ν_2. Another RF field ν_1 is frequency modulated between 3.3 and 4.3 MHz and saturates three ($\Delta m = 1$) transitions [6.17]

Hz/bar [6.204]. Suppression of the dipolar line width by saturating the ^{19}F Larmor resonance was necessary to make this tiny effect observable.

In another experiment the QCC of ^{20}F in a MgF_2 single crystal was determined [6.17,143]. Figure 6.23 shows the level scheme of the I = 2 groundstate of ^{20}F as a function of B for the crystal orientation <001>$\|$**B**. The lower part of the figure shows a ($\Delta m = 1$)-transition. In this method a special signal amplification method using two RF fields was applied (see figure caption). By this a gain in signal depth of about one order of magnitude was possible. From the measurement the QCC of ^{20}F in MgF_2 and the asymmetry parameter η were obtained. Using the QCC of the excited 197 keV-state of ^{19}F in MgF_2 and its known Q moment [6.205], Q(^{20}F) could be determined. The same procedure was applied to measure the QCC of ^{17}F in MgF_2 by MINAMISONO et al. [6.144]. The Q moment of the 197 keV state of ^{19}F was based on a molecular EFG in ClF estimated by the Townes and Dailey theory. A recent recalculation of this EFG, however, yielded a new value, larger by a factor of 1.5 [6.206]. Accordingly, all fluorine Q moments would be only two thirds of the formerly adopted values.

In a further measurement on ^{20}F in MgF_2 the sign of e^2qQ was determined by DUBBERS et al. [6.145]. To this end the β-decay asymmetry of ^{20}F and the γ-decay anisotropy of the following transition in ^{20}Ne were observed simultaneously as a

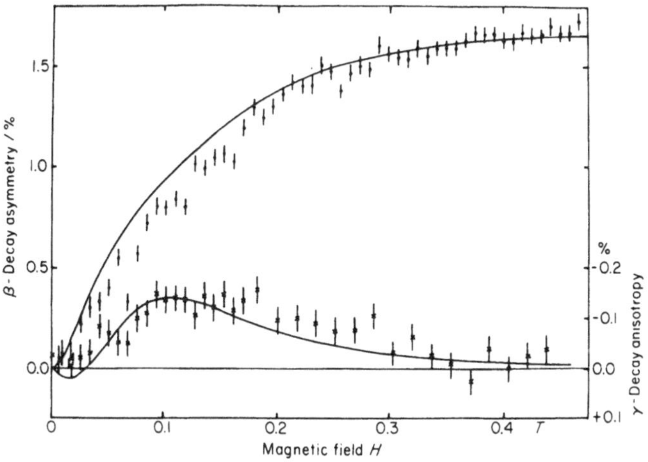

Fig.6.24

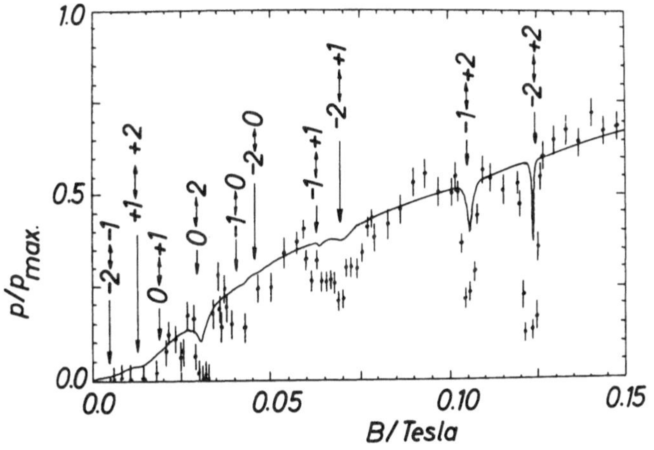

Fig.6.25

Fig.6.24. Experimental and theoretical B dependences of polarization (monitored by β-decay asymmetry) and alignment (monitored by γ-decay anisotropy) of ^{20}F in MgF_2 <001> axis ||**B** [6.145]

Fig.6.25. B dependence of the ^{20}F β-decay asymmetry in MgF_2 at low inductions (<001> axis ||**B**). The resonance like dips are due to cross-relaxation processes between the ^{20}F probes and the ^{19}F hosts [6.98]

function of B. The influence of RF irradiation on β asymmetry and γ anisotropy was studied too (Sect.6.1.2). Figure 6.24 shows β asymmetry and γ anisotropy versus B. Since the capture of polarized neutrons produces a purely dipolar polarization, the β-decay asymmetry shows its maximum value at large B, where the γ anisotropy is zero. At lower fields the dipolar polarization is partly converted by the quadrupolar interaction into an alignment, that is, the β asymmetry decreases and a γ anisotropy shows up. At zero induction the levels are completely mixed and β asymmetry as well as γ anisotropy are zero. The theoretical curves were calculated using the reorientation formalism discussed in Sect.6.1.2. The relative sign of β asymmetry and γ anisotropy yields the sign of the QCC. In a more recent measurement, the B dependence of the ^{20}F polarization in MgF_2 at low inductions was determined in greater detail [6.98,170]. At certain B values, deep resonance-like dips are observed (Fig.6.25). At these inductions transition frequencies in the ^{20}F system and the ^{19}F system of the surrounding stable nuclei coincide. By dipole-dipole interactions a mutual spin flip-flop process is energetically possible, and a loss of polarization in the ^{20}F system results due to cross-relaxation.

In a further experiment on fluorine compounds, the sign and magnitude of e^2qQ of ^{20}F in $KZnF_3$ were determined [6.146].

c) *Probe Nuclides in the Mass Number Range 24 ≤ A ≤ 39*

Several β-NMR experiments have been performed on nuclides in the mass number range 24 ≤ A ≤ 39. The experiments were mainly intended to measure nuclear moments and used both neutron activation and reactions with fast particle beams. In a NaF single crystal at RT, the isomeric 0.47 MeV state of $^{24}Na^m$ (20 ms, 1^+) was populated by polarized neutron capture [6.147]. This state decays mainly by γ emission to the ^{24}Na groundstate, but a small, not exactly known fraction undergoes a β decay to ^{24}Mg. This β decay was used to determine the magnetic moment of $^{24}Na^m$ by β-NMR.

Also using the $(\vec{n},\gamma)$-reaction, LADES produced ^{28}Al(2.4 min,3^-) in an AlP powder sample and ^{38}Cl(37.4 min, 2^-) in a NaCl single crystal [6.101]. The magnetic moments of ^{28}Al and ^{38}Cl were determined. Both measurements were performed at 1.2 K. The ^{38}Cl resonance line showed a broad threefold structure. From this splitting it was assumed that after the recoil caused by the $(\vec{n},\gamma)$ reaction, the ^{38}Cl nuclei came to rest at different lattice sites.

In a later experiment on a Al_2O_3 single crystal the QCC of ^{28}Al was determined by STÖCKMANN et al. [6.148]. Using the known QCC of ^{27}Al and Q(^{27}Al) the Q-moment of ^{28}Al was determined. This measurement also yielded a magnetic moment of ^{28}Al which did not agree with the value obtained by LADES. An independent measurement of μ(^{28}Al), already described in Sect.6.3.1b, used recoil implantation of ^{28}Al into Li metal. The obtained value was in complete agreement with that measured in Al_2O_3.

Using the selected recoil angle method, SUGIMOTO et al. determined the magnetic moment of ^{29}P (4.2 min $1/2^+$) [6.150]. The nuclide was produced through the reaction $^{28}Si(d,n)^{29}P$ in a thin silicon layer evaporated on a copper backing. The recoiled nuclides were stopped in polycrystalline red phosphorus, silicon crystals and germanium crystals. NMR was observed in the phosphorus and silicon samples.

In the following experiments the probe nuclei were generated by reactions with polarized protons. First, ^{39}Ca(0.87 s, $3/2^+$) was produced by the reaction $^{38}K(\vec{p},n)^{39}Ca$ in polycrystalline targets of KBr, $KCaBr_3$ and $KCaI_3$ by MINAMISONO et al. [6.151]. The measurements yielded the magnetic moment of ^{39}Ca. The magnetic moments of ^{31}S(2.6 s, $1/2^+$) and ^{25}Al(7.2 s, $5/2^+$) were determined by the same group

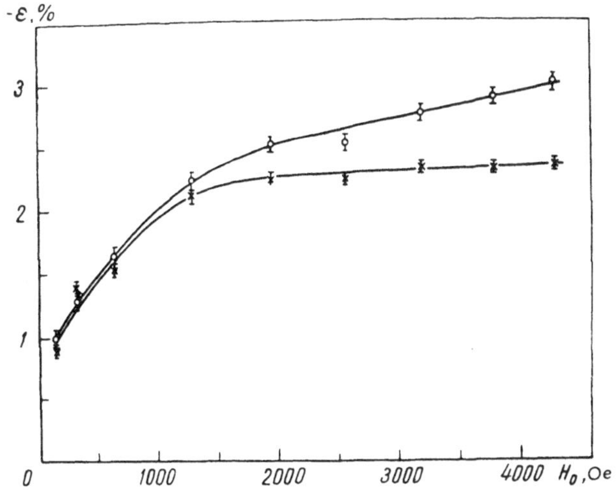

Fig.6.26. B dependence of the β-decay asymmetry ε of ^{108}Ag and ^{110}Ag in AgCl at 4.2K: (O) ^{108}Ag, (x) ^{110}Ag [6.131]

using the reactions $^{28}Si(\vec{p},\alpha)^{25}Al$ in a Si single crystal and $^{31}P(\vec{p},n)^{31}S$ in a red phosphorus target, respectively [6.36,104].

d) *Indium and Silver Compounds*

The B dependence of the nuclear polarization of ^{108}Ag (2.4 min, 1^+) and ^{110}Ag (24.4 s, 1^+) in an AgCl single crystal was measured by GUL'KO et al. [6.131,132]. Both isotopes were produced by the $(\vec{n},\gamma)$ reaction from the natural isotopes ^{107}Ag and ^{109}Ag. The B dependence showed the typical decoupling behaviour (Fig.6.26). The polarization started at low values for low B and reached its saturation value at high B.

The B dependence of ^{110}Ag was measured again by MERTENS et al. in AgCl, AgBr, and AgF powder samples [6.152]. This dependence was explained in terms of a dipolar coupling of the nuclear spin I to excited electronic states with angular momentum J. An alternative explanation was given later by DZHEPAROV and IVANTER who assumed a coupling of the activated Ag nuclei to their own paramagnetic, ionic Ag^{2+} shell produced by the (n,γ) reaction [6.153]. In a recent work, described below in greater detail, it could be shown, however, that the observed B dependence is caused by quadrupolar coupling of the nucleus to defect fields [6.171]. The magnetic moment of ^{110}Ag was also determined by MERTENS et al., performing β-NMR in powder samples of AgF, AgBr, AgCl at 8 K [6.152,154]. In a further β-NMR experiment also $\mu(^{108}Ag)$ was determined in polycrystalline AgCl [6.155].

In these experiments only a fraction of about 20% - 30% of the ^{110}Ag nuclei could be depolarized by RF irradiation. This fraction differed only slightly from compound to compound. This point was investigated further in a recent defect study in AgCl

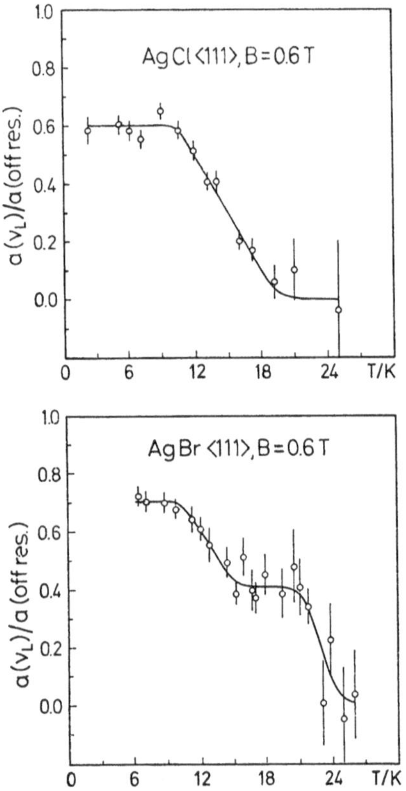

Fig.6.27. Relative part of ^{110}Ag nuclei with neighbouring defects in AgCl and AgBr, showing annealing stages in the temperature region 10...20 K [6.171]

and AgBr [6.171]. Using a statistical point charge model [6.207] it could be shown that about 70-80% of the ^{110}Ag nuclei are neighbouring a large number of Frenkel defects shifting the resonance frequency off from the Larmor frequency. As a possible explanation for the large defect number coulombic explosion was discussed, following internal conversion of the deexcition γ-cascade with high ionization of the ^{110}Ag nuclei. Defect annealing was observed already at temperatures of 10-20 K corresponding to migration enthalpies of 30-60 meV (Fig.6.27). From a comparison with ionic conductivity measurements it was concluded that Ag^+ collinear interstitialcy migration is the responsible annealing mechanism.

Similar measurements were performed by the same group on ^{116}In which was produced by the $(\vec{n},\gamma)$ reaction in powder samples of InP, InAs, and InSb [6.6,156]. The B dependence of the β-decay asymmetry as well as the magnetic moment of ^{116}In were measured. As in the case of the silver halides, a major part of the probe nuclei did not fulfill the Larmor resonance condition due to interactions with recoil defects. The detailed structure of these defects could not, however, be determined. In a recent experiment on ^{116}In in InP and InSb single crystals, this open question was settled to a large extent [6.157,219]. The defects proved to be charged and paramagnetic centres, probably F centres.

In the above-mentioned experiments on ^{108}Ag, ^{110}Ag, and ^{116}In, Q-moment ratios could also be determined from spin-lattice relaxation rates. This aspect is discussed in Sect.6.4.2.

6.4.2 Relaxation of Nuclear Orientation

Spin-lattice relaxation in diamagnetic insulators is often dominated by the quadrupolar coupling of the nuclei to the lattice phonons (Sect.6.1.3d). In a restricted temperature range ionic motion may also contribute to the total relaxation rate which then displays the characteristic maximum treated above for solid metals (Sect.6.3.2a). Contrary to the situation encountered in NMR on stable nuclides, paramagnetic impurities generally do not contribute to spin-lattice relaxation in in-beam experiments, because in the absence of spin diffusion the probe nuclei cannot communicate with paramagnetic centres beyond the near neighbourhood. Spin diffusion is hindered since the probe nucleus is surrounded by nuclei with a different gyromagnetic ratio. Paramagnetic relaxation *without* spin diffusion [Ref.6.12, Chap.IX] can, however, play a role if paramagnetic radiation defects are produced near the probes. For amorphous materials, another and hitherto not fully understood mechanism appears which is probably based on spin-phonon coupling and leads to an enhancement of the relaxation rate by at least one order of magnitude as compared to the crystalline counterpart.

Relaxation measurements in non-metals have only been done with β-emitting probes and detailed studies exist only for neutron-activated nuclides. In this case the risk of radiation damage is the least of all in-beam methods.

a) *Lithium and Fluorine Compounds*

In a recent systematic study the relaxation behaviour of ^8Li in an amorphous insulator, the silicate glass $(SiO_2)_{0.67}(Li_2O)_{0.33}$, was investigated as a function of T and of B in the range 6...340 K and 14...830 mT, respectively [6.208]. The ^8Li polarization P as function of time t was found to obey an $\exp(-\sqrt{t/T_1})$ law instead of a simple exponential (Fig.6.28). T_1, defined in this way, showed an approximately linear dependence on B (Fig.6.29) and 1/T for $T \lesssim 170$ K. Above, T_1 is influenced by classical diffusion of the alkali ions. In the low-T range both P(t) and $T_1(B,T)$ can be understood assuming relaxation by coupling to nearby defect centres with a distribution of different correlation times. Since spin diffusion is inhibited (see above), the observed polarization decay corresponds to an inhomogenous averaging of the individual exponential decays. Radiation defects cannot be responsible for the change of T_1 with B as this finding has been verified by ^7Li-NMR measurements on another sample of the same glass [6.208]. The relaxing centres have tentatively been considered as two-level systems undergoing transitions due to a multi-phonon process. Two-level systems involving one-phonon induced transitions have commonly been invoked in glasses in order to explain their anomalous thermal and acoustic properties at low temperatures [6.209].

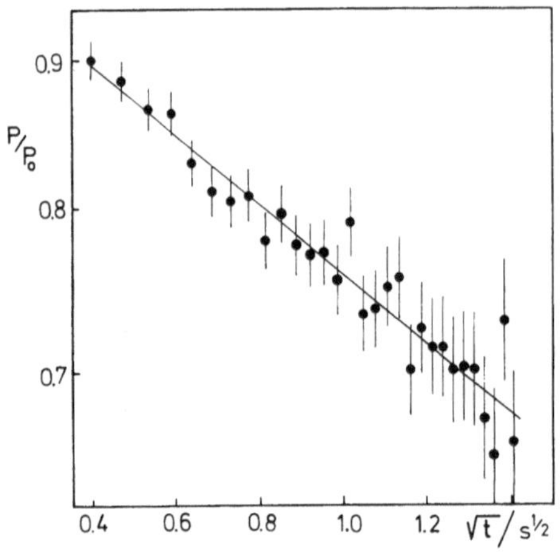

Fig.6.28. Relative polarization P/P_0 (on logarithmic scale) of ^8Li in $(SiO_2)_{0.67}(Li_2O)_{0.33}$ glass versus square root of time t. The solid line represents a fit according to $P/P_0 = \exp(-\sqrt{t/T_1})$. [T = 8 K, B = 30 mT; T_1 = 13.3(1.3)s][6.208]

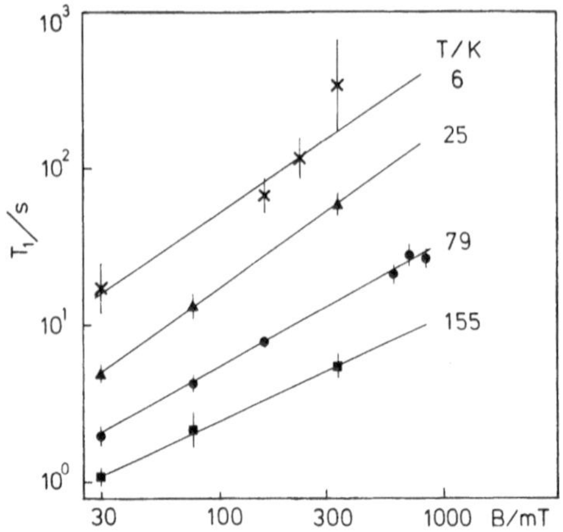

Fig.6.29. T_1 of ^8Li in $(SiO_2)_{0.67}(LiO)_{0.33}$ glass versus B for various temperatures. The lines represent fits using $T_1 \sim B^\beta$, yielding $\beta \approx 1$ [6.208]

The lamellar-type compound LiC_6 has been the subject of a preliminary study using ^8Li which has been started to complement ^7Li-NMR measurements [6.169]. Like most other graphite intercalation compounds [6.210] LiC_6 at RT has an in-plane electrical conductivity similar to a normal metal [6.211] and a small but definite Knight shift [6.212,213] so that it could well have been mentioned in Sect.6.3.2. Below RT $1/T_1$ of ^8Li was found to depend fairly linearly on T. Assuming magnetic hyperfine interaction the measured $1/T_1$ both of ^8Li and ^7Li, besides being not quite consistent, cannot be easily interpreted in terms of the experimental Knight shifts. Above RT $1/T_1$ of ^8Li shows an additional strong increase with T similar to Li metal (Fig. 6.9) which can be attributed to diffusion.

Besides these more or less systematic measurements single T_1 values have previously been published for ^8Li in $LiNbO_3$ at RT and 473 K [6.136] and in $LiIO_3$ at RT [6.102]. In the latter case ^8Li was produced by a fast particle reaction.

In an experiment on ^{20}F in CaF_2 single crystals [6.159] the phonon-induced quadrupolar relaxation rate in the case $T > \theta/2$ (θ: Debye temperature) was measured and $1/T_1 \sim T^2$ was found as predicted by theory, (6.51). In the T range 650...800 K a preliminary study of relaxation by fluorine diffusion has been made.

Single T_1 values exist for ^{20}F in MgF_2 at RT [6.17] and for ^{17}F in MgF_2 at RT and 770 K [6.144]. Comparing the two T_1 values at RT with the QCC's of the probe nuclei (Sect.6.4.1b) the ratio $T_1(^{20}F)/T_1(^{17}F)$ is found to be consistent with quadrupolar relaxation.

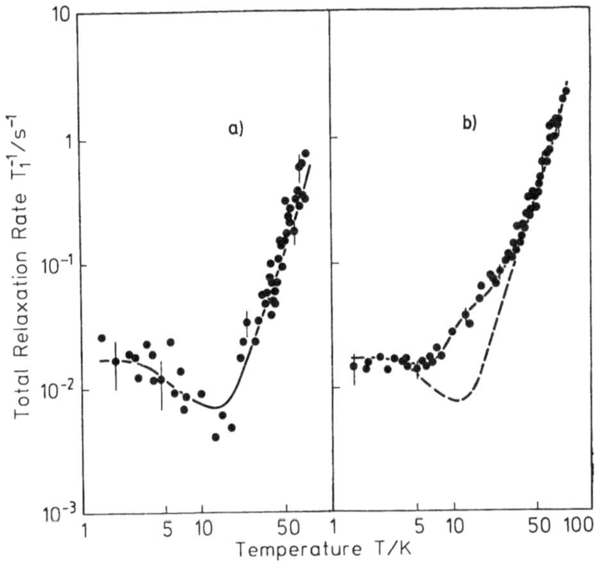

Fig.6.30a,b. Relaxation rates of resonant (a) and non-resonant (b) ^{116}In nuclei in InP versus T at B = 0.6 T. The solid lines represent fits based on (a) quadrupolar and (b) quadrupolar and additional paramagnetic relaxation [6.219]

b) *Indium and Silver Compounds*

Early systematic measurements by the neutron-activation method were done on ^{116}In (14 s) in polycrystalline InP, InAs and InSb [6.6] and on ^{110}Ag (24 s) in AgCl and AgBr powders [6.152]. The measurements extending over several decades in T_1 were confined to temperatures below LN_2 in the In and below 30 K in the Ag compounds. Above 30 and 20 K, respectively, the strong T dependence could be described by the theory of phonon-induced quadrupolar relaxation (Sect.6.1.3d) when a realistic phonon spectrum was used. At lower temperatures the relaxation rate was found to be enhanced which was tentiatively ascribed to resonant modes in the phonon spectrum induced by recoil lattice defects. Comparing the relaxation rates of ^{116}In above 30 K with those of ^{115}In from conventional NMR [6.158], the quadrupole moment ratio $|Q(^{116}In)/Q(^{115}In)|$ was deduced [6.6]. In another experiment on AgCl by the same group, both the relaxation rates of ^{110}Ag and ^{108}Ag (2.4 min) were measured and the ratio $|Q(^{110}Ag)/Q(^{108}Ag)|$ determined [6.155].

These early measurements suffered from the fact that only an averaged relaxation rate of *all* the nuclei of the same isotope was measured. In reality, two types of nuclei are produced in each case (Sect.6.4.1d), one showing a NMR signal at the Larmor frequency ω_L and another one showing no resonance effect at ω_L. In a recent re-investigation of the ^{116}In relaxation in InP and InSb single crystals [6.157,219] the behaviour of the two species was studied separately. The resonant nuclei, representing about 1/4 of all ^{116}In nuclei, showed the normal phonon-induced quadrupolar relaxation except at low T (Fig.6.30a). Describing the transverse acoustic and the optical branch of the phonon spectrum by a Debye and a Einstein spectrum, respectively, characteristic frequencies were obtained which agree with those from neutron scattering and infrared spectroscopy. By comparing the rate of the undisturbed ^{116}In nuclei with that of stable ^{115}In, a reliable value for the ratio of the quadrupole moments was obtained [6.214]. The non-resonant probes, on the other hand, showed an enhanced quadrupolar relaxation rate (Fig.6.30b) and additionally a B-dependent rate which was assigned to a nearby paramagnetic radiation defect, probably a F-centre. The spin-correlation time of the defect electron is governed by an Orbach process with an excitation energy of some meV.

Besides the In compounds, the relaxation rate of resonant and nonresonant ^{110}Ag nuclei was measured in AgCl and AgBr single crystals [6.171]. The rate of the resonant nuclei was fitted by Van Kranendonk's equation for quadrupolar relaxation (6.50) using the Debye temperature and the spin-phonon coupling constant as free parameters. Inserting a recent theoretical value for the anti-shielding factor of Ag^+ in crystals [6.215] the quadrupole moment of ^{110}Ag was estimated from the spin-phonon coupling constant.

Acknowledgement. This work was sponsored by the "Bundesministerium für Forschung und Technologie". One of the authors (P.H.) is grateful for the hospitality of the Institut Laue-Langevin, Grenoble, where part of this contribution was written.

6.5 Tabular Summary

Abbreviations used in the Tabular Summary:

p	polycrystalline solid	(applies to all host materials listed before one of these symbols)
s	single crystal	
n	non-crystalline	
ℓ	liquid	
c	fractional concentration (of alloys)	
β-NMR	β-ray detected NMR	
γ-NMR	γ-ray detected NMR	
β-asym	methods using asymmetric β radiation	
a	asymmetry of β radiation	
γ-anis	methods using anisotropic γ radiation	
μ	nuclear magnetic dipole moment, and/or Knight shift	
Q	nuclear electric quadrupole moment	
e^2qQ	electric quadrupole coupling constant	
ϑ	polar angle of the crystal EFG relative to **B**	
η	asymmetry parameter of the EFG	
$\Delta\nu$	NMR line width	
$T_1, T_1^{(2)}$	spin-lattice relaxation time (cf. Sect.6.1.3)	
MQT	multiple quantum transition	
t	time	
T	temperature	
RT	room temperature	
P	nuclear polarization	
B	static magnetic induction	
f_1, f_2	longitudinal dipolar or quadrupolar polarization	
B_{hf}	magnetic hyperfine induction	
χ	magnetic susceptibility	
E_β	energy of the β particles	
E_A	activation energy	

Probe, Reaction	Host	Method	T [K]	Results, Remarks
$^{8}Li(0.84s,2^{+})$; $(\vec{n},\gamma)$	$^{7}Li(p,\ell)$	β-asym	10-1100	$T_1(T,B) \rightarrow$ self-diffu $T_{1e}(\ell)$ related to 1
	^{7}Li-Mg, ^{7}Li-Ag(p)	β-asym	10-450	$T_1(T,B,c) \rightarrow$ self-, i
	^{7}Li-Na, ^{7}Li-Mg(ℓ)	β-asym	500-950	$T_1(T,c) \rightarrow T_{1e}$
	^{7}Li-Pb, ^{7}Li-Bi(ℓ)	β-asym	550-1500	$T_1(T,c) \rightarrow T_{1e}$; aroun 25 at% Bi departure behaviour
	LiC_6(p)	β-asym	80-420	$T_1(T)$
	$(Li_2O)_{0.33}(SiO_2)_{0.67}$ (n)	β-asym	6-340	$T_1(T,B) \rightarrow$ influence state, diffusion; P
	^{7}LiF(s)	β-NMR	RT	μ
	^{7}LiF(s)	β-asym	RT	P(B)
	^{7}LiF(s)	β-NMR	RT	sign of μ
	^{7}LiF(p)	β-NMR	RT	μ
	LiF(s)	β-NMR	RT	NMR-line shapes
	LiF(s)	β-NMR	13-295	defect induced line e^2qQ, annealing of
	Li, ^{7}Li(p)	β-asym	77,250	cross-relaxation ^{8}L
	LiF(s)	β-asym	RT	cross-relaxation ^{8}L
	$^{7}LiNbO_3$(s)	β-asym	300-820	cross-relaxation ^{8}L
	$^{7}Li_2S$, $^{7}Li_2O$(p)	β-NMR	RT	suppression of dipo precise μ
	$LiNbO_3$(s)	β-asym	RT,473	T_1
	$LiNbO_3$(s)	β-NMR	RT	e^2qQ; $Q(^{8}Li)/Q(^{7}Li)$
	$LiTaO_3$(s)	β-NMR	RT	e^2qQ; $Q(^{8}Li)/Q(^{7}Li)$

(Cont.)

Probe, Reaction	Host	Method	T [K]	Results, Remarks
^{8}Li(0.84s,2^{+}) (d,p)	Au,Pd,Pt(p)	β-NMR	RT	μ,T_1
	Be,Mg(s)	β-NMR	RT	limits for e^2qQ
^{8}Li(0.84s,2^{+}) ($\vec{d}$,p)	Li(p)	β-asym	RT	T_1
	$LiIO_3$(s)	β-NMR	RT	e^2qQ, Q(^{8}Li)/Q(^{7}Li);
^{8}Li(0.84s,2^{+}) ^{9}Be($^{7}\vec{Li}$,^{8}Li)	Au,Pd, Pt,Pb(p)	β-asym	RT	T_1
^{8}B(0.76s,2^{+}) (^{3}He,n)	Pt(p)	β-NMR	RT	μ,T_1
^{12}B(20ms,1^{+}) (d,p)	Cu,Pt,Au(p)	β-NMR	RT	μ,P(B)
	Cu,Al,Au,Pd,Pt, Be,V,Mn,Xe,C Mylar, Teflon(p); LiF,Si(s)	β-asym	RT	P(B=50 mT)
	Pd,Pt,Au(p)	β-asym	RT	P(B)
	Pd,Pt,Au(p)	β-NMR	RT	μ,T_1
	Cu,Pd,Pt,Au(p) Pd,Pt,Au(p)	β-NMR	RT 200-600	μ T_1(T)
	Al,Cu,Pt,Au(p)	β-NMR	RT	Δν → interst. sites
	Al,Cu,Rh,Pd, Ag,Pt,Au,Si,Ge, SiC(p)	β-NMR	RT	μ
	Al,Cu(s)	β-NMR	77;300; 540	Δν → octah. interst.
	Cu(s)	β-NMR	300-741	Δν → subst. sites, oc

(Cont.)

Probe, Reaction	Host	Method	T [K]	Results, Remarks
^{12}B(20ms,1^+) (d,p) (cont.)	Au(s)	β-NMR	RT	Δm = 2 transitions
	Al(s)	β-asym	68-751	$T_1(T) \rightarrow T_{1e}$, diffusion radiation damage
	Al(s)	β-asym	300-700	T_1 (orient., T)
	TiB_2,ZrB_2(p)	β-NMR	RT	e^2qQ, Q, subst. sites
	ZrB_2(s)	β-NMR	200-400	e^2qQ, Q, subst. sites
	Nb,Mo,Ta,W(p)	β-NMR	RT	e^2qQ, interst. sites
	Ta(p)	β-NMR γ-anis	RT	sign of e^2qQ
	Be(p),Be(s)	β-NMR	RT	e^2qQ, Q, site uncerta
	Be(s)	β-NMR	RT	e^2qQ, MQT
	Mg(s)	β-NMR	RT	e^2qQ
	Mg(s)	β-asym	RT	$P(B) \rightarrow e^2qQ$
	Mg(s)	β-NMR	RT	Δm = 2 transitions
	Al(p),Mg(s)	β-NMR	RT	$a(E_\beta,f_1,f_2)$
	Mg(s)	β-NMR	RT	$a(E_\beta,f_1,f_2)$
	Mg(s)	β-NMR	RT	$a(f_1,f_2)$
	Mg(s)	β-asym	RT	$a(f_1,f_2)$
	Zn(s)	β-NMR	RT	two e^2qQ's, sites unc
	Si,Ge(s)	β-NMR	RT	$\Delta\nu \rightarrow$ subst. sites
	Ge(s)	β-asym	94-745	P(T), rad. defects
	Ni(p)	β-NMR	6-730	$B_{hf}(T)$,$T_1(T,b)$,$\chi(T)$ oct./tetr. interst. s
	Fe(s),Fe(p)	β-NMR	100-1200	$B_{hf}(T)$,$T_1(T)$ oct./tetr. interst. s

(Cont.)

Probe, Reaction	Host	Method	T [K]	Results, Remarks
$^{12}B(20ms,1^+)$ $^7Li(^7\vec{L}i,^{12}B)$	Au(p)	β-asym	RT	T_1
$^{12}B(20ms,1^+)$ $^{100}Mo(^{14}N,^{12}B)$, $^{232}Th(^{14}N,^{12}B)$, $^{27}Al(^{14}N,^{12}B)$ $^{nat}Cu(^{14}N,^{12}B)$	Pt(p),Mg(s)	β-NMR	RT	$P[E_\beta(^{12}B)$,reaction-(
$^{13}B(17.3ms,3/2^-)$ (t,p)	Pd,Pt,Au(p)	β-NMR	RT	μ
	Mg(s)	β-NMR	RT	e^2qQ
$^{12}N(11ms,1^+)$ (t,n)	Al,Cu,Pt(p)	β-NMR	RT	μ,P(B)
	Al,Cu,Pt(p)	β-NMR	RT	$\Delta\nu \rightarrow$ interst. sites
	Nb,Ta(p)	β-NMR	RT	e^2qQ, interst. sites
	Be,Mg(s)	β-NMR	RT	e^2qQ, sites uncerta
	Mg(s)	β-NMR	RT	P(B),NMR $\rightarrow e^2qQ$
	Al(p),Mg(s)	β-NMR	RT	$a(E_\beta,f_1,f_2)$
	Mg(s)	β-NMR	RT	$a(E_\beta,f_1,f_2)$
	Ni(p)	β-NMR	98-773	$B_{hf}(T),T_1(T)$
$^{17}F(66s,5/2^+)$ (d,n)	CaF_2(p)	β-NMR	RT	μ
	MgF_2(s)	β-NMR	77-770	e^2qQ; η; $Q(^{17}F)/Q(^{1}$ T_1 at RT, 770 K
$^{20}F(11s,2^+)$; $(\vec{n},\gamma)$	CaF_2(s)	β-asym	RT	P(B)
	CaF_2(s)	β-NMR	RT	μ, sign of μ

(Cont.)

Probe, Reaction	Host	Method	T [K]	Results, Remarks
$^{20}F(11s,2^+)$ $(\vec{n},\gamma)$ (cont.)	$CaF_2(s)$	β-NMR	RT,4.2	μ, structured NMR-si
	$CaF_2(s)$	β,γ-NMR	10-80	defect induced quadr ting, e^2qQ,ϑ; anneal
	$CaF_2(s)$	β-asym	300-800	$T_1(T,B)$
	CaF_2,SrF_2, $BaF_2(s)$	β-NMR	10-294	defect induced line defect annealing, E_A
	$CaF_2(s)$	β-NMR	RT	pressure induced EFG
	$MgF_2(s)$	β-NMR	RT,12	e^2qQ, η, $Q(^{20}F)/Q(^{19}$ defects at 12 K; T_1
	$MgF_2(s)$	β,γ-NMR	RT	sign of e^2qQ
	$MgF_2(s)$	β-asym	5-100, RT	cross-relaxation ^{20}F annealing of defects
	$KZnF_3(s)$	β,γ-NMR	RT	e^2qQ, sign of e^2qQ
$^{24}Na(20ms,1^+)$ $(\vec{n},\gamma)$	NaF(s)	β-NMR	RT	μ,T_1
	Na(p)		10-77	
$^{25}Al(7.2s,5/2^+)$ $(\vec{p},\alpha)$	Si(s)	β-NMR	RT	μ,T_1
$^{28}Al(2.3min,3^+)$ $(\vec{n},\gamma)$	AlP(p)	β-NMR	1.5	μ
	$Al_2O_3(s)$	β-NMR	77	μ; e^2qQ; $Q(^{28}Al)/Q(^{27}$
$^{28}Al(2.3min,3^+)$ (d,p)	Li(p)	β-NMR	1.5-4.2	dynamic polarization,
$^{28}P(270ms,3^+)$ $(\vec{p},n)$	Si	β-asym	77	a ≠ 0
$^{29}P(4.2s,1/2^+)$ (d,n)	P(p)	β-NMR	RT	μ

(Cont.)

Probe, Reaction	Host	Method	T [K]	Results, Remarks
$^{29}P(4.2s,1/2^+)$ $(\vec{p},n)$	Si(s)	β-asym	RT	T_1
$^{31}S(2.6s,1/2^+)$, $(\vec{p},n)$	P(p)	β-NMR	RT	μ, T_1
$^{38}Cl(37.4m,2^-)$; $(\vec{n},\gamma)$	NaCl(s)	β-NMR	1.2	μ; structured NMR-s
$^{39}Ca(0.87s,3/2^+)$; (p,n)	$KCaBr_3$(p) $KCaI_3$ (p) KBr (p)	β-NMR	RT	T_1, μ
$^{41}Sc(0.6s,7/2^-)$ (d,n)	Pt(p)	β-NMR	4.2	μ
$^{71}Ge(20ms,9/2^+)$ (p,n)	Ga(ℓ)	γ-anis	310-1320	$T_1^{(2)}(T) \rightarrow T_{1e}^{(2)}, T_{1Q}^{(2)}$ $Q(^{71}Ge,9/2^+)/Q(^{69,7}$
	Ga(ℓ)	γ-NMR	~323	$\mu, T_1^{(2)}$
	In-Ga(ℓ)	γ-anis	350-1150	$T_1^{(2)}(T,c) \rightarrow T_{1e}^{(2)}$, T_{1} $Q(^{71}Ge,9/2^+)/Q(^{69}Ga$
$^{73}As(5.8\mu s,9/2^+)$ (α,2n)	Ga(ℓ)	γ-NMR		μ
	Ga-Te(ℓ)	γ-anis	1110	$T_1^{(2)}(c) \rightarrow T_{1Q}^{(2)}$; arou semicond. behaviour
$^{78}Br(0.12ms,4^{(+)})$ (p,n)	SeTl(ℓ)	γ-NMR	~ 613	$\mu, T_1^{(2)}$

(Cont.)

Probe, Reaction	Host	Method	T [K]	Results, Remarks
$^{81}Br(37\mu s, 9/2^+)$ (d,n)	SeTl(ℓ)	γ-NMR		μ
$^{88}Y(14ms, 8^+)$ (d,n)	Rb(ℓ)	γ-anis	298-433	$T_1^{(2)}(T,B)$, mechanisı
$^{108}Ag(2.4min, 1^+)$ $(\vec{n},\gamma)$	AgCl(s)	β-asym	4.2	P(B)
	AgCl(p)	β-NMR	8-16	$\mu; T_1(T) \to Q(^{108}Ag)/Q($
$^{110}Ag(24.4s, 1^+)$ $(\vec{n},\gamma)$	AgCl(s)	β-asym	4.2	P(B)
	AgF,AgCl,AgBr(p)	β-NMR	4.2	$T_1(T,B), P(B); \mu$; def
$^{110}Ag(24.2s, 1^+)$ $(\vec{n},\gamma)$	AgCl,AgBr(s)	β-NMR	2-25	defect induced line defect annealing, E
	AgCl(p)	β-NMR	8-16	$\mu; T_1(T,B) \to Q(^{108}Ag)/$
$^{114}Sb(0.21ms, 8^-)$ (d,2n)	Sn(ℓ)	γ-anis	510-1250	$T_1^{(2)}(T) \to T_{1Q}^{(2)}$ not or
$^{114}Sb(0.21ms, 8^-)$ (α,3n)	InSb(ℓ)	γ-anis	510-1300	$T_1^{(2)}(T) \to T_{1Q}^{(2)}$
$^{115}Sn(0.16ms, 11/2^-)$(p,n)	In(ℓ)	γ-NMR	473	$\mu, T_1^{(2)}$
		γ-anis	450-1120	$T_1^{(2)}(T) \to T_{1e}^{(2)}, T_{1Q}^{(2)}$ d
$^{115}Sn(0.16ms, 11/2^-)$(d,2n)	InSb(ℓ)	γ-anis	800-1200	$T_1^{(2)}(T)$

(Cont.)

Probe Reaction	Host	Method	T [K]	Results, Remarks
$^{116}In(14s,1^+)$; $(\vec{n},\gamma)$	InP,InAs,InSb(p)	β-NMR	4.2-70	μ;P(B);$T_1(T,B) \rightarrow Q(^{11\ldots}$ defects observed
	InP,InSb(s)	β-NMR	1.3-90	$T_1(T,B)$, resonant an… nuclei discriminated… paramagn. relax.
	InSb(s)	β-NMR	4.2	μ, T_1
	In(p)		0.45	T_1
	In(s)	β-asym	4.2	T_1
$^{117}Sb(0.34ms, 25/2^+)(\alpha,2n)$	In(ℓ)	γ-anis	430-1000	$T_1^{(2)}(T) \rightarrow T_{1Q}^{(2)}$
	In-X(ℓ)	γ-anis	260-1230	$T_1^{(2)}(T,c) \rightarrow T_{1Q}^{(2)}$ corr… enthalpy of mixing;
	X=Cu,Ag,Au,Zn, Cd,Hg,Ga,Tl,Ge, Sn,Pb,As,Sb,Bi,Te			$Q(^{117}Sb,25/2^+)/Q(^{121}\ldots$
$^{117}Sb(0.34ms, 25/2^+)(d,3n)$	In-Sn(ℓ)	γ-anis	not given	$T_1^{(2)}(c) \rightarrow T_{1Q}^{(2)}$
$^{132}Xe(8ms,10^+)$ $(\alpha,2n)$	Te(ℓ)	γ-anis	700-1050	$T_1^{(2)}(T) \rightarrow T_{1Q}^{(2)}$ not on…
	Te-X(ℓ)	γ-anis	800-920	$T_1^{(2)}(c) \rightarrow T_{1Q}$ partly … conductivity
	X=In,Bi,Sb,As,Au			
$^{136}Ce(2\mu s,10^+)$ $(^{16}O,4n)$	Sn(ℓ)	γ-anis	500-925	$\mu, B_{hf}(T), T_1^{(2)}(T)$
	Sn-Au(ℓ),Sn-Bi(ℓ)	γ-anis	460-925	$B_{hf}(T), T_1^{(2)}(T,c)$
	Pb(p)	γ-anis	300-450	$B_{hf}(T), T_1^{(2)}(T)$
	Th(p)	γ-anis	18-800	$B_{hf}(T), T_1^{(2)}(T)$

(Cont.)

Probe, Reaction	Host	Method	T [K]	Results, Remarks
$^{138}Ce(0.08\mu s, 10^+)$ $(\alpha, 4n)$	Ba(p)	γ-anis	77-900	$\mu, B_{hf}(T), T_1^{(2)}(T)$
$^{139}Ce(0.07\mu s, 19/2^-)(\alpha, 3n)$	Ba(p)	γ-anis	77-900	$\mu, B_{hf}(T), T_1^{(2)}(T)$
$^{138}Nd(0.35\mu s, 10^+)$ $(^{20}Ne, 4n)$	Sn(ℓ)	γ-anis	530-1140	$\mu, B_{hf}(T), T_1^{(2)}(T)$
	Sn-Au(ℓ)	γ-anis	795	$B_{hf}, T_1^{(2)}(c)$
	Sn-Bi(ℓ)	γ-anis	420-1000	$B_{hf}, T_1^{(2)}(T,c)$
$^{139}Pr(0.04\mu s, 11/2^-)(\alpha, 4n)$	La(ℓ)	γ-anis	1250	$\mu, B_{hf}, T_1^{(2)}$
$^{147}Eu(0.77\mu s, 11/2^-)(^{12}C, 4n)$	La(p,ℓ)	γ-anis	600-1400	$B_{hf}(T), T_1^{(2)}(T)$
$^{148}Eu(0.17\mu s)$ $(^{12}C, 3n)$	La(p,ℓ)	γ-anis	600-1400	$\mu/I, B_{hf}(T), T_1^{(2)}(T)$
$^{202}Tl(0.57ms, 7^+)$ (d,2n)	Hg(ℓ)	γ-anis	250-560	$T_1^{(2)}(T) \rightarrow T_{1Q}^{(2)}$ not on
$^{205}Pb(5.5ms, 13/2^+)$ $(\alpha, 3n)$	Hg(ℓ)	γ-anis	240-560	$T_1^{(2)}(T) \rightarrow T_{1e}^{(2)}, T_{1Q}^{(2)}$ d $Q(^{205}Pb, 13/2^+)/Q(^{201}$
$^{206}Pb(0.12ms, 7^-)$	Hg(ℓ)	γ-anis	235-530	$T_1^{(2)}(T) \rightarrow T_{1Q}^{(2)}$ diff. $Q(^{206}Pb, 7^-)/Q(^{201}Hg)$

(Cont.)

Probe, Reaction	Host	Method	T [K]	Results, Remarks
$^{206}Bi(0.88ms, 10^-)$ $(\alpha, 3n)$	Tl(ℓ)	γ-NMR	638	μ
$^{207}Bi(0.18ms, 21/2^+)(\alpha, 2n)$	Tl(p,ℓ)	γ-anis	520-1100	$T_1^{(2)}(T) \rightarrow T_{1e}^{(2)}$
	Tl-Pb, Tl-Bi, Tl-Au(ℓ)	γ-anis	500-1300	$T_1^{(2)}(T,c) \rightarrow T_{1e}^{(2)}$
$^{207}Bi(0.18ms, 21/2^+)(d, 2n)$	Pb(p,ℓ)	γ-anis	not given	$T_1^{(2)}(T) \rightarrow T_{1e}^{(2)}$
$^{208}Bi(2.5ms, 10^-)$ $(d, 2n)$	Pb(ℓ)	γ-NMR	648	μ
$^{207}Po(47\mu s, 13/2^+)(\alpha, 3n)$	Pb(p,ℓ)	γ-anis	500-1200	$T_1^{(2)}(T) \rightarrow T_{1Q}^{(2)}$ not or
	Pb-X(ℓ)	γ-anis	400-1300	$T_1^{(2)}(T,c) \rightarrow T_{1Q}^{(2)}$ corr
	X=Li,Ag,Au,In,Tl,Sn, As,Sb,Bi,Te			
$^{207}Po(47\mu s, 13/2^+)(d, 4n)$	Bi(ℓ)	γ-anis	550-1290	$T_1^{(2)}(T) \rightarrow T_{1Q}^{(2)}$ not onl
$^{212}Fr(27\mu s, 11^+)$ $(^{13}C, 4n)$	Tl(ℓ)	γ-anis	675	$T_1^{(2)}$
$^{213}Fr(0.24\mu s, 29/2^+)(^{12}C, 4n)$	Tl(ℓ) TlCl(p) Ar gas	γ-anis	575 573 RT	μ
$^{214}Ra(67\mu s, 8^+)$ $(^{12}C, 4n)$	Pb(ℓ)	γ-anis	598	$T_1^{(2)}$
	Pb(p)	γ-NMR	RT;543	μ, $T_1^{(2)}$ at RT

References

6.1 N.J. Stone: Hyperfine Interact. *2*, 45 (1976)
6.2 W.D. Brewer: J. Low Temp. Phys. *27*, 651 (1977)
6.3 D. Connor: Phys. Rev. Lett. *3*, 429 (1959)
6.4 D. Quitmann, J.M. Jaklevic, D.A. Shirley: Phys. Lett. *30*B, 329 (1969)
6.5 K. Sugimoto, A. Mizobuchi, K. Matuda, T. Minamisono: Phys. Lett. *31*B, 520 (1970)
6.6 A. Winnacker, H. Ackermann, D. Dubbers, J. Mertens, P. von Blanckenhagen: Z. Phys. *244*, 289 (1971)
6.7 D. Riegel, N. Bräuer, B. Focke, B. Lehmann, N. Nishiyama: Phys. Lett. *41*A, 459 (1972)
6.8 H. Hamagaki, K. Nakai, Y.Nojiri, I. Tanihata, K. Sugimoto: Hyperfine Interact. *2*, 187 (1976)
6.9 E. Karlsson, R. Wäppling (eds.): Hyperfine Interactions Studied in Nuclear Reactions and Decay. Physica Scripta *11*, 109-247 (1975)
6.10 R.S. Raghavan, D.E. Murnick (eds.): Proc. 4. Intern. Conf. on Hyperfine Interactions. Hyperfine Interact. *4*, 1-986 (1978)
6.11 H. Ackermann, D. Dubbers, H.-J. Stöckmann: *Adv. Nucl. Quadrupole Resonance*, Vol. III, ed. by J.A.S. Smith (Heyden, London 1978) pp.1-66
6.12 A. Abragam: *The Principles of Nuclear Magnetism* (Oxford University Press, 1961)
6.13 M.H. Cohen, F. Reif: Solid State Phys. *5*, 321 (1957)
6.14 T.P. Das, E.L. Hahn: Solid State Phys. Suppl. *1* (1958)
6.15 U. Fano, G. Racah: *Irreducible Tensorial Sets* (Academic, New York 1959)
6.16 System/360 Scientific Subroutine Package (360A-CM-03X), Version III, 4. Ed. (1968)
6.17 H.-J. Stöckmann, H. Ackermann, D. Dubbers, M. Grupp, P. Heitjans: Z. Phys. *269*, 47 (1974)
6.18 D. Dubbers: Z. Phys. A*276*, 245 (1976)
6.19 A.R. Edmonds: *Angular Momentum in Quantum Mechanis* (University Press, Princeton 1957)
6.20 F. Hartmann-Boutron: Ann. Phys. (Paris) *9*, 285 (1975)
6.21 S.R. de Groot, H.A. Tolhoek, W.J. Huiskamp: In *Alpha-, Beta- and Gamma-Ray Spectroscopy*, ed. by K. Siegbahn, Vol.2 (North-Holland, Amsterdam 1965) p.1201
6.22 U. Fano: Phys. Rev. B*133*, 828 (1964)
6.23 N. Bloembergen, E.M. Purcell, R.V. Pound: Phys. Rev. *73*, 679 (1948)
6.24 A. Abragam, R.V. Pound: Phys. Rev. *92*, 943 (1953)
6.25 A. Narath, H.T. Weaver: Phys. Rev. *175*, 373 (1968)
6.26 E.F.W. Seymour: Pure Appl. Chem. *40*, 41 (1974)
6.27 C.A. Sholl: J. Phys. F*4*, 1556 (1974)
6.28 W.W. Warren: Phys. Rev. *10*A, 657 (1974)
6.29 H. Gabriel: Phys. Status Solidi b *64*, K63 (1974)
6.30 J. van Kranendonk: Physica *20*, 781 (1954)
6.31 F.L. Shapiro: Uspekhi Fiz. Nauk *65*, 133 (1958)
6.32 H. Rauch: Z. Phys. *197*, 373 (1966)
6.33 C. Hsiung, H. Hsiung, A. Gordus: J. Chem. Phys. *34*, 535 (1961)
6.34 K. Sugimoto, A. Mizobuchi, K. Nakai, K. Matuda: J. Phys. Soc. Jpn. *21*, 213 (1966)
6.35 L. Pfeiffer, L. Madanski: Phys. Rev. *163*, 999 (1967)
6.36 T. Minamisono, J.W. Hugg, J.R. Hall, D.G. Mavis, D.L. Clark, S.S. Hanna: Phys. Rev. C*14*, 2335 (1976)
6.37 D. Riegel: Physica Scripta *11*, 228 (1975)
6.38 D. Riegel, N. Bräuer, B. Focke, K. Nishiyama, E. Matthias: Hyperfine Interact. *3*, 1 (1977)
6.39 Y. Yamazaki, O. Hashimoto, H. Ikezoe, S. Nagamiya, K. Nakai, T. Yamazaki: Phys. Rev. Lett. *33*, 1614 (1974)
6.40 N. Bräuer, B. Focke, B. Lehmann, D. Riegel: Z. Phys. *244*, 375 (1971)
6.41 N. Bräuer, B. Focke, B. Lehmann, E. Matthias, D. Riegel: Phys. Lett. *34*B, 54 (1971)
6.42 G. Schäfer, H. Hübel, C. Günther, A. Goldman, D. Riegel: Phys. Lett. *46*B, 65 (1973)

6.43 H. Hübel, C. Günther, K. Euler, N. Bräuer, D. Riegel: Nucl. Phys. A*227*, 421 (1974)
6.44 R.C. Haskell, L. Madansky: Phys. Rev. C*7*, 1277 (1973)
6.45 T. Minamisono, Y. Nojiri, A. Mizobuchi, K. Sugimoto: J. Phys. Soc. Jpn. *34*, Suppl., 156 (1973)
6.46 K. Sugimoto, K. Nakai, K. Matuda, T. Minamisono: Phys. Lett. *25*B, 130 (1967)
6.47 J. Berlijn, P. Keaton, L. Madansky, G. Owen, L. Pfeiffer, N. Roberson: Phys. Rev. *153*, 1152 (1967)
6.48 J. Wells, R. Williams, L. Pfeiffer, L. Madansky: Phys. Lett. *27*B, 448 (1968)
6.49 R. Williams, L. Pfeiffer, J. Wells, L. Madansky: Phys. Rev. C*2*, 1219 (1970)
6.50 R.E. McDonald, T.K. McNab: Phys. Rev. C*10*, 946 (1974)
6.51 K. Sugimoto, K. Nakai, K. Matuda, T. Minamisono: J. Phys. Soc. Jpn. *25*, 1258 (1968)
6.52 T. Minamisono, K. Matuda, A. Mizobuchi, K. Sugimoto: J. Phys. Soc. Jpn. *30*, 311 (1971)
6.53 K. Sugimoto, N. Takahashi, A. Mizobuchi, Y. Nojiri, T. Minamisono, M. Ishihara, K. Tanaka, H. Kamitsubo: Phys. Rev. Lett. *39*, 323 (1977)
6.54 R. Williams, L. Madansky: Phys. Rev. C*3*, 2149 (1971)
6.55 K. Sugimoto, A. Mizobuchi, T. Minamisono, Y. Nojiri: J. Phys. Soc. Jpn. *34*, Suppl., 158 (1973)
6.56 T.K. McNab, R.E. McDonald: Phys. Rev. B*13*, 34 (1976)
6.57 R.E. McDonald, T.K. McNab: Phys. Lett. *63*A, 177 (1977)
6.58 R.E. McDonald, T.K. McNab: Phys. Rev. B*13*, 39 (1976)
6.59 R.E. McDonald, T.K. McNab, J.A. Becker, J.D. Perez: Hyperfine Interact. *4*, 782 (1978)
6.60 Y. Nojiri, H. Hamagaki, K. Sugimoto: Phys. Lett. *60*A, 77 (1977); Hyperfine Interact. *4*, 465 (1978)
6.61 H. Hamagaki, Y. Nojiri, K. Sugimoto, K. Nakai: J. Phys. Soc. Jpn. *47*, 1806 (1979)
6.62 M. Hori, S. Ochi, T. Minamisono, K. Sugimoto: J. Phys. Soc. Jpn. *34*, Suppl., 161 (1973)
6.63 R.C. Haskell, F.D. Correll, L. Madansky: Phys. Rev. B*11*, 3268 (1975)
6.64 R.L. Williams, R.C. Haskell, L. Madansky: Phys. Rev. C*5*, 1435 (1972)
6.65 R.C. Haskell, L. Madansky: J. Phys. Soc. Jpn. *34*, Suppl., 167 (1973)
6.66 F.D. Correll: Hyperfine Interact. *4*, 544 (1978)
6.67 I. Tanihata, S. Kogo, K. Sugimoto: Phys. Lett. *67*B, 392 (1977)
6.68 K. Sugimoto, I. Tanihata, J. Göring: Phys. Rev. Lett. *34*, 1533 (1975)
6.69 Y. Masuda, T. Minamisono, Y. Nojiri, K. Sugimoto: Phys. Rev. Lett. *43*, 1083 (1979)
6.70 P. Lebrun, P. Deschepper, L. Grenacs, J. Lehmann, C. Leroy, L. Palffy, A. Possoz, A. Maio: Phys. Rev. Lett. *40*, 302 (1978)
6.71 H. Brändle, L. Grenacs, J. Lang, L. Roesch, V. Telegdi, P. Truttmann, A. Weiss, A. Zehnder: Phys. Rev. Lett. *40*, 306 (1978)
6.72 T. Minamisono, S. Kogo, K. Okajima, K. Sugimoto: Hyperfine Interact. *4*, 224 (1978)
6.73 R.E. McDonald, T.K. McNab: Phys. Rev. Lett. *32*, 1133 (1974)
6.74 N.L. Peterson: Solid State Phys. *22*, 409 (1968)
6.75 N.H. Nachtrieb: Ber. Bunsenges. Phys. Chem. *80*, 678 (1976)
6.76 R. Vianden: Hyperfine Interact. *4*, 956 (1978)
6.77 A. Narath: In *Hyperfine Interactions*, ed. by A.J. Freeman, R.B. Frankel (Academic, New York 1967) p.287
6.78 T. Moriya, K. Ueda: Solid State Commun. *15*, 169 (1974)
6.79 I. Weisman, L. Swartzendruber, L. Bennet: In *Measurements of Physical Properties*, ed. by E. Passaglia, Vol.VI/6 (Wiley, New York 1973) p.165
6.80 J. Winter: *Magnetic Resonance in Metals* (Oxford University Press, Oxford 1971)
6.81 J.M. Titman: Phys. Rpt. *33*, 1 (1977)
6.82 H. Ackermann, D. Dubbers, M. Grupp, P. Heitjans, R. Messer, H.-J. Stöckmann: Phys. Status Solidi b *71*, K91 (1975);
P. Heitjans: Dissertation, Universität Heidelberg (1975)

6.83 P. Heitjans, H. Ackermann, D. Dubbers, M. Grupp, H.-J. Stöckmann: Inst. Phys. Conf. Ser. *30*, 607 (1977)
6.84 T.K. McNab, J.O. Perez, R.E. McDonald: Phys. Rev. B*18*, 92 (1978)
6.85 P. Heitjans, A. Körblein, H. Ackermann, D. Dubbers, F. Fujara, M. Grupp, H.-J. Stöckmann: Proc. 19th Congr. Ampère, Heidelberg (1976) p.281
6.86 P. Heitjans, A. Körblein, H. Ackermann, D. Dubbers, F. Fujara, H. Grupp, M. Grupp, W. Hell, H.-J. Stöckmann: To be published
6.87 D.G. Hughes: Philos. Mag. *5*, 467 (1960)
6.88 J. Titman, S. Kellington: Proc. Phys. Soc. *90*, 499 (1967)
6.89 A. Körblein, P. Heitjans, H.-J. Stöckmann, F. Fujara, H. Ackermann, W. Buttler, K. Dörr, H. Grupp: To be published
6.90 D. Wolf: Phys. Rev. B*10*, 2710 (1974)
6.91 D. Wolf: Phys. Rev. B*15*, 37 (1977)
6.92 R. Messer, F. Noack: Appl. Phys. *6*, 79 (1975)
6.93 R. Messer: Private communication
6.94 C. Sholl: J. Phys. C*7*, 3378 (1974); C*8*, 1737 (1975); W. Barton, C. Sholl: J. Phys. C*9*, 4315 (1976)
6.95 A. Ott: Z. Naturforsch. Teil A: *25*, 1477 (1970)
6.96 J.W. Miller: Phys. Rev. *188*, 1074 (1969)
6.97 J. Titman, B. Moores: J. Phys. F*2*, 592 (1972)
6.98 F. Fujara, H.-J. Stöckmann, H. Ackermann, W. Buttler, K. Dörr, H. Grupp, P. Heitjans, G. Kiese, A. Körblein: Z. Phys. B*37*, 151 (1980)
6.99 H. Gutowsky, B. McGarvey: J. Chem. Phys. *20*, 1472 (1952)
6.100 H. Rauch: Z. Phys. *197*, 389 (1966)
6.101 H. Lades: Z. Phys. *252*, 242 (1972)
6.102 T. Minamisono, J. Hugg, D. Mavis, T. Saylor, S. Lazarus, H. Glavish, S. Hanna: Phys. Rev. Lett. *34*, 1465 (1975)
6.103 H.C. Torrey: Phys. Rev. *92*, 962 (1953)
6.104 T. Minamisono, J. Hugg, J. Hall, D. Mavis, H. Glavish, S. Hanna: Phys. Rev. C*14*, 376 (1976)
6.105 D. Riegel, N. Bräuer, E. Euler, C. Günther, H. Hübel: Hyperfine Interact. *2*, 273 (1976)
6.106 L.E. Drain: Metall. Rev. *12*, 195 (1967)
6.107 H. Gabriel: Hyperfine Interact. *2*, 91 (1976)
6.108 N. Bräuer, F. Dimmling, Th. Kornrumpf, M.v. Hartrott, K. Nishiyama, D. Riegel: Hyperfine Interact. *2*, 268 (1976)
6.109 F. Dimmling, M.v. Hartrott, T. Kornrumpf, K. Nishiyama, D. Riegel: 4th Intern. Conf. Hyperfine Interactions, Madison, NJ (1977); Book of Abstracts p.193
6.110 M.v. Hartrott, J. Hadijuana, K. Nishiyama, D. Quitmann, D. Riegel: Z. Phys. A*278*, 303 (1976)
6.111 K. Nishiyama, D. Riegel: Hyperfine Interact. *2*, 276 (1976)
6.112 S. Cocking, P. Egelstaff: J. Phys. C*1*, 507 (1968)
6.113 W. Schirmacher: Ber. Bunsenges. Phys. Chem. *80*, 736 (1976); Dissertation, Freie Universität, Berlin (1977)
6.114 J. Marsden, R. Havill, J. Titman: J. Phys. F*8*, 1321 (1978)
6.115 L. Varnell: Hyperfine Interact. *2*, 260 (1976)
6.116 J. Beene, O. Häusser, A. McDonald, T. Alexander, A. Ferguson, B. Herskind: Hyperfine Interact. *3*, 397 (1977)
6.117 M.v. Hartrott, K. Nishiyama, J. Rossbach, E. Weihreter, D. Quitmann: J. Phys. F*7*, 713 (1977)
6.118 M.v. Hartrott, J. Höhne, D. Quitmann, J. Rossbach, E. Weihreter, F. Willeke: Phys. Rev. B*19*, 3449 (1979)
6.119 M.v. Hartrott, K. Nishijama, J. Rossbach, E. Weihreter, D. Quitmann: Inst. Phys. Conf. Ser. *30*, 460 (1977)
6.120 J. Höhne, M.v. Hartrott, D. Quitmann, J. Rossbach, E. Weihreter, F. Willeke, W. Schirmacher: Proc. 7th Intern. Conf. Amorphous and Liquid Semiconductors, University of Edinburgh, ed. by W.E. Spear (1977) p.848
6.121 M.v. Hartrott, J. Hadijuana, K. Nishiyama, D. Quitmann: Hyperfine Interact. *2*, 271 (1976)
6.122 M.v. Hartrott, J. Höhne, D. Quitmann, J. Rossbach, E. Weihreter, F. Willeke: Hyperfine Interact. *4*, 816 (1978)

6.123 E. Weihreter, M.v. Hartrott, J. Höhne, D. Quitmann, J. Rossbach, F. Willeke: Phys. Lett. *67*A, 394 (1978)
6.124 M. Cutler: *Liquid Semiconductors* (Academic, New York 1977)
6.125 W.W. Warren: Inst. Phys. Conf. Ser. *30*, 436 (1977)
6.126 G. Cartledge, R. Havill, J. Titman: J. Phys. F*6*, 639 (1976)
6.127 F. Rossini, W. Knight: Phys. Rev. *178*, 641 (1969)
6.128 E. Claridge, D. Moore, E. Seymour, C. Sholl: J. Phys. F*2*, 1162 (1972)
6.129 D. Connor, T. Tsang: Phys. Rev. *126*, 1506 (1962)
6.130 A. Wapstra, D. Connor: Nucl. Phys. *22*, 336 (1961)
6.131 A. Gul'ko, S. Trostin, A. Hudoklin: Sov. Phys. JETP *25*, 998 (1967)
6.132 A. Gul'ko, S. Trostin, A. Hudoklin: Sov. J. Nucl. Phys. *6*, 477 (1968)
6.133 M. Bulgakov, A. Gul'ko, Y. Oratovskii, S. Trostin: Sov. Phys. JETP *34*, 356 (1972)
6.134 M. Bulgakov, S. Borovlev, A. Gul'ko, F. Dzheparov, S. Trostin: JETP Lett. *27*, 453 (1978)
6.135 A. Winnacker, D. Dubbers, F. Fujara, K. Dörr, H. Ackermann, H. Grupp, P. Heitjans, A. Körblein, H.-J. Stöckmann: Phys. Lett. *67*A, 423 (1978)
6.136 H. Ackermann, D. Dubbers, M. Grupp, P. Heitjans, H.-J. Stöckmann: Phys. Lett. *52*B, 54 (1974)
6.137 D. Dubbers, K. Dörr, H. Ackermann, F. Fujara, H. Grupp, M. Grupp, P. Heitjans, A. Körblein, H.-J. Stöckmann: Z. Phys. A*282*, 243 (1977)
6.138 T. Tsang, D. Connor: Phys. Rev. *132*, 1141 (1963)
6.139 H. Ackermann: Hyperfine Interact. *4*, 645 (1978)
6.140 K. Sugimoto, A. Mizobuchi, K. Nakai, K. Matuda: Phys. Lett. *18*, 38 (1965)
6.141 H. Ackermann, D. Dubbers, H. Grupp, M. Grupp, P. Heitjans, H.-J. Stöckmann: Phys. Lett. *54*A, 399 (1975)
6.142 H.-J. Stöckmann, D. Dubbers, M. Grupp, H. Grupp, H. Ackermann, P. Heitjans: Z. Phys. B*30*, 19 (1978)
6.143 H. Ackermann, D. Dubbers, M. Grupp, P. Heitjans, G. zu Putlitz, H.-J. Stöckmann: Phys. Lett. *41*B, 143 (1972)
6.144 T. Minamisono, Y. Nojiri, A. Mizobuchi, K. Sugimoto: Nucl. Phys. A*236*, 416 (1974)
6.145 D. Dubbers, H. Ackermann, M. Grupp, P. Heitjans, H.-J. Stöckmann: Z. Phys. B*25*, 363 (1976)
6.146 H. Ackermann, D. Dubbers, M. Grupp, P. Heitjans, H.-J. Stöckmann, K.-P. Wanczek, K. Recker, R. Leckebusch: Z. Naturforsch. Teil A: *31*, 1298 (1976)
6.147 P. Heitjans, H. Grupp, W. Buttler, F. Fujara, H. Ackermann, K. Dörr, G. Kiese, A. Körblein, H.-J. Stöckmann: Phys. Lett. *94*B, 28 (1980)
6.148 H.-J. Stöckmann, H. Ackermann, D. Dubbers, F. Fujara, M. Grupp, P. Heitjans, A. Körblein: Hyperfine Interact. *4*, 170 (1978)
6.149 S. Ochi, Y. Nojiri, T. Minamisono, K. Sugimoto: Hyperfine Interact. *10*, 1101 (1981)
6.150 K. Sugimoto, A. Mizobuchi, T. Minamisono: In *Hyperfine Interactions in Excited Nuclei*, Vol.1, ed. by G. Goldring, P. Kalish (Gordon and Breach, New York 1971) p.385
6.151 T. Minamisono, J. Hugg, D. Mavis, T. Saylor, H. Glavish, S. Hanna: Phys. Lett. *61*B, 155 (1976)
6.152 J. Mertens, H. Ackermann, D. Dubbers, P. Heitjans, A. Winnacker: Z. Phys. *262*, 189 (1973)
6.153 F. Dzheparov, I. Ivanter: Sov. J. Nucl. Phys. *23*, 280 (1976)
6.154 H. Ackermann, D. Dubbers, J. Mertens, A. Winnacker: Z. Phys. *228*, 329 (1969)
6.155 A. Winnacker, H. Ackermann, D. Dubbers, M. Grupp, P. Heitjans, H.-J. Stöckmann: Nucl. Phys. A*261*, 261 (1976)
6.156 H. Ackermann, D. Dubbers, J. Mertens, A. Winnacker, P.v. Blanckenhagen: Phys. Lett. *29*B, 485 (1969)
6.157 H. Grupp, K. Dörr, H.-J. Stöckmann, H. Ackermann, B. Bader, W. Buttler, P. Heitjans, G. Kiese, H. Lauter: Hyperfine Interact. *10*, 765 (1981)
6.158 R.L. Mieher: Phys. Rev. *125*, 1537 (1962)
6.159 M. Grupp, H. Ackermann, D. Dubbers, H. Grupp, P. Heitjans, H.-J. Stöckmann: J. Phys. (Paris) *37*, C7-538 (1976)
M. Grupp: Dissertation, Universität Heidelberg (1976)

6.160 N. Bräuer, F. Dimmling, Th. Kornrumpf, K. Nishiyama, D. Riegel: Hyperfine Interact. *2*, 265 (1976)
6.161 N. Bräuer, B. Focke, B. Lehmann, E. Matthias, K. Nishiyama, D. Riegel: Phys. Lett. *37*B, 186 (1971)
6.162 N. Bräumer, F. Dimmling, B. Focke, A. Goldmann, M.v. Hartrott, Th. Kornrumpf, K. Nishiyama, D. Quitmann, D. Riegel: In *Hyperfine Interactions Studied in Nuclear Reactions and Decay*, Vol.I, ed. by E. Karlsson, R. Wäppling (Almqvist and Wiksell International, Stockholm, New York 1974) p.254
6.163 J. Hadijuana, M.v. Hartrott, K. Nishiyama, D. Quitmann, D. Riegel, H. Schweickert: loc. cit., p.252
6.164 F. Dimmling, A. Goldmann, M.v. Hartrott, K. Nishiyama, D. Riegel: loc. cit., p.56
6.165 D. Riegel, A. Goldmann, M.v. Hartrott, K. Nishiyama, D. Quitmann: J. Phys. (Paris) *35*, C4-341 (1974)
6.166 T. Moriya, A. Kawabata: J. Phys. Soc. Jpn. *34*, 639 (1973); *35*, 669 (1973)
6.167 P. Heitjans, A. Coker, T. Lee, T.P. Das: J. Phys. (Paris) *41*, C8-403 (1980)
6.168 P. Heitjans, G. Kiese, H. Ackermann, B. Bader, W. Buttler, K. Dörr, F. Fujara, H. Grupp, A. Körblein, H.-J. Stöckmann: J. Phys. (Paris) *41*, C8-409 (1980)
6.169 H. Estrade, J. Conard, P. Lauginie, P. Heitjans, F. Fujara, W. Buttler, G. Kiese, H. Ackermann, D. Guérard: Physica *99*B, 531 (1980)
6.170 W. Buttler, H.-J. Stöckmann, F. Fujara, P. Heitjans, G. Kiese, H. Ackermann, B. Bader, K. Dörr, H. Grupp, H. Lauter: J. Phys. (Paris) *41*, C6-381 (1980); Z. Phys. B*45*, 273 (1982)
6.171 K. Dörr, H.-J. Stöckmann, B. Bader, P. Freiländer, H. Grupp, H. Ackermann, W. Buttler, P. Heitjans, G. Kiese: Hyperfine Interact. *10*, 727 (1981); J. Phys. C*15*, 4437 (1982)
6.172 J. Rossbach, M.v. Hartrott, J. Höhne, D. Quitmann, E. Weihreter, F. Willeke: J. Phys. F*10*, 729 (1980)
6.173 E.N. Kaufmann, R.J. Vianden: Rev. Mod. Phys. *51*, 161 (1979)
6.174 H.-J. Stöckmann, P. Heitjans, H. Ackermann: In *Festkörperprobleme (Advances in Solid State Physics)*, Vol.20, ed. by J. Treusch (Vieweg, Braunschweig 1980) p.19
6.175 *Proc. 5th Intern. Conf. Hyperfine Interactions*, ed. by G. Kaindl, H. Haas Hyperfine Interact. Vols. 9 and 10 (North-Holland, Amsterdam 1981)
6.176 K. Matsuda, K. Tanaka, Y. Nojiri, T. Minamisono, K. Hashimoto, K. Nagano, H. Ida, S. Kunori, Y. Toba, Y. Aoki, Y. Tagishi, K. Furuno, K. Yagi: Annual Report, Osaka University, Laboratory of Nuclear Studies, p.69 (1980)
6.177 I. König, D. Fick, R. Böttger, P. Egelhof, H. Ingwersen, S. Kossionides, K.-H. Möbius, D. Presinger, E. Steffens: AIP Conf. Proc. No.69, Polarization Phenomena in Nucl. Phys.-1980, ed. by G. Ohlsen et al., New York 1981, p.919;
I. König: Dissertation, Universität Marburg (1981)
6.178 T. Minamisono, J. Hugg, D. Mavis, T. Saylor, S. Lazarus, H. Glavish, S. Hanna: Hyperfine Interact. *2*, 315 (1976)
6.179 J.W. Hugg: Progress Report, Dept. of Physics, Stanford Univ., Stanford 1979, p.74
6.180 T. Minamisono, Y. Nojiri, S. Ochi: Phys. Lett. *106*B, 38 (1981)
6.181 H.J. Barth, K. Nishiyama, D. Riegel: Phys. Lett. *77*A, 365 (1980)
6.182 H.J. Barth, G. Netz, K. Nishiyama, D. Riegel: Phys. Rev. Lett. *45*, 1015 (1980)
6.183 M. Luszik-Bhadra, H.J. Barth, H.J. Brocksch, G. Netz, D. Riegel, H.H. Bertschat: Phys. Rev. Lett. *47*, 871 (1981)
6.184 D. Riegel: Phys. Rev. Lett. *48*, 516 (1982)
6.185 Y. Nojiri, Y. Kumata, T. Minamisono: Phys. Lett. *83*A, 85 (1981); Hyperfine Interact. *9*, 443 (1981)
6.186 A. Maio: Hyperfine Interact. *10*, 1175 (1981)
6.187 J.D. Perez, R.E. McDonald, T.K. McNab: Phys. Rev. B*19*, 163 (1979)
6.188 T. Minamisono, Y. Nojiri: Hyperfine Interact. *9*, 437 (1981)
6.189 A.G. Anderson, A.G. Redfield: Phys. Rev. *116*, 583 (1959)
6.190 R. Dupree, E.F.W. Seymour: Phys. Kondens. Materie *12*, 97 (1970)

6.191 G. Kiese, P. Heitjans, H. Ackermann, B. Bader, W. Buttler, P. Freiländer, C. van der Marel, H. Ruppersberg, H.-J. Stöckmann: In *Ionic Liquids, Molten Salts, and Polyelectrolytes*, ed. by K.-H. Bennemann, F. Brouers, D. Quitmann. Lect. Notes in Phys., Vol.172 (Springer, Berlin, Heidelberg, New York 1982) p.117
6.192 C. van der Marel, W. van der Lugt: Physica *112*B, 365 (1982)
6.193 H. Ruppersberg, J. Saar, W. Speicher, P. Heitjans: J. Phys. (Paris) *41*, C8-595 (1980)
6.194 W.W. Warren: Phys. Rev. B*3*, 3709 (1971)
6.195 J. Bosse, D. Quitmann, C. Wetzel: J. Phys. (Paris) *41*, C8-378 (1980)
6.196 C. Wetzel, J. Bosse, D. Quitmann: Hyperfine Interact. *10*, 1035 (1981)
6.197 R. Brinkmann, M. Elwenspoek, M.v. Hartrott, A. Novak, D. Quitmann: J. Phys. (Paris) *42*, C4-1055 (1981)
6.198 M.v. Hartrott, D. Quitmann, J. Roßbach, E. Weihreter, F. Willeke: Hyperfine Interact. *10*, 1027 (1981)
6.199 E. Weihreter, M.von Hartrott, D. Quitmann, J. Roßbach, F. Willeke: J. Phys. (Paris) *41*, C8-160 (1980)
6.200 A.B. Bhatia, D.E. Thornton: Phys. Rev. B*2*, 3004 (1970)
6.201a F. Willeke, M.v. Hartrott, D. Quitmann: Hyperfine Interact. *10*, 1031 (1981)
6.201b F. Willeke, M.v. Hartrott, D. Quitmann: J. Phys. F*11*, 275 (1981)
6.202 F.S. Dzheparov, A.A. Lundin: Sov. Phys. JETP *48*, 514 (1978)
6.203 F.S. Dzheparov, V.S. Smelov, V.E. Shestopal: JETP Lett. *32*, 47 (1980)
6.204 D. Dubbers, H. Vogt, A. Winnacker: Verhandl. DPG (VI) *17*, 852 (1982)
6.205 K. Sugimoto, A. Mizobuchi, K. Nakai: Phys. Rev. *134*, B539 (1964)
6.206 K. Mishra, K. Duff, T.P. Das: Phys. Rev. B*25*, 3389 (1982)
6.207 H.-J. Stöckmann: J. Magn. Res. *44*, 145 (1981)
6.208 P. Heitjans, B. Bader, K. Dörr, H.-J. Stöckmann, G. Kiese, H. Ackermann, P. Freiländer, W. Müller-Warmuth: J. Phys. (Paris) *43*, C9-143 (1982)
6.209 W.A. Phillips (ed.): *Amorphous Solids*, Topics Current Phys., Vol.24 (Springer, Berlin, Heidelberg, New York 1981)
6.210 M.S. Dresselhaus, G. Dresselhaus: Adv. Phys. *30*, 139 (1981)
6.211 E. McRae, D. Billaud, J.F. Marêché, A. Herold: Physica B*99*, 489 (1980)
6.212 J. Conard, H. Estrade: Mat. Sci. Eng. *31*, 173 (1977)
6.213 G. Roth, K. Lüders, P. Pfluger, H.-J. Güntherodt: Solid State Commun. *39*, 423 (1981)
6.214 H. Grupp, H. Ackermann, W. Buttler, K. Dörr, P. Heitjans, H.-J. Stöckmann: Nucl. Phys. A*386*, 56 (1982)
6.215 P.C. Schmidt, K.D. Sen, T.P. Das, A. Weiss: Phys. Rev. B*22*, 4167 (1981)
6.216 N. Takahashi, K. Tanaka, Y. Nojiri, T. Minamisono: *Polarization Phenomena in Nucl. Phys.*-1980, ed. by G. Ohlsen et al., AIP Conf. Proc. No.69, New York 1981, p.1115
6.217 N. Takahashi: *Polarization Phenomena in Nucl. Phys.*-1980, ed. by G. Ohlsen et al., AIP Conf. Proc. No.69, New York 1981, p.1016
6.218 T. Minamisono, Y. Nojiri, K. Tanaka, Y. Miake, N. Takahashi, K. Sugimoto: Hyperfine Interact. *9*, 53 (1981)
6.219 H. Grupp, K. Dörr, H.-J. Stöckmann, H. Ackermann, B. Bader, W. Buttler, P. Heitjans, G.Kiese: Z. Phys. B*47*, 1 (1982)

Subject Index

H. Bilz, W. Kress

Phonon Dispersion Relations in Insulators

1979. 162 figures in 271 separate illustrations. VIII, 241 pages
(Springer Series in Solid-State Sciences, Volume 10)
ISBN 3-540-09399-0

Contents: Summary of Theory of Phonos: Introduction. Phonon Dispersion Relations and Phonon Models. - Phonon Atlas of Dispersion Curves and Densities of States: Rare-Gas Crystals. Alkali Halides (Rock Salt Structure). Metal Oxides (Rock Salt Structure). Transition Metal Compounds (Rock Salt Structure). Other Cubic Crystals (Rock Salt Structure). Cesium Chloride Structure Crystals. Diamond Structure Crystals. Zinc-Blende Structure Crystals. Wurtzite Structure Crystals. Fluorite Structure Crystals. Rutile Structure Crystals. ABO_3 and ABX_3 Crystals. Layered Structure Crystals. Other Low-Symmetry Crystals. Molecular Crystals. Mixed Crystals. Organic Crystals. - References. - Subject Index.

D.C. Mattis

The Theory of Magnetism I

Statics and Dynamics

1981. 58 figures. XV, 300 pages. (Springer Series in Solid-State Sciences, Volume 17)
ISBN 3-540-10611-1

Contents: History of Magnetism. - Exchange. - Quantum Theory of Angular Momentum. - Many-Electron Wavefunctions. - From Magnons to Solitons: Spin Dynamics. - Magnetism in Metals. - Bibliography. - References. - Subject Index.

Starting with a thorough historical introduction to the study of magnetism - one of the oldest sciences known to man - through the most modern developments (magnetic "bubbles" and "soap films", effects of magnetic impurities in metals and "spin glasses"), this book develops the concepts and the mathematical expertise necessary to understand contemporary research in this field.
Volume I treats exchange forces, the theory of angular momentum, proves important theorems concerning the nature of the ground state and excited states, develops theories of magnons, vortices and solitons and gives a survey of the rapidly evolving field of magnetism in metals. The approach is thorough: all important theories are worked out in detail, using methods and notation that are uniform throughout. Footnotes and bibliograph provide a guide to the original literature, and a number of problems test the reader's skill.

Springer-Verlag
Berlin
Heidelberg
New York
Tokyo